Methods in Molecular Biology

Volume 6

Plant Cell and Tissue Culture

Methods in Molecular Biology

Edited by ***John M. Walker***

Volume I: ***Proteins,*** *1984*
Volume II: ***Nucleic Acids,*** *1984*
Volume III: ***New Protein Techniques,*** *1988*
Volume IV: ***New Nucleic Acid Techniques,*** *1988*
Volume V: ***Animal Cell Culture,*** edited by **Jeffrey W. Pollard** and **John M. Walker,** *1990*
Volume VI: ***Plant Cell and Tissue Culture,*** edited by **Jeffrey W. Pollard** and **John M. Walker,** *1990*

Methods in Molecular Biology • 6

Plant Cell and Tissue Culture

Edited by

Jeffrey W. Pollard

Albert Einstein College of Medicine, Bronx, New York

and

John M. Walker

The Hatfield Polytechnic, Hatfield, Hertfordshire, UK

Springer Science+Business Media, LLC

Originally published by The Humana Press Inc. in 1990
Softcover reprint of the hardcover 1st edition 1990

Library of Congress Cataloging in Publication Data
Main entry under title:

Methods in molecular biology.

(Biological methods)
Includes bibliographies and indexes.
Contents: v. 1. Proteins—v. 2. Nucleic acids—v. 3. New protein techniques—v. 4. New nucleic acid techniques—v. 5. Animal cell culture—v. 6. Plant cell and tissue culture.

1. Molecular biology—Technique. I. Walker, John M., 1948– II. Series.

QH506.M45 1984 574.8'8'078 84-15696

ISBN 978-1-4899-4387-3 ISBN 978-1-59259-493-1 (eBook)
DOI 10.1007/978-1-59259-493-1

Preface

For many years I performed tissue culture in large scientific institutions that had a great deal of infrastructure. When I set up a tissue laboratory outside such an infrastructure, however, I found there was a shortage of easily accessible information about the basic needs, reagents, and techniques for establishing such a facility. Much had to be done by trial and error or gleaned from original papers. Consequently, I felt that a methods book covering a wide variety of techniques from basic culture to the most sophisticated cell analysis would be a very valuable addition to scientific literature. In the interim, several useful books did appear, but none entirely fitted the bill and some are now rather dated. Then, in 1984, the first of the *Methods in Molecular Biology* volumes from Humana Press was published with its step-by-step recipe approach. This format appealed to me, and so I contacted John Walker, the series editor, about including cell culture in this series. The result was that we embarked upon a single volume covering both plant and animal cell culture. Such was the richness of the material, however, that this project soon divided itself into separate volumes on animal cell (Volume 5) and plant cell (Volume 6) culture. In this volume (Volume 6), therefore, we have aimed to describe a range of basic techniques and culture conditions appropriate to plant cells. Because of the diversity of cell types, species, and culture methods, much of the volume is devoted to the culture of particular cell types and to the regeneration of these cells into whole plants. The importance of genetic modification of plants, both for agricultural and research purposes, has also led us to include many chapters describing techniques for DNA transfer and for isolation of plant mutants via somatic cell genetics. The economic importance of plant products is also acknowledged by a section describing large-scale culture methods and possible ways of automating these cell culture processes. Finally, we have listed, in the appendix, the formulas for the most commonly used plant cell media. John Walker and I hope this volume comprehensively covers the exciting developments in these fields, as well as the more basic needs for setting up a plant cell culture laboratory.

Editors of books always have many debts of gratitude. First, I would like to thank those who taught me tissue culture and particularly somatic cell genetics. The principle influence was Dr. Cliff Stanners, who at that time was at the Ontario Cancer Institute in Toronto. Gratitude must also go to all the other members of the Toronto somatic cell genetics group, and especially to Drs. Lou Siminovitch and Vic Ling. Finally, but not least, I would like to thank Ms. Margaret Hudgell and Ms. Rita Romita for their expert and cheerful secretarial help, often at times of crisis.

Jeffrey W. Pollard

In keeping with the successful format of earlier volumes, we have again applied a step-by-step approach for describing each technique, although in some cases, particularly those involving sophisticated instrumentation, this format has necessarily been modified somewhat. Our aim, as with earlier volumes, is to provide clearly defined protocols that should allow the new worker to obtain successful results with the first attempt. We hope we have been successful in this endeavor! I would like to express my gratitude to Jeff Pollard for suggesting this volume, for his considerable efforts in helping to find appropriate authors, and for his diligent editing. Finally, as ever, my sincere thanks to my wife Jan, who once again without complaint and often at short notice, took on the role of typist, proofreader, and unofficial editor.

John M. Walker

Contents

Contributors

GEORGE W. BATES • Florida State University, Tallahassee, FL
PAMELA A. BOND • Bristol Polytechnic, Frenchay, Bristol, UK
JACQUELINE T. BROWN • University College London, London, UK
BARRY V. CHARLWOOD • University College London, London, UK
HAMISH A. COLLIN • University of Liverpool, Liverpool, UK
AGNES CSEPLO • Hungarian Academy of Sciences, Szeged, Hungary
MICHAEL R. DAVEY • University of Nottingham, Nottingham, UK
DAVID ROY DAVIES • John Innes Institute, Norwich, UK
FRANK DICOSIMO • University of Toronto, Toronto, Ontario, Canada
PHILIP J. DIX • St. Patrick's College, Kildare, Republic of Ireland
ANNE DONOVAN • University of Liverpool, Liverpool, UK
GERARD C. DOUGLAS • Agricultural Institute, Dublin, Republic of Ireland
DAVID R. DUNCAN • University of Illinois, Urbana, IL
JACQUES L. DURAND • University of Paris, Orsay, France
NICOLA E. EVANS • Department of Agriculture, Loughall, Armagh, Northern Ireland
PETER K. EVANS • University of Southampton, Southampton, UK
PETER J. FACCHINI • University of Toronto, Toronto, Ontario, Canada
FARAMARZ FAKHRAI • University of Durham, Durham, UK
HOURI KALILIAN FAHKRAI • University of Durham, Durham, UK
MICHAEL F. FAY • Royal Botanic Gardens, Kew, Surrey, UK
MATTHEW V. FORD • Royal Botanic Gardens, Kew, Surrey, UK
MICHAEL W. FOWLER • Wolfson Institute of Biotechnology, Sheffield, UK
PAUL D. GALLETTA • United States Department of Agriculture, Pasadena, CA
TIMOTHY J. GOLDS • University of Nottingham, Nottingham, UK
JEAN H. GOULD • Texas A&M University, College Station, TX
JILL GRATTON • Royal Botanic Gardens, Kew, Surrey, UK
DEBBIE GREY • Wolfson Institute of Biotechnology, Sheffield, UK

BRIAN W. W. GROUT • Plymouth Polytechnic, Plymouth, UK
NAZMUL HAQ • University of Southampton, Southampton, UK
WENDY ANN HARWOOD • John Innes Institute, Norwich, UK
RANDAL M. HAUPTMANN • Northern Illinois University, DeKalb, IL
CHRISTOPHER S. HUNTER • Bristol Polytechnic, Frenchay, Bristol, UK
SUSAN ISAAC • University of Liverpool, Liverpool, UK
DANIEL JONES • United States Department of Agriculture, Pasadena, CA
NIGEL J. KILBY • Cambridge University, Cambridge, UK
JOHN KING • University of Saskatchewan, Saskatoon, Saskatchewan, Canada
PAUL F. MCCABE • St. Patrick's College, Kildare, Republic of Ireland
MICHEL MONNIER • Université de Pierre et Marie Curie, Paris, France
JOHN MOTTLEY • North East London Polytechnic, London, UK
STEVEN J. NEILL • Bristol Polytechnic, Frenchay, Bristol, UK
JOAN M. NELSHOPPEN • University of Illinois, Urbana, IL
A. WILHELM NEUMANN • University of Toronto, Toronto, Ontario, Canada
SERGIO J. OCHATT • University of Nottingham, Nottingham, UK
CARMEL M. O'NEILL • St. Patrick's College, Kildare, Republic of Ireland
MICHAEL PARKINSON • St. Patrick's College, Kildare, Republic of Ireland
WILLIAM PIASTUCH • Florida State University, Tallahassee, FL
JEFFREY W. POLLARD • Albert Einstein College of Medicine, Bronx, New York
JOHN B. POWER • University of Nottingham, Nottingham, UK
DIETMAR RABUSSAY • Bethesda Research Laboratories, Gaithersburg, MD
LASZLO G. RADVANYI • University of Toronto, Toronto, Ontario, Canada
ELIBIO L. RECH • University of Nottingham, Nottingham, UK
ANDREW V. ROBERTS • North East London Polytechnic, London, UK
PRAVEEN K. SAXENA • University of Florida, Gainesville, FL
ANDREW H. SCRAGG • Wolfson Institute of Biotechnology, Sheffield, UK
ELAINE F. SMITH • North East London Polytechnic, London, UK
ROBERTA H. SMITH • Texas A&M University, College Station, TX
GAGIK STEPAN-SARKISSIAN • Wolfson Institute of Biotechnology, Sheffield, UK
ERIC B. SWANSON • Allelix, Inc., Mississauga, Ontario, Canada
AILEEN M. TIMMONS • St. Patrick's College, Kildare, Republic of Ireland
BRENT TISSERAT • United States Department of Agriculture, Pasadena, CA

EUGENIO ULIAN • Texas A&M University, College Station, TX
CARL E. VANDERCOOK • United States Department of Agriculture, Pasadena, CA
JOHN M. WALKER • The Hatfield Polytechnic, Hatfield, Hertfordshire, UK
JACK M. WIDHOLM • University of Illinois, Urbana, IL
LYNDSEY A. WITHERS • University of Nottingham, Nottingham, UK
STEPHEN YARROW • Allelix, Inc., Mississauga, Ontario, Canada
MOHAMED T. ZIYAD-MOHAMED • Wolfson Institute of Biotechnology, Sheffield, UK

Chapter 1

Selection of Media for Tissue and Cell Culture

Gagik Stepan-Sarkissian

1. Introduction

This chapter deals with the selection of growth media for the induction of callus and initiation of liquid suspension cultures from plants. Although the first attempts at initiating cultures of plant cells were made by the German botanist G. Haberlandt at the turn of this century, it has only been during the last three decades that rapid developments in plant cell, tissue, and organ culture have occurred. Since extensive work with microbial cultures was by then already underway, the first medium formulations used for plant culture work were inevitably based on experience of microbes. These media contained several distinct classes of compounds:

1. Carbon source
2. Inorganic salts
3. Organic supplement
4. Trace elements.

From: *Methods in Molecular Biology*, vol. 6, *Plant Cell and Tissue Culture*
Edited by Jeffrey W. Pollard and John M. Walker, ©1990 by The Humana Press

As work on plant cell cultures progressed, attention was paid to defining media more suited to the growth of plant tissue cultures. Today there are many media formulations available that, although still loosely based on microbial media, are specifically tailored to suit the requirements of plant cell cultures. The addition of undefined sources of nutrients and hormones, such as coconut milk, is now usually avoided.

The most commonly used culture media (*see* Appendix) are based on the established formulations defined by Gamborg (*1*), Heller (*2*), Linsmaier and Skoog (*3*), Murashige and Skoog (*4*), Schenk and Hildebrandt (*5*), and White (*6*). However, other media exist with variations upon these basic themes. For example, the choice of nitrogen source may vary depending upon the species being cultured and may be in the form of an ammonium or nitrate salt, amino acids (*see* Note 1), casein hydrolysate, or urea. Since so few plant cell cultures are autotrophic, the provision of an exogenous energy source is of key importance. The most common carbon source used in plant cell culture media is sucrose, although glucose is sometimes used in its place, and may support an equal or even higher growth rate in culture. However, a high rate of growth is not always of paramount importance for cultures. Often the aim of initiating a plant cell culture is to obtain secondary products that are generally accumulated during the stationary phase of growth (*see* Chapters 2 and 48) or in differentiated cultures. Therefore, high growth rates may not necessarily be desirable in all cases.

Of great importance to the growth and productivity of a cell culture is the carbon/nitrogen ratio and the form in which these media components are supplied. It is possible to investigate the effects of the carbon/nitrogen ratio by altering the concentration and/or the type of these important medium constituents. However, such medium manipulations should be undertaken only after a given culture is established and its growth kinetics are characterized.

Growth regulators or plant hormones, such as auxins, gibberellins, abscisic acid, cytokinins, and ethylene, are known to influence various stages of growth in the whole plant. Of these, only auxins and cytokinins are routinely incorporated into plant culture medium. The most commonly used auxins are 2,4-dichlorophenoxyacetic acid (2,4-D), indole 3-acetic acid (IAA), and 1-naphthaleneacetic acid (NAA). IAA is the natural auxin, whereas 2,4-D and NAA are synthetic compounds that mimic the effects of naturally occurring auxins. Other chemicals possessing auxin-like activity and used—albeit less frequently—in plant cell culture include indole-3-butyric acid (IBA) and *p*-chlorophenoxyacetic acid (CPA). 6-Furfurylaminopurine, known by the trivial name kinetin, is the most frequently used cytokinin. However, 6-benzylaminopurine (BAP) or zeatin are sometimes substituted. By careful manipulations of the relative concentrations of auxins and cytokinins, it is possible to favor either undifferentiated growth or organogenesis (root or shoot formation).

The numbers of plant species that have been cultured and the different media used are extensive (*see* ref. *7*). Once the plant species to be cultured is decided upon, a suitable medium formulation should be selected. A practical first step would be to follow up appropriate references subsequent to a literature search and/or consult a handbook dealing with plant cell culture (*7–10* and this vol.).

If a literature search is inconclusive or if access to specialized sources is limited, then it is recommended to attempt initiating callus on several of the well-known medium formulations (*see* Note 2 and the Appendix). In addition, it is advisable to experiment with different auxin and cytokinin combinations. A tentative approach is suggested in Table 1.

2. Materials

2.1. For Stock Solutions Appropriate to Selected Medium Formulation

1. Inorganic salts (*see* Appendix).
2. Organic compounds (*see* Appendix).
3. Hormones (*see* Appendix).
4. 6*M* NaOH.

2.2. For Medium Preparation

1. Inorganic stock solution.
2. Organic stock solution.
3. Hormone stock solution.
4. Carbon source (sugar).
5. Agar (*see* Note 3).
6. 60-mL sterile plastic culture pots.
7. Laminar flow cabinet (*see* Note 4) equipped with a Bunsen burner.

2.3. For Assessment of Callus Growth

Equipment for sterile transfer of callus tissue consists of scalpel, spatula, forceps, sterile Petri plates, and industrial ethanol.

Table 1
Suggested Hormone Combinations and Concentrations for Plant Cell Culture Media*

	Cytokinin, mg/L			
Auxin, mg/L	0	0.05	0.5	5
0.1	1	2	3	4
1.0	5	6	7	8
10.0	9	10	11	12

*The numbers 1–12 represent different hormone combinations resulting from three auxin concentrations and cytokinin concentrations. For ease of reference, numbers can be added after abbreviations for medium formulations (e.g., MS-6 for Murashige and Skoog's medium (4) supplemented with 1.0 mg/L 2,4-D, and 0.05 mg/L kinetin) (adapted from refs. 7 and 9).

3. Methods

The composition of the various culture media are given in the Appendix. This chapter gives guidance on how to make these media up and how to store it. It also describes an approach to establish the culture conditions for untried species.

3.1. For Stock Solutions Appropriate to Selected Medium Formulations

In order to save time, it is advisable to prepare stock solutions of the three main components of the medium, i.e., inorganic salts, organic supplement, and hormones. The inorganic salts stock solution can be prepared at a 10-fold higher concentration. Owing to small amounts of chemicals used in the organic supplement, its stock solution is prepared at a 100-fold higher concentration. In most cases, the hormone stock solutions are prepared in such a way that addition of 1 mL of stock solution to 1 L of medium gives the desired concentration. Analytical Reagent grade chemicals should be used for the preparation of stock solutions.

1. For 1 L of stock solution of inorganic salts (sufficient for 10 L of medium), place a 1-L beaker containing approximately 400 mL of distilled water on a magnetic stirrer. Weigh out individually each of the salts according to the medium formulation in 10 times larger

amounts than those given in the text (recipes are normally given for 1 L of medium) (*see* Note 5). Dissolve each salt in a minimal volume of distilled water, and add to the beaker, allowing time for the solution to become well mixed before adding the next component. In some cases, it is necessary to add the constituents in a particular order in order to prevent precipitation of certain salts (*see* Note 5).

2. When all the salts have been added, make the volume of the stock solution up to 1 L with distilled water using a volumetric flask. Invert the flask several times to allow full mixing.
3. Store the stock solution at 4°C. The solution should remain usable at this temperature for 6 mo. However, at the first sign of microbial growth, it should be replaced with a new solution (*see* Note 6).
4. The organic supplement stock solution is made in the same way as that just described for the inorganic salts, i.e., the individual components are weighed in amounts 10-fold higher than that given in the recipe, but the total volume of the stock solution is reduced from 1 L to 100 mL (sufficient to prepare 10 L of medium). There are two reasons for this reduction in the volume of organic stock solution. First, the amounts involved in this supplement are low enough to dissolve in a smaller volume. Secondly, a small volume is advantageous, since this solution needs to be stored at –20°C in order to prevent microbial growth on its rich carbon and nitrogen sources.
5. The stock solutions of hormones are normally prepared at concentrations 1000-fold higher than that required for the given medium. To obtain a concentration of 1 mg/L auxin (e.g., 2,4-D or NAA) in the medium, the stock should be prepared as a 1 mg/mL solution from which an aliquot of 1 mL will be added to the medium. To prepare most hormone stock solutions, weigh out the required amount into a 100-mL volumetric flask, and add approximately 50 mL of distilled water. Place the flask on a magnetic stirrer, and add 6*M* NaOH dropwise until the hormone is dissolved. Make up the volume to a 100 mL and store at –20°C.

3.2. For Medium Preparation

To prepare 1 L of medium:

1. Place a 1-L beaker containing approximately 500 mL of distilled water on a magnetic stirrer, and add 100 mL of inorganic stock solution, 10 mL organic supplement stock solution, the required aliquot of hor-

mone(s), the carbon source, and agar (for solid medium only) (*see* Note 7).

2. After all the medium components have been well mixed, adjust the pH to the desired value (generally between 5 and 6), and make up the volume to 1 L with distilled water using a volumetric flask (*see* Note 8).
3. Plant cell suspension cultures are routinely grown in 250-mL Erlenmeyer flasks containing 100 mL of medium. When preparing liquid media for autoclaving, dispense 100-mL aliquots into 250-mL Erlenmeyer flasks, and cover the tops with a double layer of aluminum foil measuring 15 x 15 cm (*see* Note 9). Each flask should be labeled with a water insoluble marker pen stating the medium type and date of preparation (*see* Note 10).
4. For solid medium (containing agar) used for the growth of callus cultures, there exist two alternatives for dispensing into culture vessels. One alternative is to dispense approximately 30 mL of liquid medium containing agar into 100-mL Erlenmeyer flasks, and cover the tops with aluminum foil (*see* Note 9) before autoclaving. On the other hand, the medium can be autoclaved in 500-mL lots in Pyrex bottles and then dispensed into commercially purchased sterile culture pots in a laminar flow cabinet after it has cooled to around 50°C (*see* Note 11). In either case, label containers with a water-insoluble marker pen, stating the medium type and date of preparation.
5. Autoclave the media, both solid and liquid, at 120°C and 1.06 kg/cm^2 for 20 min. If volumes of 1 L or over are to be autoclaved in a single container, the sterilization time should be increased to 40 min. To check for visible signs of microbial contamination, allow several days between the preparation and use of solid and liquid media, and store at room temperature.
6. If the medium requires addition of thermolabile components (e.g., zeatin), it is necessary to add them in a laminar flow cabinet to the individual flasks when they have been autoclaved. This is done by preparing a stock solution of the given compound at an appropriate concentration, so that addition of up to 5 mL of this solution to each flask will give the desired concentration in the medium. These additions should be made by passing the solution through a sterile 0.2-µm filter attached to a 5-mL syringe, preferably sterile. In the case of solid media, thermolabile chemicals can be added to the medium after it has cooled down sufficiently but before it has begun to set (40–50°C).

Additions can be made either before or after dispensing the medium into individual containers.

3.3. For Assessment of Callus Growth

If the callus culture of a given species has been initiated on several growth media each with a different hormone combination, it may be necessary to assess the growth and productivity (e.g., of a secondary metabolite) for each cell line. This is particularly important when space and time are limited for the maintenance of large numbers of cell lines. Assessment of callus growth and productivity provides useful data, on the basis of which a decision can be made to retain a cell line with a high growth rate, a high productivity, or a compromise between these two parameters. It should be emphasized that the method detailed below gives only a rough estimation of growth, since the procedure is based upon fresh weight rather than dry weight measurements. However, in order to increase the reliability of these growth measurements, it is necessary to standardize as many of the parameters involved as possible.

1. Under sterile conditions, transfer 25 mL of sterile agar into either Erlenmeyer flasks or sterile plastic culture pots and allow to set (*see* Note 12).
2. Aseptically transfer a known weight (e.g., 0.5 g) of callus tissue onto the fresh agar (*see* Note 13). If the callus is friable (*see* Note 14), the transfer can be made simply by using a spatula that has been sterilized by immersing it into industrial ethanol and then flamed. On the other hand, if the callus tissue is hard or partially differentiated, then it may be necessary to dissect the callus on a sterile Petri plate using a scalpel and a pair of forceps sterilized as above.
3. After a suitable length of time (e.g., 4 wk) (*see* Note 15), remove the callus from the pot with a sterile spatula and weigh using a sterile Petri plate. Record the results in a chart similar to that shown in Fig. 1. If further analysis (e.g., secondary products, starch, enzyme levels) is required, then a sample of callus tissue should be set aside for this purpose.
4. It is advisable to monitor the growth of callus lines over several passages (subcultures), in which case the callus weighed at the end of the last cycle should provide the starting material (e.g., 0.5 g) for the next cycle. In order to allow meaningful statistical analysis of the results, it is necessary to have at least 5 replicates/medium/hormone

combination for each cycle. A suitable diagrammatic way of representing these data is in the form of histograms with standard errors of the mean (*see* Fig. 2).

3.4. For Assessment of Growth in Suspension Cultures

Even when suspension cultures have been established, it is still possible to manipulate certain key components of the medium (carbon source, nitrogen source, hormone combination, or concentration) or change the medium formulation altogether.

1. Subculture cells into "new" medium in the usual way.
2. If the cells adapt to the new medium formulation, the culture should be allowed to become established over several passages (at least 5) before any analysis work is attempted (*see* Note 16).
3. The growth analysis is carried out according to the method described in Chapter 2 of this vol.
4. With suspension cultures, it is possible to take samples for product analysis at any time during the growth cycle. Ideally, samples should be taken at the same time as those for growth analysis.
5. Construct growth curves for suspension cultures according to Fig. 2 in Chapter 2 of this vol.
6. For the purposes of comparison, the growth of the culture in the "old" and "new" media should be monitored concurrently.

4. Notes

1. Care should be taken when supplying amino acids as the nitrogen source. In certain cell cultures, amino acids may be deaminated and used as a source of carbon.
2. Even when a published method for the culture initiation of a given plant species is adopted, it is still advisable to test, at the same time, other defined media formulations and hormone combinations. Occasionally, an alternative medium may emerge as the most suitable formulation for growth. An example is presented in Fig. 2.
3. Addition of agar is required only for solid media: It is omitted when preparing media for suspension cultures.
4. Although it is possible to pour agar-containing media into appropriate containers on a laboratory bench equipped with a Bunsen burner, a laminar flow cabinet is preferred for aseptic transfer of tissue material during subculturing.

'Mother' culture				'Daughter' culture					Date______ Tissue______
Medium formulation	Subculture number	Date subcultured	Weight transferred (g)	Medium formulation	Subculture number	Total tissue weight (g)	Net weight change (g)	(%)	Comments

Fig. 1. Suggested chart layout for recording growth parameters for a callus culture. At least five replicates are necessary to calculate the average fresh weight for a given cell line and to determine the percentage change.

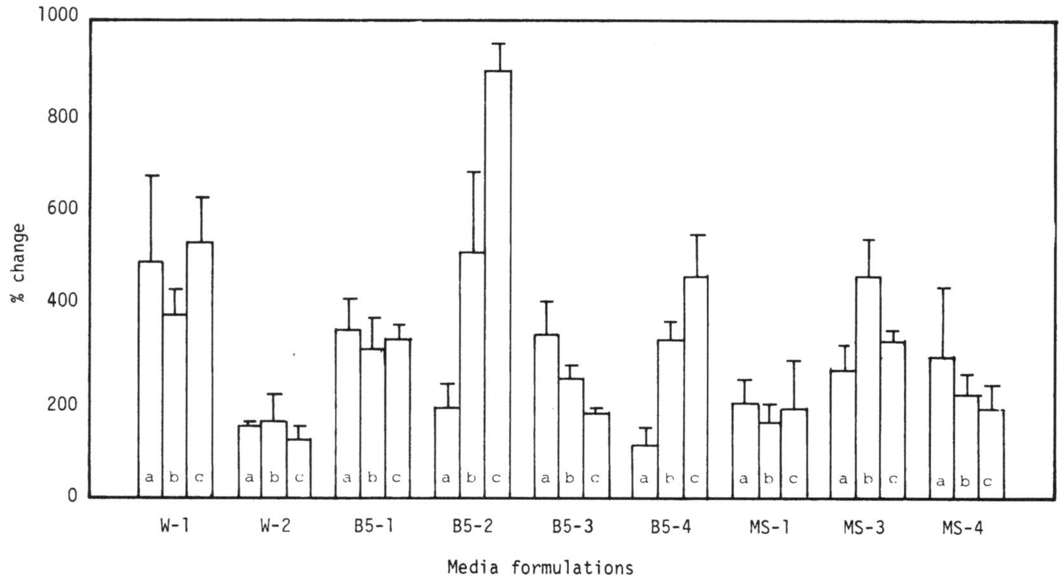

Fig. 2. Growth of potato (*Solanum tuberosum* cv. Saturna) callus on different media during three subsequent passages (a, b, c). The callus was initiated on White's medium (which was the recommended medium in the literature) as well as on two other media, namely Murashige and Skoog's and Gamborg's B5. After three successive passages, Gamborg's B5 medium emerged as the best formulation for the growth of potato callus. The growth is expressed as percentage increase in fresh weight of callus after 24 d in culture. Each histogram represents the mean percentage increase and standard error of the mean of at least five replicate cultures. Incubation conditions, medium volume, and inoculum weight were standardized (unpublished data of G. Stepan-Sarkissian and D. Grey). The abbreviations used are: W: White's medium, (b) B5: Gamborg's medium (*1*) and MS: Murashige and Skoog's medium (*4*). The numbers following medium designations represent the following auxin and cytokinin combinations: (1): 5.7 mg/L 2,4-D and 5.0 mg/L kinetin, (2): 10 mg/L 2,4-D, (3): 10 mg/L NAA, (4): 10 mg/L NAA and 10 mg/L kinetin.

5. In some media, it is necessary to add some components separately. For example, in Murashige and Skoog's medium, calcium chloride must be thoroughly dissolved before any other salt is added. To prevent precipitation of other medium components, magnesium sulfate should be added last.
6. One suitable way of prolonging the life of inorganic salt stock solution is to filter-sterilize before storage. This is particularly advisable if the solution is to be used intermittently.
7. The percentage of agar added to the medium can vary according to the type of medium and hormones used. In this laboratory, for example, when Lab M agar (London Analytical & Bacteriological Media Ltd., Salford, UK) is used, 1% (w/v) dry powder is added to Gamborg's B5 medium (*1*), whereas 1.2% (w/v) dry powder is necessary to achieve the same consistency with Murashige and Skoog's medium (*4*). In addition, it has been noted that hormones present in the medium have some effect on the setting power of agar powder.
8. For most media, autoclaving results in the release of sugar acids that cause a slight drop in the final pH of the medium.
9. An alternative way of maintaining the sterility of the flask contents is to use a nonabsorbent cotton wool bung. However, care must be taken to avoid igniting the cotton wool in the presence of a naked flame.
10. Sterilized liquid medium must be stored in a cupboard sheltered from light, dust, and extremes of temperature. Certain medium salts—especially those of Murashige and Skoog's and related media—tend to precipitate following extended periods of exposure to light. Light is also reported to have a deleterious effect on some hormones, e.g., kinetin. Liquid medium should be used within a period not exceeding 3 mo from the date of preparation.
11. If for some reason the agar is not poured into appropriate containers before it sets, then it is possible to melt the agar by placing the bottle containing the solid medium in a water bath at 80°C before pouring. Maintaining a supply of solid medium in these 500-mL lots is a convenient method of storage, since it requires less space and minimizes loss of moisture resulting from evaporation.
12. The use of a volumetric flask or a measuring cylinder with hot agar can be difficult. An alternative method of standardizing the amount of agar contained in each pot is by weight. This is done by placing the sterile culture vessel on a balance, and taring it before pouring a known weight (e.g., 25 g) of agar into the vessel.

13. When transferring a known weight of callus during growth experiments, it is important to ensure that the weight transferred consists of one single rather than several pieces of callus. It is likely that the availability of nutrients to several small pieces will be higher than to a large single piece of callus. This may affect the apparent growth rate of the callus tissue.
14. In most cases, it is possible to enhance the friability of callus tissue by either increasing the auxin or decreasing the cytokinin concentration in the growth medium.
15. A typical subculture cycle for a callus culture is 4 wk. However, if the callus exhibits rapid growth, then growth measurements should be made every 2–3 wk since it is important that the growth is not limited by nutrient depletion. On the other hand, if the callus grows very slowly, it is not advisable to extend the subculture cycle beyond 6 wk, since after this time, the agar begins to dry out and this will affect the growth rate.
16. When changing from one medium formulation to another, it is important to allow sufficient time, e.g., 5 passages, for the cells to adapt to the new medium. This is necessary for two reasons. First, the culture may survive the first passage into "new" medium because of the carryover of nutrients from the old medium, and it is necessary to eliminate this effect before a meaningful measurement of growth is undertaken. Secondly, transferring cells into "new" medium can often cause a drop in culture viability during the first few passages. It is possible that the cells may adapt to this new medium, in which case the viability will gradually return to normal.

Acknowledgments

The author wishes to thank Debbie Grey, Angela Stafford, and Laura Smith for helpful suggestions and critical review of this manuscript.

References

1. Gamborg, O. L., Miller, R. A., and Ojima, K. (1968) Nutrient requirements of suspension cultures of soybean root cells. *Exp. Cell Res.* **50,** 151–158.
2. Heller, R. (1953) Recherches sur la nutrition minerale des tissues vegetaux cultives, *in vitro. Ann. Sci. Nat. Bot. Biol. Veg.* **14,** 1–223.
3. Linsmaier, E. M. and Skoog, F. (1965) Organic growth factor requirements of tobacco tissue cultures. *Physiol. Plant.* **18,** 100–127.

4. Murashige, T. and Skoog, F. (1962) A revised medium for rapid growth and bioassays with tobacco tissue cultures. *Physiol. Plant.* **15,** 473–497.
5. Schenk, R. U. and Hildebrandt, A. C. (1972) Medium and techniques for induction and growth of monocotyledonous and dicotyledonous plant cell cultures. *Can. J. Bot.* **50,** 199–204.
6. White, P. R. (1963) *A Handbook of Plant Tissue Culture* (Jacques Cottell, Pennsylvania).
7. George, E. F., Putlock, D. J. M., and George, H. J. (1987) *Plant Culture Media.* vol. I: *Formulations and Uses* (Exegetics Limited, Westbury, Wiltshire, England).
8. Bhojwani, S. S. and Razdan, M. K. (1983) *Plant Tissue Culture: Theory and Practice* (Elsevier, Amsterdam).
9. Dixon, R. A. (1985) Isolation and maintenance of callus and cell suspension cultures, in *Plant Cell Culture. A Practical Approach* (Dixon, R. A., ed.), IRL Press, Oxford and Washington, DC, pp. 1–20.
10. Evans, D. A., Sharp, W. R., Ammirato, P. V., and Yamada, Y., eds. *Handbook of Plant Cell Culture* (1983) vol. I: *Techniques for Propagation and Breeding* (Macmillan, New York).

Chapter 2

Growth Determination and Medium Analysis

Gagik Stepan-Sarkissian and Debbie Grey

1. Introduction

The importance of plant cell suspension cultures as a research tool in various aspects of plant biology is well established. A basic requirement of work with plant cell suspension cultures is the ability to monitor growth on the basis of various parameters. Reliable and reproducible measurement of growth is essential in order to assess the performance of the culture in general and pinpoint the occurrence of certain metabolic events at given growth stages.

Plant cell cultures exhibit a growth profile not dissimilar to that of microorganisms (Fig. 1) (*1*). During an initial lag phase, following subculture into fresh medium, the cells regain the ability to divide. The next stage (exponential phase) involves rapid cell division. Depending on the cells and nutrient regime, this stage can vary in duration. However, in most cases, the exponential phase is a short one, and only 3–4 generations are passed through before the rate of cell division declines (linear and progressive deceleration phases) and eventually ceases as the cells enter stationary phase. Typically, this pattern of culture growth is observed regardless of the method employed to monitor growth.

From: *Methods in Molecular Biology*, vol. 6, *Plant Cell and Tissue Culture*
Edited by Jeffrey W. Pollard and John M. Walker, ©1990 by The Humana Press

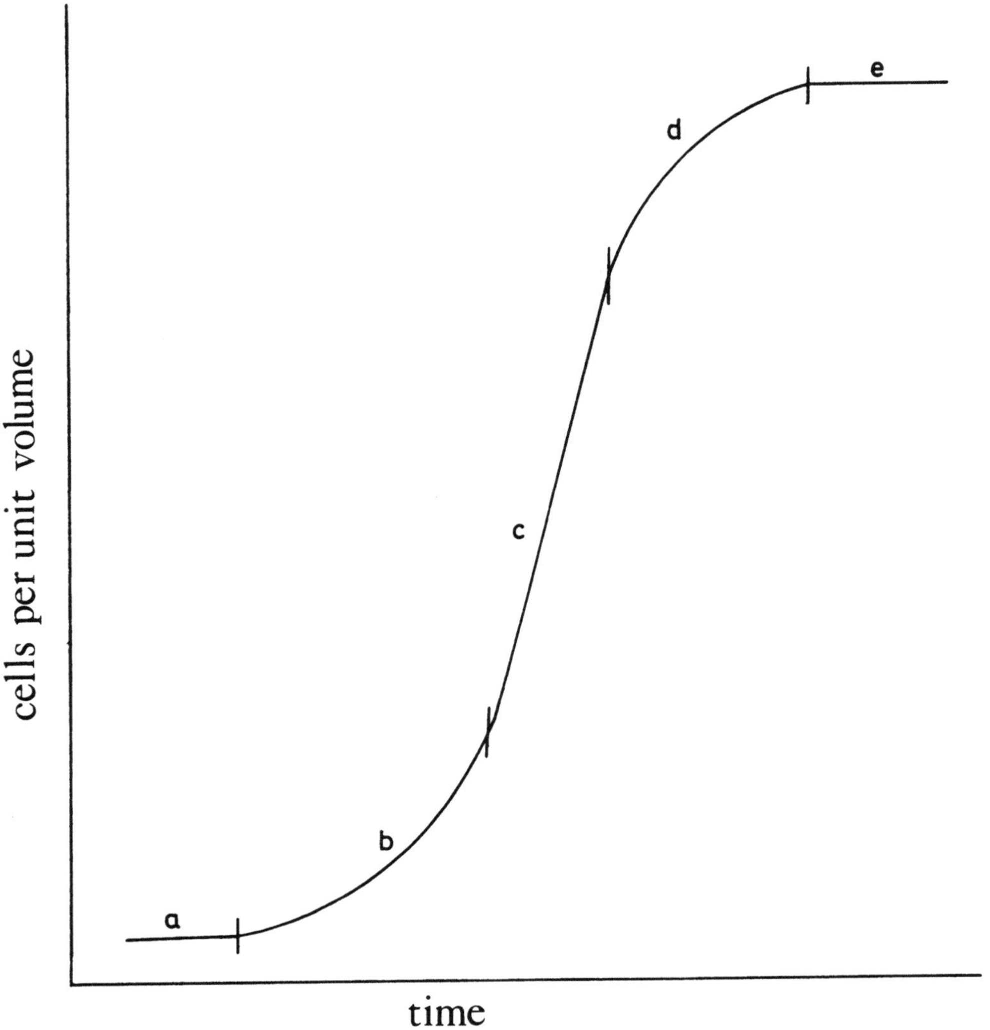

Fig. 1. Model curve relating culture growth to time in plant cell suspension cultures (from ref. *1*). The various stages of growth are: (a) lag phase, (b) exponential phase, (c) linear phase, (d) progressive deceleration phase, and (e) stationary phase.

There are several techniques available for measuring culture growth. The methods used consist of determination of cell number, fresh and dry weights, and packed cell volume. The method that gives the most accurate picture of cell growth is the determination of cell numbers. However, plant cell suspensions rarely contain a high percentage of single cells. Instead, they consist mostly of cell clumps of varying sizes. In consequence, the determination of cell numbers should include an additional process during which cell clumps are broken into single cells. Since the efficiency of this process is inversely proportional to the size of the clumps, the application of this method to suspension cultures with a high density of large clumps results in a less than accurate measurement of cell numbers. In addition, this method with its two stages of clump disruption and cell count is a time-consuming process and, therefore, not recommended for routine growth measurements. In contrast, packed cell volume measurements give very rapid results, which nevertheless are not more than rough approximations and are very rarely used for plant cell suspensions.

The most widely used method of growth determination is the measurement of fresh and dry weights of cells. It is essential to measure both parameters, since often the wet weight of the cells continues to rise after the dry weight goes into decline (Fig. 2). During the linear phase, plant cells synthesize cell wall material and starch from available carbohydrate. When the culture passes into progressive deceleration and stationary phases, the cells metabolize all available carbohydrate. There is no further starch accumulation. The decline in dry weight corresponds with the utilization by the cells of the accumulated starch. However, the wet weight of the cells continues to rise. At this stage, the cells are large and have a tendency to trap culture medium on the filter bed, which leads to a high fresh weight.

The methods just described give a measure of cell biomass, but yield little information about the health of the culture. This parameter is routinely assessed by the measurement of the percentage viability of the cell population throughout a growth cycle. There are various ways of assessing the viability of a cell culture (*see* Chapter 3, this vol.). Most of these methods involve the use of cytological stains (2). Here we describe one method based on the vital stain fluorescein diacetate (FDA). When FDA molecules come into contact with living cells, they are hydrolyzed by the action of membrane-bound esterases and the resulting fluorescein moieties accumulate in the cytoplasm of the cells. When excited by light, fluorescein molecules give a bright green fluorescence. It should be noted that the use of a single method for viability estimation will only be qualitative.

Nutrient uptake from the medium is also a measure of culture growth. Analysis of certain key components in the growth medium will indicate if a particular nutrient is limiting at any stage in the culture cycle. We describe here methods for the determination of sucrose (3), fructose (3), glucose (4), nitrate, and inorganic phosphate (5) in the culture medium of plant cells. The results from carbohydrate analysis together with biomass measurements will provide the necessary data for calculation of the carbon conversion efficiency of a cell suspension culture growing under defined conditions (*see* Note 8).

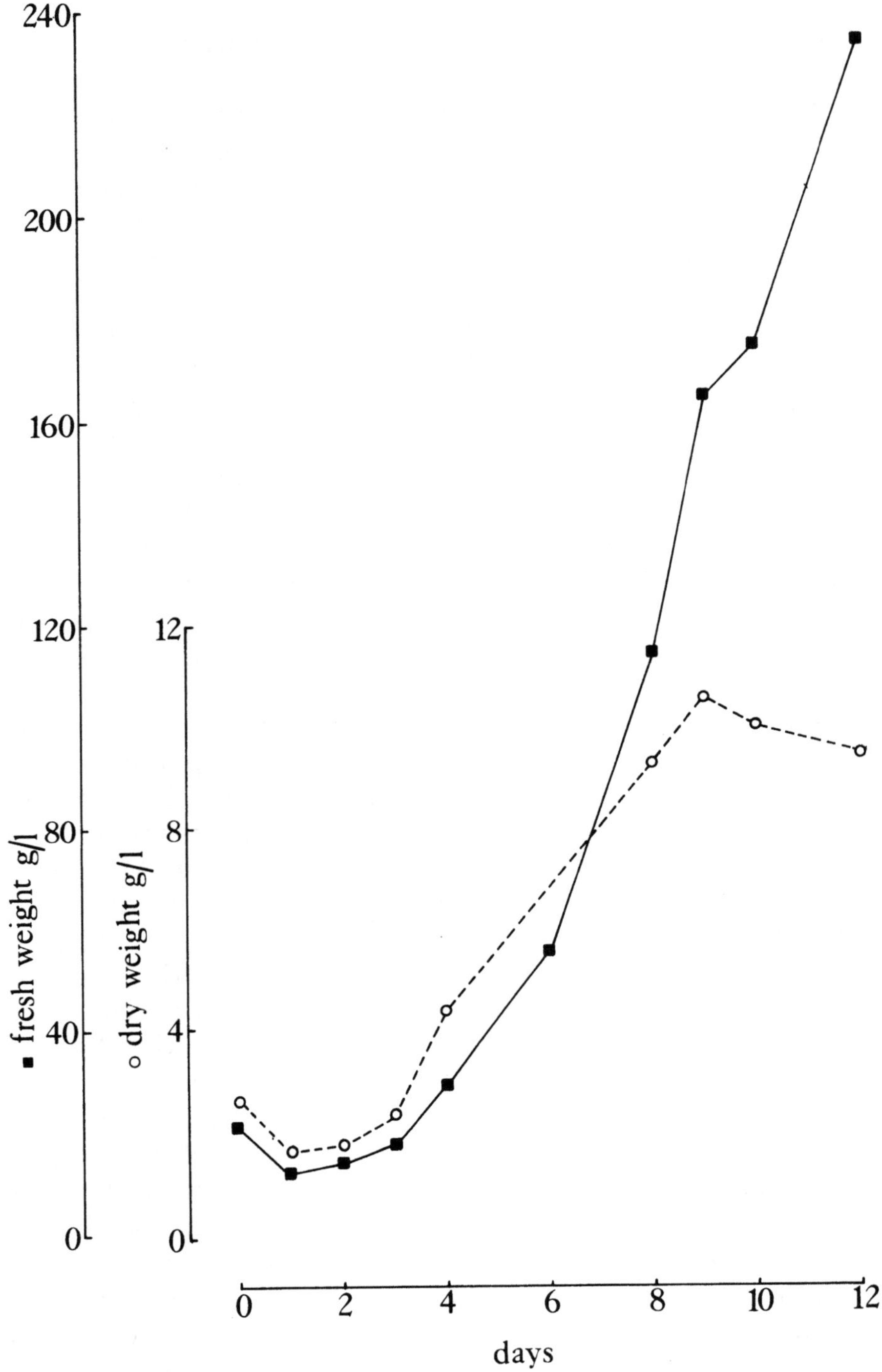

Fig. 2. Growth profile of a carrot suspension culture grown on 2% sucrose.

2. Materials

2.1. *For Fresh and Dry Weights*

1. Whatman No. 1 filter paper disks (2.5 cm).
2. Oven at 60°C.
3. Stainless-steel filter bed (*see* Fig. 3).

2.2. *For Viability*

1. Fluorescein diacetate stock solution (5 mg/mL acetone). This solution must be stored at 4°C, where it is stable for 6 mo.
2. Fluorescein diacetate working solution. This is prepared by adding 0.1 mL stock solution to 5 mL distilled water. This solution is not stable for more than 30 min.
3. Cavity microscope slides and coverslips.
4. Fluorescence microscope equipped with an exciter filter (wide spectral range) in the blue excitation region, a dichronic mirror in the range of 500–515 nm, and a barrier filter at 515 nm.

2.3. *For Sucrose Measurements*

1. 30% (w/v) KOH solution.
2. 70% (v/v) H_2SO_4 (for preparation, *see* Methods section).
3. Anthrone reagent. This is prepared by dissolving 150 mg anthrone in 100 mL 70% H_2SO_4. It is imperative that the final color of this solution be yellow, not green (*see* Note 5). This reagent is stable for 2 wk at 4°C, after which a greenish-brown coloration develops.
4. Standard sucrose solution (1 mg/mL). This reagent should be made fresh each time.

2.4. *For Fructose Measurements*

1. 70% (v/v) H_2SO_4 (for preparation, *see* Method of Sucrose Measurements).
2. Anthrone reagent (*see* reagent 3 above and Note 5).
3. Standard fructose solution (1 mg/mL). This solution should be made fresh each time.

2.5. *For Glucose Measurements*

1. Boehringer Mannheim GOD-PAP test kit for glucose (Catalog No. 166391).

2. Standard glucose solution (1 mg/mL). This solution should be made fresh each time.

2.6. For Inorganic Phosphate Determination

1. 2.5% (w/v) ammonium molybdate in 2.5*M* H_2SO_4.
2. Reducing agent stock mixture: 1-amino-2-naphthol-4-sulfonic acid (1 g), sodium metabisulfite (69 g), and sodium sulfite (5 g, anhydrous). This mixture should be blended thoroughly with a mortar and pestle, and stored in the dark at room temperature.
3. Reducing agent working solution: This is prepared by dissolving 7.5 g of the stock mixture in 50 mL of distilled water just prior to use. The solution is stable for no more than 2 h.
4. Standard phosphate solution: 4 μ*M* KH_2PO_4 (0.545 mg/mL). This solution should be made fresh each time.

2.7. For Nitrate Determination

1. Standard nitrate solution: Prepare 0.1, 1.0, 10.0, and 100.0 m*M* solutions of sodium nitrate in nitrate-free culture medium. These solutions are made up by dissolving sodium nitrate (0.85 g) in 100 mL of nitrate-free culture medium for a 100 m*M* solution, and the rest are diluted from this. Standard nitrate solutions should be prepared fresh each time.
2. A nitrate probe connected to any meter that will give readings in millivolts (mV) using the correct adaptors.

3. Methods

3.1. Fresh and Dry Weights

1. Place a supply of filter paper disks in the oven at least 24 h before use.
2. Weigh dried filter paper disk directly after removal from the oven (1 disk/sample aliquot).
3. Place the disk on the stainless-steel filter bed (Fig. 3) seated on a rubber stopper in a Buchner flask under vacuum. Wet the filter disk with distilled water (about 3 mL), and allow vacuum to operate for a constant period of time (e.g., 10 s). Reweigh the wet filter disk immediately before any evaporation has taken place.
4. Replace the disk on the filter bed, and position the filter top centrally over the disk. Filter a known volume of culture (e.g., 3 mL). To ensure

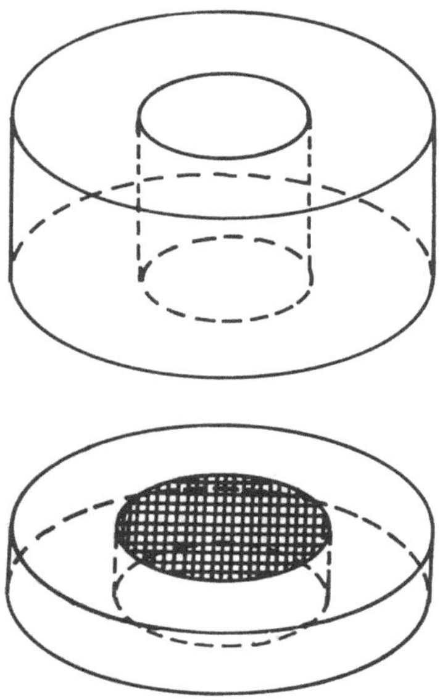

Fig. 3. Stainless-steel filter unit used in fresh and dry weight determination. The diagram is drawn to show the actual size of the unit. The filter unit consists of two parts: the filter top (A) and the filter bed (B). The filter top has a central hole. The filter bed also has a central hole, but with a slightly larger diameter than that of the filter top and covered with wire mesh to support the filter paper disk when in use.

that a representative sample is taken, the culture should be well mixed before sampling (*see* Note 1). When the cells appear "dry" allow vacuum to operate for the same period of time as above (*see* Note 2). Reweigh the wet disk cells immediately.

5. Place the disk cells on a clearly labeled weighing boat or in a Petri plate, and return to the oven to dry for at least 24 h before weighing the dry disk cells.
6. Fresh and dry weights of cells are calculated according to Equations 1 and 2:

$$\text{Wet wt (g/L)} = \frac{\text{Wt of wet disk + cells (g)} - \text{wt of wet disk (g)} \times 1000}{\text{Sample vol (mL)}} \quad (1)$$

$$\text{Dry wt (g/L)} = \frac{\text{Wt of dry disk + cells (g)} - \text{wt of dry disk (g)} \times 1000}{\text{Sample vol (mL)}} \quad (2)$$

Figure 4 shows a sample of a chart used to record data for fresh and dry weights (*see* Note 3).

Culture details	Age		Cult. vol.	Wet weight of cells				Dry weight of cells				Mean±SEM
	Day	hr	(ml)	FP+C (g)	Wet FP (g)	Net wet weight (g)	(g/L)	FP+C (g)	Dry FP (g)	Net dry weight (g)	(g/L)	(g/L)
												Wet weight
	Viability (%)		Medium pH									
												Dry weight

Fig. 4. Suggested chart layout for recording growth parameters. Three or more measurements are used to calculate the average fresh and dry weights. The abbreviations used are: FP, filter paper; C, cells; SEM, standard error of the mean.

3.2. Viability

1. Thoroughly mix the culture before placing a drop of cell suspension into the cavity of the microscope slide. Add one drop of fluorescein diacetate working solution to the cell sample and place a coverslip over it (*see* Note 4).
2. After 2 min, observe the fluorescence under a microscope. Fluorescein diacetate, a polar molecule, quickly penetrates the cells and is converted to nonpolar fluorescein, which accumulates in the cytoplasm in approximately 2 min. Viable cells exhibit a bright green fluorescence, which lasts for at least 15 min. Viability is estimated by counting the number of cells that fluoresce in a given field and expressing this as a percentage of the total number of cells in that field. For a more representative measure of culture viability, it is advisable to count the cells in several fields. After about 30 min, the fluorescein begins to diffuse from the cells into the medium, which then also fluoresces.

3.3. Sucrose Measurement

3.3.1. Preparation of 70% H_2SO_4

NOTE: The preparation of 70% H_2SO_4 from concentrated acid is an extremely exothermic reaction, and certain precautions need to be observed.

1. Wear safety glasses and gloves, and keep a container of Na_2CO_3 at hand to be used in the event of acid spillage.
2. Embed a 2-L conical flask containing a magnetic flea in ice in a plastic bowl. Place the bowl on a magnetic stirrer in a fume cupboard.
3. Pour 300 mL distilled water into the flask, and switch on the stirrer.

4. Slowly add 700 mL of concentrated H_2SO_4 to the water in the flask. Leave to stir for at least 1 h to allow the solution to cool. Store the reagent in dark.

3.3.2. Preparation of Calibration Curve

It is advisable to construct the calibration curve prior to the start of the medium analysis, since it is easier to detect whether anything is wrong with the assay at this stage (*see* Note 5).

1. Pipet 20, 40, 60, 80, and 100 µL standard sucrose solution into duplicate sets of test tubes. Make up all volumes to 100 µL with distilled water and mix well.
2. Add 100 µL of 30% KOH to each tube and mix well.
3. Cover the tubes with foil, and place them in a boiling water bath for 10 min. Remove and leave to cool.
4. Add 3 mL of anthrone reagent to the first tube and mix well. Place in a 40°C water bath, and start the timer. After exactly 2 min, repeat with the second tube. Follow this pattern with the remaining tubes.
5. After exactly 20 min from the addition to the first tube, remove this tube from the water bath and read the absorbance at 620 nm. Repeat at 2-min intervals with the remaining tubes. The absorbance at 620 nm of the tube containing 100 µL of standard sucrose solution should be approximately 1.1 ± 0.1.
6. Construct a calibration curve of sucrose concentrations (mg/mL) against absorbance readings.

3.3.3. Analysis of Medium Samples

1. Pipet into duplicate sets of tubes appropriate amounts of sample to give absorbance readings that lie within the range of standard curve. Dilution of medium samples may be necessary in most cases.
2. Repeat steps 2–5 of section above (Calibration Curve).
3. To calculate sucrose concentrations in medium samples, use the standard curve constructed above.

3.4. Fructose Measurement

This method uses 70% H_2SO_4 prepared as described above.

3.4.1. Preparation of Calibration Curve

It is advisable to construct the calibration curve prior to the start of medium analysis, since it is easier to detect whether anything is wrong with the assay at this stage.

1. Pipet 20, 40, 60, 80, and 100 µL aliquots of standard fructose solution into duplicate sets of tubes, and make up the volume in each tube to 100 µL with distilled water.
2. At timed intervals (*see* section on Sucrose Measurement above), add 3 mL of anthrone reagent, and leave at room temperature for 15 min. Read the absorbance at 620 nm.
3. Construct a standard curve of fructose (mg/mL) vs absorbance reading at 620 nm. The absorbance at 620 nm for the tube containing 100 µL standard fructose solution should be approximately 1.8 ± 0.2.

3.4.2. Analysis of Medium Samples

1. Pipet into duplicate sets of test tubes appropriate amounts of sample, which will give an absorbance reading within the standard range.
2. Make up all the volumes in the test tubes to 100 µL with distilled water.
3. Add 5 mL of anthrone reagent, mix, and leave at room temperature for 15 min. Read the absorbance at 620 nm.
4. The procedure described above determines both fructose and sucrose amounts in the given sample. It is therefore necessary to run sucrose assays on medium samples as described above. The absorbance value for a particular sample obtained in the sucrose assay is then subtracted from the corresponding absorbance value from step 3 above. The net absorbance value thus obtained is read against the fructose calibration curve in order to determine the fructose concentration in the given sample.

3.5. Glucose Measurement

3.5.1. Preparation of the Assay Reagent

1. Dissolve the contents of bottle 1, which contains buffer, enzymes, and 4-aminophenazone in 200 mL distilled water. Add to this the contents of bottle 2 containing phenol.
2. Mix well, and store in the dark at 4°C. Care must be taken with this reagent (for details refer to literature enclosed with the kit).

3.5.2. Construction of Calibration Curve

1. Pipet 20, 40, 60, 80, and 100 µL aliquots of standard glucose solution into duplicate tubes. Make up the volume in each tube to 200 µL and mix well.
2. Add 2 mL of assay reagent to each tube at timed intervals, and mix well. Incubate the tubes at room temperature in the dark for 60 min. Read the absorbance at 510 nm. The color will remain stable for up to

1 h. The absorbance at 510 nm of the tube containing 100 μL of standard glucose solution should be approximately 1.5 ± 0.2.
3. Construct a standard curve of glucose (mg/mL) vs absorbance reading at 510 nm.

3.5.3. Analysis of Medium Samples

1. Pipet 200 μL of each medium sample into duplicate sets of tubes. Dilute when necessary.
2. Add 2 mL of assay reagent to each tube at timed intervals. Mix well and read the absorbance at 510 nm, incubating tubes in the dark at room temperature for 60 min.
3. To calculate glucose concentrations in the medium samples, use the standard curve constructed above (*see* Notes 6–8).

3.6. Inorganic Phosphate Determination

3.6.1. Preparation of Calibration Curve

1. Pipet 25, 50, 100, 150, 200, 300, and 400 μL aliquots of standard phosphate solution into duplicate sets of test tubes. Make up all volumes to 500 μL and mix well.
2. Add 500 μL of 2.5% ammonium molybdate solution to each tube. Add 3.8 mL of distilled water and mix well.
3. Add 200 μL of reducing agent working solution at timed intervals to each tube and mix well. Incubate the tubes at room temperature for 10 min, and read the absorbance at 660 nm.
4. Construct a calibration curve of phosphate concentrations (μmol P_i) against absorbance readings. The absorbance at 660 nm for the tube containing 400 μL (1.6 μmol P_i) standard phosphate solution should be approximately 1.2 ± 0.2.

3.6.2. Analysis of Medium Samples

1. Pipet 500 μL of undiluted medium sample into duplicate test tubes. Repeat steps 2 and 3 above.
2. To calculate phosphate levels, use the standard curve constructed above (*see* Note 9).

3.7. Nitrate Determination

3.7.1. Construction of Calibration Curve

1. Immerse the nitrate probe in approximately 10 mL of the 0.1 n*M* nitrate standard, and set the mV reading on the meter to an arbitrary value, e.g., 65 mV. The probe may take some time to stabilize.

2. Read the values for the remaining nitrate standards beginning with the lowest concentration. Rinse and blot dry the probe between readings.
3. Plot the calibration curve on semilog graph paper—m*M* nitrate values on the log axis, mV readings on the linear axis.

3.7.2. Analysis of Medium Samples

1. Dilute the medium samples 1:10 with nitrate-free culture medium. Working volumes of 20 mL will be convenient to use.
2. Immerse the probe in the solution, and note the mV reading. Wash and blot dry the probe between readings.
3. Use the calibration curve to convert the mV readings for unknown samples into nitrate concentrations (m*M*).

4. Notes

1. A convenient way of taking samples for fresh and dry weights is to use a 5-mL automatic pipetor (e.g., Gilson). The end 5 mm of the pipet tip is cut off to widen the bore size, and ensure that larger clumps are also taken and included in the measurement. By using this method, it is also possible to withdraw samples from cell suspension cultures aseptically, since the pipet tips and filter tips can be autoclaved.
2. Often cells adhere to the upper part of the filter unit during filtering. In this case, remove the top section, scrape the cells from its inner surface, and place them on the filter disk while the latter is still on the bottom half of the filter unit. You may find that the use of a pair of forceps with blunt ends or a spatula will facilitate the handling of the filter disks.
3. For routine measurement of culture growth, samples may be taken on alternate days. To ensure that a representative average value is obtained for fresh and dry weights, triplicate samples are taken. This can be done by either removing 3 samples from 1 flask/d or taking 1 sample from 3 flasks. Ideally, in both cases 1 flask/d should be used for sampling, since a significant reduction in culture volume will affect mixing and aeration of the culture. Therefore, if the second of the two alternatives is adopted, three times as many flasks should be set up for the experiment.
4. The fluorescein diacetate (FDA) method used for viability studies is in fact a measure of membrane-bound esterase activity. Therefore, strictly speaking, this method indicates that the cell membrane is in-

tact, and this feature is taken to mean that the cells are viable. The following points are helpful hints for viability assessment:

a. If the culture being used is a very fine suspension, it is possible to substitute the cavity slides with ordinary microscope slides.
b. The presence of background fluorescence on the microscope slide is an indication of extensive cell lysis.
c. The presence of crystals on the slide indicates that the FDA has come out of solution and needs to be replaced by a fresh solution.
d. It is relatively easy to estimate the viability of a fine suspension culture. However, caution is needed when assessing the viability of a culture containing large cell clumps where the fluorescence can be attributed to cells on the surface of such clumps. Allowance should therefore be made for the possible presence of nonviable cells within the clumps.

5. The presence of even trace amounts of a reducing sugar on the glassware used to prepare the 70% H_2SO_4 solution and subsequently the anthrone reagent will cause a green coloration in the latter and make it unusable. It is essential, therefore, to acid-wash all glassware used in these determinations and keep them separate. The spatula used in weighing the anthrone reagent should also be washed in acid, rinsed with distilled water, and air-dried. It should never be wiped with tissue or cloth, since fibers of these materials affect the anthrone reagent.
6. An example of the pattern of medium carbohydrate utilization is shown in Fig. 5. In this case, sucrose is hydrolyzed externally to give fructose and glucose in the medium, as seen by the falling sucrose concentration and the concomitant rise in the levels of the two monosaccharides. The presence of glucose and fructose in the sample for d 0 is explained partly by the hydrolysis of sucrose during autoclaving and partly by its own rapid hydrolysis when the cells are subcultured into the fresh medium. The apparent discrepancy in the total level of carbohydrate may be the result of the use of different methods to determine individual sugars.
7. It is possible to determine the levels of various sugars in medium samples by HPLC analysis with the appropriate column and detector (*see* Chapter 43, this volume).
8. As mentioned in the Introduction section, it is possible to calculate the carbon conversion efficiency of a given culture using data from medium carbohydrate analysis and dry weight measurements. Carbon conversion values are calculated according to the Equation 3:

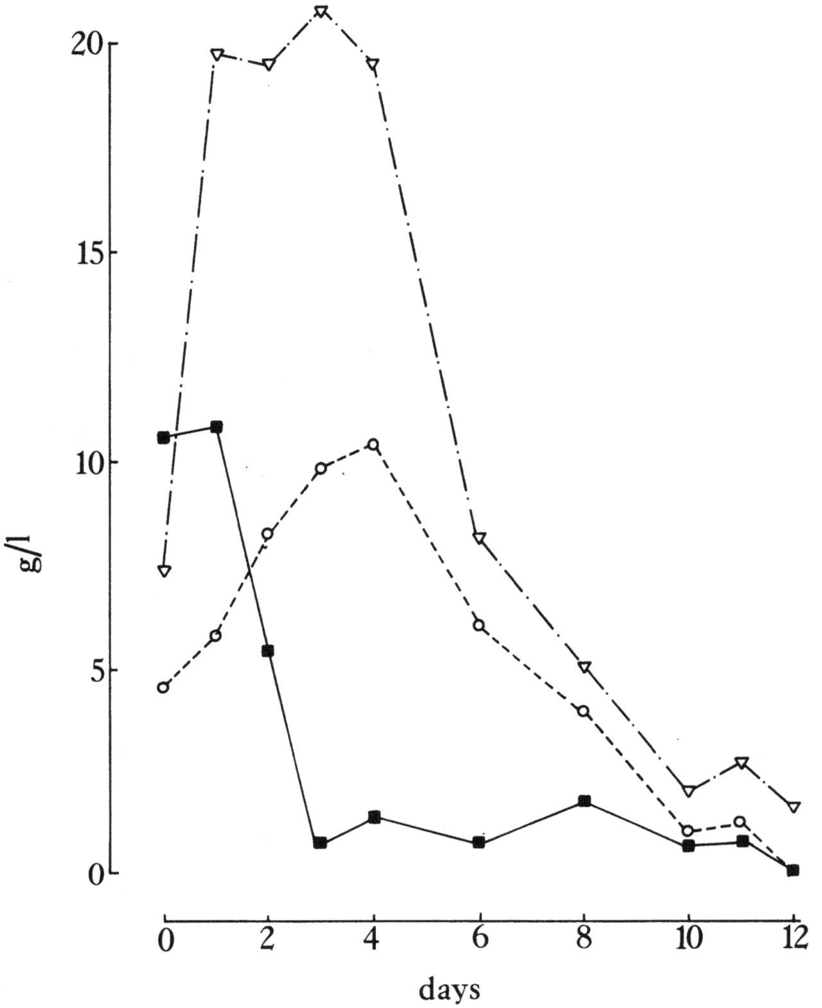

Fig. 5. Pattern of carbohydrate utilization in the medium of carrot cultures grown on 2% sucrose. The sugars analyzed are sucrose (■), glucose (∇), and fructose (O).

$$\text{Carbon conversion (\%)} = \frac{\text{Final dry wt (g/L)} - \text{Initial dry wt (g/L)} \times 100}{\text{Initial carbon level (g/L)} - \text{Final carbon level (g/L)}} \qquad (3)$$

Care must be taken in interpreting carbon conversion efficiency values, since not all the carbohydrate taken up is converted into biomass, but is also oxidized. For a discussion of carbohydrate utilization by plant cells, the reader is referred to Fowler (*6,7*).

9. During a typical growth cycle, phosphate is rapidly taken up from the medium by plant cells. In certain cases, towards the end of the growth

cycle, the levels of phosphate in the medium may rise sharply. This phenomenon can be ascribed to cell lysis and the release of intracellular phosphate pools into the medium. This rise in phosphate level may correspond to a concomitant fall in cell viability.

References

1. King, P. J., Mansfield, K. J., and Street, H. E. (1973) Control of growth and cell division in plant cell suspension cultures. *Can. J. Bot.* **55,** 1807–1823.
2. Dixon, R. A. (1985) Isolation and maintenance of callus and cell suspension cultures, in *Plant Cell Culture: A Practical Approach* (Dixon, R. A., ed.), IRL, Oxford and Washington DC, pp. 1–20.
3. van Handel, E. (1968) Direct microdetermination of sucrose. *Anal. Biochem.* **22,** 280–283.
4. Trinder, P. (1969) Determination of glucose in blood using glucose oxidase with an alternative oxygen acceptor. *Ann. Clin. Biochem.* **6,** 24–27.
5. Leloir, L. F. and Cardini, C. E. (1957) Characterization of phosphorous compounds by acid lability. *Meth. Enzymol.* **3,** 840–850.
6. Scragg, A. H. and Fowler, M. W. (1988) Large-scale culture of plant cells, this vol.
7. Fowler, M. W. (1982) Substrate utilization by plant cell culture. *J. Chem. Tech. Biotechnol.* **32,** 338–346.
8. Fowler, M. W. (1986) Industrial applications of plant cell culture, in *Plant Cell Culture Technology* (Yeoman, M. M., ed.), Blackwell Scientific, Oxford, pp. 202–227.

Chapter 3

Measurements of Viability Suitable for Plant Tissue Cultures

David R. Duncan and Jack M. Widholm

1. Introduction

Experiments using plant tissue cultures, such as measuring their tolerance to herbicides or metabolite analogs, elucidating biochemical pathways by exposing the tissues to metabolic inhibitors, or cryopreserving the cultures, require at some point the determination of the tissue's viability. Viability is the capacity of a cell or organism to live, and the state of being viable is most commonly recognized by the growth of the studied tissues (*1–3*). Measurements of growth can include measuring increases in fresh or dry weight of a cultured tissue (*2,3*) or measuring increases in cell number or size (*2–4*).

Growth, however, can be a misleading measure of cell viability (*3*). For instance, low temperatures or metabolic inhibitors may arrest growth, but not kill a tissue. The tissue may just be growing extremely slowly, and normal growth may resume once the inhibition to growth is removed (*1,3*). Removing the inhibitor, however, may take a long time, if for instance the

From: *Methods in Molecular Biology*, vol. 6, *Plant Cell and Tissue Culture*
Edited by Jeffrey W. Pollard and John M. Walker, ©1990 by The Humana Press

inhibitor has accumulated in the tissue. Thus, factors such as the rate of inhibitor removal can greatly complicate growth measurements, and may result in the data being misleading or the analysis requiring a long time to complete. Growth measurements are also usually not very quantitative, so that it is difficult to determine what proportion of the cells is viable. Tissues may also not grow, even though a small proportion of the cells is actually alive.

When feasible, growth measurements can be useful, and they can be applied to all forms of plant tissue cultures, meristem clones, callus, cell suspensions, and protoplasts. Using growth as a test of viability also has the advantage of being nondestructive, which may be of critical importance, if for instance the tissue is difficult or costly to obtain. However, because of the possibility of misinterpreting the results, growth measurements are most often best used in conjunction with other viability tests, such as those that measure some parameter of metabolic activity (*1*).

Viability tests that measure "vital metabolic functions" can be grouped into three categories:

1. Observations of cytoplasmic streaming (*3*).
2. Measurements of membrane integrity, such as electrolyte leakage, plasmolysis, and stain exclusion or retention (*1,2,5–7*).
3. Measurements of biochemical activity, such as protein synthesis (*2*), tetrazolium chloride reduction (*3,8,9*), DNA and RNA synthesis (*2,10*), fluorescein diacetate staining (*7*), and ATP content (*11*).

Probably arising from human experience with the death of other individuals, these tests are often viewed from the "nonscientific" perspective that cell death occurs at a discrete instance, at which time all vital functions stop (*1*). It is apparent from the literature that, once a lethal stress is applied to a tissue, a period of time may pass, perhaps minutes to days, before all vital cellular functions stop (for examples, *see* refs. *1* and *12*). Consequently, which viability test is used and when it is applied to a tissue are extremely important experimental parameters to consider for a clear interpretation of the viability of the tissue, and setting these parameters may strongly depend on the type of tissue being studied. For instance, to use DNA or RNA synthesis as a viability test in a differentiating culture would require doing the assay at the same physiological stage of differentiation, which may be very difficult to accomplish if experimental conditions alter the rate of differentiation. Thus, these two viability tests may be inappropriate assays for a differentiating tissue.

Most of these viability tests are also generally based on metabolic studies using animal tissues (*1,2*). The application of these tests to plant cultures is a possible point for concern. The structural and biochemical differences in plant cells as compared to animal cells may possibly invalidate or render useless a viability test. For instance, a stain used with animal cell cultures may bind to or not pass through a plant cell wall. It is also possible that a plant cell or a particular species of plant may have metabolic activities or accumulated metabolites that interfere with a viability test. Such potential problems suggest that it would be a prudent practice to verify, with each plant culture system studied, that a viability test actually measures what the literature reports it to be measuring.

The fact that most of the viability tests used today were developed from studies of animal cells also suggests that there could be many plant-specific vital functions that could be, but are rarely being, exploited for tests of viability in plant cell cultures. For instance, ethane emination (*13,14*) or polygalacturonase accumulation (*15*) may be useful for monitoring viability in heterotrophic cultures. If the cultures are photoautotrophic, then the degradation of light harvesting complex II may be a useful measure of viability (*16*).

Of the viability tests that assay vital metabolic functions, only the observation of cytoplasmic streaming is nondestructive, but it can only be applied to cell suspensions, protoplasts, and small clumps of cells through which light can pass. Even in single cells and protoplasts, the large vacuole in cultured plant cells may prevent the observation of cytoplasmic streaming (*7*).

Of the membrane integrity tests, electrolyte leakage can be used with virtually all types of cultured plant material. Some cultured tissues, however, may not be homogeneous in composition. If the different cell types in the tissue leak electrolytes at different rates, then the electrolyte loss measured may be determined more by the composition of the tissue than by its viability. Also, the conductance of distilled water containing a tissue sample is typically used to assess electrolyte leakage from the sample. Consequently, electrolyte loss is measured only from cells that are in contact with the water at a tissue's surface. Thus, variation in the surface area of the studied tissues may also affect the electrolyte loss from a tissue.

Stain exclusion assays, such as the use of phenosafranine (*7*) with observing cytoplasmic streaming, are useful only with cell suspensions and protoplasts or small clumps of cells through which light can pass. These assays can also be misinterpreted if dead or dying cells lose their cytoplasm. Such cells would not be stained and potentially could be considered as being alive. We have also noted that, although phenosafranine gives valid results with many species, the cell walls of *Zea mays* suspensions bind the dye, and thus obscure any exclusion or cytoplasmic binding.

Of the viability tests that assay metabolic activity, most can be used on any type of plant tissue culture. These assays will, however, have similar limitations as the measurement of electrolyte leakage, if the culture being studied is composed of several metabolically different types of cells (*1,3*).

This brief general introduction should serve to show that the techniques available today for assessing cell viability have limitations. This is not to say that these assays should not be used. Instead, to use these assays, some practical precautions should be taken. To list a few:

1. Several viability tests should be used to reduce interpretation error.
2. The validity of the viability test should be established for each culture system studied. Validity in this circumstance would mean that the assay results correlate to the viability of the cultured tissue, as measured, for instance, by comparing the assay results from activity growing samples to samples killed by freezing and thawing. Validity also would mean that the mode of action of the assay is the same in the plant system being studied as in the system reported in the literature.
3. The viability test should be conducted after a tissue has been removed from the experimental situation and placed back into conditions of optimal growth.

Because emphasis has been placed on using several viability tests to verify a culture's viability, the remainder of the chapter will be dedicated to discussing two widely used viability tests, the reduction of 2,3,5-triphenyltetrazolium chloride (TTC) and fluorescein diacetate (FDA) staining. These assays, in conjunction with growth under optimal conditions, can give fair assessments of cell viability in plant tissue cultures.

The theory behind the use of TTC reduction as a viability test is that, when TTC, a water-soluble, colorless solution, is within a live cell, it will be reduced to a water-insoluble red formazan by dehydrogenase activity or the mitochondrial electron transport chain (*17*). A dead cell should have no mitochondrial activity and, consequently, should not be able to reduce TTC. Thus, live tissue in a TTC solution will turn red, and dead tissue will not change color. To quantify the assay, the red precipitate can be extracted from the tissue with ethanol and its absorbance determined spectrophotometrically (*9*).

In the case of FDA, it is thought that the nonpolar, nonfluorescent molecule enters plant cells where the acetate moieties are removed by an esterase. The resulting fluorescent product, fluorescein, is consequently retained in the cell because its polar nature prevents it from crossing the plasmalemma. Dead cells should have both limited esterase activity and leaky membranes. Consequently, these cells would produce less fluorescein than live cells and would not retain the fluorescein produced. The maximum absorption of light by fluorescein is between 450 and 500 nm at biological pH values, and its emitted fluorescent light spectrum peak is near 520 nm. Thus, the fluorescence of fluorescein can be observed using a conventional light microscope fitted with the appropriate filters. Thus, live cells exposed to FDA will fluoresce, and dead cells will not.

TTC reduction requires several hours to occur, and it is difficult to visualize under a microscope. Consequently, it can be used only as a general assay of viability for a population of cells. FDA, however, is rapid and can be visualized within individual cells microscopically. Consequently, FDA can not only be used as a general viability assay, but its use also facilitates determining the number of live cells in a population.

2. Materials

2.1. 2,3,5-Triphenyltetrazolium Chloride (TTC) Reduction

1. Potassium phosphate buffer 0.05*M*, pH 7.5 (8.7 g K_2HPO_4 + 6.8 g KH_2PO_4 dissolved in deionized distilled H_2O to make 1 L and adjusted to pH 7.5 with KOH).
2. 23.9 m*M* 2,3,5-triphenyltetrazolium chloride (TTC) solution (0.4 g TTC/100 mL of 0.05*M* potassium phosphate buffer) (*18*).

2.2. Fluorescein Diacetate (FDA) Staining

1. Fluorescein diacetate (FDA) stock solution: 5 mg/mL acetone stored at –20°C.
2. Culture medium (typically Murashige and Skoog medium) (*see* Appendix) (*19*).
3. Fluorescence microscope or a light microscope equipped with barrier filters to produce an excitation wavelength of 493 nm and to detect an emission wavelength of 520 nm (*20*).

3. Methods

3.1. 2,3,5-Triphenyltetrazolium Chloride (TTC) Reduction (See *Notes 1–8*)

1. Place 10–30 mg fresh weight of callus tissue or suspension cultured cells in a microfuge tube with 1 mL of the 23.9 m*M* TTC solution.
2. Set the mixture at 28°C for 6 h in the dark. The formation of the water-insoluble (ethanol soluble) formazan in a positive reaction will yield a red pigmented tissue.
3. Remove the TTC solution with a Pasteur pipet attached to a vacuum line if callus tissue is used or, if a suspension culture is being tested, by centrifuging the solution in a Microfuge for 2 min and then removing the supernatant carefully with a Pasteur pipet.
4. Add 0.5 mL of 95% (v/v) ethanol to the tube, and pulverize the tissue with a plastic pestil that fits the microfuge tube (Direct, products for science and technology, P. O. Box 1039, Millville, New Jersey, 08332, USA). Add to the tube an additional 0.5 mL of 95% ethanol and set, covered, at room temperature for 4 h to completely solubilize the formazan.
5. Centrifuge the resultant material in a microfuge for 5 min (pellet should be white after complete extraction).
6. Determine the absorbance of the supernatant spectrophotometrically at 485 nm.

3.2. Fluorescein Diacetate (FDA) Staining (See *Notes 9,10*)

1. Dilute 0.5 mL of FDA stock with 24.5 mL of culture medium (0.01% w/v solution).
2. On a glass slide, mix one drop of cell suspension and one drop of diluted FDA.
3. Wait at least 5 min before viewing under a microscope to allow fluorescence to develop.
4. View slide with the microscope, and count fluorescing cells as viable and nonfluorescing cells as dead.

4. Notes

1. The TTC solution should be stored frozen or 4°C in the dark. If the solution has a red tinge to it, then the solution should not be used.

2. Since TTC reduction requires mitochondrial activity, factors known to slow or inhibit mitochondrial activity, i.e., low temperatures, will also affect TTC reduction. Consequently, for TTC reduction to be an effective viability test, the assay must be conducted under optimal growth conditions.
3. The procedure uses a 0.05*M* potassium phosphate buffer; however, 0.5*M* (*8*) and 0.025*M* phosphate buffers (*3*) have been used with apparent success. For special situations, we have found a need to use liquid culture medium as a TTC solvent and have seen no apparent reduction in the effectiveness of the assay. The pH (around 7.5), however, is particularly critical for dissolving the TTC.
4. The time that the tissue is exposed to the TTC solution, although listed above as 6 h, varies in the literature from 6 h (*18*) to 15 h (*8*) to 18–20 h (*3,9*). Thus, the actual time chosen for the assay should probably depend upon how rapidly the tissue to be studied reduces TTC.
5. Since TTC reduction depends upon mitochondrial activity, if the tissue is crushed well, this should stop most if not all further TTC reduction. If, however, many samples must be extracted at about the same time, the TTC reduction can be effectively stopped by rapidly cooling or freezing the samples. The samples can be stored in the cooled or frozen state until they can be extracted.
6. The time the tissue sits in 95% ethanol to dissolve the formazan can vary and should be adjusted so that all the formazan is dissolved. We have left solutions overnight in the dark, and others have heated the solution for 5 min at 80°C (*8*) or for 5–15 min at 60°C (*9*) to ensure that the formazan dissolved. If heating is used, the volumes of each assay extract will need to be adjusted to a specified final volume, since the 95% ethanol will have partially evaporated.
7. Several reports use a wavelength of 530 or 540 nm to measure the formazan absorbance (*8,17*). To avoid absorbtion by compounds other than reduced TTC, an absorbance optima at 485 nm has been widely used based on the assumption that plant cultures do not have compounds that would appreciably absorb light at that wavelength (*3,9,17*). A safe precaution would be to extract untreated cultured tissue in 95% ethanol, and examine the absorbance of the extract at 485 and 530 nm to see which wavelength has the least background absorbance.
8. Appropriate controls for TTC assays can be heat-killed or frozen and thawed tissue and actively growing tissues. The heat-killed or frozen

tissues should not react with the TTC, and the actively growing tissues should react. One should be able to plot the absorbance reading of the dead tissues as zero viability and the absorbance of the actively growing tissue as 100% viable (*18*). Such a plot facilitates determining the quantity of viable tissue in samples of comparably fresh weight.

9. Appropriate controls for FDA assays, again, can be heat-killed or frozen and thawed tissue and actively growing tissues. The heat-killed or frozen tissues should not fluoresce, and the actively growing tissues should fluoresce.
10. The diluted FDA solution is only usable for a few hours.

References

1. Malinin, T. I. and Perry, V. P. (1967) A review of tissue and organ viability assay. *Cryobiol.* **4,** 104–115.
2. Patterson, M. K., Jr. (1979) Measurements of growth and viabilty of cells in culture. *Methods Enzymol.* **58,** 141–152.
3. Zilkah, S. and Gressel, J. (1978) The estimation of cell death in suspension cultures evoked by phytotoxic compounds: Differences among techniques. *Plant Sci. Lett.* **12,** 305–315.
4. Gilissen, L. J. W., Hanisch ten Cate, C. H., and Keen, B. (1983) A rapid method of determining growth characteristics of plant cell populations in batch suspension culture. *Plant Cell Reports* **2,** 232–235.
5. Gaff, D. F. and Okong 'O-ogola, O. (1971) The use of nonpermeating pigments for testing the survival of cells. *J. Exptl. Bot.* **22,** 756–758.
6. Patterson, B. D., Murata, T., and Graham, D. (1976) Electrolyte leakage induced by chilling in passiflora species tolerant to different climates. *Aust. J. Plant Physiol.* **3,** 435–442.
7. Widholm, J. M. (1972) The use of fluorescein diacetate and phenosafranine for determining viability of cultured plant cells. *Stain Technol.* **47,** 189–194.
8. Kartha, K. K. (1981) Meristem culture and cryopreservation—Methods and applications, in *Plant Tissue Culture: Methods and Applications in Agriculture* (Thorpe, T. A., ed.), Academic, London and New York, pp. 181–211.
9. Towill, L. E. and Mazur, P. (1975) Studies on the reduction of 2,3,5-triphenyltetrazolium chloride as a viablity assay for plant tissue cultures. *Can. J. Bot.* **53,** 1097–1102.
10. Ferrari, T. E. and Widholm, J. M. (1973) A sample, rapid, and sensitive method for estimation of DNA, RNA, and protein synthesis in carrot cell cultures. *Analyt. Biochem.* **56,** 346–352.
11. Sugawara, Y. and Takeuchi, M. (1986) A rapid method for the determination of cell survival in shoot apex after freezing and thawing, in *Abstracts: VI International Congress of Plant Tissue and Cell Culture* (Somers, D. A., Gengenbach, B. G., Biesboer, D. D., Hackett, W. P., and Green, C. E., eds.), Minneapolis, Minnesota, Intl. Assn. Plant Tissue Culture, p. 233.

12. Thimann, K. V. (1987) Plant senescence: A proposed integration of the constituent processes, in *Plant Senescence: Its Biochemistry and Physiology* (Thomson, W. W., Nothnagel, E. A., and Huffaker, R. C., eds.), American Society of Plant Physiologists, Rockville, Maryland, pp. 1–19.
13. Elstner, E. F. and Konze, J. R. (1974) Effect of point freezing on ethylene and ethane production by sugar beet leaf disks. *Nature* **263,** 351,352.
14. Konze, J. R. and Elstner, E. F. (1978) Ethane and ethylene formation by mitochondria as an indication of aerobic lipid degradation in response to wounding of plant tissue. *Biochim. Biophys. Acta* **528,** 213–221.
15. Bennett, A. B. and DellaPena, D. (1987) Polygalacturonase: Its importance and regulation in fruit ripening, in *Plant Senescence: Its Biochemistry and Physiology* (Thomson, W. W., Nothnagel, E. A., and Huffaker, R. C., eds.), American Society of Plant Physiologists, Rockville, Maryland, pp. 1–19.
16. Thomas, H. and Hilditch, P. (1987) Metabolism of thylakoid membrane proteins during foliar senescence, in *Plant Senescence: Its Biochemistry and Physiology* (Thomson, W. W., Nothnagel, E. A., and Huffaker, R. C., eds.), American Society of Plant Physiologists, Rockville, Maryland, pp. 1–19.
17. Nachlas, M. M., Margulies, S. I., and Seligman, A. M. (1960) Sites of electron transfer to tetrazolium salts in the succinoxidase system. *J. Biol. Chem.* **235,** 2739–2743.
18. Gray, L. E., Guan, Y. Q., and Widholm, J. M. (1986) Reaction of soybean callus to culture filtrates of phialophora gregata. *Plant Sci.* **47,** 45–55.
19. Murashige, T. and Skoog, F. (1982) A revised medium for rapid growth and bioassays with tobacco tissue cultures. *Physiol. Plant* **15,** 473–497.
20. Slavick, J. (1982) Intracellular pH of yeast cells measured with probes. *FEBS Lett.* **140,** 22–26.

Chapter 4

Cryopreservation of Plant Cells

Lyndsey A. Withers

1. Introduction

Cryopreservation, that is, the viable storage of cells at the temperature of liquid nitrogen (–196°C), has wide relevance in many areas of pure and applied biology. Examples of its very successful use can be found in the storage of microbes and of semen (*1*). More recently, attention has been given to the development of cryopreservation methods for embryos, including those of humans, tissues, and organs for transplantation and blood components. In the context of plant research, realization of the potential applications of cryopreservation has been somewhat latent. However, several clear areas for application can be identified that involve the need to store exact genotypes in a stable state.

The tendency for plant cell and tissue cultures to change over a period of time with respect to desirable features, such as the capacity to regenerate whole plants, has been recognized for many years. The disadvantages of this are obvious. Even when particular known and well-characterized traits are retained, other subtle and cryptic changes may be taking place that render controlled experimentation difficult or impossible. However,

From: *Methods in Molecular Biology*, vol. 6, *Plant Cell and Tissue Culture*
Edited by Jeffrey W. Pollard and John M. Walker, ©1990 by The Humana Press

instability in culture is a double-edged sword. One of the most positive developments in plant cell culture technology over recent years has been the recognition of the potential advantages of in vitro-generated instability. This positive thinking has led to the coining of a name for the phenomenon: "somaclonal variation" (2). The key to using somaclonal variation without suffering its disadvantages lies in control of the factors governing instability. Some of these have been identified, such as the genotype of the parent material, the explant used, and the culture system. These factors can be manipulated to a degree, but there is a dilemma when somaclonal variation is being used to generate useful genotypes that must then be stabilized. In such circumstances, it becomes necessary to eliminate time-related factors from the culture process by effectively suspending time itself. At present, cryopreservation is the only feasible means of doing this for higher plant cultures.

Cell and callus cultures are only two of the wide range of systems in use in plant in vitro work. Their tendency to instability identifies them strongly as candidates for cryopreservation, but other systems have specific storage needs. Cryopreservation of protoplasts is mainly used to explore cryoinjury and other low temperature phenomena (3), but protoplasts are essentially ephemeral structures that would benefit from a protocol that prevents time-related change, such as wall regeneration.

Shoot cultures are the material of choice in the use of in vitro techniques for the genetic conservation of sterile plants and of unique clones that would be lost if the genotype were stored as seed (4). Shoots are chosen because of their reputation for genetic stability and for the practical reason that a whole plant may more readily be generated from them than from an unorganized culture. The cryopreservation of entire seeds and of zygotic embryos is an attractive proposition for the very stable storage of plant germplasm over extended periods of time. It may be an improvement upon the conventional way of storing most seeds and the only possible way of conserving "recalcitrant" (short-lived) seeds that quickly lose viability under normal conditions of storage. The use of somatic embryos ("embryoids") as subjects for cryopreservation is being considered seriously in the context of both vegetative and seed-propagated material, since these structures may combine physical attributes that render them amenable to cryopreservation with an inherent genetic stability. However, their suitability for conservation has yet to be tested rigorously, and shoot cultures remain the most closely examined from the conservation standpoint.

Success in the cryopreservation of shoot cultures has been variable, and in the majority of studies there has been the tendency for regeneration to be indirect and not a one-to-one resumption of growth in the original meristem region. Only in a few cases does recovery growth resume without callusing (*5,6*). Generally, shoot-tips and dormant buds taken from independently-growing plants survive at a higher frequency and show greater physical stability than shoot-tips taken from in vitro cultures.

For the short to medium term, it may be safer and more practical to store shoot cultures by slow growth. This can be achieved by culturing at a reduced temperature or in the presence of inhibitors (*6,7*). The germplasm of many important crops, including staple root and tuber crops, fruits, and ornamental species, has yielded to slow growth storage as shoot cultures. Attempts to store cell and callus cultures in a similar way have proven far less satisfactory and cannot be recommended. The risk of selection under the stress conditions inevitably associated with slow growth must be recognized and, although slow growth can retard the loss of totipotency, there is evidence for the loss of both growth potential and the capacity to synthesize secondary products in callus cultures stored for relatively brief periods (*8*).

In clear contrast to the situation in slow growth is the good record of stability in cryopreserved cell and callus cultures. Evidence for stability relates to karyotype, growth kinetics, morphogenic potential, secondary product synthesis, and other specific biochemical characters (*5,6*). This is not to say that cryopreserved cultures do not show any symptoms of damage. In the days following thawing, metabolic disturbances and symptoms of physical damage to the cell membranes, for example, are evident (*9,10*). Growth may be slow initially, and a number of passages may ensue before standard behavior is resumed (*5,7,10*). However, this should not be confused with irreversible genetic change.

The following sections detail a cryopreservation method for cell suspension cultures. It provides a model upon which cryopreservation methods for other culture systems may be based. However, the reader is advised to consult the appropriate specialist literature for details of procedures intended for other types of cultures (*5,6,11–13*). Background information on the principles of cryopreservation can be found in the latter and other texts (*1,14*).

2. Materials

1. Cultures: Certain morphological types of culture are known to be more resistant to the stresses of cryopreservation than others. It is not always going to be possible to choose a culture on the basis of its likely

amenability to cryopreservation. However, it is helpful to be able to anticipate its response and take advantage of possible means of enhancing survival potential.

In general, a suspension containing large, dense cell aggregates will freeze less well than a fine one. Small, highly cytoplasmic cells are likely to survive better than large, highly vacuolated ones. Cells in late lag phase and early exponential phase are likely to freeze well, partly because of their having a more suitable morphology. As well as selecting the most suitable growth phase, fragmenting and filtering may be used to improve the culture before cryopreservation. Callus cultures may be treated in the same way as cell cultures if first chopped into sufficiently small pieces. Also, the more actively growing cells on the surface of callus pieces can be "shaved off" and treated as a suspension.

2. Pregrowth medium: For some cultures, pregrowth in standard medium is satisfactory, but is often necessary to transfer cells to a pregrowth phase in modified medium; this is rarely detrimental and is therefore recommended as routine. Standard liquid medium (e.g., Murashige and Skoog medium; *see* Appendix) supplemented with 6% (w/v) mannitol should be prepared and autoclaved in the normal way. Other possible additives are given in Note 1. Supplemented media can be stored under standard conditions and have a similar shelf-life to plain medium.
3. Recovery medium: For recovery, standard medium solidified with agar (7 g/L) should be prepared and dispensed into 5-cm diameter Petri dishes at 5 mL/dish. Storage is under standard conditions.
4. Cryoprotectants: These should be prepared in standard liquid culture medium (Note 2). Medium prepared and autoclaved in advance may be used. Alternatively, the medium may be made up fresh. In either case, once cryoprotectants have been added, sterilization must be carried out by filtration; autoclaving would cause caramelization.

   For the procedure recommended here, a mixture of 1*M* dimethyl sulfoxide (DMSO), 1*M* glycerol, and 2*M* sucrose is prepared in liquid medium. (Final concentrations are half of these values.) Reagents of the highest available grade should be used. In the case of DMSO, purity is particularly important and a grade suitable for UV spectroscopy should be obtained. It should be purchased in small quantities (e.g., 250 mL), since its shelf-life is limited. As a guide, the author would store DMSO concentrate for no more than 1 yr. It should be stored at room temperature with the precaution that this should always be higher than its melting point (ca. 19°C).

DMSO has an unpleasant, penetrating odor and is best dispensed under a fume hood. Always use glass containers for concentrated DMSO, since it will attack plastics. Glycerol is most conveniently weighed out, because it is highly viscous and difficult to pipet. Warming glycerol will reduce its viscosity and make pouring easier. Warming the mixture to ca. 40°C with stirring will help dissolve the cryoprotectants in the culture medium, but overheating should be avoided. The cryoprotectant mixture is adjusted to the standard culture medium pH and then sterilized by filtration. Very small quantities (up to 10 mL) may be filtered by hand pressure through a small filter unit attached to a hypodermic syringe. Larger quantities must be sterilized using a compressed air or vacuum-assisted filter unit.

Cryoprotectants should normally be prepared no more than 24 h in advance and stored at 4°C. Surplus cryoprotectant mixture can be stored at –20°C, but will require refiltration as a precaution against contamination. Repeated freezing and thawing of cryoprotectant mixtures is not recommended.

5. Ampules: Presterilized polypropylene ampules (2 mL capacity) are recommended as containers for freezing cell suspensions. They should be stable at the temperature of liquid nitrogen (196°C) and should have a labeling area that is resistant to solvents. The lid of the ampule should preferably have a gasket that fits inside the lid rather than lying externally between the lower edge of the lid and a flange at the base of the threaded part of the ampule. It is convenient to purchase or have made a small rack to hold the ampules during filling, cooling, and handling after thawing. Glass ampules of either the screw-top or flame-sealed type are not recommended, since they can shatter in storage and handling.
6. For dispensing cells and cryoprotectants, use sterile, plugged Pasteur pipets or, when an exact volume is to be dispensed, graduated pipets with an opening sufficiently wide to allow unimpeded passage of cell aggregates.
7. Freezing and storage apparatus: A purpose-built, liquid nitrogen-cooled, controlled freezing apparatus is desirable, although very satisfactory work can be carried out using improvized freezing units (*15,16*). For storage, a liquid nitrogen-cooled refrigerator is indispensable. There are two basic types. One stores ampules attached to metal

canes held in cannisters (*see* Fig. 1); in the other, the ampules are placed in stacks of drawers. The former type has a narrow neck and, therefore, a low rate of loss of liquid nitrogen. The latter, having a wide neck, is more convenient to use and offers easier cataloging and retrieval of samples. However, liquid nitrogen consumption is considerably higher.

8. One or more Dewar flasks of ca. 25 L capacity are required for storage and transport of liquid nitrogen, and one of ca. 5 L capacity is useful for rapid quenching of samples at the end of a slow cooling run.
9. Requirements for thawing: A thermostated water bath containing a large beaker to hold sterile water is ideal for thawing. Alternatively, jars (ca. 500 mL capacity), half-full of sterile distilled water may be used.

3. Method

The method below is based on that of Withers and King (*16*), modified slightly on the basis of subsequent experience with its application to a range of species. The numbered stages correspond to those illustrated in Fig. 1.

1. Subculture cells into a pregrowth passage in mannitol-supplemented medium. Time the transfer, and then calculate the inoculum density to give a harvest of cells in early exponential growth at the proposed date of cryopreservation. This should normally be 3–7 d after inoculation.
2. Prepare cryoprotectant solution as detailed above; cool on ice.
3. Transfer 1 volume of cell suspension to a sterilized flask; cool on ice or in a cold room. Add an equal volume of cryoprotectant solution in three aliquots over a period of ca. 20 min, stirring well at each addition.
4. Leave for 1 h, stirring or shaking continually to mix the viscous cryoprotectants with the culture medium.
5. Allow the cells to settle and draw off sufficient supernatant to leave ca. 30% settled volume of cells. Dispense 1 mL aliquots into 2 mL capacity ampules; seal and label.
6. Cool the controlled freezer to ca. 4°C. Transfer the ampules to the controlled freezer. Cool at 1°C/min to –35°C; hold for 40 min, and then plunge into liquid nitrogen.
7. Transfer ampules to storage refrigerator. Record date, contents, and location in catalog.

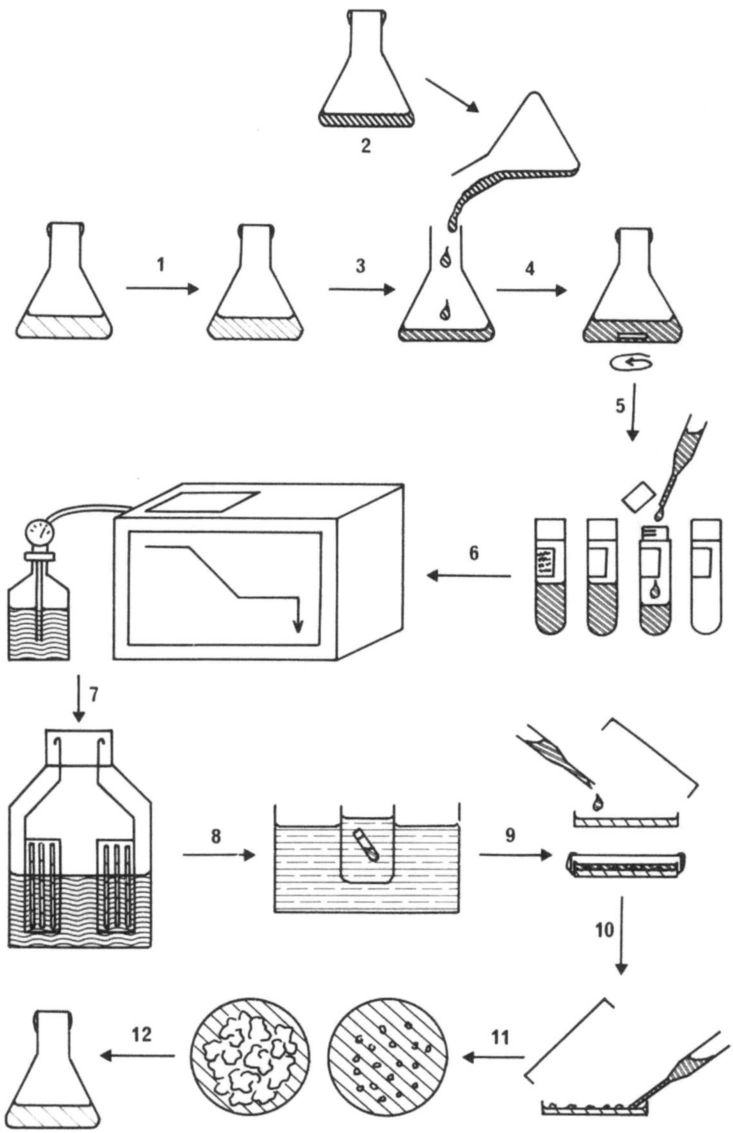

Fig. 1. Cryopreservation procedure for cultured plant cells (*see* ref. *16*). Numbers correspond to text Methods Section, where fuller details are given: (1) Transfer cells to pregrowth passage; (2) prepare cryoprotectants; cool on ice; (3) cool cells on ice; add cryoprotectants; (4) incubate cells with cryoprotectants, stirring continually; (5) transfer cells and cryoprotectant solution to ampules; (6) cool slowly to transfer temperature; (7) store in liquid nitrogen-cooled refrigerator; (8) thaw; (9) transfer cells and cryoprotectant solution to plate of recovery medium; (10) remove excess cryoprotectant solution; (11) leave cells to recover growth; (12) transfer cells to liquid medium.

8. To thaw, transfer ampules to a beaker of sterile water standing in a water bath at 40°C. Agitate until the last piece of ice has melted, and then transfer immediately to a rack at room temperature.
9. Wipe ampule with a swab of 70% ethanol; leave for a few moments to dry. Withdraw cells with a sterile Pasteur pipet, and transfer the contents of one ampule to one or two dishes of semisolid medium. Seal and place in culture room under standard conditions.
10. After 2–3 d, draw off any liquid medium that has not been reincorporated. Reseal the dish.
11. Culture under standard conditions until a vigorously growing lawn of cells has formed. Cell viability can be tested at this or earlier stages using standard viability tests (7), but recovery growth is the only categorical indicator of success.
12. Using a suitable instrument, e.g., a sterile spatula, transfer cells to a flask of liquid medium. Seal and place on shaker. Where the amount of recovering culture is not great, a satisfactory inoculum density for the species in question should be achieved by inoculation into a smaller volume of medium, rather than by bulking the contents of more than one dish.

4. Notes

The above method has been carried out successfully in the author's laboratory and by several other scientists working with a wide range of species. It provides a good starting point for developing a cryopreservation procedure for an untried subject species. If problems are encountered in achieving a satisfactory level of survival or rate of recovery, alternative treatments may be introduced. The following notes suggest some of the modifications that may be tried; others are detailed in the literature (*see 6,7,13* and references therein). It is recommended that modifications be introduced systematically into the stages of the cryopreservation procedure in the following order: pregrowth, cooling, cryoprotection, recovery. Modification of warming conditions is least likely to be beneficial.

1. Pregrowth medium additives: 3% w/v mannitol; 1*M* sorbitol (up to 20 h exposure only); 5% v/v DMSO; 10% w/v proline; $7.5 \times 10^{-5}M$ abscisic acid (ABA) (pregrowth at standard or at a reduced temperature).
2. Cryoprotectants (final concentrations): 5–10% v/v DMSO; 5% v/v DMSO plus 5% w/v glycerol; 0.5*M* DMSO plus 0.5*M* glycerol plus 1*M* proline; 10% v/v DMSO plus 8% w/v glucose plus 10% w/v poly-

ethylene glycol (PEG 6000). Cryoprotectants may be added at room temperature and in 1 vol rather than over a period of time.

3. Cooling: 0.1, 0.5, or 2°C/min, to a temperature in the range –23 to –100°C before transfer to liquid nitrogen. Longer or shorter holding times at –35°C or another temperature may be used.
4. Warming: In air at room temperature. (Effective only exceptionally; use if overdehydration during cooling is suspected and cells are very highly plasmolyzed upon thawing.)
5. Recovery: Wash cells in fresh liquid medium at room temperature. (Use only if cryoprotectant toxicity is suspected.) Use conditioned medium, medium designed to support cells at a low inoculum density, or plates with a feeder cell layer. Layer cells on a filter paper resting on semisolid medium and move to a fresh plate after hours or days. Transfer to liquid medium after only 1 or 2 d of recovery or inoculate directly into liquid medium (only if cell aggregates fail to break up when inoculated into liquid medium after a longer period of recovery).

References

1. Ashwood-Smith, M. J. and Farrant, J., eds. (1980) *Low Temperature Preservation in Biology and Medicine* (Pitman Medical, Tunbridge Wells).
2. Larkin, P. J. and Scowcroft, W. R. (1981) Somaclonal variation—a novel source of variability from cell cultures. *Theor. Appl. Gen.* **60,** 197–204.
3. Steponkus, P. L. (1984) Role of the plasmalemma in freezing injury and cold acclimation. *Ann. Rev. Plant Physiol.* **35,** 543–584.
4. Withers, L. A. and Williams, J. T. (1985) Research on long-term storage and exchange of *in vitro* plant germplasm, in *Biotechnology in International Agricultural Research* International Rice Research Institute, Manila, pp. 11–24.
5. Withers, L. A. (1985) Cryopreservation of cultured cells and meristems, in *Cell Culture and Somatic Cell Genetics of Plants,* vol 2 (Vasil, I. K., ed.), Academic, New York, pp. 254–316.
6. Withers, L. A. (1987) Long-term preservation of plant cells, tissues and organs, in *Oxford Surveys of Plant Molecular and Cell Biology,* vol 4 (Miflin, B. J., ed.) pp. 221–272.
7. Withers, L. A. (1986) Cryopreservation and genebanks, in *Plant Cell Culture Technology* (Yeoman, M. M., ed.), Blackwell, Oxford, pp. 96–140.
8. Hiraoka, N. and Kodama, T. (1984) Effects of non-frozen cold storage on the growth, organogenesis and secondary metabolite production of callus cultures. *Plant Cell, Tiss. Organ Cult.* **3,** 349–357.
9. Benson, E. E. and Withers, L. A. (1987) Gas chromatographic analysis of volatile hydrocarbons produced by cryopreserved plant tissue cultures. *Cryo-Lett.* **8,** 35–46.
10. Cella, R., Colombo, R., Galli, M. G., Nielson, E., Rollo, F., and Sala, F. (1982) Freeze-preservation of rice cells: A physiological study of freeze-thawed cells. *Physiol. Plant.* **55,** 279–284.

11. Withers, L. A. (1985) Cryopreservation and storage of germplasm, in *Plant Cell Culture: a Practical Approach* (Dixon, R. A., ed.), IRL Press, Oxford, pp. 169–191.
12. Kartha, K. K., ed. (1985) *Cryopreservation of Cultured Cells and Protoplasts* (CRC Press, Boca Raton).
13. Withers, L. A. (1985) Cryopreservation of cultured cells and protoplasts, in *Cryopreservation of Plant Cells, Tissues and Organs* (Kartha, K. K., ed.), CRC Press, Boca Raton, pp. 243–267.
14. Meryman, H. T., ed. (1966) *Cryobiology* (Academic Press, London).
15. Withers, L. A. (1984) Freeze preservation of cells, in *Cell Culture and Somatic Cell Genetics of Plants*, vol 1 (Vasil, I. K., ed.), Academic, New York, pp. 608–620.
16. Withers, L. A. and King, P. J. (1980) A simple freezing unit and cryopreservation method for plant cell suspensions. *Cryo-Lett.* **1**, 213–220.

Chapter 5

Hormonal Control of Growth and Development

Houri Kalilian Fakhrai and Faramarz Fakhrai

1. Introduction

Growth and organogenesis in vitro is highly dependent on the interaction between naturally occurring endogenous growth substances and an analogous growth regulator added to the medium. It is often necessary to alter the growth regulator composition and/or concentration for in vitro culture according to the species or variety of plant being grown, the explant origin, the type of culture, and other medium constituents. A balance between auxin and cytokinin growth regulator is most often required for the formation of adventitious shoot and root meristems. The interactions found are often complex, and more than one combination of two substances is likely to produce optimum results. Figure 1 gives an example of the relative amounts of an auxin and a cytokinin that are often required to bring about some kind of morphogenesis.

In this chapter, we have compared hormonal control of in vitro morphogenesis in tissue and cell culture of *Psophocarpus tetragonolobus* (winged bean) and the tissue culture of *Vicia faba*.

From: *Methods in Molecular Biology*, vol. 6, *Plant Cell and Tissue Culture*
Edited by Jeffrey W. Pollard and John M. Walker, ©1990 by The Humana Press

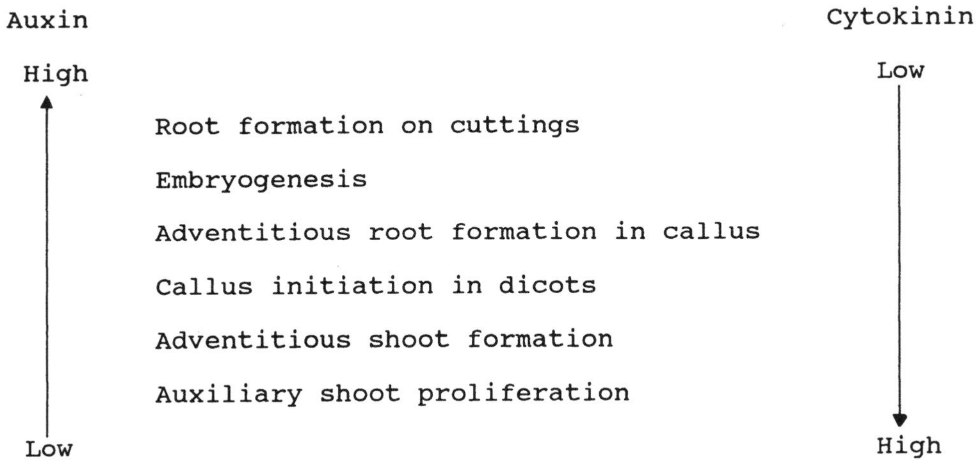

Fig. 1. Effect of auxin and cytokinin on in vitro morphogenesis (extracted from George and Sherrington, 1984) (*1*).

2. Materials

1. Mature leaves from winged bean accession No. SU623 (Southampton University) are grown in a growth room with 12-h photoperiod and 25/22°C light/dark, and 84 µE $m^{-2}s^{-1}$ light intensity.
2. Seeds of *Vicia faba* L. (cultivar Spring blaze) obtained from Cambridge University.
3. *Vicia* germination medium (VGM), containing the following ingredients (2): Ca $(NO_3)_2 \bullet 4H_2O$ 1.881 g/L, KNO_3 0.506 g/L, KH_2PO_4 0.136 g/L, $MgSO_4 \bullet 7H_2O$ 0.255 g/L Fe-citrate 0.005 g/L, agar 7.0 g/L pH 5.8.
4. MS salts and hormones (*3*).
5. UM3M medium (*4*) containing the following ingredients: MS salts and hormones, 3% sucrose, 2 g/L casein hydrolysate, 4.5 mg/L nicotinic acid, 9.5 mg/L pyridoxine. HCl, 9.9 mg/L thiamine. HCl, 2 mg/L kinetin, 3% mannitol, 0.8% agar, pH 5.8.
6. 6-Benzyl amino purine (BAP), prepared by dissolving the powder in a few drops of 1*M* HCl, and then adding H_2O to make up the volume to give 1 mg/mL. Store at 4°C, and use within 3 mo.
7. 2,4-Dichlorophenoxyacetic acid (2,4-D), prepared by dissolving the powder in a few drops of ethanol and then add H_2O to make up the volume to give 1 mg/mL. Store at 4°C. Use within 1 mo.

8. 3-Indolebutyric acid (IBA), prepared by dissolving the powder in a few drops of ethanol, and then adding H_2O to make up the volume to give 1 mg/mL. Store at 4°C. Use within 1 wk.
9. 3-Indoleacetic acid (IAA), prepared by dissolving the powder in a few drops of ethanol, and then adding H_2O to make up the volume to give 1 mg/mL. Filter sterilize by passing through 0.2-µm pore size membrane filter. Store at 4°C. Use within 1 wk.
10. Kinetin, prepared by dissolving the powder in a few drops of 1*M* HCl, and then adding H_2O to make up the volume. Store at 4°C. Use within 3 mo.
11. 1-Naphthylacetic acid (NAA), prepared by dissolving the powder in a few drops of 1*M* NaOH, and then adding H_2O to make up the volume. Store at 4°C. Use within 1 mo.
12. Agar: Bacto-agar (DIFCO)—use 0.7–0.8% in the medium.
13. Domestos: commercial bleach containing NaOCl.
14. Cereal protoplast wash (CPW) containing the following ingredients in analar distilled water: KH_2PO_4 27.2 mg/L, KNO_3 101 mg/L, $CaCl_2 \bullet 2H_2O$ 1480 mg/L, $MgSO_4 \bullet 7H_2O$ 246 mg/L, $CuSO_4 \bullet 5H_2O$ 0.025 mg/L, KI 0.15 mg/L.
15. 2% Cellulose "Onozuka" R10 and 0.4% of Macerozyme R10 (Yakuit Pharmaceutical Industry Co. Ltd., Nishinomiya, Japan). Enzymes are dissolved in CPW containing 5% mannitol, the pH adjusted to 5.8 and centrifuged at 700*g* for 10 min to remove debris. Filter sterilize the supernatant and store aliquots at –20°C. Do not freeze twice.
16. 10 m*M* 2-Mercaptoethanol.
17. 10% Ficoll (Pharmacia, Uppsala, Sweden).
18. Fluorescein-diacetate (FDA): 5 mg/mL made up in acetone diluted 1:25 with protoplast media and 1 drop with 1 drop of protoplast.
19. DAP1 (4,6-diamidino-2-phenylindole): DAPI is prepared (0.5 mg/mL) in a buffer solution containing 0.1*M* citric acid monohydrate and 0.2*M* anhydrous Na_2HPO_4 at pH 4 (*5*).

3. Method

3.1. Hormonal Control of Growth and Development in Winged Bean Tissue Culture

1. Surface sterilize leaves by immersing in 7% "Domestos" for 30 min, (*see* Notes 1–5) followed by a wash with four changes of sterile distilled water (SDW).

2. Cut surface sterilized leaves into 1-cm^2 segments, and place the lower epidermis down on 15 mL of solidified MS medium supplemented with 2 mg/L BAP and 0.2 mg/L NAA in a 60-mL glass "Beaston" jar.
3. Incubate explants in a culture room with 12-h photoperiod with a light intensity of 84 $\mu E\ m^{-2}\ s^{-1}$ derived from cool white fluorescent tubes (Crompton 75 [85] W, UK). Maintain the temperature at 27/22°C light/dark.
4. After 4 wk, pieces of callus tissue are transferred to fresh MS medium in which IAA at 0.2 mg/L was substituted for NAA (bud initiation medium).
5. Individual shoots that develop on the callus are excised above the callus and placed on half-strength MS and 1.5% sucrose, supplemented with 1 mg/L IBA and incubated in the dark. After 2 wk, roots will grow from the cut end of the stem.

3.2. Hormonal Control of Growth and Development in Winged Bean Cell and Protoplasts

1. Pieces of leaf callus, made as described before, are cut finely and transferred to MS liquid medium supplemented with 2 mg/L 2,4-D and 0.2 mg/L kinetin, and incubated on a rotary shaker at 80 rpm at 25°C. These suspensions are subcultured 50/50 with fresh medium every 2 wk.
2. Prior to protoplast isolation, cultures are stimulated to undergo rapid division by subculturing to fresh medium twice at 4-d intervals. On the third day following subculture, cells are collected and plasmolyzed with CPW5M for 1 h before adding the enzyme mixture containing 2% Cellulose "Onozuka" R10, 0.4% "Macerozyme" R10, and 10 mM 2-mercaptoethanol in CPW5M (*6*).
3. Protoplasts are released after 5 h, washed 2x with CPW5M, and cleaned with CPW5M and 10% Ficoll, resuspended in UM3M medium supplemented with 2 mg/L 2,4-D and 0.25mg/L kinetin.
4. After 2 wk in UM3M medium, protoplasts are subcultured in MS medium supplemented with 2 mg/L BAP and 0.2 mg/L NAA. The plasmolyticum level in the culture is reduced gradually by diluting the protoplast culture media.
5. After 6–8 wk from the time of isolation, cell colonies reach macroscopic size and are transferred to agar medium with the same combination of growth regulators as above.

6. Within 1 mo after transfer to agar medium, callus becomes green and compact, and about 10% of these cultures produce green nodules that give rise to regeneration of adventitious shoots after several subcultures on medium supplemented with 1–2 mg/L BAP.

3.3. *Hormonal Control of Growth and Development in* Vicia faba *Tissue Cultures* (See *Notes 6–9*)

1. Seeds are rinsed in 80% ethanol for 1 min, immersed in 7% "Domestos," and incubated on a rotary shaker at 80 rpm for 1 h. Wash 4x with SDW, and soak for 48 h in the dark at 27°C. The water is changed after 24 h. Then the seeds are decoated and sterilized for a second time in 7% "Domestos" for 5 min, washed and allowed to germinate in 300-mL glass cylindrical jars containing 35 mL of VGM. The jars are incubated at 27°C in the dark for 2–3 wk.
2. Nodal stem segments from well-developed seedlings are excised and cut to 0.5-cm long explants, and cultured horizontally on the surface of agar in MS medium supplemented with 2 mg/L BAP and 0.2 mg/L NAA. Axillary shoots and masses of green callus develop after 4 wk. The same combination of growth regulators also support the growth of axillary shoots from cotyledonary nodes and hypocotyl explants.
3. Adventitious roots are regenerated from stem explants when auxin NAA or 2,4-D (2 mg/L) is used as the sole growth regulator. Effect of 2,4-D on rhizogenecity is more pronounced than NAA.
4. Adventitious shoots develop from nodal stem explant callus after several subcultures on MS medium supplemented with 2 mg/L BAP. 2,4-D inhibits the growth of both axillary and adventitious shoot meristems from stem and cotyledonary explants.
5. Both axillary and adventitious shoots can be rooted on half-strength MS, and 1.5% sucrose supplemented with 0.1 mg/L NAA and 0.5 mg/L kinetin.
6. Callus growth is initiated on leaf explants from 2-mo-old plants growing in the greenhouse, on MS medium supplemented with 2 mg/L 2,4-D and 0.25 mg/L kinetin. When 2,4-D is replaced by NAA, no callus develops. Callus developed this way is very small (1 mm^2), white, and globular, which is scattered on the surface of the leaf. Each globular callus is excised from the rest of the leaf, and subcultured in MS medium supplemented with 1–2 mg/L BAP and 0.2–0.5 mg/L NAA. After 2 wk, masses of green and compact callus

(x10 of the original callus size) grow. This callus fails to show any organogenic response upon successive subcultures in medium supplemented with a range of concentrations of BAP and IAA or NAA.

4. Notes

1. A characteristic feature of winged bean acc. number SU623 is that shoot regeneration from callus involves a two-step process. The hormone combination given of NAA and BAP stimulates callus development on the leaf explant. Transfer to a medium in which NAA is replaced by IAA appears to trigger off the gene expression, which leads to shoot development. However, a second subculture to IAA-containing medium results in a decline in shoot regeneration. Furthermore, subculture of shoot-regenerating callus to a medium containing NAA and BAP or BAP alone initiates vigorous growth of shoot buds. It seems, therefore, that IAA is needed for the initiation of shoot bud primordia, but once it is established, IAA has a deleterious effect on further development of the shoots.
2. The combination and concentrations of growth regulators in the initial callus production medium has an effect on the size, shape, and growth characteristics of cells grown in suspension cultures. Suspension cultures made from callus grown in media supplemented with 12 mg/L BAP and 0.2 mg/L NAA contain small, round, and cytoplasmic single cells and colonies of 6–8 cells. These cells are 90% viable and 99% nucleated as determined by FDA and DAPI staining, respectively. Protoplast isolation and shoot regeneration from protoplast are achieved by using these suspension cultures. On the other hand, suspension cultures made from callus grown in media supplemented with 2 mg/L 2,4-D and 0.2 mg/L kinetin contains a very large number of elongated cells (90%) as opposed to round cells (10%). There are also fewer cell colonies in comparison to the above suspension culture. However, all these elongated cells are cytoplasmic and are 90% viable. When these cells are stained with DAPI for the presence of nuclei, only 10% of the elongated cells are observed to be nucleated, the nucleus is either masked by amyloplasts (a condition known as systrophy) or is very sparse and diffuse. Hence, their fluorescence is difficult to observe. The lack of division of these cells makes these cultures unsuitable for protoplast isolation.

3. The presence of 2,4-D (1–2 mg/L) in combination with kinetin (0.2–0.5 mg/L) is critical for subsequent growth of cells in suspension cultures. BAP- and NAA-supplemented media fail to produce a vigorous suspension culture.
4. Although 2,4-D supplemented media support the highest plating efficiency of protoplasts in the first 2 wk after the isolation, incubation of protoplasts in this media for longer periods (2–4 wk) results in the loss of shoot regeneration capacity.
5. The origin of the leaf material for callus induction is critical for subsequent growth and organogenic response of callus. Leaves from clonal plants derived from tissue culture and subsequently grown in the greenhouse are the best source of explant for culture initiation.
6. Tissue culture of *Vicia faba* is particularly difficult. It releases a substantial amount of phenolic substances in culture. Explants frequently turn brown or black shortly after isolation, and when this occurs, growth is inhibited and the tissue usually dies. Care must be taken to choose young, undamaged, and intact seeds for subsequent in vitro seedling cultures. Phenolic oxidation of the explants can often be prevented by one or several of the following approaches.
   (a) Use young tissue (seedling tissue), which is often less prone to browning upon excision.
   (b) Soak explants in sterile water 2–3 h after isolation before being transferred to culture.
   (c) Transfer explants to fresh medium after a short time interval (2 d), and again thereafter at 1 wk intervals if blackening occurs.
   (d) Either incubate in complete darkness or in a low light intensity (50–70 $\mu E\ m^{-2}\ s^{-1}$) in the first 2 wk after isolation.
7. On the basis of 32 variations of the growth regulators tested, we have observed that callus production from leaf explants is strictly dependent on the type of auxin used. With NAA, when applied alone or with combinations of kinetin over a range of concentrations, no callus is produced. With 2,4-D applied either alone or with combinations of kinetin, the explants start to callus.
8. Incubation in the dark always produces more callus from leaves for all 2,4-D supplemented medium tested (2–4-fold more callus in the dark than in 12 h photoperiod).
9. When leaf callus is separated from the original leaf tissue and subcultured onto a fresh media, it responds somewhat differently to the

addition of growth regulators. Callus subcultured in medium MS supplemented with 1–2 mg/L BAP and 0.2 mg/L IAA or NAA and incubated in a 12-h photoperiod for 2 wk produces masses of green and compact callus. Higher concentrations of IAA (0.5–2 mg/L) reduce the vitality of calluses. They become dark brown and die within 2 wk.

References

1. George, E. F. and Sherrington, P. D. (1984) Plant propagation by tissue culture, in *Handbook and Directory of Commercial Laboratories* (Exegetics Ltd. Eversley, Basingstoke, Hants. RG27 OQY).
2. Roper, W. (1979) Growth and cytology of callus and cell suspension cultures of *Vicia faba* L. Z. *Pflanzenphysiol.* **93,** 245–257.
3. Murashige, T. and Skoog, F. (1962) A revised medium for rapid growth and bioassays with tobacco tissue cultures. *Physiol. Plant* **15,** 474–497.
4. Uchimiya, H. and Murashige, T. (1974) Evaluation of parameters in the isolation of viable protoplasts from cultured tobacco cells. *Plant Physiol.* **54,** 936–944.
5. Hull, H. M., Horshaw, R. W., and Wang, Y. C. (1982) Cytofluorometric determination of nuclear DNA in living and preserved algae. *Stain Tech.* **57,** 273–281.
6. Wilson, V. M., Haq, N., and Evans, P. K. (1985) Protoplast isolation, culture and plant regeneration in the winged bean. *Psophocarpus tetragonolobus* (L) D. C. *Plant Sci.* **41,** 61–68.

Chapter 6

The Initiation and Maintenance of Callus Cultures

Jacqueline T. Brown

1. Introduction

Plant tissues grown in vitro provide an ideal research tool for the study of a wide range of aspects of plant science. For example, they have been used in the investigation of both primary and secondary metabolism, cytodifferentiation, morphogenesis, plant tumor physiology, and the formation of plant hybrids via protoplast fusion techniques. Plant tissue culture is also being increasingly adopted for the commercial propagation of plants.

Callus culture concerns the initiation and continued proliferation of undifferentiated parenchyma cells from parent tissue on clearly defined semi-solid media. Such cultures may be maintained for extended periods by subculture at 2–4 weekly intervals, and therefore represent a convenient form for the long-term maintenance of cell lines. They are also usually the material from which cell suspensions are derived and are often the form of cultures from which plant regeneration is initiated.

From: *Methods in Molecular Biology*, vol. 6, *Plant Cell and Tissue Culture*
Edited by Jeffrey W. Pollard and John M. Walker, ©1990 by The Humana Press

In vivo, callus is frequently formed as a result of wounding at the cut edge of a root or stem, following invasion by microorganisms or damage resulting from insect feeding. Its formation is controlled by endogenous auxin and cytokinin. By incorporation of these plant growth regulators into a growth medium, callus can be induced to form in vitro on explants of parent tissue. The initiation of callus material from angiosperms, gymnosperms, ferns mosses, and liverworts can by achieved in this way.

Since the majority of plant tissue cultures are photosynthetically incompetent, and hence heterotrophic, it is necessary to supply them with a carbon source, usually in the form of sugar, e.g., sucrose or glucose. Plant cells in culture also require the same macro- and micronutrients as the whole plant (usually supplied in the medium as mineral salts), amino acids, B vitamins and plant growth regulators. The presence of plant growth regulators is generally essential to promote growth, but the nature and concentrations required varies from species to species. Typically, an auxin, such as indoleactetic acid (IAA), 2,4-dichlorophenoxyacetic acid (2,4-D), or napthaleneacetic acid (NAA), and a cytokinin, such as kinetin, benzylamino purine (BAP), or zeatin, are required either singly or more commonly in combination. There are numerous plant tissue culture media documented in the literature facilitating callus formation. One commonly used is that of Murashige and Skoog (MS) (*1*). The following protocol will utilize the nutrient components of this medium. However, if there is no callus formation after incubation, the use of an alternative nutrient mixture should be investigated, (e.g., *2,3* and *see* Appendix). Such media can also be prepared and used in a similar way to that outlined below.

The growth rate of plant cells in culture is slow compared to microorganisms, and because of the rich nature of the growth medium employed, it is necessary to maintain complete sterility. Manipulations of plant tissue cultures should therefore be carried out using standard microbiological techniques in a sterile area.

2. Materials

2.1. Medium Preparation

1. Medium components—as described in Table 1.
2. Plant growth regulator stock solutions (1 mg/mL). Store at 4°C. NAA, IAA, and 2,4-D should be titrated into solution with NaOH. Kinetin, zeatin, and BAP can be dissolved in dilute NaOH or 95% aqueous ethanol.

Table 1
Murashige and Skoog Medium (1962)

Compound	Concentration in medium, mg/L	Amount in stock solution[a]	Stock volume, mL
1. NH_4NO_3	1650	8.25 g	400 mL (items 1–10)
2. KNO_3	1900	9.50 g	
3. $MgSO_4{\bullet}7H_2O$	370	1.85 g	
4. KH_2PO_4	170	0.85 g	
5. KI	0.83	4.18 mg	
6. H_3BO_3	6.20	31.00 mg	
7. $MnSO_4{\bullet}4H_2O$	22.30	111.50 mg	
8. $ZnSO_4{\bullet}7H_2O$	8.6	43.00 mg	
9. Myo-inositol	100.00	0.50 g	
10. $CaCl_2{\bullet}2H_2O$	440.00	2.20 g	
11. $FeSO_4{\bullet}7H_2O$	27.8	139.25 mg	100 mL (items 11–12)
12. $Na_2EDTA{\bullet}2H_2O$	37.3	186.25 mg	
13. $CuSO_4{\bullet}5H_2O$	0.025	12.50 mg	100 mL
14. $Na_2MoO_4{\bullet}2H_2O$	0.25	12.5 mg	10 mL
15. $CoCl_2{\bullet}6H_2O$	0.025	12.5 mg	100 mL
16. Nicotinic acid	0.50	25.0 mg	10 mL
17. Pyridoxine-HCl	0.50	25.0 mg	10 mL
18. Thiamine-HCl	0.10	5.0 mg	10 mL
19. Glycine	2.0	100.0 mg	10 mL
20. Sucrose	30 g/L		

[a]All solutions should be stored at –20°C for no longer than 3 mo.

3. 0.1M NaOH
4. Agar, e.g., Oxoid Bacteriological Agar No. 1.

2.2. Callus Initiation and Maintenance

1. Plant material—a wide range of plant organs and specialized tissues can be used to initiate callus formation. However, sterile seedlings and nonwoody stem tissue tend to faciliate ease of callus initiation. Seeds should be checked for viability before use, and stem tissue should be devoid of senescence.
2. "Domestos" solution (3% aq), or other proprietary bleach.
3. Temperature-controlled incubator 25 ± 2°C preferably with additional light-control facility.

3. Method

3.1. Media Preparation

It is ideal to have a separate area reserved for media preparation. However, if this is not possible, care should be taken to ensure that glassware remains scrupulously clean and that media do not become contaminated with trace amounts of foreign chemicals. (Note 1).

To prepare 1 L of MS medium (*see* Table 1, and Chapter 1 for consideration of choice of media):

1. Dissolve compounds 1–10 (in ascending numerical order) in 350 mL of double-distilled water. Add 1 mL each of solutions of 13,14, and 15, and make up to a final volume of 400 mL. This is stock solution A.
2. Dissolve 11 and 12 in 50 mL of double-distilled water, and make up to a final vol of 100 mL. This is stock solution B.
3. Dissolve sucrose (30 g) in double-distilled water (600 mL). Add solution A (80 mL) and solution B (20 mL). Stir well and dilute to 970 mL.
4. Add 0.2 mL each of solutions of 16,17,18, and 19.
5. Adjust the pH of the salt solution to 5.5 with 0.1*M* NaOH, and dilute to 1 L.
6. Add the chosen plant growth regulators to the appropriate concentration. When preparing media for the initiation of a new callus line, it is wise to prepare several batches of media containing a variety of combinations and concentrations of auxin and cytokinin (*see* Chapters 1 and 5 for greater detail). The following supplements to MS medium have proven useful for the initiation of callus in a number of species: 5 mg/L BAP and 1 mg/L NAA; 1 mg/L kinetin and 1 mg/L NAA; 0.2 mg/L kinetin and 1 mg/L NAA; 0.2 mg/L kinetin and 1 mg/L 2,4-D.
7. Add agar (10 g) and stir well.
8. If the medium is to be used in Erlenmeyer flasks, then the agar should be premelted either in a water bath or with steam, and the medium dispensed into the culture vessels (30 mL/100 mL flask). After stoppering, the flasks should be sterilized by placing in an autoclave for 15 min at 120°C (1.06 kg/cm^2) (*see* Notes 2 and 3).
9. If the medium is to be used in pre-sterilized containers the medium should be autoclaved, allowed to cool to approximately 40°C and then dispensed as appropriate under sterile conditions. The medium should be allowed to set before moving the vessels (*see* Note 4).

3.2. Callus Initiation and Maintenance

1. Sterilization of seed material:
   a. Place seeds in ethanol for 30 s.
   b. Remove the seeds, and place in "Domestos" solution for 30 min, shaking occasionally.
   c. Remove the seeds and wash five times, with fresh, sterile, double-distilled water.
   d. Place the seeds on appropriate media, and arrange them so that they are not in contact with each other.
   e. Incubate in the dark for 1 wk, and then transfer to the desired incubation conditions. Once germinated, the young seedling provides ideal tissue for callus initiation. This may take place from either the plumule or the radicle.
   f. View daily for callus formation (*see* Fig. 1).
2. Sterilization of stem explants:
   a. Remove nonwoody stem from the plant and cut into 5 cm lengths.
   b. Wash in double-distilled water, and then remove the leaves and axillary buds.
   c. Place the explants in "Domestos" solution for 5 min. Remove and wash five times, with fresh, sterile, double-distilled water (*see* Note 5).
   d. Trim away the end 2–3 mm of the explant, and cut the remaining tissue into 2 cm lengths. Cut these in half lengthwise, and place the cut side down on the agar (*see* Note 6).
   e. Incubate the explants, and watch daily for callus formation (*see* Fig. 1). This often takes place along the cut edge of the explant in contact with the agar, or at the site of axillary bud removal.
   f. Once sufficient callus growth has taken place, it should be carefully removed from the explant and transferred to fresh medium. It should now be possible to maintain the callus line by subculture at regular intervals, e.g., every 2–4 wk, depending on the growth rate of the callus (*see* Note 7). The callus should be subdivided and transferred onto fresh medium. Care should be taken not to reculture sensecent tissue and that the transferred material is of a sufficient size to be able to maintain growth (ca. 0.5 cm^3) (*see* Notes 8 and 9).

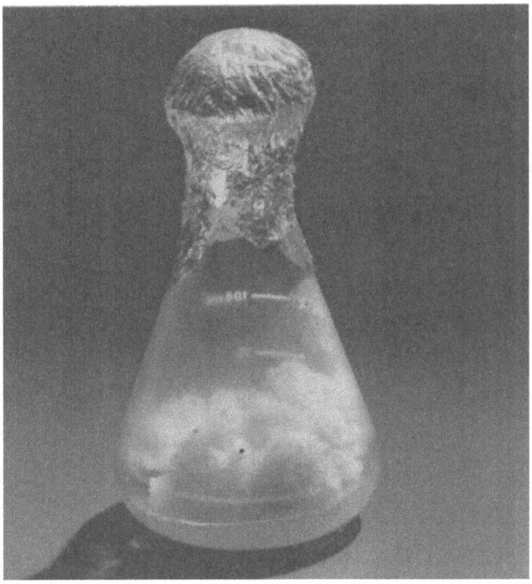

Fig. 1. Callus culture of *Nicotiana glutinosa* initiated and maintained on Murashige and Skoog medium supplemented with 5 mg/L BAP and 1 mg/L NAA.

4. Notes

1. Many of the commonly used plant tissue culture media (*see* Appendix) can be obtained in prepared powder form (minus agar, sucrose, and plant growth regulators). This provides a quick and convenient way of preparing basic media. However, this does not allow for manipulation of media components (but *see* Chapter 1 for protocols describing how to choose different media).
2. Some media components are heat labile and, therefore, should not be added to the medium prior to autoclaving, e.g., kinetin (and *see* Chapter 1). Such compounds should be filter sterilized using a 0.22 µm membrane filter and added to precooled sterile medium (40°C) prior to dispensing.
3. The method of sterilization of any one medium component should remain consistent, since its final concentration in the medium depends on its method of sterilization.
4. To prevent media desiccation, it is advisable to store media at 4°C and then return to room temperature prior to use.
5. If microorganisms persist on the explant after the sterilization procedure, then the length of the sterilization period and/or the concentration of the "Domestos" solution may be increased. However, harsh

treatments may lead to the death of the plant tissue or no seed germination. This may be overcome by subjecting the explant to a series of weaker sterilization procedures.

6. If the stem tissue available is too narrow to be cut in half lengthwise, then a cut should simply be made along the length of the explant and the cut placed on the surface of the agar.
7. It is preferable to initiate and maintain callus lines in either continuous light or continuous dark, such that if required the effect of photoperiod can be easily investigated.
8. The morphology of callus cultures can be controlled by the manipulation of the plant growth regulators in the medium. Increasing auxin concentration will increase the friability of the culture, which may be important if the callus is to be used to initiate suspension cultures.
9. Similarly, organogenesis can be induced or halted by manipulating the ratio of auxin to cytokinin in the medium. For example, a high ratio of cytokinin to auxin can induce the formation of shoots, and a high ratio of auxin to cytokinin can induce the formation of roots, in dicotyledonous callus (4).

References

1. Murashige, T. and Skoog, F. (1962) A revised medium for rapid growth and bioassays with tobacco tissue cultures. *Physiol. Plant* **15,** 473–497.
2. White, P. R. (1954) *The Cultivation of Animal and Plant Cells.* (The Ronald Press, New York).
3. Gambourg, O. L., Miller, R. A., and Ojima, K. (1968) Nutrient requirements of suspension cultures of soybean root cells. *Exp. Cell Res.* **50,** 151–158.
4. Brown, J. T. and Charlwood, B. V. (1986) The control of callus formation and differentiation in scented Pelargoniums. *J. Plant Physiol.* **123,** 409–417.

Chapter 7

Organogenesis in Callus Culture

Jacqueline T. Brown and Barry V. Charlwood

1. Introduction

Plant cells can be totipotent, i.e., each cell may be capable of developing into an entire plant when provided with the correct environmental stimuli. Research during the last 30 yr has demonstrated that successful organogenesis in callus cultures can be achieved by the correct choice of medium components, selection of a suitable inoculum, and control of the physical environment (*1*). The manipulation of plant growth regulator concentration is probably the most widely used technique for the induction of organogenesis, and this methodology has formed the basis of the propagation of commercially important plants via tissue culture in recent years (*2*).

In 1957, Skoog and Miller performed their classic work demonstrating that shoot and root initiation in callus cultures of *Nicotiana tabacum* could be regulated by manipulation of the ratio of auxin and cytokinin

From: *Methods in Molecular Biology*, vol. 6, *Plant Cell and Tissue Culture*
Edited by Jeffrey W. Pollard and John M. Walker, ©1990 by The Humana Press

present in the growth medium (3). It has since proven possible to induce organogenesis in callus cultures of a large number of dicotyledons using this method. In general, a high ratio of cytokinin to auxin results in shoot formation, and a high ratio of auxin to cytokinin gives rise to root formation (*see* Chapter 5, this vol.). However, there are a few species that differ in their developmental behavior in response to auxin and cytokinin (4). This chapter demonstrates how to develop a protocol for the control of organogenesis in callus cultures of a dicotyledon by manipulation of plant growth regulators.

2. Materials

2.1. Medium Preparation

1. Using the formula defined for the maintenance of callus tissue (Chapter 6, this vol.) as a reference, a number of media should be prepared, which vary in both the concentration and ratio of plant growth regulators used. For example, Table 1 describes a range of plant growth regulator supplements to Murashige and Skoog medium (MS) (5) used to investigate the control of organogenesis in callus tissue of *Pelargonium tomentosum* (6). Medium preparation should be carried out as described in Chapter 6 (*see* Note 1).
2. Culture vessels: to aid tissue manipulation, it is preferable to carry out this type of study in vessels with wide neck openings. Wide-necked 100-mL Erlenmeyer flasks containing 30 mL of medium and stoppered with cotton wool bungs covered with a single layer of aluminum foil are ideal for this purpose.

2.2. Organogenesis

1. Callus material that has been initiated and maintained on a defined growth medium as described in Chapter 6.
2. Temperature-controlled incubator $25 \pm 2°C$ with additional control of light intensity (10–50 $\mu E\ m^2\ s^{-1}$) and photoperiod.

3. Methods

3.1. Organogenesis

1. Subculture the maintenance callus (*see* Notes 2–4) under sterile conditions, as described in Chapter 6, onto the newly prepared media

Table 1
Experimental Approach to the Formulation of Media
for Organogenesis in Callus of *P. tomentosum*

Plant growth regulator supplement, mg/L		Morphogenic response		
BAP	NAA	Shoot	Root	Callus
0.05	0.05	†	†	
0.05	0.50	†(23)	†(22)	
0.05	1.00	†(8)	†(41)	
0.25	2.00		†	†
0.50	0.05	†(50)		
0.50	0.15	†(5)		
1.00	1.50			†
1.50	1.00			†
1.50	2.25			†
2.25	1.50			†
5.00	1.00*			†

*Callus maintenance medium.
†Growth response after 4 wk incubation. The figures in parenthesis are the average numbers of shoots/roots formed per callus mass (vol 2 cm^3) at subculture.

containing the different concentrations of plant growth regulators described in Table 1.

2. Incubate in continuous light, and view daily for organogenesis.
3. Subculture after the usual interval, taking care to transplant any tissue showing a morphogenic response.
4. A typical pattern of organogenesis controlled by plant growth regulator regime is described in Table 1 and Fig. 1 for *P. tomentosum* (*see* Note 5). High levels of the auxin naphthaleneacetic acid (NAA) and the cytokinin, benzylaminopurine (BAP) (i.e., in excess of 1.00 mg/L) result in the continued production of callus irrespective of the concentration ratios of the regulators. When the BAP level is maintained below 1 mg/L, a decrease in auxin results in shoot formation. Conversely, root formation is observed at low levels of BAP (0.25 mg/L and below) with higher NAA levels (*see* Note 6).

   When both BAP and NAA levels are maintained at 0.05 mg/L, both shoot and root formation are observed, suggesting that the effects of endogenous hormones are significant.

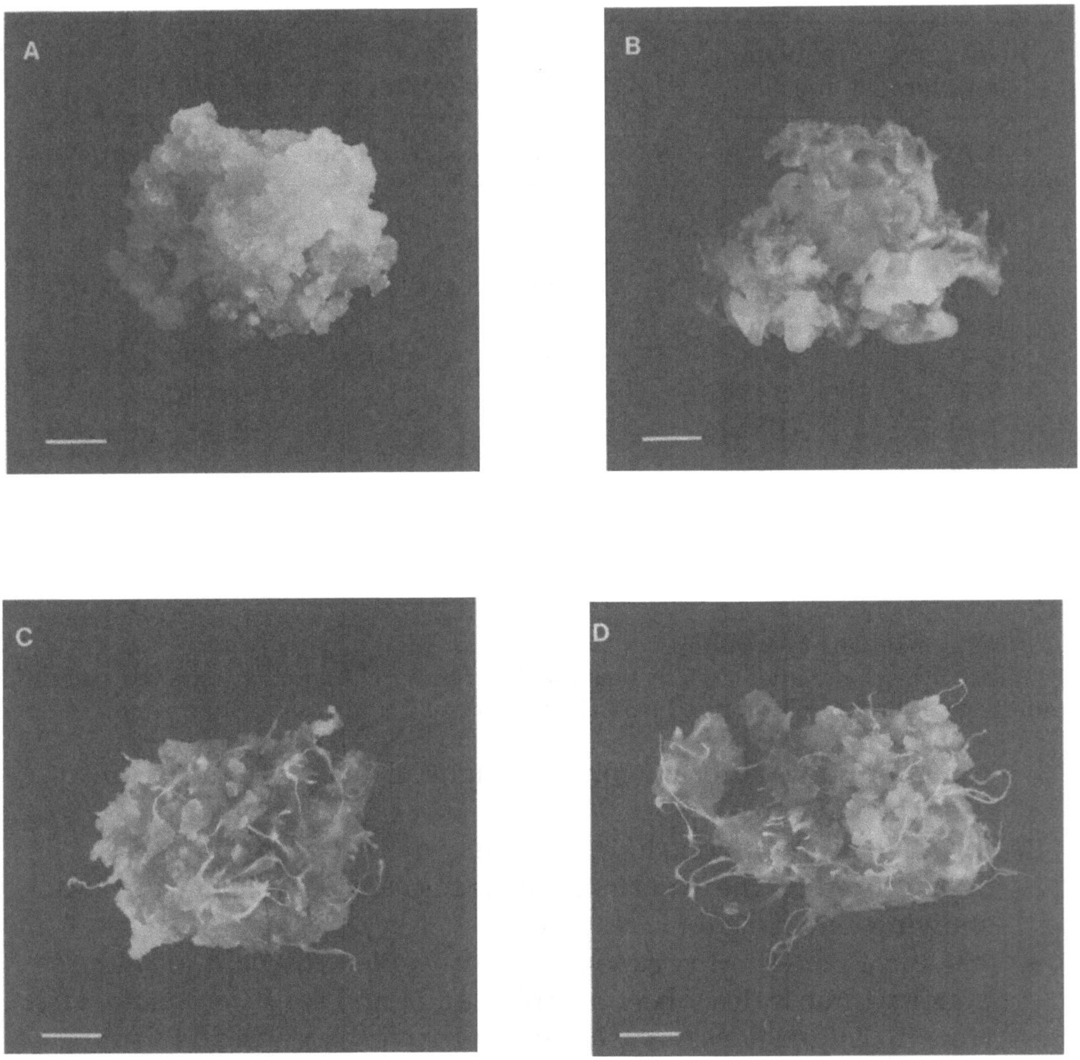

Fig. 1. Control of organogenesis by plant growth regulator manipulation in callus cultures of *P. tomentosum* (bar = 0.5 cm). (A) Maintenance callus on MS supplemented with 5.00 mg/L BAP and 1.00 mg/L NAA. (B) Shoot formation on MS supplemented with 0.50 mg/L BAP and 0.05 mg/L NAA. (C) Root formation on MS supplemented with 0.05 mg/L BAP and 1.00 mg/L NAA. (D) Shoot and root formation on MS supplemented with 0.05 mg/L BAP and 0.05 mg/L NAA.

5. Both root and shoots initiated in the above manner may be isolated from the callus material and maintained on the appropriate medium independently.

4. Notes

1. In order to save on the amount of medium needed for an experiment of this kind, 1 L of medium can be used to make a number of medium types by division into aliquots before addition of the plant-growth regulators.
2. An increased efficiency of organogenesis and a faster response to the supplement is usually observed in newly initiated callus compared to callus that has been taken through a large number of subcultures.
3. The genetic instability of callus cultures increases with time in culture. Therefore, newly initiated callus should be used for organogenesis if the organs regenerated by the above protocol are to be representative of the explant.
4. There is evidence to suggest that the use of 2,4-dichlorophenoxyacetic acid (2,4-D) in plant tissue culture media may be responsible for the loss of the ability of some callus cultures to undergo organogenesis. It is therefore advisable to avoid the use of callus material that has been maintained on a medium containing 2,4-D for this type of study.
5. Initial experiments should be carried out in continuous light. The use of a photoperiod, however, may be necessary for the induction of organogenesis or to increase its frequency in some species.
6. Although the manipulation of auxin and cytokinin levels is usually sufficient to induce organogenesis in callus cultures of most dicotyledonous species, the addition of other growth regulators, e.g., gibberellic acid, may be necessary in some cases.

References

1. Thorpe, T. A. (1980) Organogenesis in vitro: Structural, physiological and biochemical aspects. *Intl. Rev. Cytol.* Suppl. IIA, 71–111.
2. George, E. F. and Sherrington, P. D. (1984) *Plant Propagation by Tissue Culture* (Exegetics Ltd., Reading, U.K.).
3. Skoog, F. and Miller, C. O. (1957) Chemical regulation of growth and organ formation in plant tissues cultivated in vitro. *Symp. Soc. Exp. Biol.* **11,** 118–130.
4. Foskett, D. E. (1976) Hormonal control of morphogenesis in cultured tissues, in *Plant Growth Substances* (Skoog, F., ed.), Springer Verlag, Berlin, pp. 362–369.

5. Murashige, T. and Skoog, F. (1962) A revised medium for the rapid growth and bioassays with tobacco tissue cultures. *Physiol. Plant.* **15,** 473–497.
6. Brown, J. T. and Charlwood, B. V. (1986) The control of callus formation and differentiation in scented Pelargoniums. *J. Plant Physiol.* **123,** 409–417.

Chapter 8

Manipulation of Shoot Formation in Cultured Explants

Gerard C. Douglas

1. Introduction

Regeneration of buds, shoots, and roots in cells and explants in vitro has provided useful developmental systems to analyze the processes of cell differentiation and morphogenesis. Interest in these studies was greatly stimulated by the demonstration of chemical regulation of morphogenesis in 1957 by Skoog and Miller (*1*). They showed that a high ratio of cytokinin to auxin-stimulated formation of buds when these chemicals were provided in the culture medium. Subsequently, their observations have been applied, extended, and found general application with numerous genera and species.

Although we now have a better understanding of methods for manipulating morphogenesis in cultured explants, the underlying regulatory mechanisms are still poorly understood. Several different approaches have been used, and these have identified causal events (*2*), chemical interactions (*3*), and biochemical modifications (*4*) that precede and accompany morphogenesis.

From: *Methods in Molecular Biology*, vol. 6, *Plant Cell and Tissue Culture*
Edited by Jeffrey W. Pollard and John M. Walker, ©1990 by The Humana Press

Utilization of this information has been essential for applications, such as propagation of plants in vitro, and for recovery of plants from cells that have been genetically modified by mutagenic treatments or by other genetic manipulations. Further studies will hopefully provide more general methods for regeneration of plants from explants and cells of species that are currently difficult or recalcitrant.

This chapter describes simple systems for regenerating shoot buds in stem explants of *Populus* (*5*). With *Populus*, an exogenous supply of auxins or cytokinins is not required. Manipulation of explant size and other factors, such as the chemical constituents of the medium, may, however, be used to modulate and study morphogenic responses (*6*). Regeneration in leaf explants of African violet *Saintpaulia ionantha* Wendl. is also described (*7,8*). In this case, an exogenous supply of auxin and cytokinin is required in the culture medium for bud regeneration.

2. Materials

2.1. Materials 1, *Populus*

1. Collect poplar shoots (preferably 1*M* in length for optimal storage) in the dormant season from tree suckers or stool beds of either *P. nigra*, *P. deltoides*, *P. tacamahaca*, *P. wilsonii*, *P. maximowiczii*, or hybrids derived from crosses of one or more of these species. Shoots should be collected from wood produced in the previous growing season.
2. Refrigerated store 3–5°C in darkness.
3. Potting compost consisting of equal volumes of fertilized peat, loam, and sand.
4. Growth room or phytotron with cool white fluorescent lights (Philips, 40 W cool white 33) giving a photon fluence rate of 45–75 µmol m^{-2} s^{-1} at plant level with a 16-h photoperiod. Temperature 12–20°C.
5. A filtered solution of calcium hypochlorite 7% w/v prepared by placing 70 g of chemical in 1.0 L of double-distilled water and stirring continuously for 15 min, followed by decanting and filtration through filter paper.
6. Basal medium of Murashige and Skoog (*see* Appendix) (*9*) (MS) with 3% w/v sucrose and modified by addition of 40.0 mg/L Sequestrene Fe. 330 (Giegy) instead of iron EDTA; vitamins of B5 medium (*see* Appendix) (*10*). Prior to autoclaving (for 18 min at 1.1kg/cm^2), agar is added and the pH adjusted to 5.8 with 0.1*M* KOH. Agar solidified medium (Difcobacto agar, 8.0 g/L) is dispensed 6 mL/(50 x 20 mm) Petri plate.

7. Culture room with cool white fluorescent lights (Thorn 65–80 W, cool white 4300) mounted 90 cm above the culture bench providing a photon fluence rate of 63 $\mu mol\ m^{-2}\ s^{-1}$ in a 16 h photoperiod. Temperature 22–25°C.

2.2. *Materials 2, African Violet* (Saintpaulia ionantha *Wendl.*)

1. Plants of *Saintpaulia ionantha* Wendl. (*Gesneriaceae*) in a healthy condition purchased locally.
2. Solutions of 0.5% and 1.0% sodium hypochlorite, prepared by dilution of household bleach with water 1/9 and 1/4, respectively. Hypochlorite solutions should be supplemented with 3 drops/L of Tween 80 as wetting agent.
3. A basal medium prepared with macronutrients, micronutrients, vitamins, and sucrose according to MS formulation (*see* Appendix) (*9*). For shoot induction, medium is supplemented with naphthalene acetic acid (0.1 mg/L) and benzyladenine (5.0 mg/L), i.e., MS 1. For root induction, the basal medium is prepared at half-strength and naphthalene acetic acid is included (0.01 mg/L) with activated charcoal 0.6% (w/v), i.e., MS2. Difco agar (8.0 g/L) is used throughout and medium is dispensed 6 mL/50-mm Petri plate.
4. Culture room as in Method 1 point 7.

3. Methods

3.1. *Method 1, Populus*

3.1.1. *Manipulation of Shoot Formation in Stem Explants of Poplar* Populus *spp.*

1. Remove shoots from refrigerated store, and excise cuttings 20–25 cm in length. Shorter cuttings 15–20 cm may be made from thickest end of the shoot (2-cm diameter). Insert cuttings into potting compost (5/22.5-cm diameter pot), and place in a growing room or phytotron.
2. After 2 wk of growth, all or most of the buds on each cutting should show growth. Remove all buds except one; this bud becomes a shoot after a further 5–6 wk (Fig. 1 [1]).
3. Prepare culture plates as described, Materials, 1, point 6.
4. Harvest the shoots and excise leaves at the base of the leaf blade

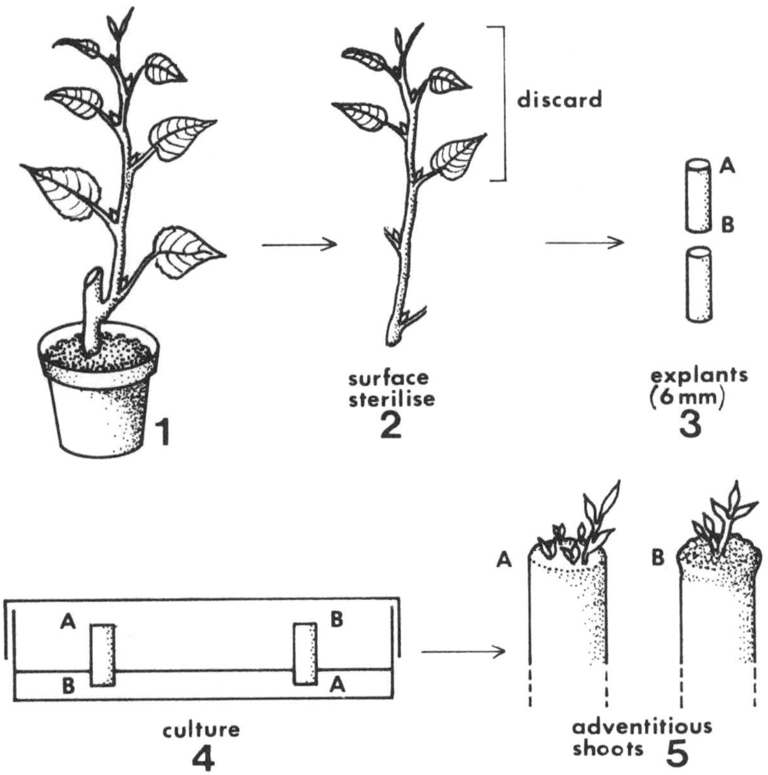

Fig. 1. (1) Poplar plant obtained from a single bud on original cutting. Preparation of internodal explants (2,3). Culture of explants in vitro (4) and production of shoots at the physiological apical end of the explant 2(5A) or physiological basal end (5B).

leaving most of the petiole attached. Take care not to injure or damage the stems during steps 3–5 (Fig. 1 [2]).

5. Counting from the shoot apex, excise and discard the first, second, third, and fourth internode (Fig. 1 [2]). Use the remainder of the shoot for experimentation.
6. Excise stem pieces approximately 9 cm in length, taking care to make the excision above a node at the apical end and just below a node at the basal end. Transfer shoots to glass or plastic jars.
7. Pour in a solution of calcium hypochlorite to cover the shoots, and replace the lid. Transfer the jar to a laminar-flow cabinet, and shake the jar occasionally.

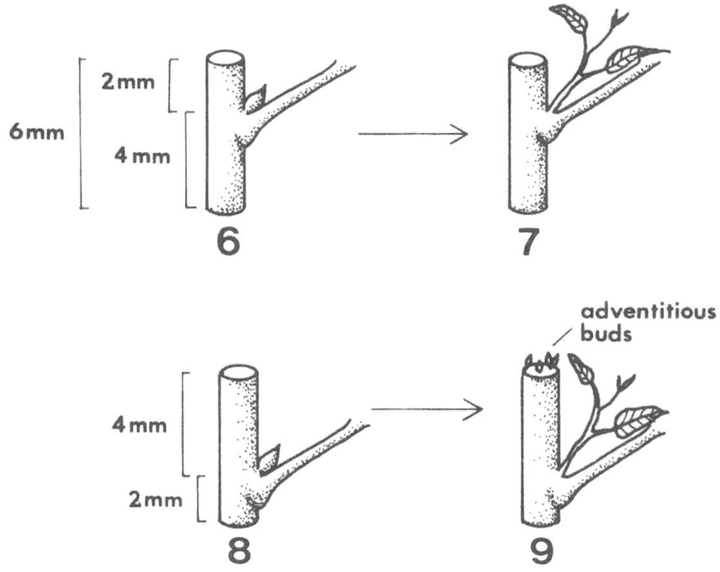

Fig 2. (6,8) Preparation of nodal explants. Growth of axillary buds (7,9) and adventitious buds, (9) in cultured explants.

8. After 20 min, drain off the calcium hypochlorite, and replace with sterile double-distilled water. Rinse in water for 5 min, and repeat rinsing twice more.
9. Mark the outside of the base and lid of a 9-cm Petri plate at intervals of 6 mm with a marker.
10. Transfer a sterilized shoot to the marked Petri dish, and aseptically excise internodal explants 6 mm in length. Take great care to note the physiological apical (Fig. 1 [3A]) and basal end (Fig. 1 [3B]) of each explant. Nodal explants may also be prepared and cultured at this time if required (Fig. 2 [6–9]).
11. Transfer the explants vertically to agar-solidified MS medium, and place the physiological basal end into the medium. Inversion of the explant and placement of the physiological apical end in the medium will give a different physiological response (Fig. 1 [4,5]). Place 1–4 explants/50-cm Petri dish (Fig.1 [3,4]). Wrap the Petri dishes with Parafilm.
12. Observe the explants from d 5–35 after culture for growth of cambial cells, initiation of meristematic areas, differentiation of shoot meristems, and for growth and development of shoots. Shoots produced at the physiological apical end of the explant arise directly from the

explant (Fig. 1 [5A]). Those produced at the physiological basal end arise from callus tissue (Fig. 1 [5B]). In the former case, adventitious buds are visible after 10–15 d.

13. Count the number of adventitious buds and shoots present after 4–5 wk of culture.
14. For root formation, remove adventitious shoots that have 2–3 developed leaves and a clearly differentiated stem axis (3–5 mm in length). Culture shoots on MS medium with 1% sucrose and 0.1 mg/L NAA. Shoots without well-developed leaves and well-developed stem axes generally fail to continue growth or to produce roots when recultured.
15. Transfer explants from which shoots have been removed onto fresh medium. This will ensure continued growth of adventitious buds and their differentiation into shoots.

3.2. Method 2, African Violet

3.2.1. Manipulation of Shoot Formation in Leaf Explants of African Violet (Saintpaulia ionantha *Wendl.*)

1. Excise fully expanded leaves with 2–4 cm of petiole, immerse in 70% ethanol for 5 s, and immediately transfer to jars containing 0.5% hypochlorite solution. Place cap on jar, and shake contents periodically for 15 min. Continue with further manipulation aseptically. Decant and replace solution with a solution of 1.0% hypochlorite, recap jar, and shake contents periodically for 20 min. Decant, and rinse leaves four times with sterile water.
2. Transfer leaves to a 9-cm Petri plate, and allow them to dry (5 min) in the airflow cabinet. Excise pieces of lamina approximately 1-cm square. Each explant should include a portion of the main vein for optimal results.
3. Transfer leaf explants to MS1 medium, one explant /Petri plate. Place the explants in an upright position with one-quarter inserted in the medium. Wrap plates and transfer them to the culture room. Adventitious shoots appear along the cut surfaces within 3–5 wk of culturing.
4. Transfer small shoots approximately 0.5 cm in length to MS2 medium for rooting. About 10–20 shoots may be placed/Petri dish, and rooting occurs within 3–4 wk.
5. Rooted plants may be transferred to soil by rinsing off the agar, and transfer to a mixture of 1 part perlite, 2 parts sand, and 5 parts peat. Cover the plants with light plastic film for 1–2 wk to facilitate the transition to autotrophic growth.

4. Notes

4.1. Method 1

1. Stem explants of all species listed (Materials 1) consistently produce adventitious shoots when donor shoots are grown and explants are made as described. There is some variability in the frequency of shoot formation between species and hybrids (*5*). To reduce variability among explants of any given genotype, it is important to choose donor plants (Fig. 1 [2]) of uniform size and developmental stage. Since variability in responsiveness occurs in explants taken from different locations in the donor shoot and also with explants of different size (*5*), care should be taken to discard at least the upper three internodes (Fig. 1 [2]) and to make explants of equal size (6 mm), respectively. For production of the maximum number of shoots/explant, best results are obtained by placing the physiological basal end of the explant in the medium (Fig. 1 [3,4,5A]).
2. After 4–5 wk of culture, explants consist of 1–2 adventitious shoots and several adventitious buds (Fig. 1 [5A]). The presence of shoots supress further development of adjacent buds. Buds may be stimulated to develop further by removal of shoots and reculture of the entire explant. Generally, only adventitious shoots are capable of rooting.
3. The influence of an axillary bud on formation of adventitious shoots in the stem may be studied. Presence of an axillary bud located 2 mm from the responsive surface of the stem supresses production of adventitious buds (Fig. 2 [6,7]). Presence of the axillary bud 4 mm from the responsive surface of the stem does not completely inhibit formation of adventitious buds (Fig. 2 [8,9]). In each case, the axillary bud grows out.
4. The influence of growth regulators on shoot formation in explants may be readily observed by their inclusion in the culture medium. Cytokinins, such as zeatin (0.01–5.0 mg/L), in the culture medium increase the number of shoots produced in explants of various lengths, indoleacetic acid (5.0 mg/L) and abscisic acid (1.0 mg/L) are inhibitory (*6*).
5. The anatomical features of the process of bud formation may be readily observed by making thin hand-sections of internodes at 3, 5, 10 and 15 d from initiation of cultures. Xylem, phloem, phloem fibers, cambium, and cortical tissues are readily distinguished by staining

sections for 20–30 s in a solution of Toluidine Blue 0 (0.05% w/v in a solution of 0.125% w/v benzoic acid with 0.145% sodium benzoate). Sections are then rinsed, mounted on a slide, and observed in the light microscope.

4.2. Method 2

1. Because of the hairy leaf surface of *Saintpaulia,* it may be difficult to obtain sterile cultures. To obtain sterile explants more easily, it is best to water donor plants from the base via a small dish for 3–4 wk prior to explanting. Although buds are usually visible within 3 wk of culture, some delay may occur for explants taken in winter.
2. The number of adventitious buds and shoots produced/explant may be easily regulated by varying the concentration of cytokinin (benzyladenine) in the culture medium.
3. Explants may be taken from leaves of regenerated plantlets, and thereby, a rapid propagation cycle may be demonstrated. In addition, plantlets, produced in vitro may be readily stored asceptically for up to 1 yr by placing each plant in a 100-mL jar containing 50 mL MS2 medium under normal growth room conditions. Plant regeneration is also possible from explants of leaf petioles.

References

1. Skoog, F., and Miller, C. O. (1957) Chemical regulation of growth and organ formation in plant tissues cultured *in vitro. Symp. Soc. Exp. Biol.* **11,** 118–131.
2. Christianson, M. L. (1986) Causal events in morphogenesis, in *Plant Tissue and Cell Culture, Plant Biology* (Green, C. E., Somers, D. A., Hackett, W. P., and Biesboer, D. D. eds.), Alan R. Liss, New York, vol 3, pp 45–55.
3. Street, H. E. (1979) Embryogenesis and chemically induced organogenesis, in *Plant Cell and Tissue Culture* (Sharp, W. R., Larsen, O. P., Paddock, E. F., and Raghavan, V., eds.), Ohio State University Press, Ohio, pp 123–154.
4. Thorpe, T. A. (1980) Organogenesis in vitro, structural, physiological, and biochemical aspects. *Intern. Rev. Cytol. Suppl.* **11B,** 77–111.
5. Douglas, G. C. (1984) Formation of adventitious buds in stem internodes of *Populus* spp. cultured *in vitro* on basal medium: influence of endogenous properties of explants. *J. Plant Physiol.* **116,** 313–321.
6. Douglas, G. C. (1985) Formation of adventitious buds in stem internodes of *Populus* hybrid TT32 cultured *in vitro*: effects of sucrose, zeatin , IAA and ABA. *J. Plant Physiol.* **121,** 225–231.
7. Vasquez, A. M., Davey, M. R., and Short, K. C. (1977) Organogenesis in cultures of *Saintpaulia ionantha. Acta Horticulturae* **78,** 249–258.
8. Harney, P. M. (1982) Tissue culture propagation of some herbaceous horticultural

plants, in *Applications of Plant Cell and Tissue Culture to Agriculture and Industry* (Tomes, E. T., Ellis, B. E., Harney, P. M., Kasha, K. J. and Peterson, R. L. eds.), Univ. Guelph Press, Ontario, Canada, pp 187–208.

9. Murashige, T. and Skoog, F. (1962) A revised medium for rapid growth and bio-assays with tobacco tissue cultures. *Physiol. Plant* **15,** 473–497.
10. Gamborg, O. L., Miller, R. A., and Ojima, K. (1968) Nutrient requirements of suspension cultures of soybean root cells. *Exp. Cell Res.* **50,** 151–158.

Chapter 9

Meristem-Tip Culture

Brian W. W. Grout

1. Introduction

The essence of meristem-tip culture is the excision of the organized apex of the shoot from a selected donor plant for subsequent in vitro culture. The conditions of culture are regulated to allow only for organized outgrowth of the apex directly into a shoot, without the intervention of any adventitious organs (*1–3*). The excised meristem-tip is typically small (often less than 1 mm in length) and removed by sterile dissection under the microscope (Fig. 1). It comprises the apical dome and a limited number of the youngest leaf primordia, and excludes any differentiated provascular or vascular tissues. A major advantage of working with such a small explant is the potential that this holds for excluding pathogenic organisms present in the donor plant from the in vitro culture (*see below*). A second advantage is the genetic stability inherent in the technique, since plantlet production from adventitious organs can be avoided (*3–7*). However, if there is no requirement for such benefits, then the related technique of shoot-tip culture (*see* Chapter 8, this vol.) may be more expedient for plant propagation. In this procedure, the explant is still a dissected shoot apex, but a much larger one, containing a relatively large number of developing leaf primordia. Typically, the explant is between 3–4 mm and 2 cm in length, and development in vitro is still regulated so as to be confined to outgrowth of an organized shoot, without adventitious propagation.

From: ***Methods in Molecular Biology***, vol. 6, ***Plant Cell and Tissue Culture***
Edited by Jeffrey W. Pollard and John M. Walker, ©1990 by The Humana Press

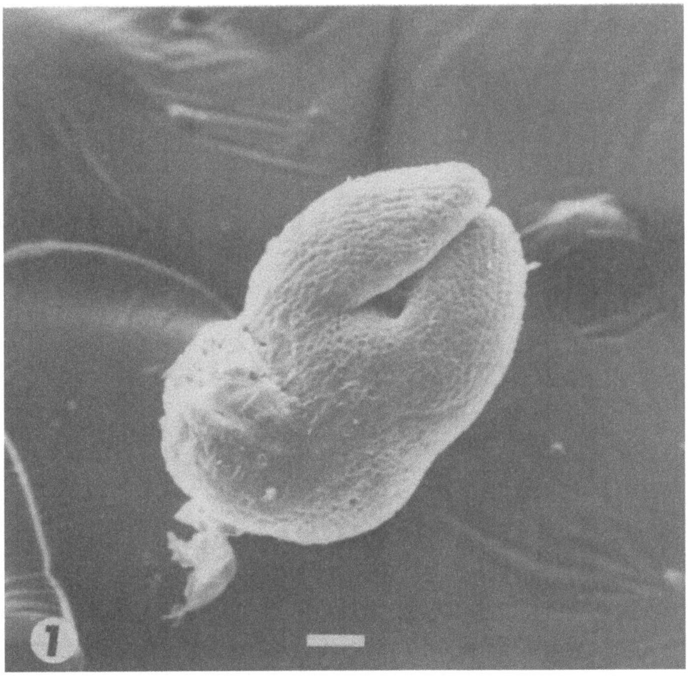

Fig. 1. A freshly excised meristem tip from an axillary bud of the potato *Solanum tuberosum*. The two smallest emergent leaf primordia are present. Scale bar represents 50 μM.

The axillary buds that develop on in vitro plantlets derived from meristem-tip cultures may also be used as a secondary propagule. When the in vitro plantlet has expanded internodes, it may be divided into segments, each containing a small leaf and an even smaller axillary bud (Fig. 2). When these nodal explants are placed on fresh culture medium, the axillary bud will grow into a new plantlet, at which time the process can be repeated. This technique adds a high propagation rate to the original meristem-tip culture technique, and together the techniques form the basis of micropropagation, which is so important to the horticulture industry (*1–3; see* Table 1 and Chapter 10, this vol.).

If the culture conditions are designed appropriately, the excised meristem will develop and elongate in essentially the same manner as if it were still part of the original parent plant. The production of a shoot from the meristem does not rely on any callus tissue formation or adventitious organogenesis, and so, problems of genetic instability and somaclonal variation should be minimized.

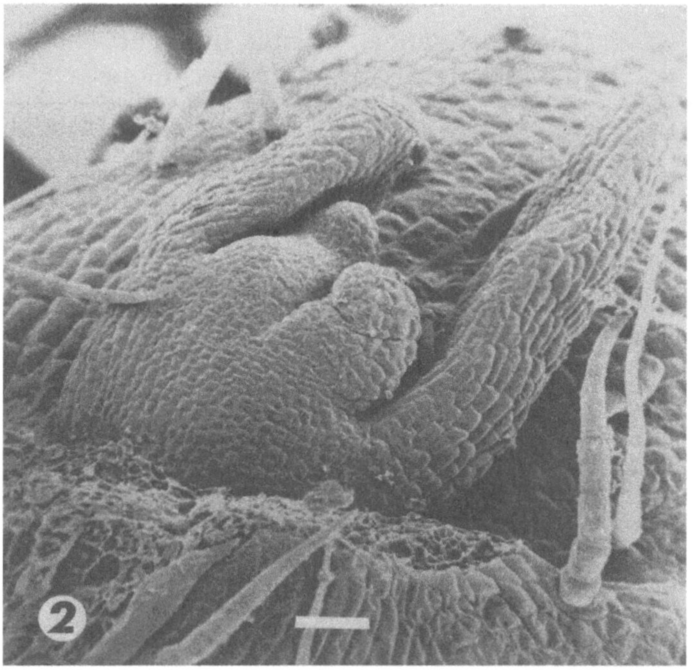

Fig. 2. An axillary bud from a freshly excised nodal segment, taken from a meristem-derived potato plantlet in vitro. A number of small leaf primordia are apparent, emphasizing the effectively normal structure of this very small axillary organ. Scale bar represents 50 μ*M*.

It is possible, however, that callus tissue may develop on certain portions of the growing explant, particularly at the surface damaged by excision (Fig. 3). The only acceptable situation under such circumstances is that the callus be sufficiently slow growing and localized to be readily identified, and that it and any organized development from it be excised at the first available opportunity. Recent studies from the author's laboratory on cauliflower plants regenerated from wound callus and floral meristems in vitro showed normal phenotypes for all the plants produced by meristem culture, whereas those of callus origin included rosette and stunted forms, and both glossy and serated leaves as examples of phenotypic deviance. These abnormal plants, derived from callus tissue, and normal plants from meristem culture showed the same DNA levels, as measured by scanning microdensitometry, but considerable differences when the isozyme of acid phosphatase was examined by gel electrophoresis. This suggests a range of unacceptable variation at the gene level, introduced by the adventitious origin of the shoots.

Table 1
The Propagation Potential Inherent in the Meristem-Tip Culture Technique[a]

Time from culture initiation, mo	No. of nodes available[b] for further subculture, equivalent to plantlet number
3	5
4	25
5	125
6	625
7	3125
8	15625

[a]This data is for a potato species, *Solanum curtilobum*, and was calculated from propagation rates achieved in the author's laboratory.

[b]Assumes 5 nodes available/plantlet.

A further advantage of meristem culture is that the technique preserves the precise arrangement of cell layers necessary if a chimeral plant genetic structure is to be maintained. In a typical chimera, the surface layers of the developing meristem are of differing genetic background, and it is their contribution in a particular arrangement to the plant organs that produces the desired characteristics, e.g., the flower color in some African violets. As long as the integrity of the meristem remains intact and development is normal in vitro, then the chimeral pattern will be preserved. If, however, callus tissues were allowed to form and shoot proliferation, subsequently, was from adventitious origins, then there would be a risk that the chimeral layers of the original explant may not all be represented in the specially required form in the adventitious shoots.

The technique of meristem culture may be exploited in situations where the donor plant is infected with viral, bacterial, or fungal pathogens, whether or not symptoms of the infection are expressed. The basis of eradication is that the terminal region of the shoot meristem, above the zone of vascular differentiation, is unlikely to contain pathogenic particles. If a sufficiently small explant can be taken from an infected donor and raised successfully in vitro, then there is a real possibility of the derived culture being pathogen-free. Such cultures, once screened and certified, can form the basis of a guaranteed disease-free stock for further propagation. The meristem-tip technique can be linked with heat therapy to improve the efficacy of disease elimination (*8,9*).

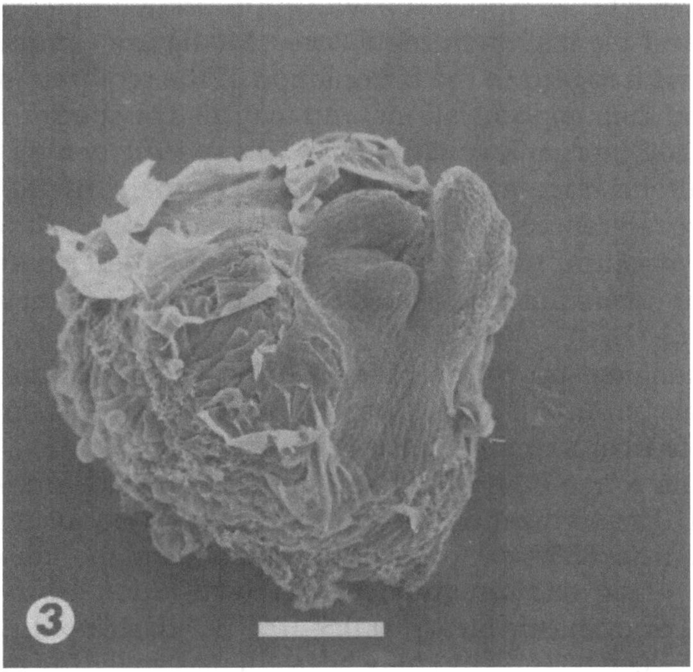

Fig. 3. A meristem-tip culture of potato 6 d after initiation. A distinct swelling can be seen at the wound surface, and larger callus cells are beginning to proliferate from it. Scale bar represents 500 μ*M*.

The major advantages of meristem culture therefore are:

1. Clonal propagation in vitro with maximal genetic stability.
2. The potential for removal of viral, bacterial, and fungal pathogens from an infected donor plant (*8–10*).
3. That the meristem tip is sufficiently small and homogeneous in tissue type to provide a practical propagule for cryopreservation and other techniques of culture storage.
4. An essential technique for accurate micropropagation of chimeral material.
5. That cultures are often acceptable for international transport with respect to quarantine regulations.

2. Materials

1. Culture vessels containing an appropriate plant culture growth medium, the most widely used formulation being that of Murashige and

Skoog (1962; 11; *see* Appendix) with appropriately chosen organic additives and plant growth regulators. Media are commonly species specific with regard to the formulation of the required growth regulators and their concentrations, and may also be specific for the other organic and inorganic constituents. This specificity may even extend to the cultivar/race level with respect to optimal growth. A variety of extensive reference sources can be consulted to select the appropriate growth medium, using the formulation for the closest taxonomic relative if no medium is detailed for the particular plant under investigation (*3,12,13*).

For meristem-tip culture, the medium usually contains 0.8–1.0% w/v agar, but liquid media may be used in conjunction with paper bridges or fiber supports in the culture vessel.

2. A suitable range of sterilant solutions for surface-sterilizing plant material. Presterilization is commonly carried out using an ethanol solution (from 85% v/v to absolute ethanol), and the subsequent sterilization using sodium hypochlorite (a dilution of a proprietary chemical or domestic bleach at 0.5–10% [v/v]; *see* text). A detergent such as Tween 80 (1%v/v) may be needed as a wetting agent. Sterile distilled water for rinsing is also required.
3. A growth room that provides a controlled environment to raise the cultures. Constant temperatures of 25 ± 1°C are common, but may vary somewhat with species.

3. Methods

1. Select a suitable donor plant (*see* Note 1) from which to remove apical meristem explants. Excise from the donor a terminal portion of stem and/or stem segments containing at least one node.
2. Remove mature and expanding foliage to expose the terminal and axillary buds. Cut the donor segments to an appropriate size and presterilize by immersion in, for example, 90% v/v ethanol or industrial methylated spirit for 10–30 s.
3. Sterilize by immersing the donor tissues in a sodium hypochlorite solution, the concentration and immersion times depending upon the size and nature of the donor tissues. A 0.5–10.0% v/v solution with 15 min immersion would represent the typical protocol range. Domestic bleach may also be used at a concentration empirically determined for the tissues involved. A few drops of a detergent, such as Tween 80, are added to the hypochlorite solutions to improve wetting of the

tissues, which is particularly important when the tissue surfaces are waxy or coated with epidermal hairs. However, caution is needed if domestic bleaches already containing high levels of detergent are to be used (*see* Note 2).

4. Following surface sterilization, the donor tissues are rinsed in sterile distilled water.
5. The tissues are then mounted on the stage of the dissection microscope with a magnification of at least 15x. A system will have to be developed to hold the tissue in place, and a piece of expanded polystyrene covered in white plastic film is ideal. The tissue can then be held in place by traditional dissection pins.
6. The tips of hypodermic needles are used to dissect away progressively smaller, developing leaves to expose the apical meristem of the bud, with the few youngest of the leaf primordia (*see* Note 3). This tissue is excised, and should comprise the apical dome and the required number of the youngest leaf primordia (*see* Note 4; ref *14*).
7. After excision of the explant, it should be transferred directly onto the selected growth medium. After closure, culture vessels with a relatively high rate of water loss to the external atmosphere, such as Petri dishes, are sealed with Parafilm tape.
8. The completed meristem-tip culture is transferred to a suitable growth room at, typically, 25 ± 1°C and illumination from warm white/cool white fluorescent lamps. A 12–16-h photoperiod of intensities up to (4000 lux) would be typical of the optimal conditions reported for a wide range of species (*see* Note 5).
9. If the explant is viable, then enlargement, the development of chlorophyll, and some elongation should be visible within 7–14 d (Fig. 4).
10. The developing plantlet should be maintained in vitro until the internodes are sufficiently elongated to allow dissection into nodal explants (Fig. 5).
11. The plantlets should be removed from their culture vessels, under sterile conditions, and separated into nodal segments. Each of these is transferred directly onto fresh growth medium, of the original formulation, to allow axillary bud growth (Fig. 5). Extension of this bud should be evident within 7–10 d of culture initiation (*see* Note 6).

4. Notes

A number of factors are likely to affect the success of attempts at meristem-tip culture. These are:

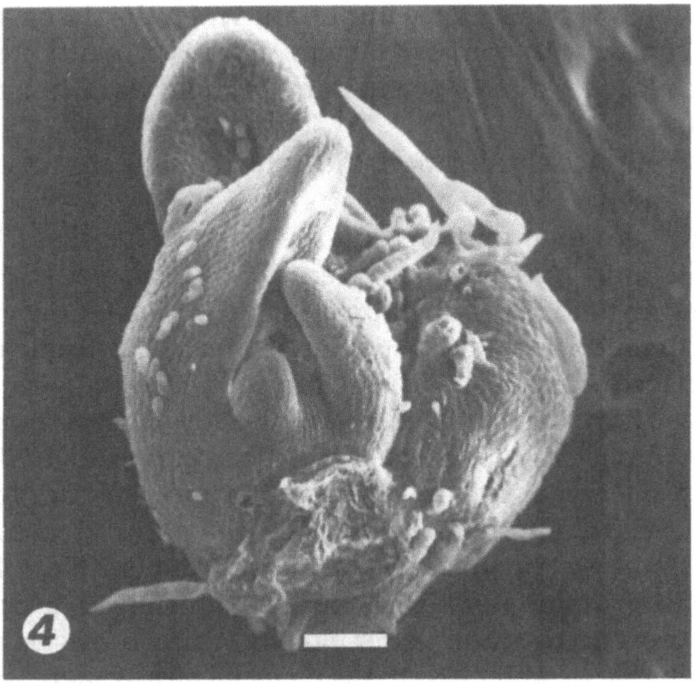

Fig. 4. A meristem-tip culture of potato 10 d after initiation. When excised, this explant had only two obvious leaf primordia, and has both developed and extended considerably during the culture period. Scale bar represents 500 µ*M*.

1. Donor plants should be selected for general health and vigor. Donor tissues should be taken from young, preferably actively growing regions of the plant to ensure the best chance of success. Seasonally dormant plants are to be avoided.
2. It is often easy to oversterilize during surface sterilization, and so fatally damage tissues and therefore the meristems in an attempt to eradicate surface pathogens. Use of donor plants that have been glasshouse raised and only watered from below will help minimize infection problems.
3. It is only with practice that reproducible, high levels of culture success will be achieved, and early attempts at dissection will often result in damaged tissue. Hypodermic needles used as dissecting tools often needed to be discarded after 2–3 meristem excisions.
4. It seems that the presence of leaf primordia is essential for successful culture growth, and so removal of too small an explant may restrict success.

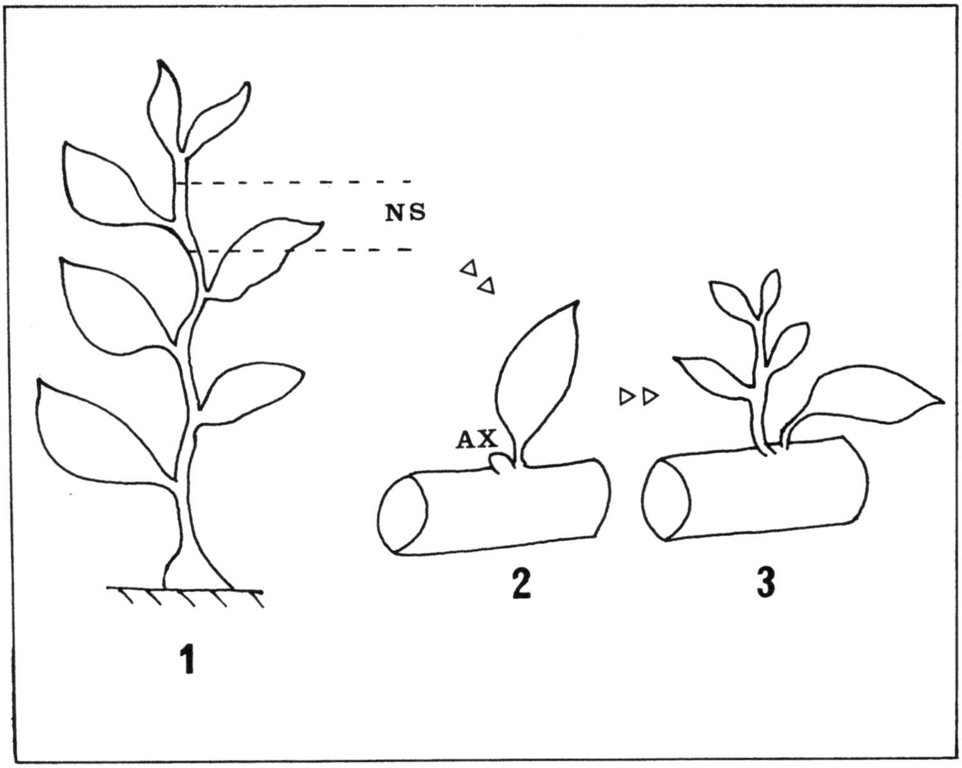

Fig. 5. Propagation from meristem-tip derived plantlets by the technique of nodal culture. (1) A plantlet showing extension growth in vitro. The nodal segment to be excised is indicated (NS). (2) The excised nodal segment as it is transferred onto fresh culture medium, showing the axillary bud (AX) that will be responsible for subsequent growth. (3) The pattern of development of a successful nodal segment culture, showing extension growth of the new plantlet.

5. A consequence of dissection may be the production by the tissue of toxic, oxidized polyphenolic compounds (3). These are more prevalent in cultures of woody species. Their effects may be minimized by rapid serial transfer of the excised meristem tip to fresh medium as soon as significant browning of the medium occurs, or by reducing light available to the culture in the early stages of development. Attempts can be made to reduce oxidation of the secreted polyphenolic compounds by incorporating antioxidants into the growth medium. These might include ascorbic acid, dithiothreitol, and polyvinyl pyrrolidone.

6. Where no previous reports of a successful medium can be found, either for the plant in question or a close relative, then an empirical study becomes necessary. The modified medium of Murashige and Skoog (1962) (*11*) has an inordinately wide application, and may be used with inorganic and organic constituents as a sensible first step in almost all cases. The energy source is typically sucrose at a 2–4% w/v level. Most problems will be found in determining the appropriate type and concentration of the necessary growth regulators to ensure culture survival and the required pattern of development.

References

1. Murashige, T. (1978) The impact of plant tissue culture on agriculture, in *Frontiers of Plant Tissue Culture* (Thorpe, T. A., ed.), IAPTC, Calgary, pp. 15–26.
2. Murashige, T. (1978) Principles of rapid propagation, in *Propagation of Higher Plants through Tissue Culture* (Hughes, K. W., Henke, R., and Constantin, M., eds.), US Dept. of Energy CONF-7804111, US Technical Information Centre, Washington, DC, pp. 14–24.
3. Hu, C. Y. and Wang, P. J. (1984) Meristem, shoot-tip and bud culture, in *Handbook of Plant Cell Culture* (Evans, D. A., Sharp, W. R., Ammirato, P. V., and Yamada, Y., eds.), Macmillan, New York, pp. 177–227.
4. Murashige, T. (1974) Plant propagation through tissue culture. *Ann. Rev. Plant Physiol.* **25,** 135–166.
5. Ancora, G., Belli-Donini, M. L., and Cozzo, L. (1981) Globe artichoke plants obtained from shoot apices through rapid *in vitro* micropropagation. *Sci. Hort.* **14,** 207–213.
6. Reisch, B. (1984) Genetic variability in regenerated plants, in *Handbook of Plant Cell Culture* Vol. 1, (Evans, D. A., Sharp, W. R., Ammirato, P. V., and Yamada, Y., eds.), Macmillan, New York, pp. 748–769.
7. Scowcroft, W. R., Brettell, I. R. S., Ryan, S. A., Davies, P., and Pallotta, M. (1987) Somaclonal variation and genomic flux, in *Plant Tissue and Cell Culture* (Green, C. F., Somers, D. A., Hackett, W. P., and Bisboev, D. D., eds.), A. Liss Inc., New York, pp. 275–288.
8. Walkey, D. G. A. (1980) Production of virus-free plants by tissue culture, in *Tissue Culture Methods for Plant Pathologists* (Ingram, D. S. and Hegelson, J. P., eds.), Blackwell Scientific, UK, pp. 109–119.
9. Long, R. D. and Cassells, A. C. (1986) Elimination of viruses from tissue cultures in the presence of antiviral chemicals, in *Plant Tissue Culture and Its Agricultural Applications* (Withers, L. A. and Alderson, P. G., eds.), Butterworths, UK, pp. 239–248.
10. Kartha, K. K. (1986) Production and indexing of disease-free plants, in *Plant Tissue Culture and its Agricultural Applications* (Withers, L. A. and Alderson, P. G., eds.), Butterworths, UK, pp. 219–238.
11. Murashige, T. and Skoog, F. (1962) A revised medium for rapid growth and bioassay using tobacco tissue cultures. *Physiol. Plant* **15,** 473–497.
12. George, E. F., Puttock, D. J. M., and George, H. J. (1987) *Plant Culture Media—Formulations and Uses* (Exegetics Ltd., UK), p. 567.

13. The information committee on plant tissue culture and the institute of physical and chemical research (1987) *Bibliographic information on Plant Tissue Culture,* Japanese Scientific Societies Press, Tokyo.
14. Huang, S. C. and Millikan, D. F. (1980) *In vitro* micrografting of apple shoot tips. *Hortrscience* **15,** 741–743.

Chapter 10

Micropropagation

Axillary Bud Multiplication

Nicola E. Evans

1. Introduction

Micropropagation techniques are being used by an increasing number of research workers and commercial firms. The main use has been that of mass production of plants ranging from nursery stock species (such as rhododendron or rose), through ornamentals (such as fuchsia or carnation) to fruits (such as apples or raspberries) and vegetables and field crops (such as cauliflower, potato, or lupins) (*1–5*). As well as enabling the mass production of selected genotypes, micropropagation techniques provide a method whereby viral and bacterial pathogens can be eliminated from infected varieties (*1*). These techniques are described in Chapter 9, this volume.

Any micropropagation system must produce large numbers of uniform plants that are genotypically and phenotypically the same as the original plant from which they were produced. To satisfy this criterion, the most appropriate technique is that of axillary bud multiplication. The advantages of this technique are that large numbers of uniform, genetically stable plants are produced under sterile conditions with a minimum of stress imposed on the growing plants. Environmental and nutritional

From: *Methods in Molecular Biology*, vol. 6, *Plant Cell and Tissue Culture*
Edited by Jeffrey W. Pollard and John M. Walker, ©1990 by The Humana Press

conditions can be easily manipulated and strictly controlled. The main disadvantage of such a system is that the scope for physiological experimentation is restricted because the plants produced are generally juvenile in character, and there is a tendency for the amount of epicuticular waxes to be reduced, palisade cells to be smaller, mesophyll air sacs to be larger, and stomatal physiology to be disrupted (*1*).

The purpose of this chapter is to describe a system that can be used to provide uniform, sterile plants for use as the basic material in genetic manipulation systems. The system described has been published by many authors in one form or another, but needs to be adapted to particular species and varieties as required (*1–6*).

Micropropagation is conducted in four main stages:

1. Selection and sterilization of elite plants.
2. Establishment of axillary buds in culture.
3. Multiplication in culture.
4. Rooting of in vitro plants and transfer to compost.

2. Materials

2.1. Stage 1: Selection and Sterilization of Elite Plants

1. Elite stock plants are selected that are visibly free from any sign of disease, stress, or surface blemishes. This material should preferably have been grown in an environmentally controlled growth cabinet or clean glasshouse, and should have been tested for the presence of specific viruses (Note 1a) using such techniques as that of ELISA (*7*) when the identity of such viruses are known.
2. 70% ethanol plus four drops of surfactant ("manoxol")/100 mL solution at room temperature (Note 1b).
3. 2% chloros plus four drops of surfactant ("manoxol")/100 mL solution at room temperature.

2.2. Stage 2: Establishment of Axillary Buds in Culture (See Note 2)

1. 50-mm diameter sterile Petri dishes containing approximately 10 mL of medium 1 (Table 1) or 85-mm diameter sterile Petri dishes containing approximately 25 mL of medium 1 (Table 1).
2. Incubator or environmentally controlled growth chamber allowing

environmental conditions of 20–24°C (depending on variety or species), a 16-h photoperiod, and a light intensity of 4000–6000 lux.

2.3. Stage 3: Multiplication in Culture

1. 250-mL sterile jars containing approximately 30 mL of medium 2 or medium 3 (Table 1).
2. Environmental conditions as described for Stage 2.

2.4. Stage 4: Rooting of *In Vitro* Plants and Transfer to Compost

1. 250-mL sterile jars containing approximately 50 mL of medium 4 (Table 1).
2. Propagation trays, plastic bags, or seed trays covered with glass sheets or a misting or fogging system.
3. 50-mm diameter pots containing compost (3:2, peat:sand) supplemented with nutrients. For potato *Solanum tuberosum* these are:

Osmocote	– 57 g
Sulphated potash	– 21 g
Epsom Salts	– 28 g
Frit 252 A	– 14 g
Lime	– 170 g
Phosphate	– 85 g

per 300 L peat: sand mix.

3. Methods

3.1. Stage 1: Selection and Sterilization of Elite Plants

1. Using a clean sharp blade, carefully excise axillary buds of the desired variety (Fig. 1, *see* p. 6; Note 3) and store in distilled water until enough buds have been obtained.
2. Working in a laminar flow bench, put a maximum of 20 buds into a sterile test tube.
3. Fill to the brim with ethanol solution and leave for 1–1.5 min.
4. Decant the ethanol.
5. Fill to the brim with chloros solution, replace the top, and agitate at 120 strokes/min for 12 min, either manually or with a shaker.

Table 1
Amounts of Chemicals (g L^{-1} or mg L^{-1}) in Media Used in Axillary Bud Culture of Solanum Species Based on Either Murashige and Skoog Medium (*8*) or Gambourg B5 Medium (*9*)

	Medium 1 (M&S)	Medium 2 (M&S)	Medium 3 (B5)	Medium 4 (M&S)
NH_4NO_3	1.65 g	1.65 g	–	1.65 g
KNO_3	1.9 g	1.9 g	2.5 g	1.9 g
$CaCl_2•2H_2O$	0.44 g	0.44 g	0.15 g	0.44 g
$MgSO_4•7H_2O$	0.37 g	0.37 g	0.25 g	0.37 g
KH_2PO_4	0.17 g	0.17 g	–	0.17 g
$(NH_4)_2SO_4$	–	–	13.4 g	–
$NaH_2PO_4•H_2O$	–	–	0.15 g	–
FeNa EDTA	36.7 mg	36.7 mg	40.0 mg	36.7 mg
H_3BO_3	6.2 mg	6.2 mg	3.0 mg	6.2 mg
$MnSo_4•4H_2O$	22.3 mg	22.3 mg	10.0 mg	22.3 mg
$ZnSo_4•7H_2O$	8.6 mg	8.6 mg	2.0 mg	8.6 mg
KI	0.83 mg	0.83 mg	0.75 mg	0.83 mg
$Na_2MoO_4•2H_2O$	0.25 mg	0.25 mg	0.25 mg	0.25 mg
$CuSO_4•5H_2O$	0.025 mg	0.025 mg	0.025 mg	0.025 mg
$CoCl_2•6H_2O$	0.025 mg	0.025 mg	0.025 mg	0.025 mg

Nicotinic Acid	0.50 mg	0.50 mg	1.0 mg	0.50 mg
Thiamine•HCl	0.10 mg	0.10 mg	10.0 mg	0.10 mg
Pyridoxine•HCl	0.50 mg	0.50 mg	1.0 mg	0.50 mg
Glycine	2.0 mg	2.0 mg	–	2.0 mg
Sucrose	20.0 g	20.0 g	20.0 g	20.0 g
Inositol	0.1 g	0.1 g	0.1 g	0.1 g
Glutamine	0.1 g	0.1 g	0.1 g	–
BAP (6-benzyl-amino purine)[4]	0.25 mg	0.25 mg	0.25 mg	0.25 mg
GA3 (gibberellic acid)[4]	0.1 mg	–	–	–
NAA (1-napthalene--acetic acid)	–	–	–	0.05 mg L^{-1}
pH[2]	5.64	5.64	5.64	5.64
Agar (Difco Bacto)[4]	8.0 g	8.0 g	8.0 g	8.0 g

[a]Commercial preparations of M&S and B5 salt (minus sucrose glutamine and hormones) are available in packets that make up 1 L of media.

[b]pH adjusted using .1M HC1 and 0.1M KOH begore agar added and prior to autoclaving.

[c]The amount of agar needed varies depending on the brand used.

[d]Varying the hormones allows the above media to be used for axillary bud culture of a wide range of species.

Fig. 1. Axillary bud being excised from a sprout of *S. tuberosum*.

6. Decant the chloros solution and refill with sterile distilled water.
7. Rinse in sterile distilled water 3x.
8. Store in sterile distilled water until ready to continue (not longer than 2 h).

3.2. Stage 2: Establishment of Axillary Buds in Culture

1. Empty the water plus axillary buds into an empty sterile Petri dish, for ease of handling.
2. Place up to 4 buds in a 50-mm Petri dish or 10 buds in an 85-mm Petri dish containing medium 1. Make sure that the base of the bud is stuck firmly in the medium, but take care that the bud is not buried (*see* Note 4).
3. Seal the Petri dish with laboratory sealing film, ensuring adequate gaseous exchange by puncturing the film 2–4x with a fine sterile needle (*see* Note 5).
4. Transfer to growth room or incubator, and leave for between 1 wk and 2 mo depending on the variety.

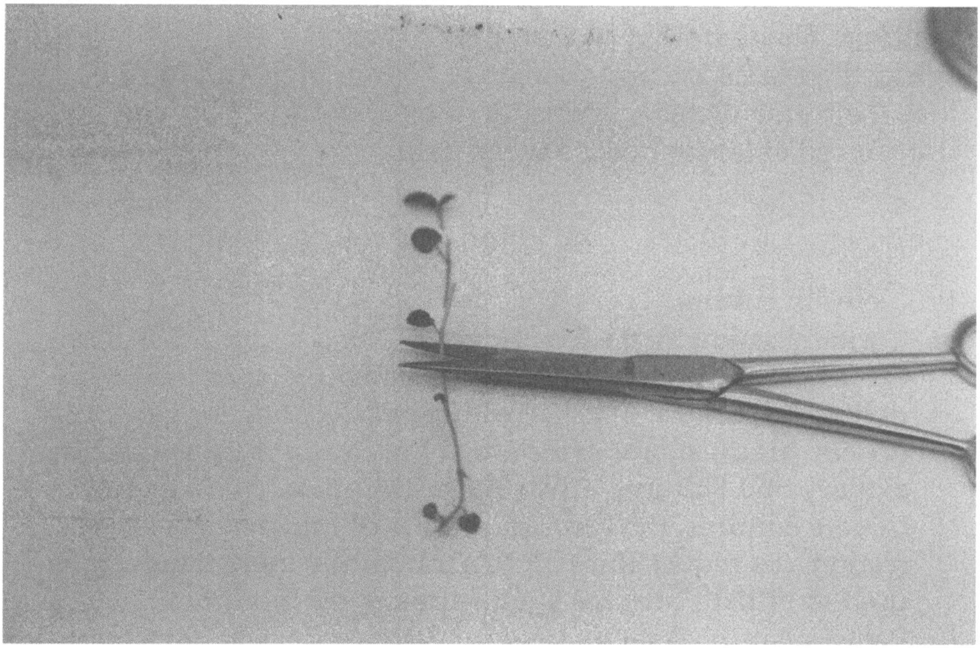

Fig. 2. Subculturing of *in vitro* grown shoot of *S. tuberosum* by internodal cutting.

3.3. Stage 3: Multiplication in Culture

When shoot extends, cut internodes, and transfer apical cutting and internodal cuttings (Fig. 2) to sterile jars containing medium 2 or 3. Make sure that the basal portion of the stem is firmly pressed into the medium without burying the explant.

3.4. Stage 4: Rooting of In Vitro Plants and Transfer to Compost

1. Cut off the top 5 mm of in vitro plant (apex plus 2 or 3 nodes) and transfer to medium 4. (12 plants/250 mL jar).
2. Leave for 3–5 d until 3 or 4 small, sturdy roots approximately 5 mm long are visible.
3. Remove each plantlet carefully from the jar, removing as much agar as is possible without harming the root structure, and transfer to damp compost in 50-mm pots (*see* Note 6).
4. Keep in high humidity conditions 12–24 h (propagator, plastic bag, misting, or the like).

5. Transfer to the glasshouse, preferably placing on top of capillary matting. Shade from direct sunlight.
6. When the plants are approximately 70 mm high and have begun to lose their juvenile characteristics, they should be hardened off and transferred to larger pots or to the field.

4. Notes

1. Problems in culture:
   a. Contamination by viral pathogens: The danger can be minimized by testing for infections by known viruses before initiating the culture. Once in culture, viral pathogens generally do not produce any symptoms, but are rapidly spread by axillary bud cultures. The smaller the piece of tissue used to initiate cultures, the less risk there is of transferring viral infection. To reduce this risk to a minimum, meristems can be used to initiate cultures (*see* Chapter 9, this volume).
   b. Bacterial and fungal pathogens: Surface sterilization should remove most contamination. Frequent scrutiny of cultures should allow healthy buds to be transferred to fresh medium if occasional contaminated buds are seen. Alternatively, placing fewer explants in smaller Petri dishes reduces the problem and infected plates can then be discarded. The use of antibiotics to suppress bacterial or fungal infections is generally ineffective.
2. Precautions for sterile culture: Use a laminar flow bench, and swab down the bench, roof, and sides with disinfectant. Always wear gloves (thin latex gloves are best), and swab gloves and bench surface frequently during any experimental work. Sterilize and flame all instruments before use. Finally, avoid passing anything over the top of opened containers.
3. Stage 1: Choose as juvenile a tissue as possible to initiate cultures. Do not try to sterilize too many different genotypes at once. Start with 1 or 2, and build up to about 6. Problems of contamination and toxic effects of the sterilant increase with increased numbers of genotypes being treated simultaneously.
4. Stage 2: Condensation can be a problem in Petri dishes. This can be minimized by stacking the Petri dishes and rotating the position of each dish in the stack daily. The intensity of light reaching each culture does not appear to be critical at this stage, although light is necessary.

Fig. 3. Axillary bud cultures of *S. sparsiphilum* grown in (A) Medium 3 and (B) Medium 2.

Gibberellins tend to produce elongated, abnormal looking shoots. These revert to normal growth when transferred to medium that does not contain gibberellins.

Callusing will occur around the base of the bud; only shoots that can be clearly seen to have derived from extension of the preexisting axillary bud should be transferred. Adventitious shoot formation in callus tissue generally produces high levels of variation and is the basis for producing somatic variants.

5. Stage 3: Murashige and Skoog medium and Gambourg B5 medium (*see* Appendix) are those most frequently used in the culture of axillary buds. They often need to be modified, mainly by varying the vitamins and growth regulators, depending on the species or variety being used. One of the effects of changing the medium can be seen in Fig. 3. Fig. 3A shows *Solanum sparsipilum* growing on medium 3, whereas Fig. 3B shows the same species growing on medium 2.

If the intent is to use the axillary bud cultures as a source of protoplasts, then all media need to be made up using distilled deionized water or water of equal purity other additions may also have to be made to the culture medium. For example, *Solanum tuberosum* can be quite satisfactorily multiplied in vitro using basic Murashige and Skooge medium without any hormones or glutamine; however, if protoplasts are to be isolated from in vitro culture, then the addition of BAP and glutamine increases the number of viable protoplasts obtained.

Vitrification can be a serious problem in in vitro cultures. In this condition, the shoots tend to elongate and look "glassy," the leaves become reduced or malformed, and if not treated the culture will die. The causes of vitrification are not fully understood, but in my experience the risk of vitrification is greatly reduced if there is adequate gaseous exchange between the culture and the external environment. This can be achieved by piercing laboratory film seals as described earlier or by loosening jar lids slightly.

6. Stage 4: Rooting and transfer of plants to compost can be difficult. If the auxin level is too high, subsequent root development after transfer to compost is reduced. Some species can be transferred to compost without a rooting step. Some fruit tree species need a dark treatment as well as additions of auxin for successful rooting to take place (5). The transition from culture to compost is critical. High humidity must be maintained, but care must be taken not to waterlog the compost. If plants are transferred to the field, they need protection from wind damage, drought, and birds.

References

1. Penell, D. (1987) Micropropagation in horticulture. *Grower Guide No. 29* (Grower Books, London).
2. Hartman, H. T. and Kester, D. E. (1983) *Principles of Tissue Culture for Micropropagation (4th ed.)* Ch. 16, (Prentice Hall, New Jersey).
3. Hartman, H. T. and Kester, D. E. (1983) *Techniques of In Vitro Micropropagation in Plant Propagation (4th ed.)* Ch. 17 (Prentice Hall, New Jersey).
4. Schcifer-Menuhr, A., ed. (1985) Propagation of lupins in in vitro techniques. *Propagation and Long Term Storage,* Martinus Nijhoff and Junk, Dordrecht, Boston, Lancaster, pp. 23–28.
5. Boxus and Druart (1985) Mass propagation of fruit trees in in vitro techniques. *Propagation and Long Term Storage* (Schcifer-Menuhr, A., ed.), Martinus Nijhoff and Junk, Dordrecht, Boston, Lancaster, pp. 29–33.
6. George, R. A. T. (1986) Technical guidelines in seed potato micropropagation and multiplication. F.A.O. Plant Production and Protection Paper No. 71. F.A.O., Rome.
7. Gaastra, W. (1984) Enzyme-linked immunosorbent assay (ELISA), in *Methods in Molecular Biology, Vol. 1,* **38,** (Walker, J. M., ed.), Humana Press, New Jersey, pp. 349–355.

8. Murashige, T. and Skoog, F. (1962) A revised medium for rapid growth and bioassays with tobacco tissue cultures. *Physiol. Plant* **15,** 473–497.
9. Gamborg, O. L., Miller, R. A., and Ojima, K. (1968) Nutrient requirements of suspension cultures of soybean root cells. *Exp. Cell Res.* **50,** 151–158.

Chapter 11

Grafting In Vitro

Michael Parkinson, Carmel M. O'Neill, and Philip J. Dix

1. Introduction

Graft formation in plants involves the severing of the vascular system with consequent loss of water and solute transport throughout the plant. This transport must be restored to prevent death resulting from nutrient starvation or dessication.

This restoration is achieved by an initial phase of rapid division of cells at the cut surfaces to close the wound and form a graft union, followed by a phase of coordinated differentiation of cells both in and around the graft union (*1*). This reunites the severed ends of the original vascular tissue, and thus restores the flow of water and nutrients through the plant. The pattern of this differentiation is related to the compatibility relations of the two graft partners (2).

Grafting is therefore a useful experimental system to investigate the factors controlling cell differentiation in plants. Such studies should also provide a valuable insight into the important problem of graft incompatibility.

From: *Methods in Molecular Biology*, vol. 6, *Plant Cell and Tissue Culture*
Edited by Jeffrey W. Pollard and John M. Walker, ©1990 by The Humana Press

Grafting is traditionally practiced with woody species; however, grafting in stem internodes of herbaceous plants provides a much more convenient experimental system. Large populations of these grafts can be easily constructed in a small area, and their development can be investigated readily over a relatively much shorter time span (1–2 wk) (*3*).

Such a system is enhanced by the capacity to graft under sterile conditions in culture, where the environment of the graft can be carefully controlled and manipulated, and where substances of interest can be maintained in the graft union at high and defined concentrations.

We have developed a method for grafting in cultured, explanted internodes from a number of species within the family *Solanaceae*, obtained either from greenhouse-grown plants (*4*) or from shoot cultures, and present the protocol below.

2. Materials

1. Aluminum box sections. These are cut to 86 mm in length from a length of aluminum curtain track and the base filed to fit the curved base of the Petri dish.
2. "Sterilization-tube." An open-ended 2.5-cm diameter Pyrex™ tube covered at one end by a single layer of muslin, which is held in place by an elastic band.
3. 12-mm lengths of "Versilic" silicone-rubber tubing of the same internal diameter as the internodal explants to be grafted.
4. Seed: *Lycopersicon esculentum* var. Ailsa Craig., *Lycopersicon peruvianum, Nicandra physaloides, Datura stramonium, Solanum nigrum, Solanum tuberosum* var. Kerrs Pink, *Nicotiana plumbaginifolia*.
5. RM Medium (per liter): 4.6 g Murashige and Skoog Plant Salt Mixture (*see* Appendix); 30 g sucrose, pH adjusted to 5.65–5.75; 7 g Difco Bacto-agar.
6. Medium A: As RM, but with the addition, prior to pH adjustment, of (per liter) 100 mg meso-inositol and 0.2 mg kinetin (1 mL of a 20 mg/L stock made by dissolving the plant growth regulator in the minimum volume of molar Potassium Hydroxide [a few drops], followed by dilution with distilled water).
7. Medium B: As Medium A, but with the addition of 2 mg/L of IAA from a filter-sterilized stock (*see* Methods).
8. 20% (v/v) "Domestos" or other proprietary bleach plus 0.04% (v/v) Tween 20.

9. Culture-room (25°C, 16 h/d, 300 µmol/M^2/s light intensity).
10. Clearing solution: Dissolve 1g Basic Fuchsin in 100 mL distilled water at 80°C. Add 6 g Potassium Hydroxide in 1-g aliquots. Filter through two filter papers and collect eluate (clearing solution). Store clearing solution in a black bottle in the refrigerator (*see* Note 1).
11. Dehydration solutions: 50%, 70%, and 100% ethanol v/v.
12. Concentrated Hydrochloric Acid.
13. Xylene.
14. Canada Balsam.

3. Methods

The procedure involves four separate techniques that follow each other sequentially:

1. The preparation of a "split-agar Petri dish" for the culture of the assembled grafts.
2. The preparation of the internodes to provide sterile explants for grafting.
3. The assembly of the sterile, internodal explants to form grafts and the culture of those grafts.
4. Treatment of the cultured grafts, so that graft compatibility may be assessed.

3.1. Preparation of "Split-Agar Petri Dish"

1. Mark an arrow across the underside of a 9-cm Petri dish and divide the dish in two with an aluminum box-section perpendicular to the arrow.
2. Prepare Medium B and autoclave for 15 min at 15 psi (121°C) and cool to 40°C in a water bath.
3. Prepare Medium A from an aliquot of Medium B from the water bath by the addition of IAA to a final concentration of 2 mg/L from a 50-mg/L stock (*see* Materials). Return to the water bath.
4. Pour 15 mL of Medium A into the dish on the side nearest to the arrowhead, and 15 mL of Medium B into the opposite side of the dish (*see* Note 2).
5. Allow the agar to solidify, and then carefully remove the box-section (*see* Note 3).

3.2. Preparation of Internodes for Grafting

3.2.1. Greenhouse-Grown Plants

1. Sow seed in trays of John Innes No. 1 compost and germinate in greenhouse conditions. After emergence of the first pair of true leaves, transfer singly to the same compost in 9-cm plastic plant pots. Internodes are selected for grafting after emergence of the first five true leaves (approximately 14 d).
2. Remove the entire first internode wih a scalpel 5 mm below the seed leaves and above the first true leaf, remove the leaves, cap the cut surfaces with molten parawax (at 60°C), and insert into the "sterilization-tube" so that the top of the internode rests against the muslin.
3. Place the "sterilization-tube" for 10 min in 150 mL of 20% "Domestos" plus 0.04% v/v Tween 20 in a 250-mL wide-necked Erlenmeyer flask (*see* Note 4).
4. Remove the sterilant by successively transferring the "sterilization-tube" for 5 min each through three changes of 150 mL distilled water.
5. Tip the internodes out of the "sterilization-tube" into a 9-cm Petri dish. If this operation is performed right-handed, the physiological apex of the internodes will lie on the right side of the dish.
6. Immediately before graft assembly, the wax-covered ends of the internode are trimmed off and the internodes cut into 7 mm lengths by cuts perpendicular to the long axis of the internode.

3.2.2. Shoot Cultures

1. Prepare RM Medium, autoclave, and pour into sterile Petri dishes (20 mL/dish) and "Plantcon" containers.
2. Surface-sterilize seeds using the same procedure as for internodes (*see* previous section). For *N. plumbaginifolia, see* Chapters 7 and 8, for seed sterilization and initiation of shoot cultures.
3. Place five seeds in each dish of RM Medium and incubate in the constant temperature room.
4. After 4–6 wk, establish shoot cultures from the plantlets by dissecting out individual nodes, retaining 1-cm of internode above and below the node, and transferring singly to fresh RM Medium in sterile containers. The basal surface should be cut at an angle and inserted into the culture Medium. A fresh shoot will develop rapidly from the axillary bud.
5. Maintain the shoot cultures by subculturing in the same way every 4–6 wk.

6. For grafting, dissect out the internodes and trim to 7 mm lengths. No sterilization is needed.

3.3. Graft Assembly and Culture

1. Select two internode segments of the same diameter for graft partners and insert in the appropriate orientation into a 12-mm length of "Versilic" silicone rubber tubing of the same internal diameter as the diameter of the graft. Hold the left end of the tubing with forceps, and push the right graft partner into the tube until it extends into the middle of the tube with approximately 1 mm of internode protruding. Then hold the right end of the tubing, and push in the left graft partner until both graft partners meet firmly in the middle of the tube.
2. Place the assembled grafts between the two surfaces of a "split-agar Petri dish" as shown in Fig. 1.
3. Seal the dish with Parafilm and hang the dish vertically on the wall of a culture room for 14 d.

3.4. Measurement of Graft Incompatibility

1. Remove the grafts from the "split-agar Petri dish" and trim to within 2 mm of the graft union with a scalpel, marking the apex of the graft with an oblique cut.
2. Divide exactly in half with a cut perpendicular to the long axis of the grafts, and remove the cut silicone-rubber tubing.
3. Place the two halves together in a test tube containing 10–15 mL of clearing solution and leave in a waterbath overnight at 60°C (*see* Note 5).
4. Replace the solution with 50% ethanol and leave for 10 min. Repeat 4x.
5. Replace the solution with 70% ethanol and leave for 10 min. Repeat a further 5x.
6. Add 5 mL of concentrated Hydrochloric Acid to the 10–15 mL of 70% ethanol and leave until all the vascular tissue has turned purple (30–120 s).
7. Replace the solution with 70% ethanol and leave for 10 min. Repeat (*see* Note 6).
8. Replace the solution with absolute ethanol every 30 min for 6 h, and then leave overnight (*see* Note 7).
9. Replace the solution with xylene and leave for 1 h. Repeat. Leave overnight in xylene (*see* Note 8).

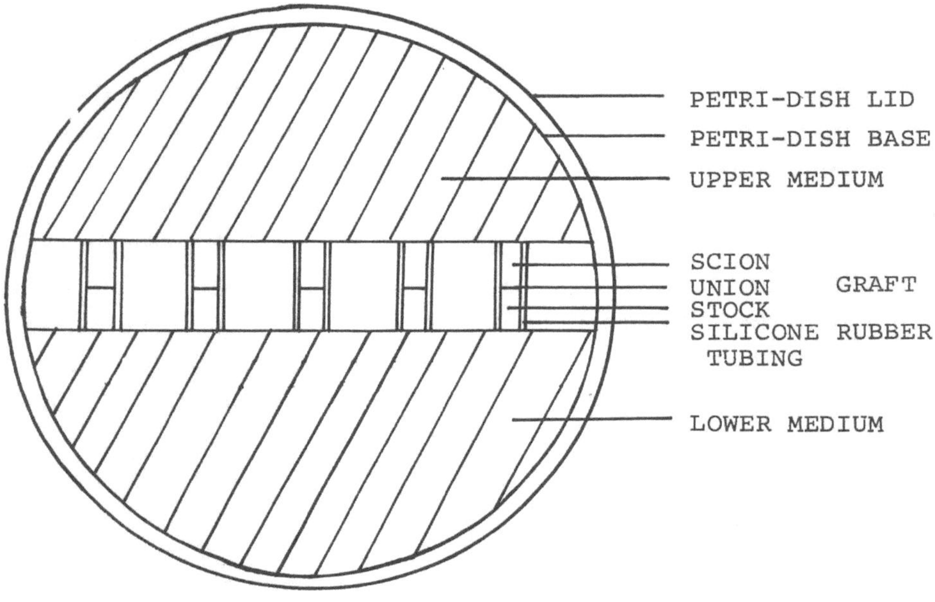

Fig. 1. A "split-agar" Petri dish containing five grafts.

10. Remove the grafts from the xylene, place both halves of the graft on a microscope slide and place a coverslip over the grafts. Press firmly with two thumbs for a few seconds, and then remove the coverslip.
11. Remove surplus xylene, then mount the graft in Canada Balsam, and seal under a coverslip (*see* Note 9).
12. The graft union will be clearly visible between the two cut ends of the original vascular tissue. The vascular tissue will lie in one plane of focus, permitting counts of individual xylem elements within the graft union. These will be aggregated into the following units:
    a. Strands: Strands of linked xylem elements entering, but failing to completely cross the graft union.
    b. Connections: Strands that completely cross the graft union.
    c. Free: Xylem elements not in strands or connections.

Compatible grafts will have the majority of xylem elements in connections (50–100% of all elements), whereas in incompatible grafts, very few

(0–20%) of the elements will be in connections, the majority being in strands or free.

4. Notes

1. The solution may be kept for up to 14 d under these conditions.
2. It is important that the box-section is cool prior to pouring. This prevents agar from running under the box-section into the opposite side of the dish.
3. The prepared dishes may be stored, at 4°C in the dark, for up to 7 d prior to use.
4. Perform this and all subsequent operations in a laminar-flow hood.
5. The grafts are delicate once they are placed in the clearing solution, so care must be taken in their handling.
6. Care must be taken in decanting the mixture, since the grafts will float in the acid mixture.
7. The grafts should appear frosty-white when fully dehydrated. If not, replace the ethanol and leave overnight.
8. The grafts should be perfectly transparent apart from the xylem elements, which are stained purple. If the grafts appear cloudy, they are not fully dehydrated and must be left in absolute ethanol (repeat step 8).
9. Crystals of KOH may appear in the graft at this stage. This is the result of incomplete removal of the clearing solution (*see* Step 4). This cannot be remedied. Air bubbles may also be present in the grafts. These will be absorbed by the Canada Balsam if the grafts are left overnight.

References

1. Stoddard, F. L. and McCully, M. E. (1979) Histology of the development of the graft union in Pea roots. *Can. J. Bot.* **57 (14),** 1486–1501.
2. Yeoman, M. M., Kilpatrick, D. C., Miedzybrodzka, M. B. W., and Gould A. R. (1978) Cellular interactions during graft formation in plants, a recognition phenomenon? *S. E. B. Symp.* **32,** 139–160.
3. Lindsay, D. W., Yeoman, M. M., and Brown, R. (1974) An analysis of the development of the graft union in *Lycopersicon esculentum*. *Annals of Botany* **38,** 639–646.
4. Parkinson, M. and Yeoman, M. M. (1982) Graft formation in cultured, explanted internodes. *New Phytol.* **91,** 711–719.

Chapter 12

Flower Organ Culture

Brent Tisserat and Paul D. Galletta

1. Introduction

A variety of factors contribute to flower induction in nature. These same factors are assumed to be responsible for in vitro flowering (*1*). Flower formation in tissue culture has been observed in several plant species, and arises from a variety of explant sources (*1*). However, the factors responsible for flowering in culture have not been extensively studied (*1*). Reasons for studying flower formation in tissue culture can be summarized as follows:

1. Provides a model system for studying whole flower development or development from an excised part of the flower (e.g., ovary culture) (currently performed).
2. Provides a means for conducting microbreeding (potential application).
3. Provides a source of biochemicals and pharmaceuticals (potential application).

Using appropriate plants (e.g., *Amaranthus*), the possibility of completing a plant's entire life cycle in vitro exists. This leads to a model system useful in studying microclimates or nutritional effects on the plants'

From: *Methods in Molecular Biology*, vol. 6, *Plant Cell and Tissue Culture*
Edited by Jeffrey W. Pollard and John M. Walker, ©1990 by The Humana Press

vegetative and reproductive processes (Fig. 1) (2). Flowering in culture can be induced from explants that are in either the vegetative or the reproductive state. In vitro flower induction from vegetative structures is described in this chapter for a woody plant (*Citrus*) and for an herbaceous plant (*Amaranthus*). Flower induction may be stimulated by both chemical and physical means.

2. Materials

1. *Citrus* explants. Shoot tips from *Citrus limon* (L.) Burm. f. cv. "Eureka" are obtained from fruit-bearing trees grown in the greenhouse or in the field.
2. *Amaranthus* explants. *Amaranthus caudatus* L. cv. "Pan," *A. gangeticus* L. (Chinese Spinach), *A. hypochondriacus* L., *A. retroflexus* L., and *A. viridis* L. seeds are obtained from a commercial source (J. L. Hudson, Seedsman, Redwood City, CA) (*see* Note 1).
3. Surface sterilant. A 2.63% sodium hypochlorite solution (containing two drops of Tween-20 emulsifier/100 mL solution) is employed as the disinfectant. This solution is prepared by diluting household bleach solution (containing 5.25% sodium hypochlorite) with water.
4. Nutrient medium. The nutrient medium contains Murashige and Skoog (MS) inorganic salts (*see* Appendix) with the following (in m*M*): thiamine•HCl, 0.001; meso-inositol, 0.667; sucrose, 8.76; naphthalene acetic acid (NAA), 0.005. Phytagar (Flow Laboratories Inc., Inglewood, CA) is added to the medium at an 8000 mg/L concentration. In some cases, 0.1 mg/L benzyladenine is included in the medium. The pH is adjusted to 5.7 ± 0.1 with 0.1*N* HCl or NaOH before the addition of Phytagar. The medium is dispensed in 25-mL aliquots into 25 x 150-mm culture tubes and capped with Bellco kaputs. Tubes are autoclaved for 15 min at 1.05 kg/cm^2 and 121°C and then slanted at a 45° angle while cooling.
5. Culture conditions. *Amaranthus* cultures are incubated in a temperature-controlled environmental chamber at 26 ± 1°C under a 16-h daily exposure to 2.2 $W{\bullet}m^{-2}$ using cool, white fluorescent lamps. *Citrus* buds are initially cultured at 16°C under a 16 h daily exposure to 2.2 $W{\bullet}m^{-2}$ using cool, white fluorescent lamps in a Hotpack incubation chamber for 8 wk. Buds are then transferred to a temperature-controlled environmental chamber maintained at 26 ± 1°C.

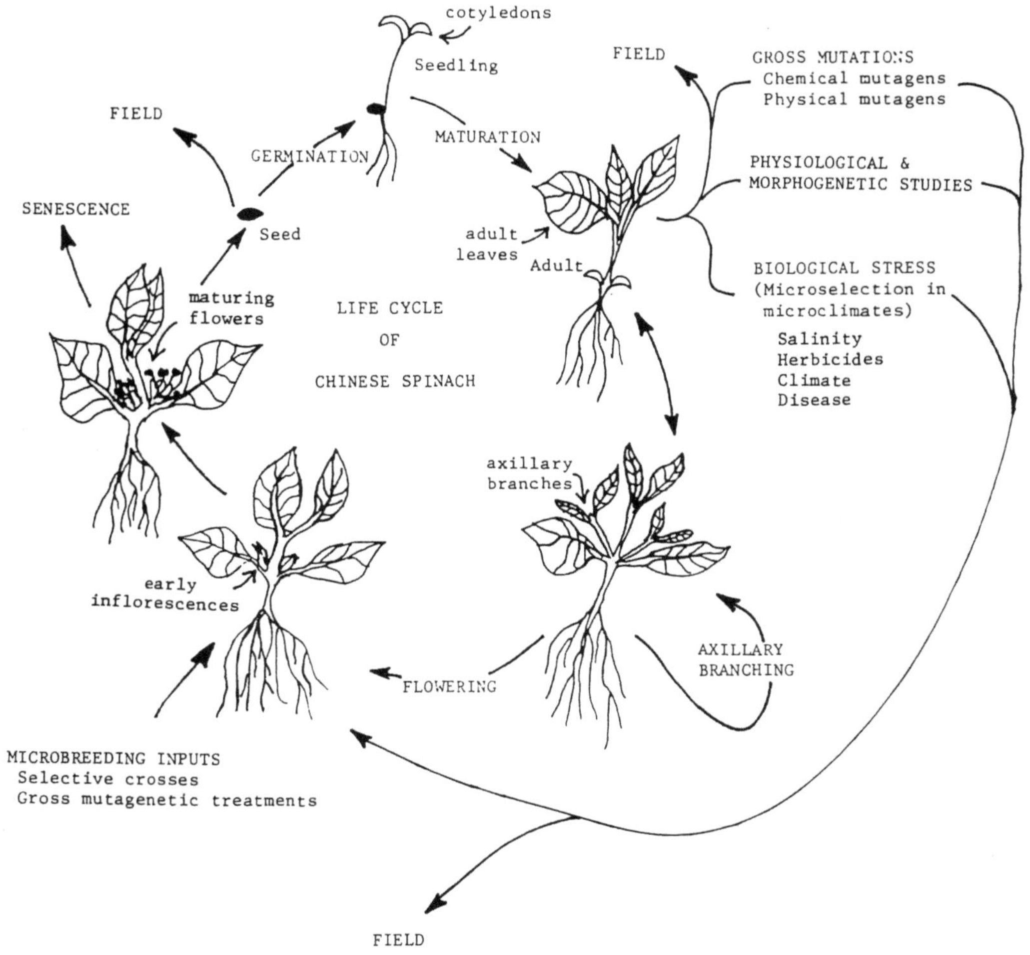

Fig. 1. Potential study applications using the *Amaranthus* culture system.

3. Methods

3.1. Flowering in Amaranthus

1. Seeds are removed from packages, wrapped in sterile cheesecloth, and surface sterilized in sodium hypochlorite solution for 15 min. A subsequent sterile water rinse is not administered.
2. Following surface sterilization, cheesecloth packages are removed to a sterile Petri dish, and unwrapped using an 11" (22 cm) bayonet fine-

tipped serrated forceps and surgeon's scalpel fitted with a # 11 blade. Exposed seeds are then planted on the surface of nutrient agar.

3. Cultures are incubated in a temperature-controlled environmental chamber at 26°C and recultured every 8 wk to fresh medium. Seeds germinate at the end of 8 wk in culture. Hereafter, only the terminal shoot tip of 1–3 cm in length is recultured. Shoot tips readily produce lateral buds through axillary outgrowths. They also readily form adventitious roots.
4. At the end of 16–32 wk in culture, shoots begin to initiate flowers from their lateral bud nodes (Fig. 2). Buds produce flowers indefinitely in culture (*see* Notes 2–3).

3.2. *Induction of Flowering in* Citrus

1. Newly initiated branches of 10–30-cm lengths are cut from *Citrus* trees and taken to the laboratory for further preparation. Using a surgeon's scalpel fitted with a # 10 blade, all leaves and petioles are removed from branches. Explants are trimmed to uniform 0.5-cm lengths, with each stem piece containing one inactive nodal bud.
2. Stem explants are then washed in detergent for a few minutes to remove excess debris, and rinsed several times in tap water. Stem explants are surface sterilized in sodium hypochlorite solution for 15 min. A subsequent sterile water rinse is then administered.
3. Following surface sterilization, explants are transferred en masse to a sterile Petri dish. Using a 22 cm long bayonet, fine-tipped, serrated forceps, stem pieces are planted upright in medium with the node above the surface of nutrient agar.
4. Cultures are maintained in the Hotpack incubation chamber at 16°C for 16 wk, and recultured every 8 wk to fresh medium.
5. At the end of 16 wk in culture, explants initiate flowers from their nodal buds. Reculture of explants to this environment stimulates continued flowering (Fig. 3).
6. Reculture stem explants to a temperature-controlled environmental chamber maintained at 26°C.
7. Culturing explants initially at 26°C on nutrient medium supplemented with 0.1 mg/L benzyladenine stimulates numerous axillary buds (Fig. 3). Flowering does not occur under these conditions (*see* Notes 4–6).

Fig. 2. Flowering responses obtained from *Amaranthus gangeticus* shoots tips after 16 wk in vitro. Note the early formation of inflorescences at the base of the petioles.

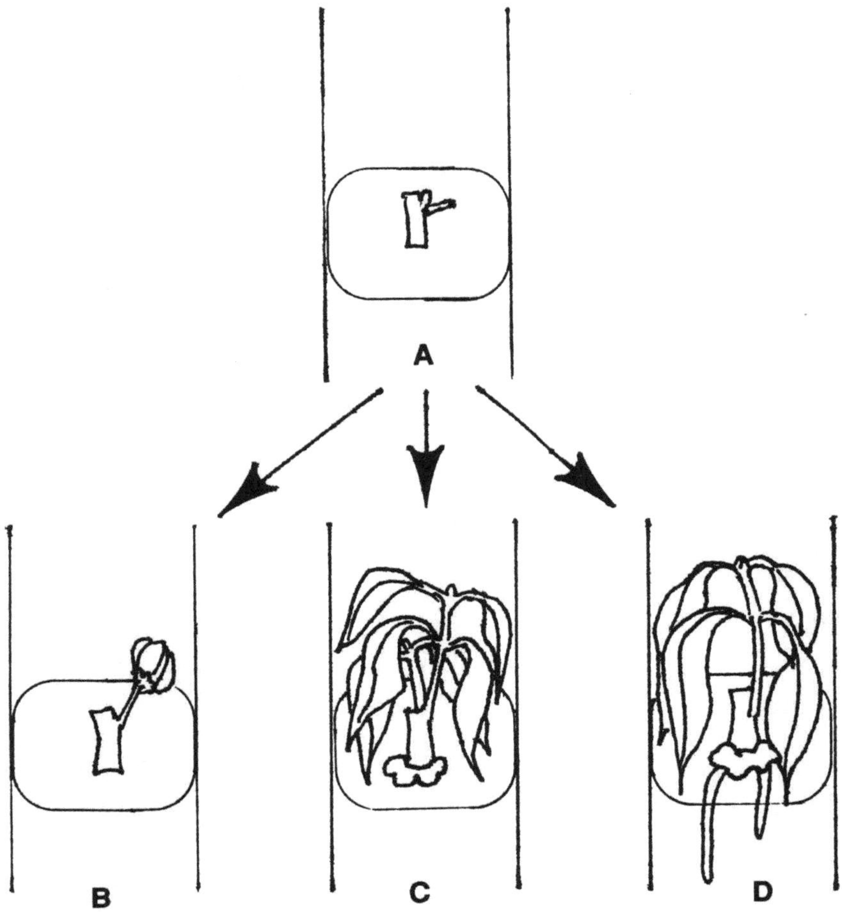

Fig. 3. Culture responses obtained from *Citrus* buds in vitro. **A.** Initial bud explant, which when subjected to culture treatments produced growth forms characteristically found in mature tree stages. **B.** Buds cultured at 16°C for 24 wk produced flowers. **C.** Buds cultured on at 26°C on nutrient medium supplemented with 1.0 mg/L benzyladenine produced axillary buds. **D.** Buds cultured at 26°C on basal medium with 0.1 mg/L NAA produced roots and elongated vegetative shoots.

4. Notes

1. A variety of plants can flower using the same culture conditions as described for *Amaranthus* (e.g., *Compositae* family members, such as sunflowers and asters) (Fig. 4).
2. For *Amaranthus*, a complete life cycle can be achieved within 32 wk in vitro. Therefore, the possibility of studying the whole life cycle of a plant in sterile culture exists.

Fig. 4. Example of growth responses obtained from 24-wk-old aster shoot tips cultured in vitro.

3. Inflorescences of *A. gangeticus* and *A. retroflexus* produced seeds that dropped to the surface of the nutrient medium and then germinated. It was not uncommon for the mother plant and germinated seedlings to coexist on the same nutrient medium. These seedlings, after 16–32 wk in culture, became flowering plants themselves. Interestingly, in some cases seeds germinate within the fruit while still attached to the parent plant. These seedlings usually died (unless rescued), since their roots failed to reach the agar surface.
4. Attainment of about 100% flowering from cultured *Citrus* stem explants is possible after 32 wk in culture. In nature, *Citrus* trees flower when they are subjected to either cold or drought stresses. Similarly, in order to induce *Citrus* buds to flower in culture, a cold stress is administered.
5. In *Amaranthus*, both reproductive and vegetative activities occurred simultaneously from shoot tips within the same culture passage. However, in *Citrus*, only flowering occurred in the cold treatment (i.e.,

16°C), whereas in the warm treatment (i.e., 26°C) only vegetative processes occur.

6. It is of interest to note that the occurrence of inflorescences is usually confined to plantlets whose size does not exceed 8 cm in length. In nature, flowering occurs from plants that are several times this size. The relationship between plantlet size and flowering is unknown. Presumably, the nutrient medium satisfies the nutritional requirements for flower induction and its prolonged initiation.

Disclaimer

Mention of a trademark name or proprietary product does not constitute a guarantee or warranty of the product by the US Department of Agriculture and does not imply its approval to the exclusion of other products that may be suitable.

References

1. Scorza, R. (1982) *In vitro* flowering. *Hort. Rev.* **4,** 106–127.
2. Tisserat, B., and Galletta, P. D. (1988) *In vitro* flowering in *Amaranthus. Hort. Sci.* **23,** 210–214.

Chapter 13

Fruit Organ Cultures

Brent Tisserat, Paul D. Galletta, and Daniel Jones

1. Introduction

The culture of fruit tissues as whole organs or isolated tissue sections has been conducted with various species (*1*). Whole, isolated ovaries have been successfully cultured to give rise to mature fruits (e.g., strawberry). Typically, however, when an isolated portion of the fruit tissue is introduced into a sterile environment, it immediately loses structural integrity and degenerates into a rapidly dividing callus mass (2). Loss of structural integrity is correspondingly associated with an alteration of physiology that is subsequently reflected in the production of an altered metabolism. Therefore, a meaningful study of fruit development using callus derived from fruit tissues is often not possible. Recently, we studied the parameters involved in the maintenance of citrus fruit tissue integrity (2). In this paper, the culture of isolated fruit tissues, as well as half and whole fruit culture, is demonstrated using the lemon fruit (Figs. 1–3).

Currently, the immediate use of fruit culture is to serve as a bioassay system to study fruit maturation events within a controlled environment. Knowledge gained from such studies may then be extrapolated for the

From: *Methods in Molecular Biology*, vol. 6, *Plant Cell and Tissue Culture*
Edited by Jeffrey W. Pollard and John M. Walker, ©1990 by The Humana Press

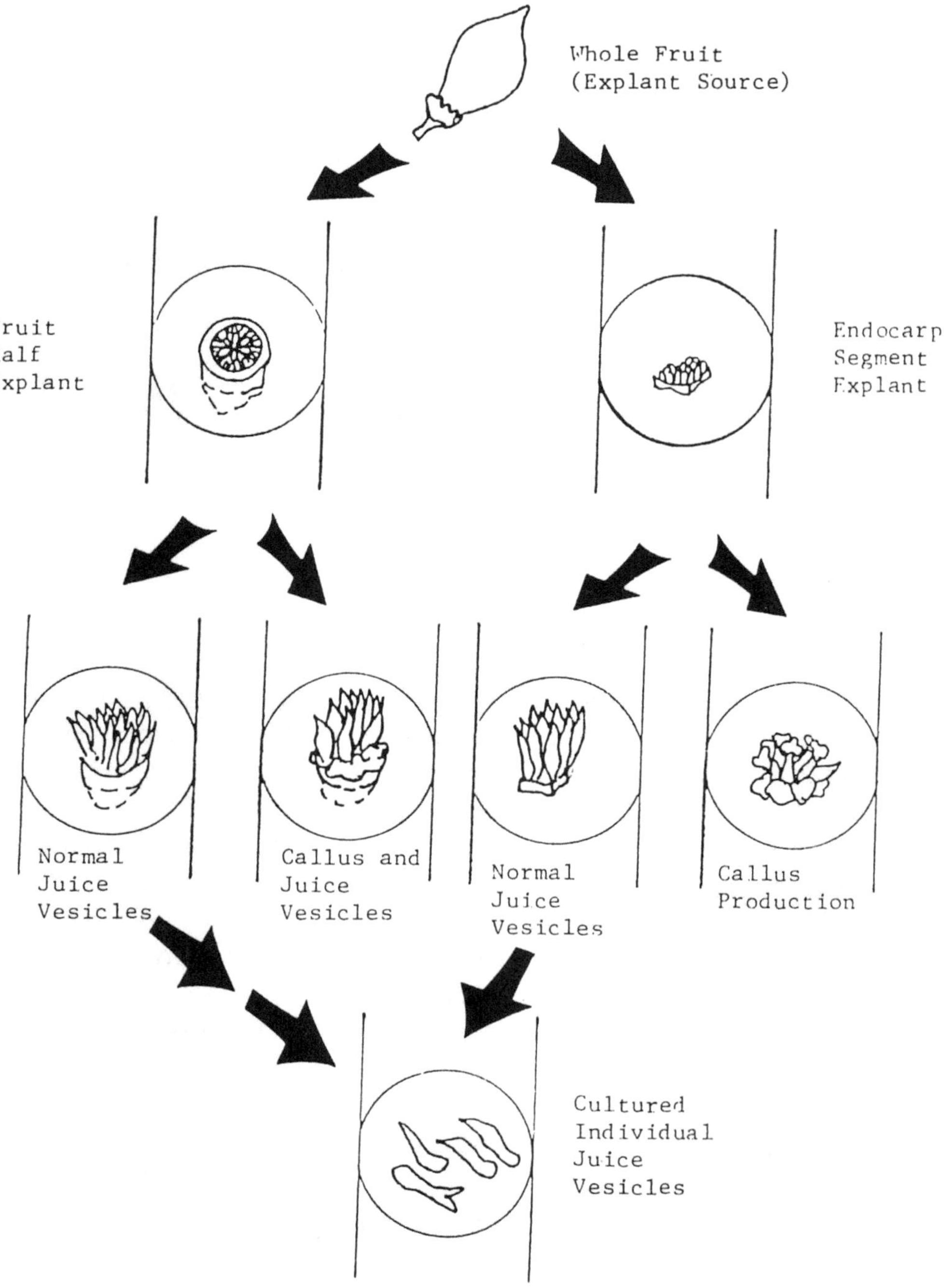

Fig. 1. Diagrammatic representation of the growth responses obtained from cultured juice vesicles in vitro.

Fig. 2. Examples of growth responses obtained from endocarp segments cultured in vitro for 30 d. Top row: juice vesicle 1-cm^2 sections; bottom row: longitudinal fruit halves.

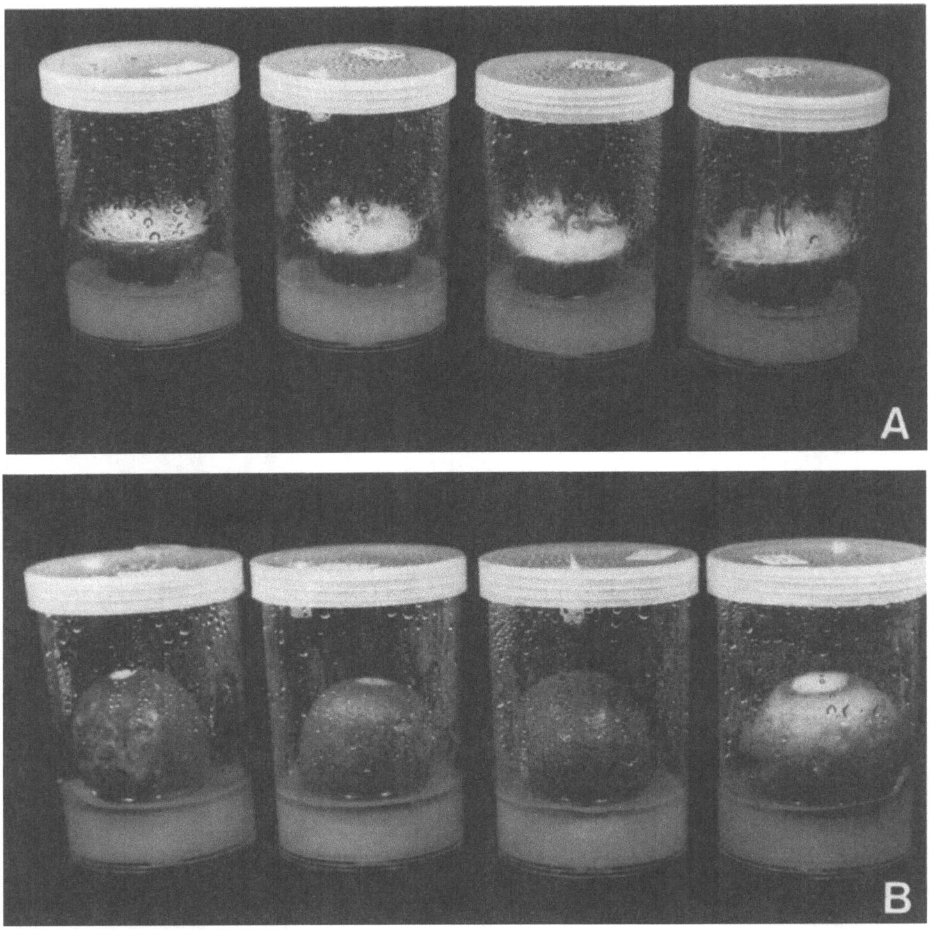

Fig. 3. Examples of 30-mm fruits of lemons cultured in vitro. A. Equatorial fruit halves after 60 d in culture. B. Whole fruits after 60 d in culture. Note that the flavedo remains viable in culture.

improvement of field grown crops. Both fruit and flower cultures have failed to attract the attention that micropropagation methods have over the last 20 yr. Micropropagation techniques have probably reached a saturation level; practically any plant can be mass cloned in vitro. Interest is now growing in means to obtain biochemicals and pharmaceuticals through tissue culture. The preponderance of extractable plant biochemicals and pharmaceuticals is derived from flowers and fruits. However, as fruit culture technology improves, the possibility of obtaining edible products from cultured fruits on demand occurs (*1*).

2. Materials

1. *Citrus* explants. Fruit, 10–50 mm diameter from *Citrus limon* (L.) Burm. f. cv. "Eureka" is obtained from bearing trees grown in a greenhouse or in the field (*see* Note 1).
2. Surface sterilant. 2.63% sodium hypochlorite solution (containing two drops of Tween-20 emulsifier/100 mL solution). This solution is most often prepared by diluting household bleach solution (containing 5.25% sodium hypochlorite) to half strength with water.
3. Nutrient medium (*see* Note 2): Medium contains the following constituents, in m*M*: KNO_3, 4.95; $Ca(NO_3)_2•3H_2O$, 2.12; $MgSO_4•7H_2O$, 0.61; $MnSO_4•H_2O$, 0.03; $CuSO_4•5H_2O$, 0.002; $ZnSO_4•7H_2O$, .014; $MgCl_2•6H_2O$, 0.88; KI, 0.003; $CaCl_2•2H_2O$, 1.02; $CoCl_2•6H_2O$, 0.004; H_3BO_3, 0.04; $Na_2MoO_4•2H_2O$, 0.001; KH_2PO_4, 0.37; EDTA, 0.12; $FeSO_4•7H_2O$, 0.01; sucrose, 8.76; thiamine•hydrochloride, 0.001; meso-inositol, 0.667, and naphthaleneacetic acid, 0.005. Agar, at an 8000 mg/L concentration, is added. The pH value is adjusted to 5.7 ± 0.1 with 0.1*N* HCl or NaOH before the addition of agar. The medium is dispensed in 25-mL aliquots into 25 x 150-mm culture tubes and capped with Bellco kaputs. Liquid medium (with pH value adjusted to 5.0 and agar excluded) is dispensed in 25-mL aliquots into 50-mL polystyrene flasks. Medium is also dispensed in 50-mL aliquots into 38 x 200-mm culture tubes, 275-mL polypropylene or polycarbonate specimen containers and capped with Bellco kaputs, polyethylene, or polypropylene lids, respectively. Media vessels are autoclaved for 15 min at 1.05 kg/cm^2 and 121°C. Tubes are slanted at a 45° angle while cooling.
4. Cultures are incubated in a temperature-controlled environmental chamber at 26 ± 1°C under a 16-h daily exposure to 2.2 W • m^{-2} using cool, white fluorescent lamps.

3. Methods

1. Fruit is washed in detergent for a few minutes to remove excess debris, and then rinsed several times in tap water. Fruit is surface sterilized in sodium hypochlorite solution for 30 min, and then rinsed three times with sterile water.
2. Using a surgeon's scalpel fitted with a #11 blade and a 21.6 cm long bayonet, fine-tipped, serrated forceps, the extreme stylar and stem ends are severed. Fruits are bisected equatorially or cut into 1-cm^2

endocarp pieces, and then planted on the surface of nutrient agar medium. These endocarp pieces consist of undamaged juice vesicles attached to a 1-mm thick layer of anchoring endocarp tissue. Also, whole fruits can be cultured directly on medium without any additional preparation.

3. Cultures are incubated in the incubation chamber and recultured every 8 wk to fresh medium (*see* Note 3) (Figs. 1 and 2).
4. Juice vesicles begin to enlarge immediately after being introduced to the sterile environment. Vesicles cultured as isolated endocarp segments (obtained from 10–20-mm diameter fruits) enlarge to several times their original size by the end of 4 wk in culture. Cultured vesicles undergo the developmental stages found within the maturing fruit. Eventually, vesicles reach their mature stage of development and begin to accumulate the characteristic juice. Older, more mature cultured vesicles (obtained from 30–50-mm diameter fruits) undergo less enlargement, since most of their development has already occurred within the fruit prior to culture (Figs. 1 and 2).
5. At the end of 8 wk in culture, fruits enlarge considerably from their original size. Juice vesicles grow out from the carpel cavities of equatorial fruit halves and are prominent. The normal maturation patterns found for fruit grown on the tree occurs for cultured fruit (*see* Note 4).

4. Notes

1. A variety of *Citrus* explant types and species sources can be employed in the culture of fruit tissues in vitro using the described system. A partial list includes: navel orange, sweet orange, sour orange, blood orange, Valencia orange, lime, rough lemon, citron, pummelo, and grapefruit.
2. Several media have been employed to culture whole and half fruit explants satisfactorily. The best medium to date for culturing endocarp sections is described herein. However, media improvement to attain normal development and to accelerate fruit development is continuing.
3. Whole fruit can remain viable for up to 8 mo in culture. However, smaller excised pieces remain alive indefinitely. Smaller explant pieces produce more callus. Callus formation is much reduced by using fruit halves and whole fruit cultures (Fig. 3).
4. A considerable number of intact juice vesicles can be used for bio-

assay testing using the described systems. For example, a lemon contains about 3000 individual juice vesicles. Fruit halves then contain about 1500 juice vesicles, of which about 90% or more continue to grow normally and do not produce callus in culture. Depending on the diameter (i.e., age) of the fruit, endocarp segments may contain about 100 or more juice vesicles.

Disclaimer

Mention of a trademark name or proprietary product does not constitute a guarantee or warranty of the product by the US Department of Agriculture and does not imply its approval to the exclusion of other products that may be suitable.

References

1. Nitsch, J. P. (1963) The *in vitro* culture of flowers and fruits, in *Plant Tissue and Organ Culture—A Symposium* (Maheshwari, P. and Rangaswamy, N. S., eds.), Univ. of Delhi, Delhi, pp. 198–214.
2. Tisserat, B. and Galletta, P. D. (1987) *In vitro* culture of lemon juice vesicles. *Plant Cell Tissue and Organ Culture* **11,** 81–95.

Chapter 14

Culture of Zygotic Embryos of Higher Plants

Michel Monnier

1. Introduction

The culture of embryos has several fundamental purposes. It allows observation of morphogenesis of isolated embryos at regular intervals and direct analysis of the various steps in embryogenesis. In this way, it gives more information than does fixation followed by histological analysis. Also, cultured embryos can undergo surgical operations similar to those performed with great success on animal embryos. In addition, embryo culture is a means of determining the nutritional factors favorable to normal embryogenesis. It is possible, by using appropriate media, to modify the pattern of embryo development. Indeed, if a modification of the experimental conditions of zygotic embryo culture provokes a more rapid growth, a higher survival, and an improved differentiation, it can be assumed that the new conditions are closer to those that exist in the embryo sac. The culture of zygotic embryos therefore permits us to determine the conditions in which the embryo grows in the ovary.

The culture of embryos also offers several applications, especially in the production of hybrids. In horticultural practices, crosses between

From: *Methods in Molecular Biology*, vol. 6, *Plant Cell and Tissue Culture*
Edited by Jeffrey W. Pollard and John M. Walker, ©1990 by The Humana Press

distantly related plants are generally unfruitful because of the abortion of embryos on the mother plant. This failure of the hybrid zygote to produce a viable embryo is commonly because of the degeneration of the endosperm. The endosperm is then unable to ensure proper nutrition for the embryo, and thus, seed formation is rare. The culture of embryos is a means of supplying nutrients when the endosperm does not play its nutritional part. The precocious excision of the embryos and their deposit on a synthetic medium can provide the nutrients that are lacking in vivo. By using this technique, a great number of hybrid plants have been obtained, and several genetic characteristics transferred (*1*).

2. Materials

1. Growth chamber with temperature and lighting regulated according to the kind of plants chosen. For *Capsella bursa-pastoris*, plants are grown in a growth chamber at 18°C. The lighting is continuous, and must be bright enough to allow plants to grow quickly. Twelve Grolux fluorescent tubes (F36W/GRO) are placed close to the plants.
2. Culture room for embryo culture at 25°C with continuous light provided by Grolux fluorescent tubes (one tube by the shelf).
3. Glass Petri dishes: the dishes must be ordinary glass, not Pyrex™. The lid and the bottom of a Petri dish made of Pyrex™ are not even enough to permit observation of the embryos. It is useful to engrave the bottom to spot the embryos and to number the dishes (Fig. 1).
4. Glass slides covered with silicon (slides are prepared in advance by spraying with an aerosol containing liquid used to waterproof fabrics). These slides are wrapped with filter paper before sterilization. All the glassware is sterilized in the autoclave (115°C, 20 min).
5. Medium:
   a. The effect of different solutions on the growth of embryos has been examined, and it was observed (*2*) that the solution that ensured the best growth was Murashige and Skoog solution (*3*, and *see* Appendix). But Murashige and Skoog solution was toxic for the smaller embryos. Modification of the concentration of different salts or elements gave a new solution that produced the same growth as Murashige and Skoog's, but an enhanced survival of small embryos (*4*).

      Here is the composition of this solution (concentration in mg/L). Generally stock solutions are prepared in advance at 100x concentration:

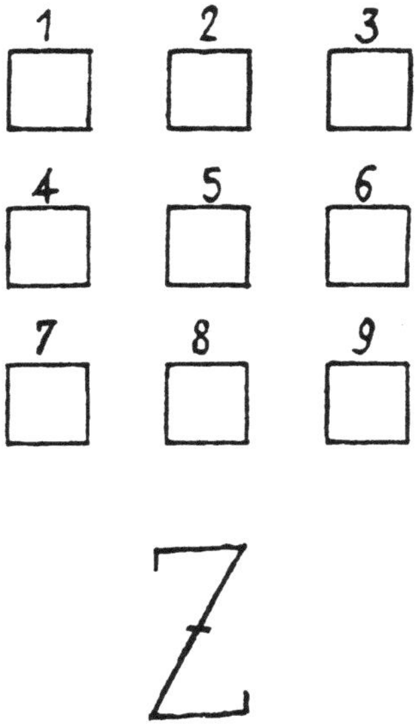

Fig. 1. The bottom of the Petri dish is engraved to locate the embryos. Three embryos are placed in the same square. Each square has a 0.5-mm side.

KNO_3	=	1900	H_3BO_3	=	12.4
$CaCl_2{\bullet}2H_2O$	=	880	$MnSO_4{\bullet}H_2O$	=	33.6
NH_4NO_3	=	825	$ZnSO_4{\bullet}7H_2O$	=	21.0
$MgSO_4{\bullet}7H_2O$	=	370	KI	=	1.66
KCl	=	350	$Na_2MoO_4{\bullet}2H_2O$	=	0.5
KH_2PO_4	=	170	$CuSO_4{\bullet}5H_2O$	=	0.05
Na_2EDTA	=	14.9*	$CoCl_2{\bullet}6H_2O$	=	0.05
$FeSO_4{\bullet}7H_2O$	=	11.1*			

* 2 mL of a stock solution containing 5.57 g $FeSO_4{\bullet}7H_2O$ and 7.45 of Na_2EDTA/L.

Some other items must be put into the medium:
Sucrose: 120 g/L (*see* Note 1), glutamine: 400 mg/L, vitamin B_1: 1 mg/L, vitamin B_2: 1 mg/L, Difco agar: 7 g/L.

The action of the different mineral solutions was tested on the survival of embryos of various sizes (Fig. 2). It can be seen that the survival of embryos increases with their size. The survival of 100 µm long embryos varies particularly with the kind of solution used. The above solution supplies the best survival for these embryos (*see* Note 2).

b. Dissection medium: To avoid dessication of fruits, they have to be dissected in a solution containing 120 g/L sucrose

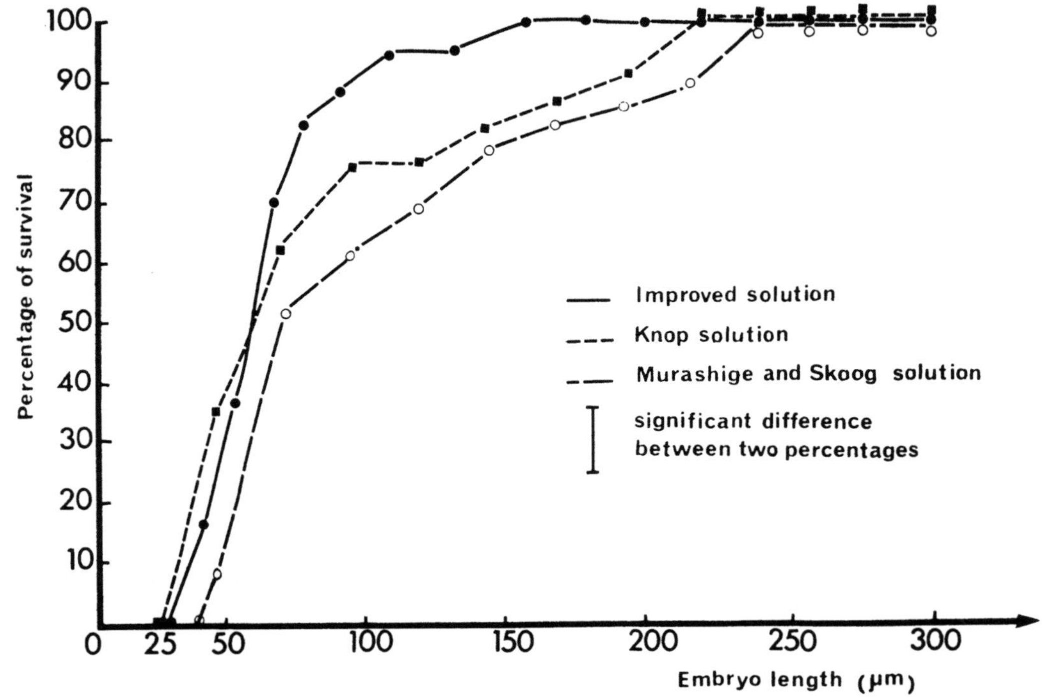

Fig. 2. The action of different mineral solutions on the survival of embryos according to their length. The effect is particularly evident when the embryos are 100 µm long. The improved solution allows small embryos to grow.

and 1.75 g/L soft agar to immobilize the ovules. Only 100 mL of this solution is needed and after sterilization, it is placed in a sterilized bottle dropper.

c. Sterilization of the medium: It has been observed that the temperature of sterilization has an effect on the growth and survival of embryos (5). A low temperature does not alter the elements, and ensures a better growth and survival of embryos. The best method to sterilize the medium is Tyndallization. This involves heating the medium at 100°C twice, with an interval of 24 h separating these two heatings. The first heating lasts 5 min, and the second 15 min. Since these times are short, it is necessary to put the medium in a large Erlenmeyer flask, so that the medium is in a thin layer before heating. The medium, when hot, is poured into the Petri dishes, which have been previously sterilized. This operation is performed in a sterile flow cabinet.

6. Calcium hypochlorite: Prepared freshly by adding 70 g of calcium hypochlorite to 1 L H_2O. After stirring for 15 min, decant the solution and filter it.

3. Methods

1. The fruits of the plant are chosen at the right stage, and then sterilized by plunging them into the calcium hypochlorite solution for 10 min. They are washed three times in sterile water. When the fruits are obtained from plants grown in a growth chamber, they do not need to be sterilized.
2. After sterilization, the fruits are put into a sterile Petri dish on a filter paper infused with water to prevent their dessication. All further operations are made under a binocular dissecting microscope in the sterile flow cabinet.
3. Several drops of the dissection medium containing sucrose are put on the glass slide (Fig. 3). The coating of the slide with silicon avoids flattening of the drops.
4. The two valves of a fruit are removed with tweezers. The placenta with attached ovules is placed in one of the drops. When the fruit has not been previously sterilized, the placenta is washed in two of the drops to prevent infection. Dissection is achieved with microinstruments sterilized by alcohol followed by quickly passing them over a Bunsen flame.
5. One ovule is detached from the placenta and put in the next drop. The ovule is immobilized by placing the joined ends of the tweezers on the ovule. With the microscalpel, an aperture is made in the area opposite to the micropyle. By pressing the micropylar region, the embryo is ejected. When the endosperm attached to the embryo has been pulled off, the embryo is taken with the micropipet and put in the engraved Petri dish.
6. Generally, three embryos are put in one square. The Petri dish is sealed with parafilm.
7. The Petri dishes are put in the dark at 25°C. When embryos are cultured on a high concentration of sucrose, exposure to light delays the appearance of chlorophyll during germination. The development of the embryos is observed through the agar at the bottom of the Petri dish (*see* Note 3).
8. The length of the embryos can be measured after 6 d with a micrometer gage. Generally, when we want to examine the effect of a factor on the development of embryos, it is necessary to inoculate two

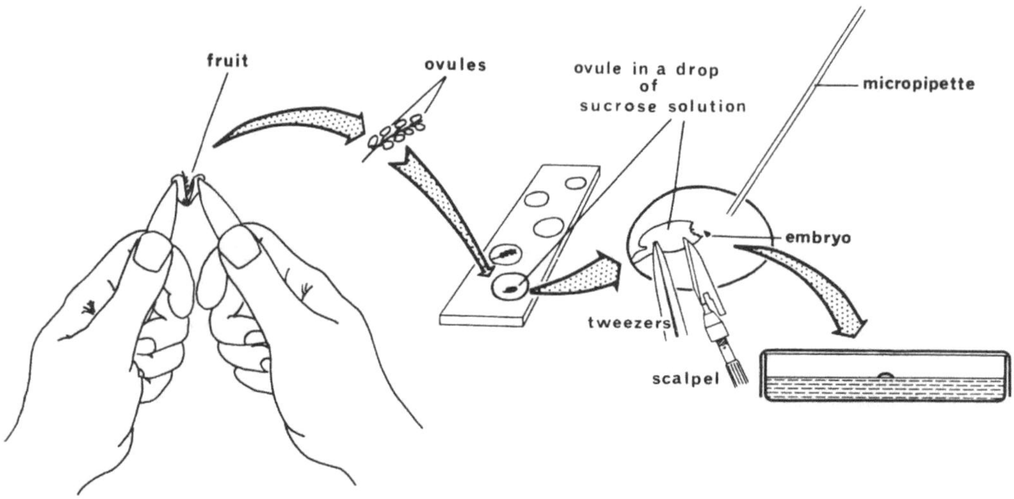

Fig. 3. Inoculation of the embryos. The two valves of a fruit are removed, and the placenta with the ovules is placed in a drop of sucrose solution. An aperture is made at the top of the ovule, and the embryo is put on the medium with a micropipet.

types of embryos: large embryos whose growth follows statistical laws and small embryos that are very fragile, and whose survival varies according to the toxicity of the medium. For example, in studies on *Capsella bursa-pastoris* embryos, we inoculate 160 µm long embryos to examine growth and 65 µm long embryos to have information about survival. Growth is measured by calculating the increase of size according to the initial length, that is to say the ratio L_6-L_0/L_0, where L_6 = length after 6 d, L_0 = length at the inoculation time. This ratio is multiplied by 100 to convert to a percentage of growth. The percentage of survival is obtained by the ratio: number of embryos grown/number of transplanted embryos x 100.

9. After a 2-wk period of culture, embryos are transferred into tubes containing the same medium, but without glutamine and, only 2% of the sucrose, in order for germination to start. Nevertheless, because of the transfer, a great number of embryos die, so we have evolved an apparatus to avoid an osmotic shock during the transfer. This device ensures a continuously variable composition of the medium (*see* Fig. 4, Note 2).

4. Notes

1. The culture medium for embryos contains a higher concentration of sucrose than the general medium used for plant tissue culture. With

a smaller amount of sucrose, the growth of embryos is reduced and they may germinate precociously (6). With a high concentration of sucrose, it is necessary, at the end of the culture, to decrease this concentration, so that the embryos can germinate. In fact, the concentration of sucrose has to be chosen according to the age of the embryos, i.e., the length of embryos. In Fig. 5, the optimal concentration for each type of embryo inoculated is plotted. As embryos grow in the medium, there is a continuous change in the requirements of embryos, so it is useful to adjust the concentration of elements over the course of time. The device drawn in Fig. 4 ensures variation of the composition of the medium during culture.

2. Two media can be used to cultivate the embryos in the apparatus shown in Fig. 4. One (central disc medium) is composed of a solution that allows the development of very small embryos with a high percentage of survival, and the other (external ring medium) is a medium that ensures a rapid growth of well-developed embryos (Table 1). A diffusion between these two media can continually adjust the composition of the medium to the requirements of the embryos in the course of time (Fig. 4).

   A heavy glass container is placed in the center of a Petri dish before the medium is poured (Fig. 4). Its diameter is 30 mm, and it has a central rod to ease handling. The medium of the "external ring," which is liquid after sterilization, is poured into the Petri dish around the central container. After the agar medium cools and solidifies, the central container is removed. Into the hole, obtained in that way, is poured a second medium of a different composition (central disk medium). The embryos are cultured on the central disk medium, and, as a result of diffusion, they are subjected to the action of a variable medium over time. At first the Petri dish is placed in the dark, and after 6 d, it is placed in the light. This method makes the transition from heterotrophy to autotrophy easier for the embryo . In a nonvariable medium, the growth of embryos stops very quickly, but in the medium with a variable composition, growth keeps going at a steady pace (Fig. 6). In that last case, after 1 mo of culture, small green plantlets can be obtained, and transfer into pots is easy. Use of this system has made possible the culture of very young embryos of about 50 µm long, up to germination for the first time. At the beginning of the culture, the embryo is globular, and during culture, the differentiation of cotyledons (a significant event) can be observed and studied (7).

3. Growth of immature embryos can also be sustained by culturing the

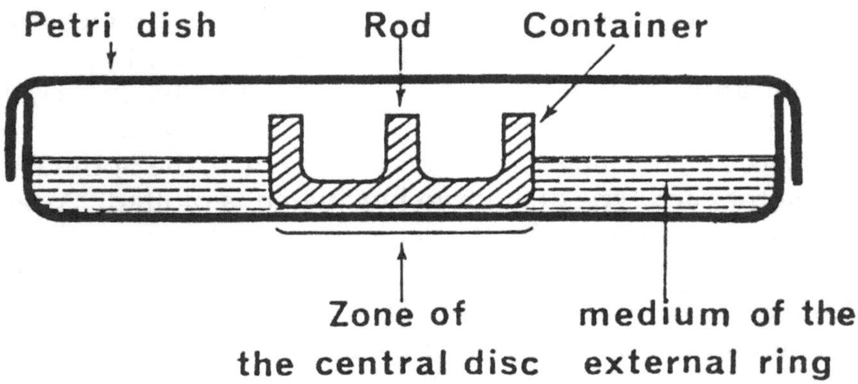

Fig. 4. Device allowing the juxtaposition of two media with different composition. The first medium is poured around the central glass container, and after cooling, the container is removed. A second medium is poured into the hole. The embryos cultured in the center are subjected to the action of variable medium because of diffusion.

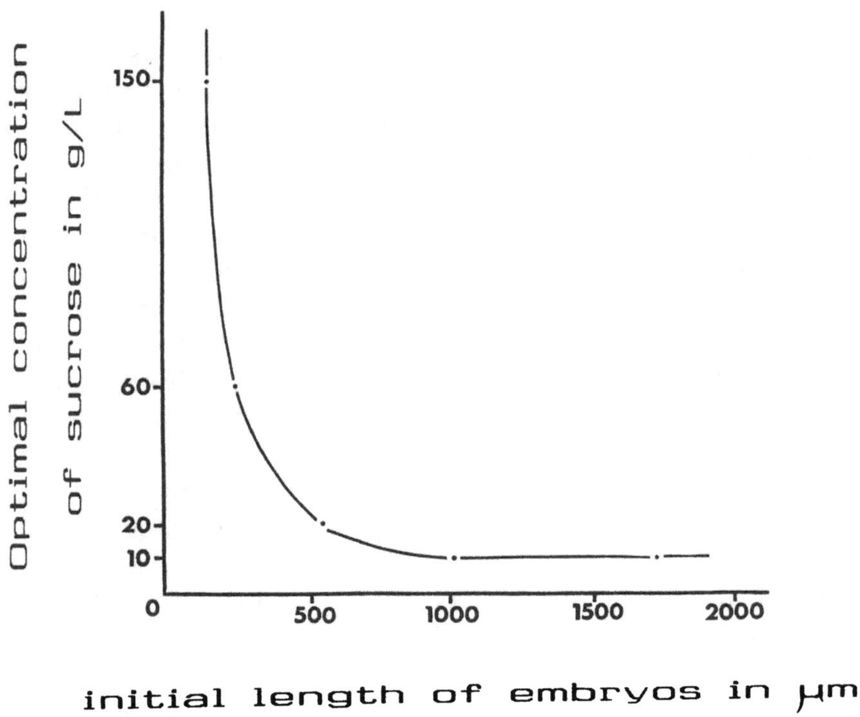

Fig. 5. Variation of the optimal concentration of sucrose according to the length of the embryos.

Table 1
Composition of the Two Media Used to Fill the Apparatus

Elements	Central disk medium mg/L	External ring medium mg/L
KNO_3	1900	1900
$CaCl_2 \cdot 2H_2O$	1320	484
NH_4NO_3	825	990
$MgSO_4 \cdot 7H_2O$	370	407
KCl	350	420
KH_2PO_4	170	187
Na_2EDTA	0	37.3
$FeSO_4 \cdot 7H_2O$	0	27.8
H_3BO_3	12.4	12.4
$MnSO_4 \cdot H_2O$	33.6	33.6
$ZnSO_4 \cdot 7H_2O$	21	21
KI	1.66	1.66
$Na_2MoO_4 \cdot 2H_2O$	0.5	0.5
$CuSO_4 \cdot 5H_2O$	0.05	0.05
$CoCl_2 \cdot 6H_2O$	0.05	0.05
Thiamine	1	1
Pyridoxine	1	1
Glutamine	600	0
Saccharose	180,000	0
Agar Difco	7,000	7,000

whole fertilized ovule. The ovule is put on the medium and the embryo, surrounded by endosperm and integuments, grows *in ovulo*. This technique has two advantages in comparison with culture of embryos directly on the medium; the growth rate and the survival frequency are better. The drawback is that, because the embryo cannot be directly observed, it is difficult to determine its initial length at the time of inoculation. Nevertheless, knowing the size of the inoculated ovule, this initial length can be deduced from a curve that establishes the relationship between the size of the ovule and the length of the enclosed embryo. For *Capsella* embryos, we obtained the equation of the curve:

$$y = e^{(0.0069.x - 0.31)} \tag{1}$$

where e is the base of naparian logarithm, x the length of ovules, and y the length of embryos (*8*). The medium to culture the ovules is

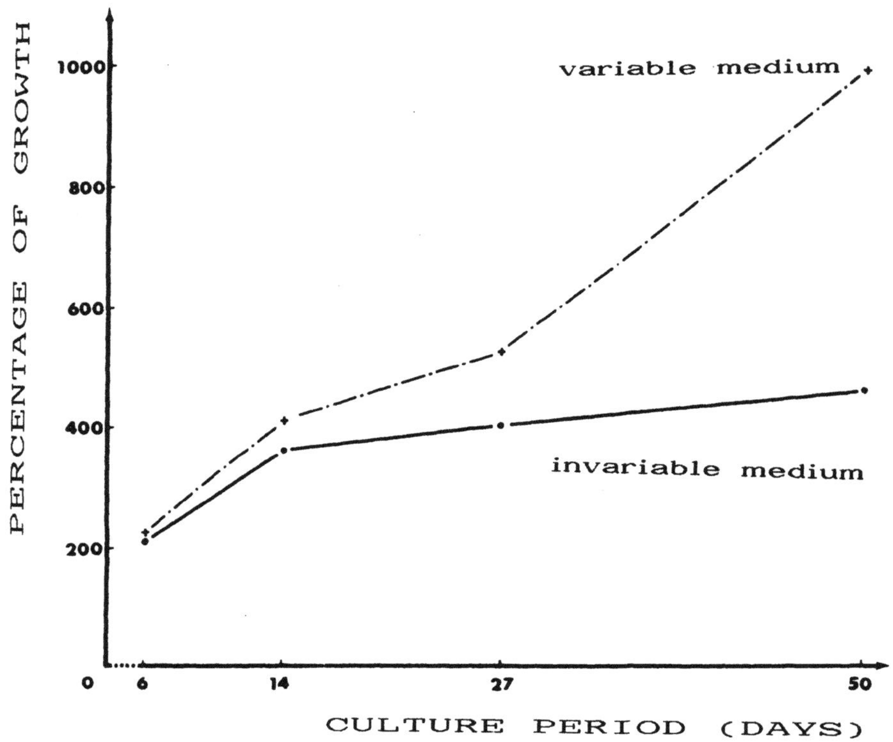

Fig. 6. The control medium has an invariable composition, and the growth of embryos slackens slowly. When the medium has a composition that varies according to time, the growth rate remains the same.

slightly different from that to culture the embryos. The optimal concentration of sucrose is 80 g/L, but the mineral solution is the same. With this method, 25 µm long embryos of *Capsella* can survive with a survival rate of 50%. However, the influence of the medium's composition is less, because the endosperm plays a part in the nutrition of embryo.

References

1. Chueca, M-C., Cauderon, Y., and Tempé, J. (1977) Technique d'obtention d'hybrides Blé tendre x Aegilops par culture "in vitro" d'embryons immatures. *Ann. Amélior. Plantes* **27,** 539–547.
2. Monnier, M. (1976) Culture *in vitro* de l'embryon immature de *Capsella bursa-pastoris* Moench. *Rev. Cyt. Biol. vég.* **39,** 1–120. Thesis.

3. Murashige, T., and Skoog, F. (1962) A revised medium for rapid growth and bioassays with tobacco tissue cultures. *Physiol. Plantarum* **15,** 473–497.
4. Monnier, M. (1978) Culture of zygotic embryos. 4th Interna. Cong. Plant Tissue and Cell Cult, *Frontiers of plant tissue cult.* Calgary (Canada), pp. 277–286.
5. Monnier, M. (1971) Action des conditions de stérilisation sur la valeur nutritive des milieux utilisés pour la culture des embryons isolés de *Capsella bursa-pastoris. Rev. gén. Bot.* **78,** 57–60.
6. Monnier, M. (1976) Variations des besoins nutritifs des embryons immatures de *Capsella bursa-pastoris* au cours de leur culture *in vitro.* Nat. Cong. soc. sav. *Lille Sciences* **1,** 595–606.
7. Monnier, M. (1984) Survival of young immature *Capsella* embryos cultured *in vitro. J. Plant Physiol.* **115,** 105–113.
8. Lagriffol, J. and Monnier, M. (1983) Etude de divers paramètres en vue de la culture *in vitro* des ovules de *Capsella bursa-pastoris. Can. J. Bot.* **61,** 3471–3477.

Chapter 15

Induction of Embryogenesis in Callus Culture

Michel Monnier

1. Introduction

The capacity of a cell to produce an embryo is not limited to the development of the fertilized egg. Cells of differentiated tissues (somatic tissues), which are first submitted to the action of an auxin and secondly transferred to a new medium without auxin, are capable of producing embryos (Fig. 1). It seems that the influence of a hormone, such as auxin, rejuvenates the cell so much that it reacquires the capacity of a zygote to produce an embryo.

Embryos that arise from the vegetative cells of the plant are called somatic embryos. This asexual reproduction is an alternative to sexual reproduction. Initially, these structures were termed embryoids, but they are now referred to as somatic embryos because of their similarity to zygotic embryos (*see* Chapter 14, this vol.). These somatic embryos are diploid, and must not be confused with haploid embryos, which are derived from the development of the pollen grain and are called androgenetic embryos.

Somatic embryos are very different from buds. Embryos that have a bipolar structure (axis stem-root) are without a vascular connection with

From: *Methods in Molecular Biology*, vol. 6, *Plant Cell and Tissue Culture*
Edited by Jeffrey W. Pollard and John M. Walker, ©1990 by The Humana Press

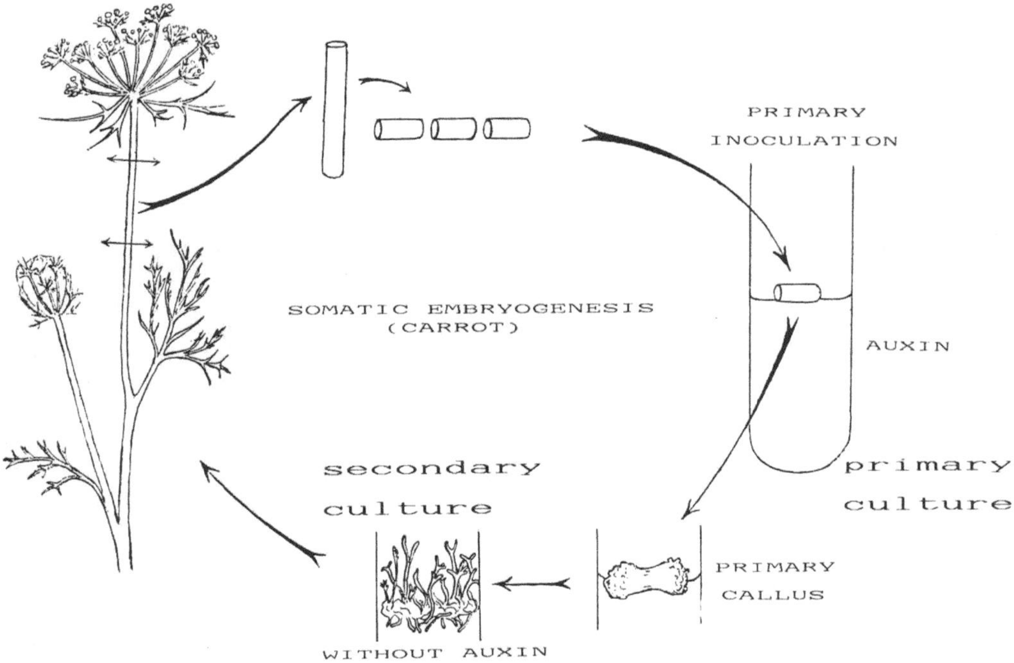

Fig. 1. Diagram representing the technique of obtaining somatic embryos from wild carrot. Initiation of somatic embryos occurs in primary culture with auxin. The embryogenic potentiality is expressed in a medium without auxin (secondary culture).

the mother tissue, and can, for that reason, be easily separated from the tissue. In addition, the first leaves of the embryo have characteristics that are typical of cotyledons.

The embryogenic callus can be cultured routinely on a solid medium composed of agar and nutrients. This procedure has several advantages compared to liquid-medium culture. Culture on solid medium requires only a normal room with shelves and little space, whereas suspension culture needs a mobile apparatus and is more difficult to maintain.

2. Materials

1. Growth chamber for plant culture. Temperature (25°C) and lighting are regulated. Lighting must be bright (5 W/m^2), discontinuous (16 h of lighting/d), and provided by Grolux fluorescent tubes (F36w/GRO). This growth chamber is not needed if plants can be collected from outdoors.

2. Culture room for tissue culture. Temperature has to be regulated at approximately 25°C. Continuous or discontinuous lighting (1 W/m^2) can be supplied by ordinary fluorescent tubes (one tube per shelf).
3. Culture medium: Three chemical factors are considered to have a significant effect on the induction of embryogenesis. A substantial amount of reduced nitrogen (NH^+_4) is required (*1*), and it is beneficial to supplement the medium with an amino acid, such as glutamine. Among the other ions, K^+ is very effective (2). In addition , an auxin must be added at a concentration of 1 or 0.5 mg/L (3).

   The mineral solution of Murashige and Skoog can be used for culture of somatic embryos (*4*). The solution described below, which was evolved for the culture of zygotic embryos (5), also provides satisfactory results for induction of somatic embryos, especially because of its low toxicity.

   Here is the composition of this solution (concentration in mg/L). The basic medium is identical to that described in the Methods section in Chapter 14, this vol. In addition, some other items must be put into the medium: Sucrose: 30 g/L, glutamine: 400 mg/L, vitamin B_1: 1 mg/L, vitamin B_6: 1 mg/L, 2,4-D (2,4-dichlorophenoxyacetic acid): 1 mg/L, agar: 7 g/L. It is not necessary to adjust the pH.

   At first, agar is put into distilled water to swell, and then dissolved in hot water by continually stirring it. After adding the chemicals, the medium is poured into tubes and sterilized by autoclave (10 min, 110°C).
4. Solution of calcium hypochlorite. The solution must be freshly prepared by putting 70 g of calcium hypochlorite powder (220°C) into a liter of water. After stirring the solution for 15 min and decanting it, it is filtered.
5. Use sterile water to wash stems after sterilization by calcium hypochlorite.
6. Subsequent plantlet development requires a filter-paper bridge apparatus and small pots containing soil mixture.

3. Methods

Embryogenesis occurs after two successive cultures (Fig. 1). The first one consists of culturing the tissue (primary culture) on a medium with auxin where there is a loss of differentiation (embryogenic induction). The second (secondary culture) consists of culturing the dedifferentiated tissue on an auxin-free medium where the embryogenic capacity is expressed.

3.1. Primary Culture

1. The stems of the plant are sterilized by plunging them into the calcium hypochlorite solution for 15 min. They are then washed 3x in sterile water.
2. In a sterile flow cabinet, the stems, after sterilization, are put into a sterile Petri dish. Scalpels and forceps are sterilized with alcohol and passed quickly through a flame. The ends of the stems that were bleached by hypochlorite are discarded, and the remaining stem is cut into pieces (Fig. 1). Each piece is put either horizontally or vertically on the medium in a tube.
3. After a few weeks in culture, the surface of the explant becomes rough, and the ends of the piece of stem form nodule-like structures, which are proembryonic masses. The tissue grows very slowly, and is sufficiently large to be transferred only after $1^1/2$ mo of culturing. The tissue (callus) is divided into fragments that serve as inocula for the new medium (secondary culture). Only healthy callus tissue can be used; brownish tissue is a sign that localized necrosis has occurred.

3.2. Secondary Culture

The medium is the same as described for primary culture, except that the auxin and glutamine are omitted.

1. After the transfer of healthy callus tissue bearing protuberances to the new medium, the nodules grow into embryos after a few weeks.
2. The tissue consists of an intertwining of embryos, sometimes aberrant with green blades. In the callus, all the intermediary stages from embryos to typical leaves are visible (Fig. 2). For observation, the callus can be put into a Petri dish with a small quantity of medium and dissociated with needles. Hundreds of embryos appear, showing the recognizable steps in embryogenesis. These are globular, heart-shaped, torpedo-shaped, mature, and germinating embryos. In Fig. 3, embryos of wild carrot were chosen and classified. In some plants (*Ranunculus*), second-generation embryos can be seen on the hypocotyl of the primary somatic embryo (Fig. 4).
3. The mature embryos can be transferred to a filter-paper bridge dipping into a liquid medium (Fig. 5). The solution of Murashige and Skoog (4 and *see* Appendix) at half strength and sucrose (5 g/L) is necessary for germination and growth.
4. The tubes are placed on a well-lit shelf in the growth chamber for plant

Fig. 2. Secondary culture of the callus. Appearance of numerous intertwining somatic embryos among abnormal embryos, which resemble leaves (wild carrot).

culture. When the roots are sufficiently developed, the plantlets are transferred into pots in a soil mixture. The pots must be covered by plastic for a few days to prevent dessication of the plants (*see* Notes section).

4. Notes

This procedure is successful with wild carrot, but with some varieties of cultivated carrots, the results are erratic. In a number of cases with other species, this method fails. Different alternatives are proposed here to finally obtain somatic embryos.

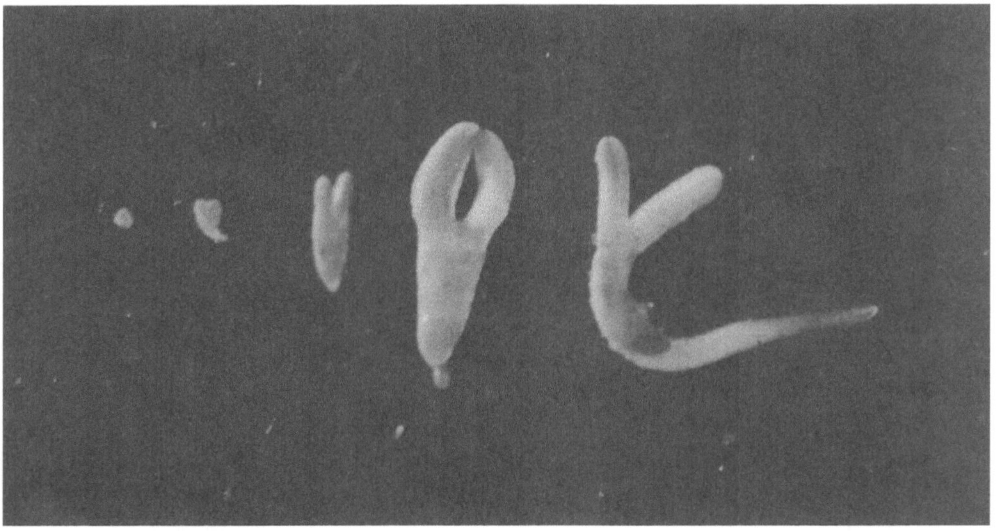

Fig. 3. Somatic embryos from a dissociated callus (wild carrot). They have been set in order of development from left to right and they range as follows: globular, heart-shaped, torpedo-shaped, mature embryo, and germinating embryo.

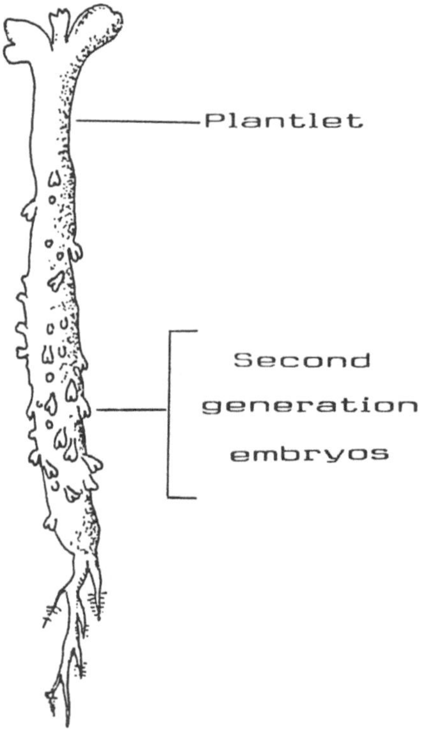

Fig. 4. Second generation embryos appear on the hypocotyl of the primary somatic embryo (*Ranunculus sceleratus*) (7).

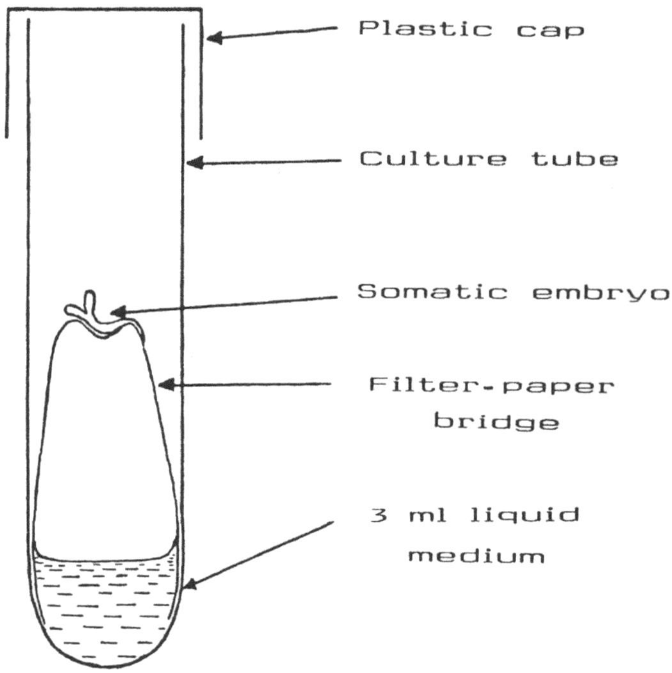

Fig. 5. Culture of mature somatic embryo on filter-paper bridge. The explant is carefully placed in the central depression at the top of the filter-paper bridge.

1. Change of the plant: It is preferable to work with some families of plants known to produce somatic embryos in culture, especially members of the Umbelliferae and Ranunculaceae. Nevertheless, in these families, some species successfully produce embryogenic callus, whereas others do not. In the same species, it can be useful to change the cultivar, because different genotypes have shown variable potentialities for somatic embryogenesis.
2. Change of the explant: For *Daucus carota,* any part of the plant taken at any stage of development successfully produces somatic embryos in culture. However, for other species, only certain regions of the plant may respond in culture. Tissues that are younger and less differentiated than stems can be used. For instance, zygotic embryos can develop into a callus from which a multitude of embryos may be produced (*6*). Reproductive tissue has proven to be an excellent source of embryogenic material (*7*). Immature inflorescences are suitable, too.
3. Change of the medium:
   a. Organic substances. Glutamine is the most important amino acid but others, such as serine and proline, can be added to

glutamine, at a lower concentration (100 mg/L) (*8*). Casein hydrolysate (800 mg/L) can promote somatic embryo formation in some cases. Coconut milk (20%) has been proven beneficial.

b. Growth regulators. Embryogenic culture has been produced with NAA (naphthaleneacetic acid) at the dose of 1 mg/L or more (*9*). Uncommon auxins like picloram (*10*) can be tested. Cytokinins, at the concentraton of 1 mg/L, like kinetin or BAP (6-benzylaminopurine) can be either associated with auxin in the primary culture or used just in the secondary culture.

c. Activated charcoal. The addition of activated charcoal (20 g/L) to the medium has been proven very useful for somatic embryo development of woody plants. It absorbs many inhibitors, especially oxidized polyphenols, which are exuded into the medium by the woody tissues.

References

1. Halperin, W. and Wetherell, D. F. (1965) Ammonium requirement for embryogenesis *in vitro*. *Nature* **205,** 519, 520.
2. Brown, S., Wetherell, D. F., and Dougall, D. K. (1976) The potassium requirement for growth and embryogenesis in wild carrot suspension cultures. *Physiol. Plant.* **37,** 73–79.
3. Halperin, W., and Wetherell, D. F. (1964) Adventive embryony in tissue cultures of the wild carrot *Daucus carota*. *Amer. Jour. Bot.* **51,** 274–283.
4. Murashige, T., and Skoog, F. (1962) A revised medium for rapid growth and bioassays with tobacco tissue cultures. *Physiol. Plant.* **15,** 473–497.
5. Monnier, M. (1976) Culture *in vitro* de l'embryon immature de *Capsella bursa-pastoris Moench*. *Rev. Cyt. Biol. veg.* **39,** 1–120.
6. Vasil, V., and Vasil, I. K. (1981) Somatic embryogenesis and plant regeneration from tissue cultures of *Pennisetum americanum* and *P. americanum* x *P. purpureum* hybrid. *Am. J. Bot.* **68,** 864–872.
7. Konar, R. N., and Nataraja, K. (1964) *In vitro* control of floral morphogenesis in *Ranunculus sceleratus* L. *Phytomorphology* **14,** 558–563.
8. Armstrong, C. L., and Green C. E. (1985) Establishment and maintenance of friable embryogenic maize callus and the involvement of L-proline. *Planta* **164,** 207–214.
9. Ammirato, P. V., and Steward, F. C. (1971) Some effects of the environment on the development of embryos from cultured free cells. *Bot. Gaz.* **132,** 149–158.
10. Beyl, C. A., and Sharma G. C. (1983) Picloram induced somatic embryogenesis in *Gasteria* and *Haworthia*. *Plant cell tissue organ cult.* **2,** 123–132.

Chapter 16

Induction of Embryogenesis in Suspension Culture

Michel Monnier

1. Introduction

The initiation and development of embryos from differentiated tissue (somatic tissue) was first observed by Steward et al. (*1*) in 1958, but earlier researchers had already observed these formations in culture, without recognizing them as embryos.

These embryos are termed somatic embryos, and their origin is different from zygotic embryos, which result from the fusion of gametes. This phenomenon is also called adventive or asexual embryogenesis. Somatic embryos (*see* Chapter 15, this vol.) closely resemble zygotic embryos in their structure, as well as in their physiological properties, especially with wild carrot (2). Nevertheless, the organization of these somatic embryos is generally more variable than that of zygotic embryos. Abnormalities in shape and size are frequent, and because of this variation they were, at first, called embryoids.

In contrast to buds, somatic embryos have a bipolar axis with an apical meristem and a root, and as they have no vascular connection with the mother callus, they are easily detached by the swirling action of the agitated medium.

From: *Methods in Molecular Biology*, vol. 6, *Plant Cell and Tissue Culture*
Edited by Jeffrey W. Pollard and John M. Walker, ©1990 by The Humana Press

Three types of medium are required to induce somatic embryogenesis in suspension. The first medium is solidified by agar and contains auxin; this dedifferentiates the somatic tissue and provokes initiation of embryogenic cells (primary culture). The second medium, which is liquid and also contains auxin, ensures the multiplication of these cells. The third medium, which is also liquid but without auxin, allows the cells to express their embryogenic potential.

Culture in a liquid medium is not easy to carry out. When the tissue is inoculated, it sinks and rapidly dies because of lack of oxygen. To prevent asphyxia, the liquid must be agitated to dissolve air in the medium. The culture of somatic embryos in a liquid medium, however, has numerous advantages. The swirling medium naturally separates the embryos, which are then easily observed. Thus, the embryos can be fractionated according to their stages. They can be obtained in great quantity and used as a basis for a large-scale micropropagation.

2. Materials

1. Growth chamber for plant culture. Temperature (25°C) and lighting are regulated. Lighting must be bright (5 W/m^2), discontinuous (16 h of lighting/d), and provided by Grolux fluorescent tubes (F36w/GRO). This growth chamber is optional, since plant material can also be collected from outdoors.
2. Culture room for tissue culture. Temperature has to be regulated at approximately 25°C. Continuous light (1 W/m^2) can be supplied by ordinary fluorescent tubes placed above the rotary shaker.
3. Actively growing callus in tubes.
4. Culture medium: The mineral solution of Murashige and Skoog can be used for culture of somatic embryos (*3*). The solution described in Chapter 15 was evolved for the culture of zygotic embryos, and is particularly useful for induction of somatic embryos because of its low toxicity (*5*). It must have a pH of 5.5. In addition, it contains NH_4^+ and K^+ ions, which are considered to be promoters of somatic embryo growth (*6*).

   For mature embryo growth, use half-strength M and S medium (*4* and Appendix, this vol.) containing sucrose at 5 g/L.
5. Sieves (1-mm grid) made of stainless screens (tea sieves from hardware store) to fraction suspension culture (Fig. 1) and transfer only the smaller embryos into the new flask. Sieves can also be made with nylon mesh filtration cloth (industrial nylon mesh filter) (Fig. 1).

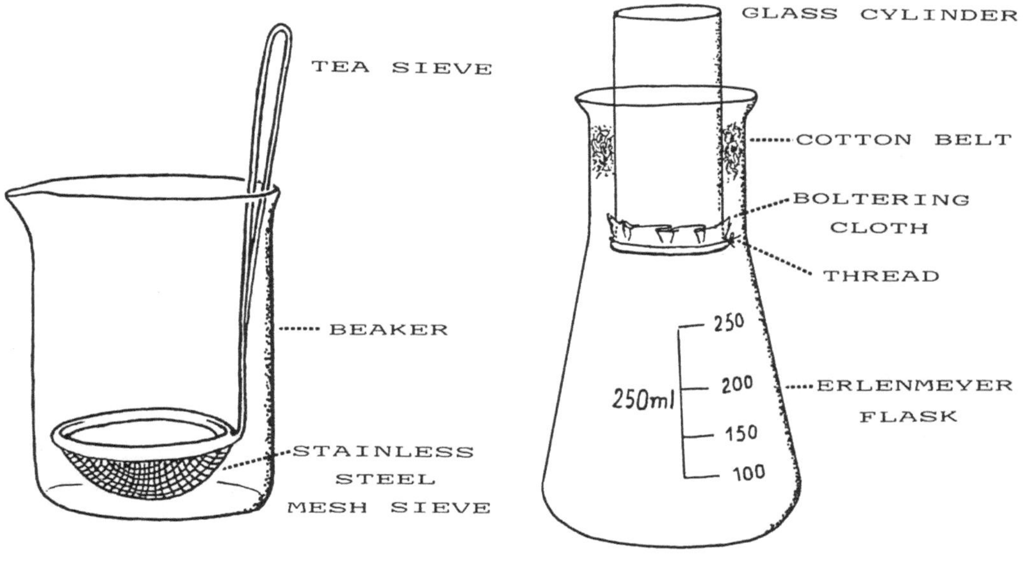

Fig. 1. The metal sieve, on the left, is used principally to remove the larger embryos before transferring the embryonic population into a fresh medium. The nylon sieve, on the right, is used to obtain a homogeneous population of embryogenic cells and synchronize the different stages of embryo development.

6. Subsequent plantlet development requires filter-paper bridge apparatus (*see* Fig. 5 of Chapter 15) and small pots containing soil mixture.

3. Method

Generally, three steps are needed to obtain embryogenic suspension culture. First, embryogenic tissue is initiated by culture on an agar medium with auxin; it is then transferred to liquid medium with auxin, and eventually transferred into a liquid medium without auxin.

3.1. Initiation of Embryogenic Tissue

Growth initiation is better on solidified medium containing 2,4-D (1 mg/L). The detailed procedure is described in Chapter 15. When the callus has developed the potentiality for embryogenesis, that is to say 5 wk at least after explant isolation (4), it is transferred to a liquid medium.

3.2. Establishment of Embryogenic Suspension Culture

1. The callus should be removed from the culture tube with sterile forceps and transferred to a Petri dish containing filter paper. Discard brownish parts of the callus, and inoculate a fragment of about 2 g in a 250-mL Erlenmeyer flask containing 50 mL of medium. Smaller Erlenmeyer flasks can be used, but the volume of liquid in relation to the size of the flask must be, for adequate aeration, about 20% of the volume of the flask. These flasks are placed on a gyrotory or orbital shaker and are agitated at 100–150 rpm.
2. When the plant material is first placed in the medium, there is an initial lag period prior to cell division. This is followed by an exponential rise in cell number. Finally, the cells enter a stationary stage. In order to maintain the viability of the culture, the cells should be subcultured at the beginning of this stationary phase. It is reached in 2–3 wk, and the suspension has to be transferred to fresh medium at regular intervals within this period. In addition, a minimum density must be achieved when cells are transferred to fresh medium, to maintain embryogenic potential (7).

   To subculture embryogenic suspension, wait several minutes until the cells are settled at the bottom of the flask and decant almost all the medium. Resuspend the suspension by gently rotating the flask, and transfer one-fourth of the entire population to fresh medium.

   All these operations can also be carried out, under safer conditions, using a micropipet. In that case, when the cells have settled, aspirate almost all the supernatant above the cells. Then, aspirate one-fourth of the cell population and transfer it into a new medium.
3. A population of cells typically shows a wide range of proembryonic stages visible under the dissecting microscope. It is a mixture of single cells, proembryonic cell clusters, and new embryogenic centers and older embryos (Fig. 2). To obtain a certain degree of uniformity, it is necessary to sieve the inoculum when transferring the suspension.

   The proembryo suspension is passed through a 200-μm sieve and then a 100-μm sieve. In the second sieve, differentiated embryos from the late globular stage are retained and may be examined, while proembryonic cells pass through. When the suspension, which has passed the sieves, is settled, decant most of the medium to obtain a suspension with a high density of cells.

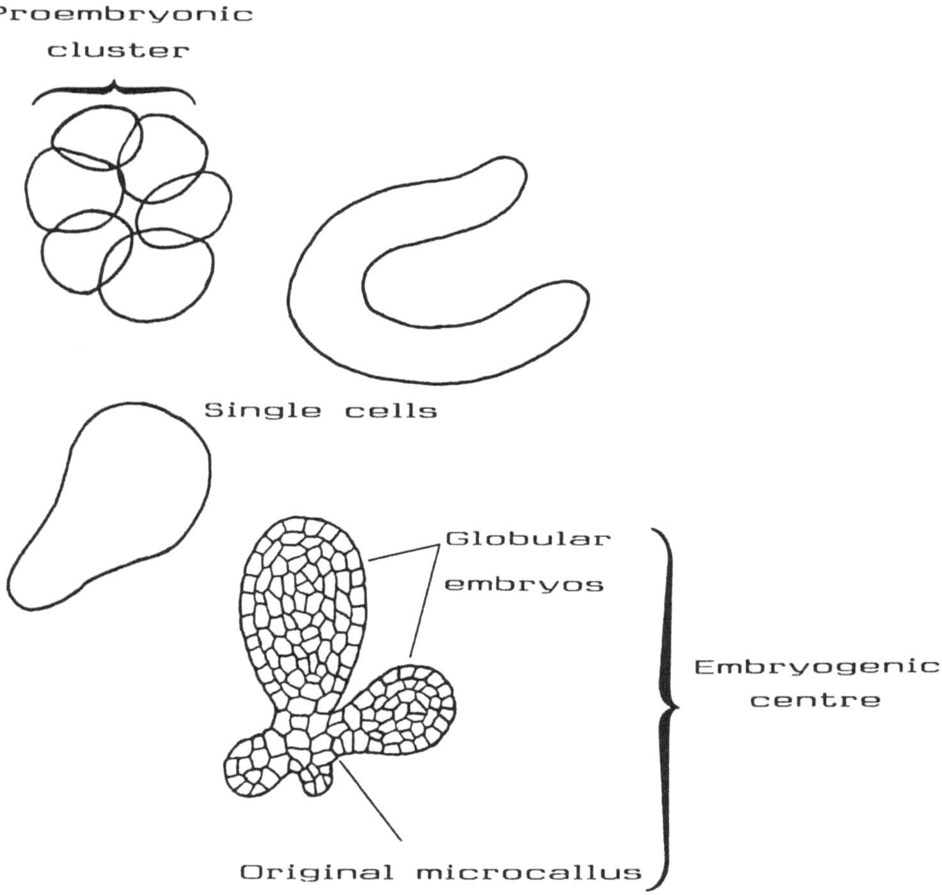

Fig. 2. The embryogenic cells divide in the liquid medium and constitute proembryonic clusters. A certain number of new embryogenic centers will give rise to globular embryos of different sizes (wild carrot).

3.3. *Maturation of Embryos*

When embryo expression and maturation are needed, the inoculum is transferred to a liquid medium having the same composition, but lacking auxin and glutamine. Cells express their embryogenic potentiality, and if the cells have not been carefully homogenized by sieving, a mixture of somatic embryos of different stages appear (Fig. 3). They reach mature size in 2 wk. These embryos can be dispersed, in a Petri dish, on an agar medium without auxin, but with small amounts of cytokinins (0.1 mg/L of BAP or zeatin).

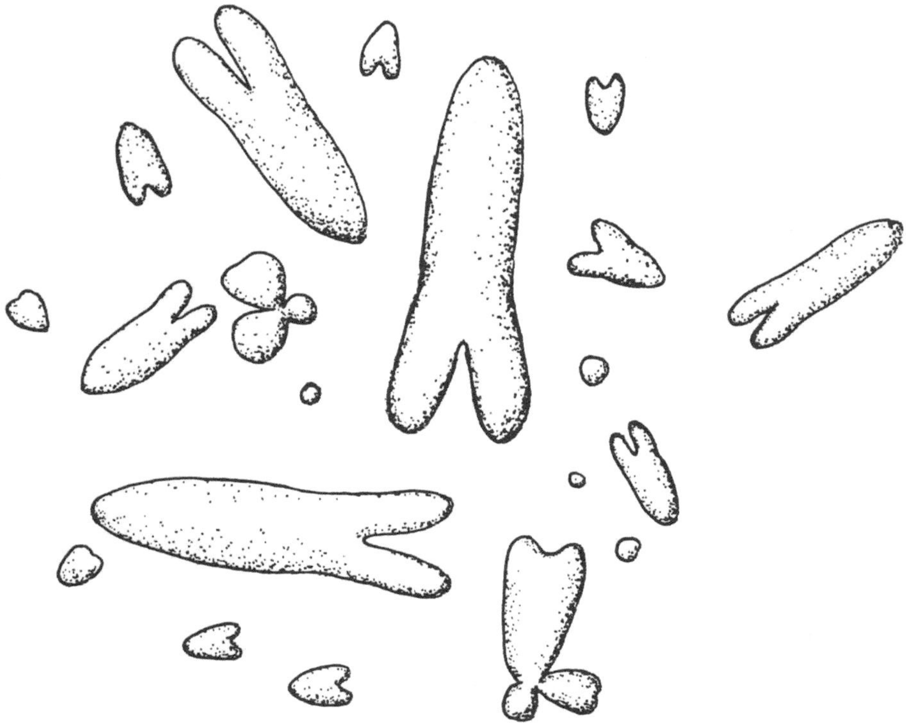

Fig. 3. During maturation without filtration, a heterogeneous population of somatic embryos appears, showing a large range of different stages (wild carrot).

When the embryos are mature, they are transferred to tubes on filter-paper bridges (Fig. 5 of Chapter 15). The medium is composed of Murashige and Skoog mineral solution (4) diluted by half and containing a low concentration of sucrose (5 g/L). To stimulate the formation of leaves, the tubes must be well illuminated. When the plantlets are tall enough, with a large number of leaves and roots, they are transferred into pots with a mixture of peat, soil, and vermiculite (*see* Chapter 15 for details) for subsequent development. After their transfer, plantlets have to be protected from dessication by covering the pots with plastic film for a few days.

4. Notes

4.1. Simplification of the Technique

1. Suspension culture can be started by inoculating the liquid medium containing 2,4-D (1 mg/L) with an explant of differentiated tissue iso-

lated directly from the plant. The initiation of multiplication of cells requires the inoculation of a large amount of tissue—about 3 g for 100 mL of medium (*8*). The dividing cells will gradually free themselves from the inoculum because of the swirling action of the liquid.

4.2. Different Causes of Failures (and Remedies)

1. Darkening of the medium: In a number of cases, especially with explants isolated from woody plants, the liquid medium becomes dark after a few days. The radicle ends of the embryos turn brown, and the embryos rapidly stop growing and show necrosis. Several procedures can help to reduce this process.

   More frequent transfer into fresh medium can prevent browning. After transfer to liquid, for the first days, when the proembryos are particularly sensitive, the culture must be performed in complete darkness. After this period, the culture can again be exposed to light. This lethal browning is the result of formation of oxidized polyphenols, so antioxidants, such as ascorbic acid and cysteine, may be used, but the results are variable.
2. No appearance of embryos: Not all plant species respond to culture and produce somatic embryos. Some plants give only adventitious buds, and others, no organogenesis at all. If a good experimental model is sought, it is preferable to select some species in families like Umbelliferae and Ranunculaceae, since they offer an easy way to obtain somatic embryogenesis.

   The explant has to be carefully chosen, since different regions of the plant body do not have the same embryogenic potential. Young material like zygotic embryos and reproductive tissue is an excellent source to initiate embryogenesis. For example, the diploid tissue of an immature anther of grapevine develops into somatic embryos (*9*) (Fig. 4).
3. Insufficient number of embryos: If a culture produces few embryos, suspending the cells in a plasmolyzing solution containing a high concentration of sucrose (1*M*) for a short time (45 min) is recommended. When these are replaced in normal conditions, somatic embryogenesis is efficiently promoted (*10*). The effect may be because of the rupture of intercellular connections and isolation of cells that are reinstated into the condition of the zygote.

   Alterations of the composition of the medium can be beneficial. Some authors demonstrated that NAA (naphtaleneacetic acid) is

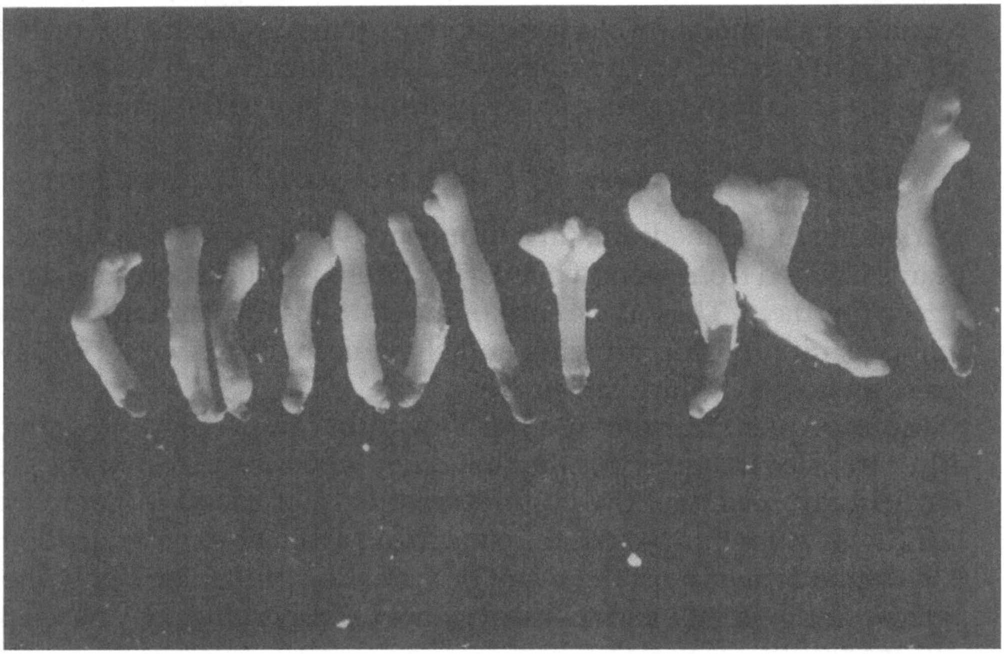

Fig. 4. Somatic embryos can be slightly different from zygotic embryos. The cotyledons are generally shorter. These embryos were obtained from the diploid tissue of the immature anther of grapevine.

sometimes better to initiate somatic embryogenesis than 2,4-D (*11*). Cytokinins like zeatin can help the embryos to mature. The addition of some amino acids like serine, proline, and glycine is generally considered to stimulate embryo growth.

References

1. Steward, F. C., Mapes, M. O., and Mears, K. (1958) Growth and organized development of cultured cells. II. Organization in cultures grown from freely suspended cells. *Am. J. Bot.* **45,** 705–708.
2. Wetherell, D. F. and Halperin, W. (1963) Embryos derived from callus tissue cultures of the wild carrot. *Nature* **200,** 1336,1337.
3. Murashige, T. and Skoog, F. (1962) A revised medium for rapid growth and bioassays with tobacco tissue cultures. *Physiol. Plant.* **15,** 473–497.
4. Reinert, J., Backs-Hüsemann, D., and Zerban, H. (1971) Determination of embryo and root formation in tissue cultures from *Daucus carota,* in *Les Cultures de Tissus de Plantes.* Colloques internationaux du C.N.R.S., Paris, **193,** 261–268.

5. Monnier, M. (1973) Croissance et développement des embryons globulaires de *Capsella bursa-pastoris* cultivés *in vitro* dans un milieu à base d'une nouvelle solution minérale. *Bull. Soc. bot. France, Mémoires,* 179–193.
6. Halperin, W. and Wetherell, D. F. (1965) Ammonium requirement for embryogenesis *in vitro*. *Nature* **205,** 519, 520.
7. Halperin, W. (1967) Population density effects in embryogenesis in carrot-cell cultures. *Exp. Cell Res.* **48,** 170–173.
8. Helgeson, J. P. (1979) Tissue and cell suspension culture, in *Nicotiana. Procedures for Experimental Use* (Durbin, R. D., ed.), Department of Agriculture Tech. Bull., Washington, D.C., **1586,** pp. 52–59.
9. Rajasekaran, K. and Mullins, M. G. (1979) Embryos and plantlets from cultured anthers of hybrid grapevines. *J. Exp. Bot.* **30,** 399–407.
10. Wetherell, D. F. (1984) Enhanced adventive embryogenesis resulting from plasmolysis of cultured wild carrot cells. *Plant Cell Tissue Organ Culture* **3,** 221–227.
11. Ammirato, P. V. and Steward, F. C. (1971) Some effects of the environment on the development of embryos from cultured free cells. *Bot. Gaz.* **132,** 149–158.

Chapter 17

Microspore Culture in *Brassica*

Eric B. Swanson

1. Introduction

Pioneering research in *Brassica* microspore culture (*1,2,3*) rapidly led to the realization that microspores provide a powerful alternative to protoplast culture as a single-celled culture method in plants. These two single-celled systems are fundamentally different, both in tissue origin and in genetic variability. The microspore system improves significantly on the protoplast system by virtually eliminating the large somaclonal variation associated with protoplast selection, by utilizing a true haploid cell system, and by resulting in a more synchronized embryo development that facilitates accurate mutation and selection methods. Most critical, however, are the observations that plant regeneration frequencies in excess of 80% can be readily obtained and that the entire sequence from microspore isolation to plantlet development may take place in as little as 4 wk (*4*).

Microspore culture has been particularly effective in both winter and spring cultivars of *Brassica napus*, with a strong genotypic effect (*5*). Following suggestions from Keller (personal communication), we de-

From: *Methods in Molecular Biology*, vol. 6, *Plant Cell and Tissue Culture*
Edited by Jeffrey W. Pollard and John M. Walker, ©1990 by The Humana Press

veloped improved microspore-responding spring genotypes via inter- and intra-varietal selections. We have also achieved higher microspore success rates in winter types by using F1 hybrid material.

Developmental research on microspore culture in *Brassica* has focused on problems associated with donor plant physiology, selection of immature buds, cytological staging of microspores, medium composition, genotype, isolation methods, and physical culture conditions. One of the more interesting aspects of *Brassica* microspore culture is the observation that, although all laboratories acknowledge the importance of donor plant conditioning, the necessary conditions vary in stringency with different institutions. The methods outlined in this review should result in excellent success in most laboratories.

Microspore culture has several advantages over anther culture, including the production of a haploid single-celled population and the removal of any undesirable anther-microspore interactions. However, until recently, large-scale isolation was often a laborious and time-consuming effort. We developed a rapid and simple large-scale mechanical microspore isolation method that resulted in the production of hundreds of thousands of normal microspore-derived embryos with high plant regeneration rates (*4*). This technology was subsequently applied to microspore mutagenesis and selection, to produce a number of herbicide-tolerant plants (*6,7*). It has been our experience that microspore-derived plants of *Brassica* are remarkably uniform when embryo development has been normalized. This feature is being utilized in our own and a number of other plant breeding programs (*8*). The improved control over spontaneous somaclonal variation is a result, at least in part, of the normalization of embryo development. One of the critical observations on the quality of embryos developed through microspore culture was that torpedo-shaped embryos are more likely to develop and form plants (*3,9,4*). The use of overnight incubation and embryo shaking greatly improved the percentage of torpedo-shaped embryos, and resulted in rates of plant regeneration in excess of 80% (*4*). Keller (personal communication) observed that growing the donor plants at lower temperatures (day 10°C/night 5°C) resulted in better embryo development. We have also observed that a reduction in the growth temperature of the donor plant can extend the period of good bud sampling. It was recently reported (*10*) that BA levels of 0.01–0.255 mg/L in the embryo induction medium improved embryo yields. It should be pointed out, however, that hormone supplements are not essential, and in the simplest system, hormone-free medium may be used throughout all stages of microspore isolation, embryo development, and plant regeneration. We have observed that placing embryos briefly on dry filter paper or transferring the large cotyledonary embryos to sterile, damp Pro Mix with a filter paper support may improve plant regeneration (*11*); however, more research is needed in this area.

2. Materials

1. Plants of *B. napus* (cv. Topas) have been used by several laboratories to produce a high percentage of microspore-derived embryos. Numerous other genotypes also will respond well; however, it is recommended that a "tester" genotype, such as the cultivar Topas, be used initially to ensure that the proper technique has been implemented. The methodology described will produce excellent results from the cultivar Topas.
2. Donor plant conditioning is a critical variable, and plants should be grown in an environmental chamber or growth room for optimum results. Lighting is usually provided by a mixture of incandescent bulbs and cool, white fluorescent tubes. The donor plants should be grown in a 16-h photoperiod of not less than 300 $\mu E\ m^{-2}\ s^{-1}$ at 18°C, with an 8-h dark interval at 13°C.
3. For efficient microspore isolation from the bud, we recommend the use of a microblender (Micro S/S Eberbach Co., Ann Arbor, Michigan). Other blenders may also be used; however, the microblender provides for more consistent and efficient blending of the buds regardless of whether the researcher is interested in isolating microspores from a single bud or 100 buds at a time. The large base of this blender is designed to maintain a more constant internal blending temperature, has mainly metal parts for autoclaving, and will remain cool for several blendings if desired.
4. Autopsy baskets (Ingram and Bell, Weston, Ontario, M9M 1M6) for bud sterilization.
5. Two incubators (one at 32°C in the dark, and the other at 25°C containing a rotary shaker) will be required for microspore-derived embryo development.
6. The medium used during microspore isolation and all washing steps is B5 medium (*12* and *see* Appendix) with no phytohormones and with 13% (w/v) sucrose. It is important *not* to use the microspore culture medium described in step 7 below for isolation and washing.
7. A microspore culture medium that is used with minor modification by many laboratories is provided in Table 1. This medium is a modi-

fication of the medium first used for microspore development in *Brassica* by Lichter (*1,13*), which was a previous modification of the original basal medium developed by Nitsch and Nitsch (*14* and *see* Appendix). BA (0.05 mg/L) may optionally be included in the microspore medium to support normal embryo development.

8. Nitex with a pore size of 45 μ*M* (B.SH. Thompson Co., Toronto, Canada).
9. Falcon (3002) 60 x 15 mm Petri dishes. Use either the indicated brand of Petri dish, or a similar brand that is coated to reduce surface tension and permit low medium volumes to be plated.
10. Commercial bleach (6% hypochlorite).

3. Method

1. The donor plants are germinated in Pro Mix C (Plant Products, Toronto, Canada) in 8-in fiber pots. Fertilizer (20–10–20 [N:P:K]) is applied with routine watering (3–4 x/wk) once plants are past the 3–5 leaf stage. It is very important not to overwater the pots, which can result in precocious bud opening and poor microspore response (i.e., the soil should be well aerated). Transferring the plants to a lower temperature (10°C day/5°C night) during actual bud harvesting may increase the successive number of bud harvests, and improve embryo yields and development. If donor plant growth space is limited, this modification may be highly desirable. The donor plants, when grown under the conditions described, should have developed excellent vegetative growth (large green leaves) to support proper bud development from the emerging racemes.
2. For the initial isolations, the terminal racemes are removed from the donor plants just as they begin to bolt and usually before any visible flower development. The buds, which contain the responding microspores (primarily at the uninucleate stage, although some may be at the binucleate stage), are approximately 1.5–3.5 mm in length. However, size alone does not result in the proper bud selection. Depending on the growth conditions of the donor plant, the genotype used, the age of the plant, and the placement of the raceme (either terminally or laterally) on the plant, the raceme structure may differ. The approximate locations of suitable buds from a number of different raceme types from the cultivar Topas are indicated in Fig. 1. Responsive buds are generally found in a whorled arrangement on a large raceme, and frequently may be identified by

Table 1[a]
Culture Medium for the Development
of Microspore-Derived Embryos in *Brassica Napus*

Compound	Final conc. g/L	
KNO_3	0.125	
$MgSO_4 \bullet 7H_2O$	0.125	
$Ca(NO_3)_2 \bullet 4H_2O$	0.500	
KH_2PO_4	0.125	
Fe EDTA (Na salt)	0.040	
Amino acid supplements		
Glutamine	0.800	
Glutathione	0.030	
Serine	0.100	
Vitamins	**Stock conc. (100x) g/L**	
Myo-inositol	10.00	Use 10 mL stock/L in microspore medium
Nicotinic acid	0.50	
Pyridoxine HCl	0.05	
Thiamine HCl	0.05	
Folic acid	0.05	
Glycine	0.20	
Biotin	0.005	
Freeze vitamin stock in 100-mL bags		
Micronutrients	**Stock conc. (1000x) g/L**	
$MnSO_4 \times 4H_2O$	22.3	Use 10 mL stock/L in microspore medium
H_3BO_3	6.2	
$ZnSO_4 \times 7H_2O$	8.6	
$Na_2MoO_4 \times 2H_2O$	0.250	
$CuSO_4 \times 5H_2O$	0.025	
$CoCl_2 \times 6H_2O$	0.025	
Freeze micronutrient stocks in 100-mL bags		
Other supplements		
BAP	0.05 mg/L	(Optional)
SUCROSE	130 g/L	
pH 6.0		

[a]Table derived from *1,13,14*.

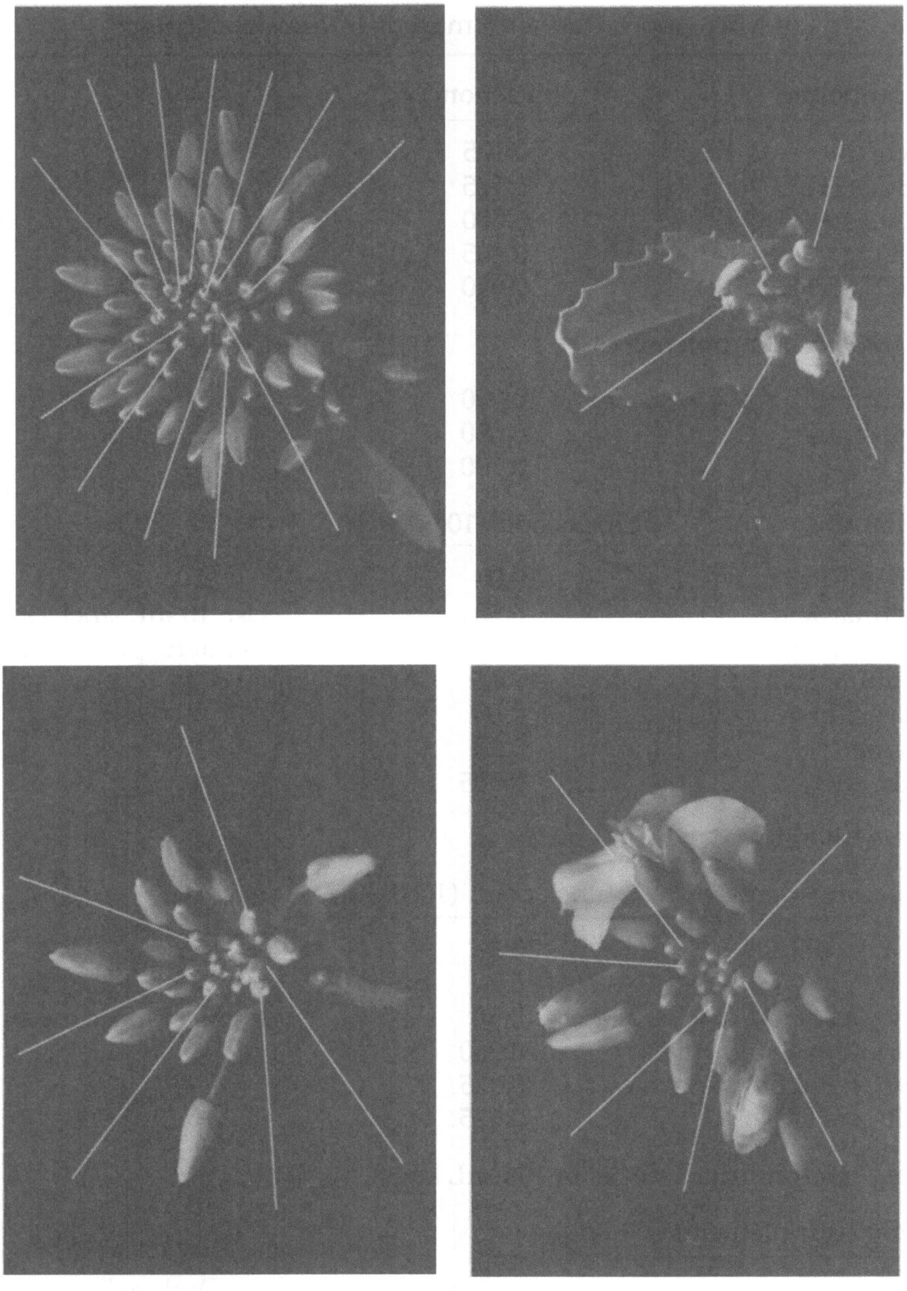

Fig. 1. Several different raceme types with buds producing microspore-derived embryos indicated.

having recently begun to elongate from the raceme. The responding buds may be as few as 1 or as many as 15–20/raceme.

3. The selected buds are placed in small wire autopsy cages (1–100 buds/cage) and immersed in commercial bleach (6% hypochlorite) for 15 min. The cages are then dipped in sterile distilled water (DiW) and rinsed for three 5-min washes in DiW.
4. The buds are placed in the cool microblender (10°C), using approximately 0.5–1.0 mL of cool (10–12°C) (not cold) B5 medium/bud, and blended at high speed for 5–7 s. Do *not* overblend (this can be monitored by ensuring no pieces of cell debris are in the final slurry), since this will reduce the frequency of microspore response (*see* Note 1).
5. The slurries from the blender are passed through two layers of nitex (45 µ*M*), and collected in conical centrifuge tubes. The blender and nitex are washed with additional B5 medium, and the microspores collected as before.
6. The filtrates are centrifuged at 250*g* for 10 min in a swing bucket centrifuge. The supernatant is discarded, and the pellet resuspended in B5 wash (for 30–50 buds use 40–50 mL B5 wash) and recentrifuged.
7. The washing procedure (step 6) should be repeated at least once more (total of three B5 washes). After the final wash, the microspore pellet should be resuspended in microspore culture medium (Table 1).
8. The density of the microspore preparation can be determined using a Fuchs Rosenthal Ultra Plane Corpuscle Counting Chamber and the microspore density adjusted to approximately 100,000/mL. The final suspension will contain only microspores.
9. The microspores are then left overnight at 32°C.
10. The microspores are centrifuged (350*g* for 10 min) the following morning and resuspended in fresh microspore medium.
11. The final concentration of microspores should fall between 75,000–100,000 microspores/mL (allowing for some loss of microspores caused by the final recentrifugation step).
12. The microspores (75–100,000 microspore/mL) are plated at 2.5–3.0 mL/60 x 15 mm Petri dish.
13. The microspores are placed in an incubator at 32°C until numerous globular and young torpedo-shaped embryos can be observed in the plates (Fig. 2) (usually around 10–14 d). The microspores are then placed in an incubator at 25°C, and a small amount of fresh microspore medium is added per plate.

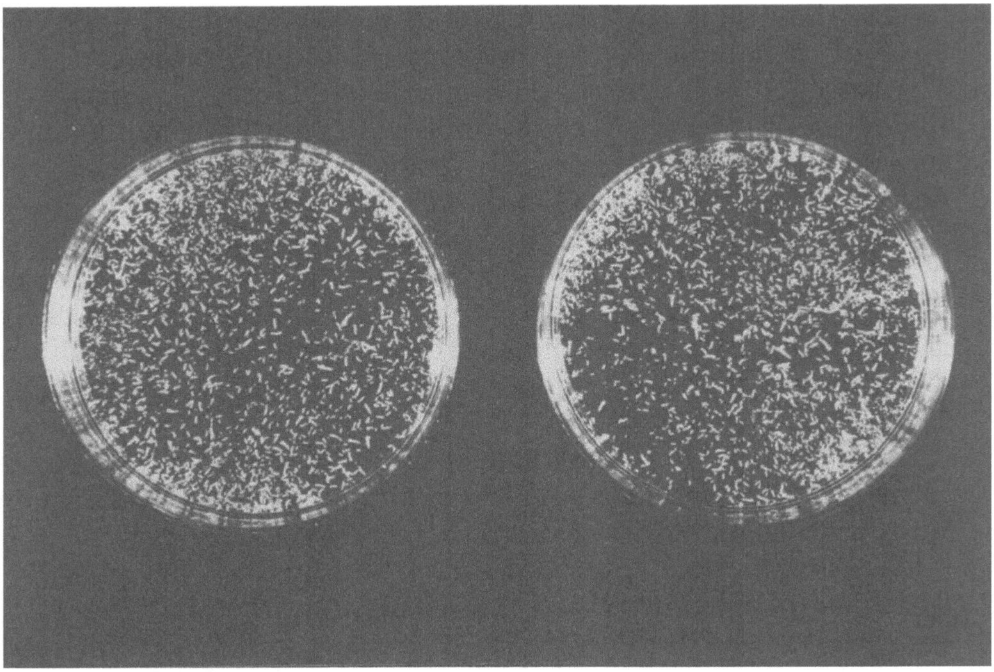

Fig. 2. Microspore-derived embryos developing from isolated microspores of five buds (day 14 from isolation).

14. The embryos are placed on a slow rotary shaker (40–50 rpm) and left until torpedo-shaped embryos have developed.
15. The Petri dishes containing torpedo-shaped embryos are transferred to a rotary shaker in the light at 25°C.
16. To induce better quality cotyledonary embryos, the number of embryos in step 14 or 15 should be reduced to approximately 250/Petri dish.
17. The cotyledonary embryos are transferred to B5 medium (0.8% agar) with 2% sucrose (GA3 at 0.1 mg/L has been added to the B5 medium in some labs; however, in our hands this treatment has varied with the genotype used and is not essential). The use of strong cotyledonary embryos before transferring to solid B5 will help considerably in rapid development of plantlets.
18. After 2 wk on the B5 medium without hormones, embryos that have not produced true leaves are transferred to fresh B5 medium with 0.05 mg/L benzyladenine. Plant regeneration rates of 90% can be routinely achieved from these cotyledonary embryos.
19. Placing the cotyledonary embryos on filter paper, and either recult-

uring them to B5 medium providing a cold shock of 5–10°C for 48 h, or planting them directly into peat, may facilitate more rapid plant development.

20. The plants with true leaves are micropropagated by cuttings (if rooting has not been initiated) and then transferred to Multiwell transfer flats (72 well) containing moist Pro Mix, and maintained under high humidity until strong plant development is obtained (usually 1–2 wk).
21. Colchicine doubling of haploid plants can be achieved by rinsing well-rooted small cuttings with water and placing them in a beaker containing 0.2% (w/v) colchicine for appoximately 5–6 h (*see* Note 2). The plantlets should have a wilted appearance. The roots are rinsed in water and repotted to larger pots. The ultimate size of the pot will be determined by how much seed set is desired from each double haploid. If large seed quantities are desired (for example from a transformed or an in vitro selected microspore population), then the large 8-in fiber pots should allow the researcher to harvest more than 1000 seeds/plant. Additional seed may be obtained if the plant is subsequently cut back and allowed to regrow. Once a colchicine double plant has begun to flower, unproductive non-doubled branched sections (i.e., identified by having sterile flowers) should be pinched off to encourage more seed production from the doubled sectors of the plant. Further, if seed production is very low on a particular branch, care must be exercised to ensure that this is a doubled sector. Normal pod filling and seed development should be obtained on properly doubled branches.

4. Notes

1. We utilized the mechanical isolation methods described in this chapter to produce the first successful use of in vitro microspores mutagenesis and selection in plants (*6,11*). We have since produced a number of herbicide-tolerant plants with stable, genetically inherited herbicide-tolerant mechanisms (*7*). We are also using microspore mutagenesis and selection to produce disease and quality mutants.
2. Colchicine treatments may not be required if the researcher follows a cryopreservation protocol after microspore isolation (*15*). These authors obtained over 88% spontaneous diploid plants from microspores frozen in the microspore culture medium. Although the pro-

cedure does reduce the final number of embryos, it may permit the elimination of the colchicine step, and thus reduce the overall time required to produce fertile plants.

3. A useful modification to the microspore isolation method described involves the use of Percoll density gradients to partially separate the embryogenic fraction of microspores from the unresponsive types (*16*). Following this technique, we have enriched the responsive microspores, which are floating at the interface between the 32% Percoll/NLN and the NLN medium that is carefully loaded on the top. In general, we observed that the oblong microspores and also several different categories of other microspores sink through the gradient, but the rounder enlarged microspores remain floating in a band at the interface. Use of this technique will result in the loss of some portion of the responsive microspores; however, microspore development is subsequently faster and more synchronized.
4. Anthers of *Brassica* yield from 15,000–20,000 microspores/anther, or 90,000–120,000 microspores/bud. The frequency of microspore-derived embryos can routinely reach 1% from some genotypes (Topas is one of these genotypes). Higher levels (1–10%) may also be obtained by better genotype selection, by more careful bud selection, and by altering donor plant conditioning; however, in our hands a level of 1% has been satisfactory. When large bud isolations are used, the overall response percentage may drop. A satisfactory microspore-derived embryo response from large-scale isolations would be 50,000 embryos from 100 buds.
5. Numerous laboratories are currently applying a number of genetic engineering systems towards microspore transformation, including microinjection (*17*). We are examining the possibility of using microspore culture techniques with *Agrobacterium* and direct DNA uptake methods, including the use of a particle acceleration apparatus (*18*), to produce transformed plants.

References

1. Lichter, R. (1982) Induction of haploid plants from isolated pollen of *Brassica napus*. *Z. Pflanzenphysiol.* **105,** 427–434.
2. Keller, W. A. and Armstrong, K. C. (1983) Production of *Brassica napus* haploids through anther and microspore culture. *Proc. of Sixth International Rapeseed Congress,* Paris, France, pp. 239–245.
3. Choung, P. V. and Beversdorf, W. D. (1985) High frequency embryogenesis through isolated microspore culture in *Brassica Napus* L. and *B. Carinata* Braun. *Plant Sci.* **39,** 219–226.

4. Swanson, E. B., Coumans, M. P., Wu, S.-C., Barsby, T. L., and Beversdorf, W. D. (1987) Efficient isolation of microspore and the production of microspore-derived embryos in *Brassica napus* L. *Plant Cell Reports* **6,** 94–97.
5. Choung, P. V., Deslauriers, C., Kott, L. S., and Beversdorf, W. D. (1988) Effects of donor genotype and bud sampling in microspore culture of *Brassica napus*. *Can. J. Bot.* **66,** 1653–1657.
6. Swanson, E. B., Coumans, M. P., Brown, G. L., Patel, J. D., and Beversdorf, W. D. (1988) The characterization of herbicide tolerant plants in *Brassica napus* L. after *in vitro* selection of microspores and protoplasts. *Plant Cell Reports* **7,** 83–87.
7. Swanson, E. B., Herrgesell, M. J., Arnoldo, M., Sippell, D., and Wong, R. S. C. (1989) Microspore mutagenesis and selection: The development of canola plants with field tolerance to the imidazolinones. *Theor. Appl. Genet.*, in press.
8. Lichter, R. (1985) From microspores to rape plants: A tentative way to low glucosinolate strains in world crops, in *Production, Utilization, Description*, vol. II (Sorensen, H., ed.), Nijhoff/Junk Dordrecht, Boston, Lancaster, pp. 268–277.
9. Klimaszewska, K. and Keller, W. A. (1983) The production of haploids from *Brassica hirta* (*Sinapis alba*) anther cultures. *Z. Pflanzenphysiol.* **109,** 235–238.
10. Charne, D. G. and Beversdorf, W. D. (1988) Improving microspore culture as a rapeseed breeding tool: 1. The use of auxins and cytokinins in an induction medium. *Can. J. Bot.* **6,** 1671–1675.
11. Beversdorf, W. D., Swanson, E. B., and Coumans, M. P. R. (1987) *Microspore-Based Selection Systems and Products Thereof.* US Patent Serial #030987.
12. Gamborg, O. L., Miller, R. A., and Ojima, K. (1968) Nutrient requirements of suspension cultures of soybean root cells. *Exp. Cell Res.* **50,** 151–158.
13. Lichter, R. (1981) Anther culture of *Brassica napus* in a liquid culture medium. *Z. Pflanzenphysiol.* **103,** 229–237.
14. Nitsch, C. and Nitsch, J. P. (1967) The induction of flowering *in vitro* in stem segments of *Plumbago indica* L. 1. The production of vegetative buds. *Planta (Berlin)* **72,** 355–370.
15. Charne, D. G., Pukacki, P., Kott, L. S., and Beversdorf, W. D. (1988) Embryogenesis following cryopreservation in isolated microspore cultures of rapeseed (*Brassica napus* L.) *Plant Cell Reports* **7,** 407–409.
16. Fan, Z., Armstrong, K. C., and Keller, W. A. (1988) Development of microspores *in vivo* and *in vitro* in *Brassica napus* L. *Protoplasma* **147,** 191–199.
17. Spangenberg, G., Neuhaus, G., Gland, A., and Schweiger, H.-G. (1986) Transgenic plants of *Brassica napus* obtained by microinjection of microspore-derived embryos. *Eur. J. of Cell Biol.* **53,** 56.
18. Klein, T. M., Fromm, M. E., Weissinger, A., Tomes, D., Schaaf, S., Sleeten, M., and Sanford, J. C. (1988) Transfer of foreign genes into intact maize cells using high velocity microprojectiles. *Proc. Natl. Acad. Sci. USA* **85,** 4305–4309.

Chapter 18

Growth of Ferns from Spores in Axenic Culture

Matthew V. Ford and Michael F. Fay

1. Introduction

In this chapter, a method by which many fern species can be successfully grown from spores in axenic culture will be described. Unlike the conventional method of sowing the spores on compost, this method allows spore populations free from contamination by spores of other species to be sown. The method can be used for the production of mature sporophytes or to provide a controllable system for biosystematic studies of, or experimentation with, fern gametophytes (*1,2*).

For many years, it has been well known that spores of most ferns require exposure to light for germination and growth (*3,1,4*). However, dark germination has been recorded in *Pteridium aquilinum* (*5,6*), *Blechnum spicant* (*7*), and *Onoclea sensibilis* (*8,9*), among others. This ability to germinate in the dark may vary with specimen age, exposure to different temperatures, and plant hormone treatment (*3,1,10*). In general, the percentage of dark germination is low, and subsequent growth is abnormal or retarded.

Many workers have investigated the relative effectiveness of different spectral regions of light on spore germination, and have obtained

From: *Methods in Molecular Biology*, vol. 6, *Plant Cell and Tissue Culture*
Edited by Jeffrey W. Pollard and John M. Walker, ©1990 by The Humana Press

results indicating the involvement of phytochrome (*11,12,13*). In general, it was found that exposure to red light induced germination (*14,15*). This was reversed by exposure to far-red light and blocked by blue light (*16*). Light quality also influences prothallial growth (*17,18,19,20*). The use of spectral filters has not been found necessary for the germination of the majority of ferns using the method described later, since white light seems to provide an adequate spectral balance to satisfy the requirements for germination. Light intensity, however, has been reported to have a marked influence on germination and subsequent prothallial growth (*21*). This phenomenon can be observed in nature in the various habitat preferences of ferns.

Spore sowing density can prove to be critical at its extremes (*22*), with slower germination at lower densities and reduced germination at higher densities. This appears to be the result of volatile substances, possibly including ethylene, which are produced by germinating spores. This may also lead to aberrations in prothallial growth. The problem may also be one of population density, involving competition for light and mineral resources. The production of large amounts of antheridiogens, often found in high-density gametophyte populations, can also prove problematic, since they induce a predominance of maleness in the cultures (*23*), hence preventing fertilization and sporophyte formation.

It can be seen from the brief résumé given above that much work has been carried out on the physiological aspects of spore germination and gametophyte growth. The following method takes into account much of the knowledge now available, but has been designed to be simple and effective.

2. Materials

2.1. Media

Murashige and Skoog (MS) medium (*24* and *see* Appendix) is obtained as a ready mixed powder, without sucrose, growth regulators or agar. The medium is made up at half the normal strength, using 2.35 g MS powder and 15 g sucrose dissolved in 1 L of deionized water. Adjust the pH to 5.6 using dilute HCl or NaOH as appropriate. Dissolve 9 g of Oxoid No. 1 Agar in the medium using a microwave oven (1 min at high setting for every 100 mL of medium), and stir thoroughly. Dispense the medium into the desired vessels, and sterilize in the autoclave for 15 min at 121°C and 1.05 kg/cm^2. If the medium is not to be used immediately, store the vessels in plastic bags at 4°C.

2.2. Compost

Ferns, such as *Cyathea* and *Angiopteris*, are fairly sturdy when they are brought out of culture. These can be weaned in the first compost given below. However, some ferns (*Adiantum, Cheilanthes, Nephrolepis,* and so on) are quite sensitive when first out of culture, and thus require a much finer compost, such as the second one given below. In all cases, good drainage must be provided. For the more robust ferns: 2 parts leafmold; 2 parts coarse peat; 1 part charcoal; 1 part coarse bark. For more delicate ferns: 1 part loam; 1 part grit.

3. Methods

3.1. Spore Collection and Storage

1. For each species a frond, or piece of frond, with mature sporangia is collected and kept warm and dry in a folded white card, or paper, awaiting dehiscence (3). Depending upon the maturity of the sporangia collected, this usually takes place within 48 h. More than enough spores can be obtained from a single frond. It has been calculated that the approximate spore content of an individual frond can range from 750,000–750,000,000 depending upon the species (*10*). Spore viability declines with age and depends on storage conditions, but this also varies between species. Ferns with chlorophyllous spores (those in the *Osmundaceae, Gleicheniaceae, Grammitidaceae* and *Hymenophyllaceae*) should be sown as soon as possible, because they lose their viability after only a few days (*25,10*). Nonchlorophyllous spores, in general, retain their viability for substantial periods when kept at room temperature, but there are exceptions (e.g., spores from the tree-fern family, *Cyatheaceae,* lose viability rapidly after only a few weeks) (*10*). If storage of the spores is unavoidable, they are best kept in sealed packets in a desiccator, or in sealed tubes at 4°C.

3.2. Sterilization

1. Ensure that the spore sample is reasonably free from chaff and other debris. This can be done either manually with fine forceps, for the larger pieces, or by giving the white card a short, sharp, but gentle tap. This should separate chaff from spore.
2. Place a quantity of spores in the filter paper packet, and seal with a metal staple (remembering that too small or too great a quantity may result in density problems).

3. Soak the spore packets for 5–10 min, and carefully expel any air within them by gently squeezing with forceps.
4. Remove the packets from the water, put them in a plastic beaker, and place on a magnetic stirrer. Add enough 10% NaOCl solution to more than cover the packets, and set the speed on the stirrer to provide gentle agitation. Leave to agitate for 10 min.
5. Move the beaker into a laminar airflow bench. Standard aseptic techniques should now be followed.
6. Remove each packet from the solution and rinse in separate tubes of sterile deionized water 3x.

3.3. Sowing

1. Carefully take the packet from the water with large forceps, and squeeze gently to expel any excess water.
2. Remove the staple end of the packet with scissors.
3. With the Petri dish open, using two pairs of forceps, unfold the packet and wipe the spores in a swirling motion over the surface of the medium.
4. If any obvious clumping of spores is observed, gently spread them out with the flat end of a spatula.
5. Seal the Petri dish with Parafilm or Nescofilm and place in the growth room (temp. 22 +/– 2°C; light intensity 400–4200 lx; photoperiod 16 h) or into an incubator.

3.4. Germination and Gametophyte Growth

1. It is generally believed that a period of 3–96 h in the dark after sowing is required for imbibition before the spores become light receptive (*26*). This preinduction phase of germination is not directly accommodated in this method because the photoperiod provided in the growth room has proved adequate in supplying the necessary environment. If, however, synchronous or uniform germination is desired, then a dark period before exposure to light is advisable.
2. The next induction phase of germination will go largely unnoticed. It is followed by the postinduction phase, where dark processes are triggered, and which results in the protrusion of a rhizoid and the protonema (*27*). The time from sowing to this stage varies greatly between species, and can range from a few days to many months. Some spores will inevitably be killed by the surface sterilization process, but a good percentage of germination should still result.

3. After a few weeks, the characteristic thalloid gametophyte form will be noticed. From this stage, the rate of growth varies between species and with the light conditions to which the gametophyte has been subjected. If growth to mature gametophyte appears particularily retarded, try moving the dish to a higher or lower light intensity.
4. During growth, the following problems may occur:
   a. The population is too dense.
   b. The medium begins to dry and split.
   c. The gametoyphytes begin to grow brown around their bases and discolor the medium.
   d. Mass sporophyte formation begins.

The solution to problem a. is to gently ease the gametophytes from the medium and move to fresh medium in either jars or dishes, depending upon the size of the plants. A spatula with a flattened end bent to a 90° angle is useful for this operation. By gently pulling the latter across the surface of the media, the rhizoids (vital for the uptake of nutrients and moisture and anchorage to the substrate) are eased out of the media with minimal physical damage. Remove any media adhering to the plant during transfer.

If problem b. occurs, movement to fresh media is required as soon as possible. Gametophytes have been shown to be sensitive to drying (*28*). This can lead to a rapid death of the population if left too long.

If problem c. happens, then either (1) the medium is not suitable for this plant and it needs to be transferred to another medium (*see* Note 1), or (2) the nutrients are being rapidly exhausted, and frequent transfer to fresh medium is necessary.

Situation d. quite often occurs where the level of moisture in the environment is sufficient to facilitate fertilization. This becomes a problem if the container is too small or the sporophytes arise too close to their neighbors. This can lead to a high degree of competition and stunted growth. It is, therefore, necessary to transfer the plants to a fresh medium, ensuring adequate space is provided for unrestricted growth.

If the gametophytes form clumps, then gently tease them apart and move to a fresh medium.

3.5. Sporophyte Formation and Potting

1. Many ferns are homosporous, producing spores that germinate to form bisexual, haploid gametophytes. These have both antheridia (male) and archegonia (female) parts, and produce gametes that

fuse and form the diploid sporophyte (*29*). Other ferns are heterosporous, producing micro- and megaspores that germinate to form microgametophytes (male) and megagametophytes (female), respectively. These spores must be sown together if the ultimate aim is to produce sporophytes (e.g., in *Marsilea* and *Platyzoma*).

2. Some species will produce sporophytes readily, and others will do so only after several transfers to fresh medium. A few require supplementary moisture. This can be provided simply in one of two ways:
   a. Rinsing in sterile deionized water between transfers.
   b. Adding a small amount of sterile deionized water to the jar, swirling gently, and pouring off the excess (care has to be taken here not to add so much water that the structure of the medium is damaged). Do not attempt this if the gametophytes are not firmly anchored to the medium.

   In the early stages of the life of the sporophyte, it is dependent upon the gametophyte for moisture and nutrients, which are transferred to the sporophyte via a foot.
3. Subsequently roots will be formed and, when these are large enough, the sporophytes may be pricked out onto fresh medium and cultured separately. It is often advisable to leave the parent gametophyte attached to the sporophyte, reducing the risk of physical damage. The gametophytes may then senesce and die, or continue growing, in which case they can then be removed and cultured separately, if so desired.
4. As the sporophyte grows, more frequent transfers to larger vessels may prove necessary. When a reasonable amount of roots have been formed, the plant is ready for potting. This can be done using one of two methods:
   a. Direct potting of the fern in the compost described in the Materials section. Adhering media is carefully removed by rinsing in tepid deionized water before potting. Good drainage must be provided, and a covering layer of gravel helps to conserve water and prevents the growth of algae. The potted ferns are grown in shaded frames in a glasshouse at temperatures between 20–30°C and high relative humidity (90–98%). They can then be gradually acclimatized to the recommended growth conditions for the species in question.
   b. Transferring to sealed jars of sterile compost. Some ferns are

This intermediate step can help to encourage root growth and reduce the potential damage to roots when the plant is finally potted as in (a).

4. Notes

1. Some ferns may prefer a different media to that given here, e.g., *Angiopteris boivinii* will grow on 1/2 strength MS, but will grow better on Knudson's C medium (*28*) if frequent transfers are made. Useful media variations for experimentation are: Knudson's C; Knudson's C modified (with 1 mL microsalt solution /L); and Moore's media. (personal communication, Moore, London University.)

NH_4NO_3	1.0 g/L
KH_2PO_4	0.2 g/L
$MgSO_4 \cdot 7H_2O$	0.2 g/L
$CaCl_2 \cdot 2H_2O$	0.1 g/L

Dissolve separately and add solutions a. and b.

a. Microsalt solution 1 mL/L as follows:

$B_2O_3 \cdot H_2O$	6.20 g/L
$Na_2MoO_4 \cdot 2H_2O$	0.25 g/L
$ZnSO_4 \cdot 7H_2O$	8.60 g/L
KI	0.83 g/L
$CuSO_4 \cdot 5H_2O$	0.025 g/L
$CoCl_2 \cdot 6H_2O$	0.025 g/L

Dissolve separately. Store in the dark at 0°C.

b. Ferric citrate solution (2 mL/L), as follows:

$C_6H_5O_7Fe \cdot 5H_2O$ 1 g/L

Store medium in the dark at 0°C.

Adjust pH to 5.6, and add 9 g of agar.

2. The time period and the strength of the sterilant can be varied. Increasing the time or concentration will improve the sterilization of those specimens that are repeatedly contaminated. Reducing the time or concentration will decrease the percentage of spore fatality in especially delicate samples.
3. For some species with short spore viability, manual dissection of the sporangia may be required if dehiscence has not occurred within 48 h. This is carried out in a draught-free environment using a clean blade and fine forceps.
4. If fungal contamination is severe, the culture must be thrown away. If mild, sections of the population may be saved by moving the uncontaminated parts to a nonsugar-based medium (such as Moore's)

or to 1/2 MS supplemented with 1% Benylate. Growth on the former may be slow and pale, but the plants should reach maturity. Transfer back to a sugar-based medium should be possible at a later stage, but the fungus may manifest itself again, after lying dormant in the absence of sugar.

5. Abnormal growth may occur at either the gametophyte or sporophyte stage.
   a. Elongation of the prothallus may occur. This likely to be a density problem and should be treated as such.
   b. Sporophytes are sometimes formed without the fusion of gametes taking place, i.e., apogamously. This sometimes occurs in old cultures of gametophytes or in cultures of species that are particularily susceptible to it. Ethylene has been implicated as a promoter of this process (*30*). The apogamous (haploid) sporophyte may arise from the prothallial cushion, apical region, or on the end of an apical protuberance or podium (*31*).
   c. The sporophytes grow down into the media as well as upwards. This is geotrophic confusion, possibly brought on by either ethylene or too much light below the culture. This can be averted by thinning out the population and blocking light from below the cultures.
6. The pH used in this method was chosen because it suits a range of ferns (*32*). However, for obligate calcicoles or calcifuges, the pH may have to be adjusted. If lower pH values are desired, then a higher concentration of agar is needed to maintain a sufficient gel strength (12 g or more). If the pH values greater than 7.0 are required, it is better to use $Ca(OH)_2$ rather than NaOH, since high concentrations of sodium can prove toxic.
7. Withering of fronds during transfer occurs in particularly sensitive ferns or if the transfer takes a long time. This is the result of desiccation caused by the stream of air in the flow bench. A screen, using one or more sterile dishes set up in the bench, helps to prevent this problem.

Acknowledgment

We would like to thank Ms. L. Goss for her advice on compost and glasshouse conditions.

References

1. Miller, J. H. (1968) Fern gametophytes as experimental material. *Bot. Rev.* **34,** 362–441.
2. Windham, M. D. and Haufler, C. H. (1986) Biosystematic uses of fern gametophytes derived from herbarium specimens. *Am. Fern J.* **76(3),** 114–128.
3. Dyer, A. F. (1979) The culture of fern gametophytes for experimental investigation, in *The Experimental Biology of Ferns* (Dyer, A. F., ed.), Academic Press, London, pp. 254–305.
4. Williams, S. (1938) Experimental morphology, in *Manual of Pteridology* (Verdoorn, Fr. Martinus Nijhoff, ed.), The Hague, pp. 105–140.
5. Ragavan, V. R. (1970) Germination of bracken fern spores. Regulation of protein and RNA synthesis during initiation and growth of the rhizoid. *Expl. Cell Res.* **63,** 341–352.
6. Weinberg, E. S. and Voeller, B. R. (1969) Induction of fern spore germination. *Proc. Nat. Acad. Sci. USA* **64,** 835–842.
7. Orth, R. (1937) Zur keimungsphysiologie der farnsporen in verschidenen spektralvezirken. *J. Wis. Bot.* **84,** 358–426.
8. Miller, J. H. and Greany, R. H. (1974) Determination of rhizoid orientation by light and darkness in germinating spores of *Onoclea sensibilitis*. *Am. J. Bot.* **51,** 329–334.
9. Miller, J. H. and Miller, P. M. (1970) Unusual dark-growth and antheridial differentiation in some gametophytes of the fern *Onoclea*. *Am. J. Bot.* **57,** 1245–1248.
10. Page, C. N. (1979) Experimental aspects of fern ecology, in *The Experimental Biology of Ferns* (Dyer, A. F. ed.), Academic Press, London, pp. 552–589.
11. Ragavan, V. R. (1971) Phytochrome control of germination of the spores of *Asplenium nidus*. *Plant Physiol.* **48,** 100–102.
12. Ragavan, V. R. (1973) Blue light interference in the phytochrome-controlled germination of the spores of *Cheilanthes farinosa*. *Plant Physiol.* **51,** 306–311.
13. Sugai, M., Takeno, K., and Furuya, M. (1977) Diverse responses of spores in the light-dependent germination of *Lygodium japonicum*. *Pl. Sci. Letters* **8,** 333–338.
14. Towill, L. R. and Ikuma, H. (1973) Photocontrol of the germination of *Onoclea* spores I. Action spectrum. *Plant Physiol.* **51,** 973–978.
15. Towill, L. R. and Ikuma, H. (1975) Photocontrol of the germination of *Onoclea* spores II. Analysis of germination processes by means of anaerobis. *Plant Physiol.* **55,** 150–154.
16. Furuya, M. (1985) Photocontrol of spore germination and elementary processes of development in fern gametophytes, in *Biology of Pteridophytes* (Dyer, A. F. and Page, C. N., eds.), *Proc. Royal Soc. Edin.* **86B.** pp. 13–19.
17. Greany, R. H. and Miller, J. H. (1976) An interpretation of dose-response curves for light induced cell elongation in the fern protonemata. *Am. J. Bot.* **63,** 1031–1037.
18. Howland, G. P. and Edwards, M. E. (1979) Photomorphogenesis of fern gametophytes, in *The Experimental Biology of Ferns* (Dyer, A., F. ed.), Academic Press, London, pp. 394–434.
19. Miller, J. H. and Miller, P. M. (1967) Action spectra for light induced elongation in fern protonemata. *Pysiologia Pl.* **20,** 128–138.
20. Miller, J. H. and Miller, P. M. (1974) Interaction of photomorphogenetic pigments in fern gametophytes: Phytochrome and a yellow-light absorbing pigment. *P. Cell Physiol. Tokyo,* **8,** 765–769.

21. Ford, M. V. (1984) Growth responses of selected fern prothalli to various light regimes (unpublished).
22. Smith, D. L. and Robinson, P. M. (1971) Growth factors produced by germinating spores of *Polypodium vulgare* (L.). *New Phytol.* **70,** 1043–1052.
23. Bell, P. R. (1979) The contribution of the ferns to an understanding of the life cycles of vascular plants. IV. Sexuality in gametophytic growth, in *The Experimental Biology of Ferns* (Dyer, A. F., ed.), Academic Press, London, pp. 64, 65.
24. Murashige, T. and Skoog, F. (1962) A revised medium for rapid growth and bioassays with tobacco tissue cultures. *Physiol. Pl.* **15,** 473–497.
25. Lloyd, R. M. and Klekowski, E. J., Jr. (1970) Spore germination and viability in the Pteridophyta: Evolutionary significance of chlorophyllous spores. *Biotropica* **2(2),** 129–137.
26. Jarvis, S. J. and Wilkins, M. B. (1973) Photoresponses of *Matteuccia struthiopteris* (L.) Todaro. *J. Exp. Bot.* **24,** 1149–1157.
27. Brandes, H. (1973) Spore Germination. *Ann. Rev. Plant Physiol.* **24,** 115–128.
28. Knudson, L. (1946) A new nutrient solution for the germination of orchid seed. *Am. Orchid Soc. Bull.* **15,** 214–217.
29. Hyde, H. A., Wade, A. E., and Harrison, S. G. (1978) *Welsh Ferns, Clubmosses, Quillworts, and Horsetails,* National Museum of Wales, Cardiff.
30. Elmore, H. W. and Whittier, D. P. (1973) The role of ethylene in the induction of apogamous buds in *Pteridium* gametophytes. *Planta* **111,** 85–90.
31. Walker, T. G. (1985) Some aspects of agamospory in ferns—the Braithwaite System, in *Biology of Pteridophytes* (Dyer, A. F. and Page, C. N., eds.), *Proc. Royal Soc. Edin.* **86B,** pp. 59–66.
32. Otto, E. A., Crow, J. H., and Kirby, E. G. (1984) Effects of acidic growth conditions on spore germination and reproductive development in *Dryopteris marginalis* (L.). *Ann. Bot.* **53,** 439–442.

Chapter 19

Clonal Propagation of Orchids

Daniel Jones and Brent Tisserat

1. Introduction

This chapter will deal with methods of clonal propagation for members of the two major morphological groups of orchids. The first group, sympodials, includes such genera as *Cymbidium, Cattleya, Dendrobium,* and *Oncidium*. They are characterized by a multi-branching rhizome that can supply an abundance of axillary shoots for use as explants. They were among the first orchids to be successfully propagated, and techniques for their in vitro initiation (i.e., establishment) and subsequent proliferation are well established (*1–8*). The second group, monopodials, include *Phalaenopsis* and *Vanda*, and are characterized by a single, unbranched axis of growth that possesses few readily available axillary shoots for use as explants. Significantly different in their morphologies, the two groups require different approaches to explant selection and subsequent culturing. The successful large scale micropropagation of monopodials is, in fact, a relatively recent achievement (*9*): the culmination of a wide variety of studies using different media compositions and supplements (*10–18*).

Since the introduction of orchid meristem culture in 1960, a divergence in orchid micropropagation techniques has developed. This divergence manifests itself in the utilization of two distinctly different culture schemes. In one method, the initial explant is induced into (and is subsequently maintained in) an undifferentiated callus or protocorm-

From: *Methods in Molecular Biology*, vol. 6, *Plant Cell and Tissue Culture*
Edited by Jeffrey W. Pollard and John M. Walker, ©1990 by The Humana Press

like body (PLB) state. PLBs are spherical tissue masses that resemble an early stage of orchid embryo development. Proliferation occurs via PLB multiplication, and differentiation into plantlets is permitted only after the desired volume of callus is achieved. The alternate method minimizes the role of callus, and encourages the differentiation of cultures into plantlets early in the procedure. Consequently, proliferation is accomplished by the induction of axillary shoots from plantlets derived from the original explant. Though not unique to orchid propagation, these two methods and the attempts to optimize them constitute the bulk of all recent propagative studies.

This chapter describes four methods of propagation: one callus and one axillary shoot technique for each of the two morphological groups of orchids. Advantages and disadvantages associated with these two methods will be discussed and possible variations pointed out (*see* Notes).

2. Materials

1. Disinfesting solution: Use a solution of 1% sodium hypochlorite (household bleach diluted five times with distilled water) with 0.1% Tween 80 emulsifier. Prepare fresh solution for each treatment and use within the hour.
2. Post disinfestation holding solution: Prepare a 100 m*M* phosphate buffer with 8.76 m*M* sucrose. Adjust the pH to 5.8 and autoclave for 15 min at 1.05 kg cm^{-2} and 121°C.
3. Culture media: Use Murashige and Skoog inorganic salts (*see* appendix) with: 8.76 m*M* sucrose, 0.003 m*M* thiamine-HCl, 0.005 m*M* pyridoxine-HCl, 0.008 m*M* nicotinic acid, 0.555 m*M* i-inositol, and 100 mg/L casein hydrolysate. This basic formulation is supplemented as indicated in Table 1. An initiation media (IM) and a differing proliferation media (PM) are required for each method. The proliferation steps of Methods 3 and 4 necessitate the preparation of a series of media (Table 1) to determine optimum yields. The pH of all solutions is adjusted to 5.7 with 1.0*N* HCl or NaOH before being dispensed into 25 x 150-mm culture tubes (for agar media) or 250-mL flasks (for liquid media) in 25-mL aliquots. Culture tubes and flasks, capped with appropriate closures, are autoclaved for 15 min at 1.05 kg cm^{-2} and 121°C.
4. Culture conditions: Cultures receive 13 $\mu EM^{-2}\ s^{-1}$ from wide spectrum fluorescent tubes for 16 h daily. Temperature is maintained at 27°C.

Table 1
Supplements to the Basal Medium
Employed in the Four Orchid Micropropagation Methods[a]

		Supplements			
Methods	Media	BA, μ*M*	NAA, μ*M*	Agar, mg/L	Charcoal, mg/L
1	IM	7.83	1.07	–	–
	PM	0.78	13.43	6000	–
2	IM	7.83	1.07	–	–
	PM	11.75	2.68	6000	–
3	IM	19.58	5.37	6000	–
	PM	0.39	2.68	6000	–
		0.39	5.37	6000	–
		0.39	8.05	6000	–
		3.92	2.68	6000	–
		3.92	5.37	6000	–
		3.92	8.05	6000	–
4	IM	19.58	5.37	6000	–
	PM	7.84	0.54	6000	–
		15.67	0.54	6000	–
		31.34	0.54	6000	–
		7.84	5.37	6000	–
		15.67	5.37	6000	–
		31.34	5.37	6000	–
1,2,3,4	RM	–	–	6000	–

[a]IM—initiation medium; PM—proliferation medium; RM—rooting medium; BA—benzyladenine; NAA—Naphthaleneacetic acid.

5. Sympodial orchid explants (*see* Table 2). Developing shoots (less than 1/2 their mature size) are severed from the rhizome (Fig. 1).
6. Monopodial orchid explants (*see* Table 2). Inflorescences (preferably with flowers still intact) should be severed near their point of emergence from leaf bases.

3. Methods

3.1. Clonal Propagation of Sympodials by Callus Production

1. Leaves surrounding developing shoots are removed, exposing underlying lateral buds and shoot tips. Though the form of the leaves may vary from one sympodial genus to another, the basic

Table 2
Some Commonly Encountered Sympodial and Monopodial Orchid Genera

Sympodials		
Cattleya	*Laelia*	*Broughtonia*
Brassavola	*Sophronitis*	*Epidendrum*
Dendrobium	*Oncidium*	*Encyclia*
Miltonia	*Odontoglossum*	*Cochlioda*
Brassia	*Masdevalia*	*Cymbidium*
Monopodials		
Phalaenopsis	*Vanda*	*Rhynchostylis*
Ascocentrum	*Aerangis*	*Angraecum*
Aerides	*Doritis*	*Cyrtorchis*
Neofinetia	*Renanthera*	*Vandopsis*

structure of the developing shoot is as depicted in Fig. 1.

2. Excised shoots are sterilized for 20 min in disinfesting solution.
3. Replace disinfesting solution with buffered holding solution. Tissues may remain in this solution while they await final trimming.
4. Excise explants for culturing by maneuvering the blade of a scalpel behind the lateral buds. Remove the buds by cutting approximately 1 mm into the underlying stem tissue. The shoot tip is removed by cutting just below the base of the elliptical leaf primordia. All dissecting operations are performed while the tissues are immersed in the holding solution. Final trimming of explants involves the removal of one or two of the tunicate leaf scales of the lateral buds or leaf primordia surrounding the shoot tips. Finally, carefully remove any fibrous stem tissue that may have adhered to the lateral buds or shoot tips when dissected from the stem.
5. Place explants in flasks containing liquid IM (Table 1) and agitate on a gyrotory shaker at 100 rpm.
6. Within days, explants should turn green and begin to swell. After 4–6 wk, the explants should be large and plump. At this time, the explants are removed from culture, lightly scarred with two or three superficial scalpel incisions, and placed on PM (Table 1). If after the intital 4–6 wk any cultures appear to have undergone little or no growth, reculture to fresh IM and observe for suitability for transfer to PM in another few weeks. In addition, if at any time the medium of an initiating culture begins to discolor (i.e., darken because of phenolic compound accumulation), it should be replaced with fresh medium immediately (*see* Note 1).

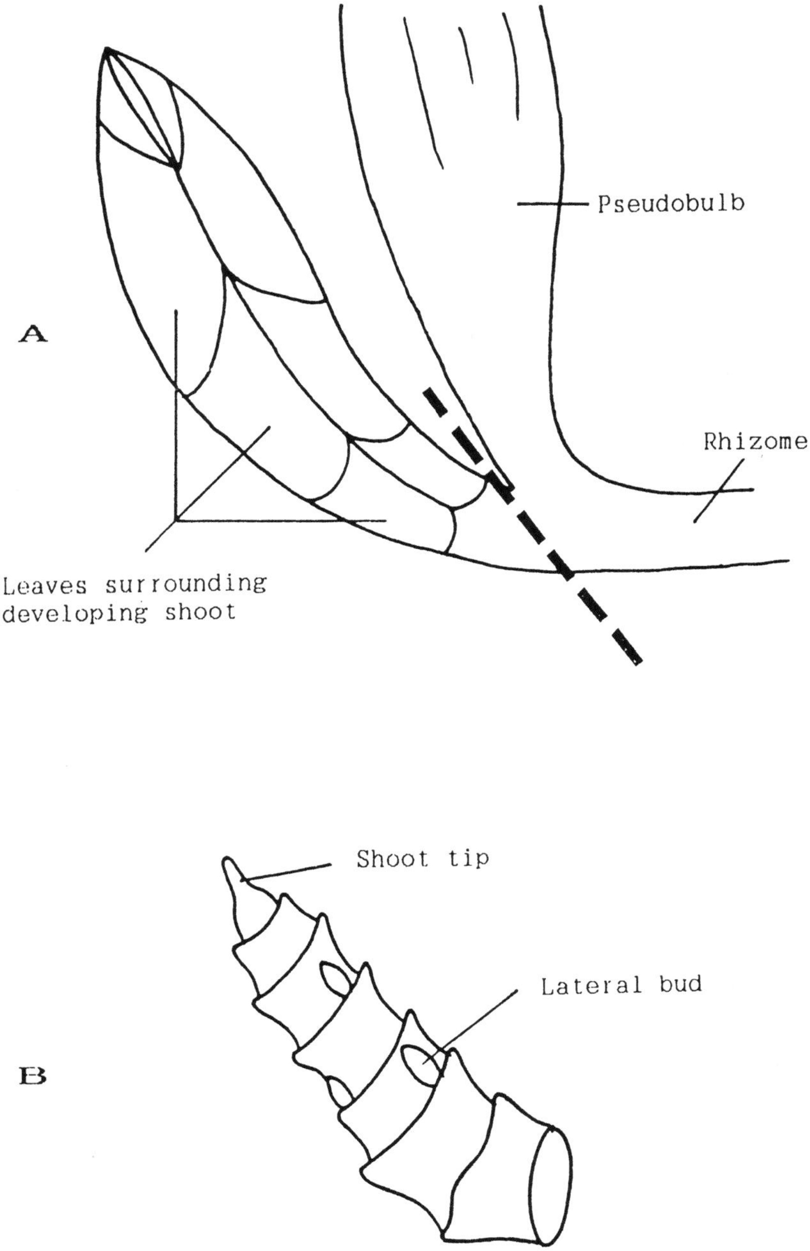

Fig. 1. Explant selection for sympodial orchids. Dotted line in (A) indicates point of removal of developing shoot. (B) Example of growth after removal of leaves. Lateral buds and shoot tip, used for culturing, are revealed.

7. Upon transfer to PM, explants begin to produce callus or PLBs on their surface and from the incisions made before transfer. Large PLB clusters should be broken up and subcultured to fresh PM to enhance multiplication. Continue sectioning and subculturing callus masses until the desired volume of PLBs is achieved.
8. To induce differentiation into plantlets with roots, transfer PLBs to the rooting medium (RM) in Table 1.

3.2. Clonal Propagation of Sympodials by Axillary Shoot Formation

1. Follow steps 1–5 as described in Method 1. Step 6 is executed using Method 2 PM instead of Method 1 PM (Table 1).
2. When transferred to PM, cultures will start to produce well-differentiated axillary shoots and/or callus masses, which quickly differentiate into shoots. From this point onward, most shoots should give rise to additional axillary shoots without passing through a callus stage. Shoot clumps should be separated to enhance proliferation, and individual shoots subcultured to fresh PM. This step is repeated until the desired number of plants is obtained (*see* Note 2).
3. To establish a root system, shoots are transferred to RM (Table 1).

3.3. Clonal Propagation of Monopodials by Callus Production

1. Cut the inflorescence into sections, each piece possessing one dormant node with approximately 2 cm of the stem attached above and below the node (*see* Note 3).
2. With fine-tipped forceps, remove the bracts surrounding the dormant buds (Fig. 2). All traces of the bract, including any old, papery, or scarred tissues that may be present at their bases, should be removed before disinfestation.
3. Wash nodal sections by agitating in dilute household detergent for 5 min.
4. Sterilize in disinfesting solution for 20 min. Agitate frequently.
5. Replace disinfesting solution with buffered holding solution. Agitate vigorously to ensure deactivation of the remaining sterilant.
6. With a scalpel, recut the two ends of the nodal sections, making sure that each fresh cut is at least 5 mm below the margins created by the bleached (damaged) areas at each end of the section.
7. Place nodal sections on Method 3 IM so that the bud is just above the medium surface. Within a week, buds turn green and begin to swell.

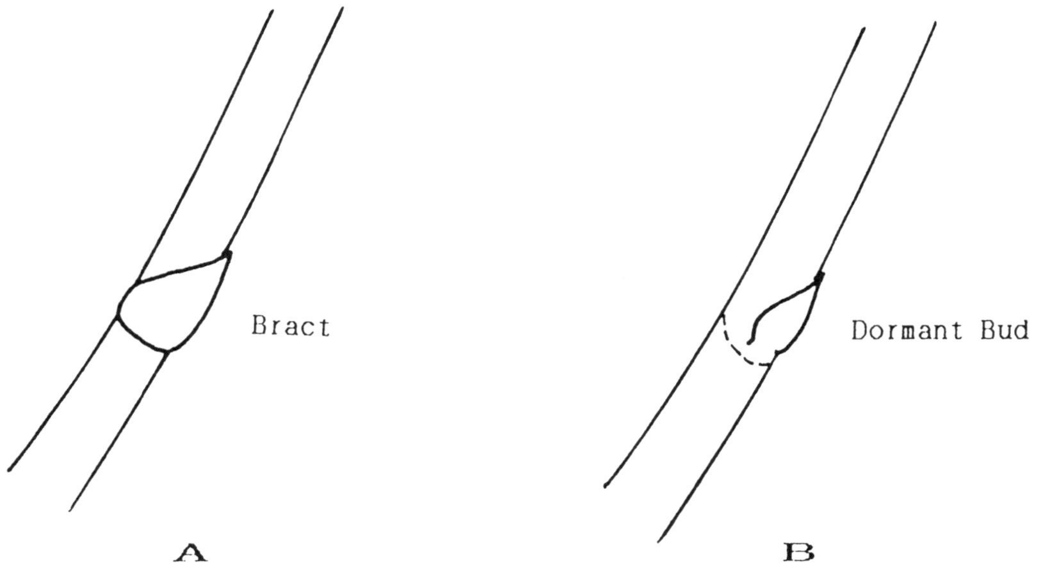

Fig. 2. Explant selection for monopodial orchids. (A) Monopodial inflorescence stem section. (B) Section after removal of the bract to reveal the underlying bud. Dotted line indicates location where bract attached to the stem.

After 12–16 wk, explants possess multiple shoots with well-developed leaves. Occasionally, this shoot cluster may also possess peripheral callus masses that contain newly developing plantlets in various stages of development (Fig. 3).

8. If callus masses described in step 7 are present on the explant, they should be severed from the rest of the node culture, sectioned into smaller pieces, and transferred to the PM series of Method 3 (Table 1).
9. Cultures not possessing visible callus are treated as follows: Sever the nodal plantlets from the stem, being sure to leave at least a 3–5-mm stub of the nodal base still attached to the stem. Severed shoots may be used for Method 4 (Fig. 3). Reculture the stem section, devoid of its nodal shoots, on fresh IM. The cut surface and periphery of the nodal stub will produce callus, which can be handled as described in step 8 (Fig. 3).
10. The PM (from Table 1) deemed optimum for a particular callus should be employed for further subculturing (*see* Notes 4–6). Continue sectioning callus clusters and transferring to fresh PM until the desired volume of callus is achieved.
11. To induce differentiation into plantlets with roots, callus or PLBs should be transferred to RM (Table 1).

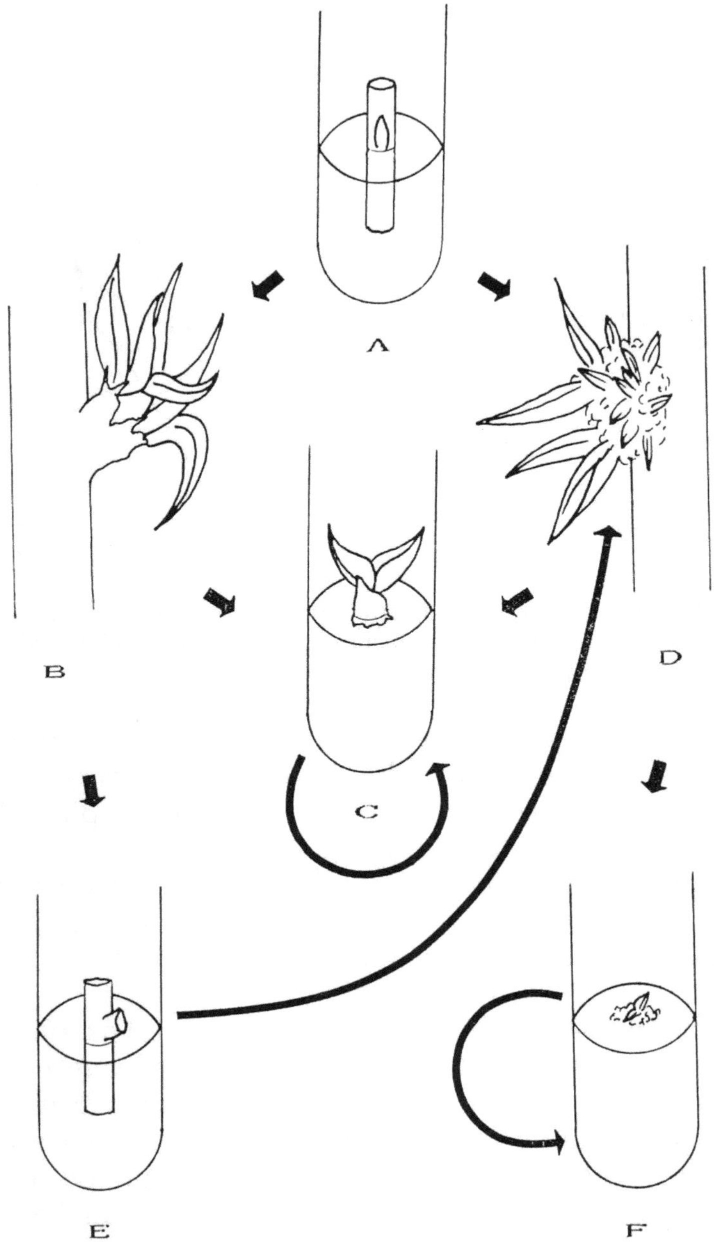

Fig. 3. Steps in the micropropagation of monopodial orchids. (A) Newly planted inflorescence node culture. (B) Multiple shoots developing from a node. (C) An individual shoot separated from the shoot cluster derived from either B or D. (D) Multiple shoots, with accompanying callus, developing from node. (E) Recultured inflorescence node with shoots removed to induce callus from the remaining stub. (F) Isolated callus and shoots obtained from culture D.

3.4. Clonal Propagation of Monopodials by Axillary Shoot Formation

1. Follow steps 1–7 as described for Method 3.
2. Multiple nodal shoots are separated, so that each plantlet has a sufficient base to support its leaves above the medium surface (Fig. 3). Culture plantlets on the entire Method 4 PM series (Table 1).
3. The medium giving the highest axillary shoot yield for a particular plant is used for all subsequent proliferation. Continue separating shoot clusters into individual plantlets and subculturing on fresh PM until the desired number of plantlets is achieved (*see* Note 7).
4. To establish a root system, shoots are transferred to RM (Table 1).

4. Notes

1. Exudation of phytotoxic phenolics from wounded orchid tissue poses a serious problem for in vitro culture. The most effective way of overcoming this problem in orchids is frequent media replenishment. It is important that culture vigor not be compromised by phenolic accumulation, because once attenuated, the productivity of a culture is difficult to restore.
2. It should be apparent that the IM formulations of all four presented methods are closely allied to the PM formulations of Methods 2 and 4. The primary reason for indirectly approaching even the callus proliferation techniques by first employing a shoot proliferation analog is to achieve greater initial explant survival. The presence of cytokinins in the IM improves the survival rate of explants, probably owing to the antisenescence properties of the cytokinin.
3. In monopodial propagation, it is improtant that inflorescences be young and healthy. Ideally, flowers at the apex should still be fresh. Older inflorescences, or ones from which the flowers have faded, will give adequate, but not optimum results.
4. In monopodial callus propagation, loss of chlorophyll, and an accompanying decline in growth, may be observed in some cultures. In such instances, transfer of the callus to PM lacking sucrose can often restore color and growth to cultures. All subsequent proliferation steps should, therefore, employ sucrose-free media. When differentiation into plantlets is desired, use sucrose-containing RM (Table 1).
5. Generally, callus propagation tends to be swifter than axillary shoot propagation. However, culturing orchids for extended periods in

an undifferentiated callus state, can result in significant occurrences of mutation. Genetic aberration has long been accepted as a risk in callus proliferation. However, as the demand for orchid clones increases, the problem of mutation in callus is only likely to worsen. Conversely, axillary shoot proliferation is very resistant to genetic aberration and its development was largely in response to increasing concern over the occurence of mutation (3,5).

6. Though NAA and BA perform satisfactorily, experiments to substitute other growth regulators for them should be beneficial. In particular, cytokinins, such as isopentenyladenine and zeatin riboside (the latter being very expensive), hold great promise in orchid micropropagation.
7. Both axillary shoot methods presented are capable of producing approximately 1000 individual plants in the first year of culture, and as many as 100,000 by the end of the second year. The two callus methods described are capable of producing 100,000 or more plants within the first year of culture.

References

1. Churchill, M. E. (1973) Tissue culture of orchids. *New Phytol.* **72,** 161–166.
2. Lindemann, E. G. P. (1970) Meristem culture of *Cattleya*. *Amer. Orchid Soc. Bull.* **39,** 1002–1004.
3. Morel, G. (1960) Producing virus free *Cymbidiums*. *Amer. Orchid Soc. Bull.* **29,** 495–497.
4. Morel, G. (1964) A new means of clonal propagation of orchids. *Amer. Orchid Soc. Bull.* **33,** 473–478.
5. Sagawa, Y. (1966) Clonal propagation of *Cymbidiums* through shoot meristem culture. *Amer. Orchid Soc. Bull.* **35,** 188–192.
6. Sagawa, Y. (1967) Clonal propagation of *Dendrobiums* through shoot meristem culture. *Amer. Orchid Soc. Bull.* **36,** 856–859.
7. Scully, R. (1967) Aspects of meristem culture in the *Cattleya* alliance. *Amer. Orchid Soc. Bull.* **36,** 103–108.
8. Wimber, D. (1963) Clonal multiplication of *Cymbidiums* through tissue culture of the shoot meristem. *Amer. Orchid Soc. Bull.* **32,** 105–107.
9. Kuhn, L. (1982) Orchid propagation in the EYMC. *Marie Selby Bot. Gard. Bull.* **9,** 32–34.
10. Griesbach, R. J. (1983) The use of indoleacetylamino acids in the in vitro propagation of *Phalaenopsis* orchids. *Sci. Hort.* **19,** 363–366.
11. Intuwong, O. (1974) Clonal propagation of *Phalaenopsis* by shoot tip culture. *Amer. Orchid Soc. Bull.* **43,** 893–895.
12. Huang, L. C. (1984) Alternative media and method for *Cattleya* propagation by tissue culture. *Amer. Orchid Soc. Bull.* **53,** 167–170.
13. Rotor, G. (1949) A method of vegetative propagation of *Phalaenopsis* species and hybrids. *Amer. Orchid Soc. Bull.* **18,** 738,739.

14. Sagawa, Y. (1961) Vegetative propagation of *Phalaenopsis* stem cuttings. *Amer. Orhid Soc. Bull.* **30,** 803–809.

15. Scully, R. (1966) Stem propagation of *Phalaenopsis. Amer. Orchid Soc. Bull.* **35,** 40–42.

16. Lay, M. and Fan, F. (1978) Studies on the tissue culture of orchids. *Orchid Rev.* **Oct,** 308–310.

17. Tanaka, M. (1977) Clonal propagation of *Phalaenopsis* by leaf tissue culture. *Amer. Orchid Soc. Bull.* **46,** 733–737.

18. Zimmer, K. and Pieper, W. (1978) Clonal propagation of *Phalaenopsis* by excised buds. *Orchid Rev.* **July,** 223–227.

Chapter 20

Tissue Culture and Top-Fruit Tree Species

Sergio J. Ochatt, Michael R. Davey, and John B. Power

1. Introduction

The commercial cultivation of rosaceous fruit trees (e.g., pear, apple, cherry, peach, plum) relies heavily upon the quality and performance of the rootstocks. This is even more the case now that self-rooted scions produce larger trees with a longer juvenile phase (*1*). It would, therefore, be of special interest for the fruit breeder to have general purpose rootstocks with a wide ecophysiological adaptation and high compatibility coupled with early cropping. In addition, many of the older and highly adapted scion varieties of fruit trees could benefit greatly from the introduction of stable, yet minor changes in their genome. Fruit trees are generally highly heterozygous, outbreeding, and thus are asexually propagated (*see* Chapter 10, this vol.). Consequently, genetic improvement is likely to be based on protoplast technology, and achieved mainly through somatic methods, such as somaclonal variation or somatic hybridization.

From: *Methods in Molecular Biology*, vol. 6, *Plant Cell and Tissue Culture*
Edited by Jeffrey W. Pollard and John M. Walker, ©1990 by The Humana Press

In this context, knowledge of the conditions for sustained plant propagation from true callus (of either complex explant, cell, or protoplast origin), as distinct from in vitro cultured organized tissues, is a prerequisite. For most fruit tree species, this still remains to be accomplished. In addition, although many herbaceous species will regenerate plants from protoplasts (*see* Chapter 24, this vol.), woody species are still regarded as recalcitrant in this respect (*1*), with relatively few examples (*2–8*).

Since micropropagation is covered in Chapter 10 (this vol.), and for many woody species tissue culture is interpreted as being at this level, this chapter will be restricted to the methodology of callus initiation and maintenance as the essential intermediate stage for the establishment of a protoplast-to-tree system.

Among those few protoplast-to-tree systems established for woody species, only three belong to rosaceous fruit trees (*3–5*). However, limited differentiation has been reported in two other cases for members of this family (*9,10*), and a general strategy for the isolation of mesophyll protoplasts of temperate fruit and nut tree species is now available (*11*). This background, coupled with the detection of phenotypic variation among protoplast-derived wild pear plants (*12*), the isolation of protoplast-derived cherry variants with the ability to tolerate severe salt and water stress both at the tissue and whole plant levels (*13*), and the enhancement of growth (*14*) and regeneration responses (*15*) after electroporation of isolated cherry and pear protoplasts, shows that the use of somaclonal variation and protoplast technology is indeed feasible in the context of top-fruit tree breeding.

2. Materials

2.1. Initiation and Maintenance of Proliferating Callus and Cell Suspensions

Callus can be induced from almost any explant using standard tissue culture procedures and employing a medium based on MS (*16*) salts (*see* Appendix) supplemented with 2.0 mg/L NAA and 0.5 mg/L BAP (MSP1 medium). However, when embryos (immature and taken only a few weeks after pollination) or seedling explants (e.g., hypocotyls or cotyledons) are the starting materials, better results are obtained, in terms of callus induction and proliferation, using MS medium with 0.1 mg/L 2,4-D and 0.1 mg/L BAP (*10*). Once initiated, calluses can be transferred to MSP1 medium for further proliferation.

Prior to culture, explants are surface sterilized by

1. Gentle agitation in a bleach solution (0.5–2.0% active chlorine; 20–30 min)
2. Immersion in 70% v/v ethanol (5 s) and
3. Several rinses in sterile tap water.

However, disinfectants acting through protein precipitation (such as 0.1% $HgCl_2$; 15 min), rather than oxidation, are preferred when handling tissues of woody species that are very prone to phenolic oxidation, such as most fruit tree genotypes (apple, pear, and cherry). Several other approaches can be adopted to alleviate the browning/oxidation of the initial explants, but in general, maintenance of the tissues (prior to explant excision) in a sterile solution of citric acid (0.4 g/L) is helpful. In addition, filter-sterilized citric acid solution added to a liquid culture medium, coupled with constant agitation (40 cycles/min on a rotary shaker) during the first week of culture both enhances tissue growth responses and negates the tendency for browning. Alternatively, antioxidants acting by adsorption, such as 1.0% (w/v) PVP-10 (polyvinylpyrrolidone, MW 10,000), can be used successfully for this purpose for several *Prunus* species. Initial explants, taken from in vitro-grown plants (*see* Chapter 10, this vol.) will not only bypass the need for surface sterilization but, for most woody species, can also be more readily induced to proliferate as callus. In this respect, in vitro rooted plants are better sources than axenic shoot cultures. In turn, root segments or leaf disks should be used as explants for callus initiation.

Callus can be induced on such explants within 2–4 wk of culture, and following removal from the explants, can be maintained by monthly subculture. Such tissue will proliferate rapidly at 25°C under continuous illumination (1000 lx, daylight fluorescent tubes). Cultures maintained in the dark generally exhibit a slower growth rate, with associated tissue browning.

To initiate cell suspension cultures, the most friable portions of fast-growing callus are taken (5.0 fresh wt of callus to 50.0 mL of liquid in a 250-mL Erlenmeyer flask) and are dispersed by constant agitation on a rotary shaker (90–120 cycles/min). The resulting cell suspensions should be provided with constant illumination, and subcultured every 7–21 d (depending on the genotype) by transfer of approximately 10.0 mL of a settled (20 min) cell suspension or 1.5 mL packed cell volume to 75.0 mL of fresh medium. Liquid MSP1 medium has proved particularly useful for suspension cultures of apple, Colt cherry (*5*), and other cherry genotypes.

2.2. Tissue Sources for Protoplast Production

2.2.1. Leaf Tissues

A comparison of leaves from in vitro and field-grown plants for mesophyll protoplast production (3) suggests that the yield, viability, and survival of protoplasts from in vivo grown plants are erratic, and that such cultures are prone to persistent bacterial contamination. Leaves from axenic shoot cultures or in vitro-rooted plants always provide more consistent and reproducible results; the latter source of protoplasts is superior for most species. In either case, the largest yields of protoplasts ($>5 \times 10^6$ protoplasts/g fresh wt of digested tissues) coupled with high viability (>60%) are obtained from the youngest fully expanded leaves, i.e., those taken from the upper one-third of the shoot. This applies not only to apple, cherry, pear, and peach, but also to other nonrosaceous temperate fruit and nut tree species, including kiwifruit, walnut, hazelnut, and trifoliate orange.

2.2.2. Callus / Cell Suspension Cultures

Actively growing cells should be used in their exponential growth phase, with cell viabilities being monitored as described in Chapters 3 and 24 (this vol.). Thus, for callus subcultured every 4 wk, the largest yield ($\geq 10^7$ protoplasts/g fresh wt) and viability (approaching 100%) are obtained 15–20 d after transfer of cells to fresh medium. This same principle applies to cell suspensions, which generally are found to be a better source of protoplasts. However, root cell suspension cultures, particularly those of *Prunus* species, tend to be slow growing. Consequently, a longer period (20–30 d) should be allowed after subculture before cells are used for protoplast production.

3. Methods

3.1. Plant Regeneration from True Callus Cultures

Plant regeneration should be attempted as early as possible after establishment of callus, since undifferentiated cultures of most rosaceous fruit trees gradually lose their totipotency, which is often concomitant with browning of tissues. The addition of gibberellic acid at 0.05 mg/L or less to the media can retard this aging response for some genotypes (e.g., *Prunus* species). Cultures are maintained at 25°C with continuous illumination or with a 16/8-h light/dark photoperiod. In general, continuous illumination induces a larger number of calluses to regenerate, coupled with an increased number of buds/callus. Hard, compact callus regenerates more readily. The composition of the media used for plant regeneration from true callus tissues is genotype dependent (Table 1).

3.2. Isolation of Protoplasts

1. Tissues (leaf or cultured cell origin) are plasmolyzed for 1 h in the osmoticum used in the enzyme mixture (generally at a ratio of 1.0 g fresh wt tissues/10.0 mL osmoticum; *see* Table 2). When large, pieces of callus tissue should be finely sliced or chopped prior to plasmolysis. Leaves are cut into thin strips (1.0 mm wide) with the central and main veins discarded, and/or the lower epidermis removed by peeling with fine forceps, or bruising with a soft brush or the cutting edge of the scalpel.
2. The largest yields of protoplasts of the highest viability are obtained by digesting the tisue (8–18 h, depending on genotype), at 25°C, with constant illumination (100 lx) and slow agitation (40 cycles/min), utilizing 1.0 g fresh wt or 1.0 mL packed cell volume of tissue to 10.0 mL of enzyme solution (*see* Note 1).
3. For some genotypes, protoplasts can be separated from debris by floatation on CPW 21S solution (*see* Chapter 24, this vol.) or a Percoll gradient (*see* Chapter 25, this vol.). In some cases (e.g., apple mesophyll protoplasts), serial sieving will provide a debris-free protoplast preparation as follows:
   a. Digested tissues are sieved through a nylon mesh (64 µm pore size) and the filtrate retained in centrifuge tubes.
   b. Tubes are centrifuged (100 x *g*, 10 min) to pellet the cells/protoplasts, and the supernatant discarded.
   c. The pellet is layered on top of CPW 21S solution, or on a Percoll gradient (30, 25, 20, 15, 10, 5% in the appropriate osmoticum), and centrifuged (100 x *g*, 5–10 min). Protoplasts can be collected at the surface of the CPW 21S solution, or at the 20 and 25% Percoll interphase. For nonfloating protoplast systems, the pellet is resuspended in the osmoticum, and serially sieved through nylon meshes of 53, 45, 30, and 20 µm pore sizes.

3.3. Culture of Protoplasts

1. For most fruit tree protoplast systems, semi-solid media with agarose are recommended because of the long lag phase (≥7 d) prior to the

Table 1
Culture Media Used for Plant Regeneration from True Callus of Top-Fruit Tree Species

Species	Callus source	Basal medium[e]	Growth regulators, mg/L					
			IBA	NAA	BAP	Kin	Z	GA_3
Malus Xdomestica	Stem	(1/4 strength MS)			5.0			
	Stem	MS[a]			0.5			
	Endosperm	MS[a]			0.5			
Prunus avium	Leaf	MS[b]		0.025	2.5		0.025	0.05
P. avium x *pseudo-cerasus*	Leaf	MS[a]			0.5			
	Root	MS[b]	0.1		2.0			0.1
	Mesophyll protoplasts	MS[b]		0.1	0.75		0.1	
	Cell suspension protoplasts	MS[a]		0.05	5.0		0.05	
P. cerasus	Mesophyll protoplasts	MS		0.025–0.05	1.0–5.0		0.1–0.5	
P. dulcis	Seedling	MS		1.0		1.0		
P. lannessiana	Leaf	MS			2.0			
P. persica	Embryo	MS		0.01	0.88			
Prunus spp.	Root	MS			1.0			0.1
Pyrus communis	Nucellus	Yehia[c]			2.0			
	Mesophyll protoplasts	(1/2 strength MS)[d]	0.2		2.0			0.2
P. communis var *pyraster*	Mesophyll protoplasts	MS	0.1		1.0			

[a]Casein hydrolysate added at 100 mg/L or pantothenate.
[b]50 mg/L.
[c]*See* (*1*), also with 10 mg/L L-glutamine, 2.5 mg/L Ca.
[d]10.0 mg/L Ca pantothenate added.
[e]*See* Appendix for media composition.

Table 2
Enzyme Solutions for the Isolation of Protoplasts from Top-Fruit Tree Species

Species	Source	Enzyme Mixture[a] (%)	Osmoticum
Malus Xdomestica	Leaf	(1.0)ONO, (1.0)HEMI, (0.1)PEC	CPW13*M* + 1.0% PVP–10 + 5 m*M* MES
		(2.0)ONO, (0.5)MAC	CPW9*M*
	Callus/cell suspension	(2.0)CELL, (0.5)MASE	13% mannitol
		(2.0)ONO, (0.5)MAC	CPW9*M*
Prunus avium	Leaf	(1.0)ONO, (1.0)HEMI, (0.1)PEC	CPW13*M* + 1.0% PVP–10 + 5 m*M* MES
P. avium x *pseudo-cerasus*	Leaf	(1.0)ONO, (0.2)MAC, (0.1)DRI	CPW13*M* + 1/0% PVP–10 + 5m*M* MES
	Cell suspension	(2.0)MEI, (2.0)RHO, (0.03)MAC	CPW13*M*
P. cerasus	Leaf	(1.0)ONO, (1.0)HEMI, (0.1)PEC	CPW13*M* + 1/0% PVP–10 + 5 m*M* MES
	Callus	(1.5)ONO, (0.2)MAC	11% mannitol
P. dulcis	Cell	(2.0)CELL, (4.0)PASE	11% mannitol
	suspension	(1.0)ONO, (0.5)HEMI, (0.5)MAC, (2.0)DRI	0.35*M* mannitol + 0.35*M* sorbitol 4 m*M* $CaCl_2 \bullet 2H_2O$ + 8 m*M* KCl + 1.4 m*M* $NaH_2PO_4 \bullet H_2O$ + 0.5m*M* MES + 0.5% bovine serum albumin + 6 μ*M* ascorbic acid
P. persica	Cell suspension	(2.0)ONO, (0.1)PEC, (0.5)DRI	0.25*M* mannitol + 0.25M sorbitol
Pyrus communis	Leaf	(1.0)ONO, (1.0)HEMI, (0.1)PEC	CPW13*M* + 1.0% PVP–10 + 5 m*M* MES
	Callus	(2.0)MEI, (2.0)RHO, (0.03)MAC	CPW13*M*
P. communis var *pyraster*	Leaf	(0.5)ONO, (0.1)MAC, (0.1)DRI	0.1MS (without NH_4NO_3) + 1.0% PVP–10 + 3 m*M* MES + 50 mg/L casein hydrolysate + 0.35*M* sucrose

[a]Abbreviations: CELL: Cellulysin; DRI: Driselase; HEMI: Hemicellulase, MEI: Meicelase; MAC: Macerozyme R-10, MASE: Macerase; ONO: Cellulase Onozuka R-10; PASE: Pectinase; PEC: Pectolyase Y-23; RHO: Rhozyme HP-150.

onset of division, coupled with the requirement for a high initial plating density.

2. A temperature of 25 ± 2°C and illumination (1000 lx, daylight fluorescent tubes), either continuously or with a 16-h photoperiod, is required for all protoplast systems in the genus *Pyrus*, whereas for other genera (e.g., *Prunus, Malus*), culture in the dark is conducive to division.

   Protoplast-to-protoplast relationships are particularly important for fruit tree species, and therefore, a high initial plating density range ($5 \times 10^4 - 1 \times 10^6$ protoplasts/mL medium) must be employed. The choice of protoplast culture medium is discussed in Note 2.

3. Mannitol, at 0.5–0.7*M*, is required for most fruit tree protoplasts as osmoticum. Once division has been initiated, the osmotic pressure of the medium must be reduced (*see* Chapter 24, this vol.).

   In this respect, in order to avoid browning during this process, protoplast cultures are diluted initially (original medium; osmoticum-free medium) in a ratio 3:1, after 10 d for members of the genera *Pyrus* and *Malus*, and after 20–25 d for *Prunus* species. This is repeated at 7–10-d and 15-d intervals, respectively, until colonies are 1–2 mm in diameter.

3.4. Plant Regeneration from Protoplast-Derived Cultures

1. MSP1 medium supports the growth of protoplast-derived calluses of most fruit tree species. It should be noted that, in contrast to protoplast-derived calluses of herbaceous species, plant regeneration can only be attempted after individual calluses have attained a minimum mass (typically 200 mg fresh wt).
2. The information currently available relating to growth regulator requirements for differentiation of protoplast-derived callus is scant. A high cytokinin to auxin ratio will generally stimulate organogenesis. A supplement of organics to this regeneration media (50–100 mg/L casein hydrolysate) may also be stimulatory (*see* Table 1).
3. In this context, researchers should not be discouraged if only rhizogenesis can be induced, since an alternative pathway to plant regeneration is now available. Roots are detached and transferred to MS medium with 0.01 mg/L NAA and 2.0 mg/L BAP, whereupon shoot buds can be differentiated from the basal region. All cultures are maintained at 25°C in the light (1000 lx, daylight fluorescent tubes).

Once caulogenesis is induced, buds must be detached as soon as possible and transferred to the appropriate medium that supports multiplication/internode elongation (*see* Chapter 10). Regenerated shoots of woody species will root only after they have attained a minimum size (2–3 cm) on half-strength hormone-free MS (*16*) medium (*Prunus* species), or this medium supplemented with 3.0 mg/L IBA (1 wk) followed by half-strength, hormone-free MS medium (3 wk) (e.g., *Pyrus, Malus* species).

4. The plants are finally transferred to soil, and the transition from heterotrophic to autotrophic growth conditions can be achieved with little loss by placing individual plants in small pots containing commercial compost or a 3:1 (v/v) mixture of soil:peat (*see* Chapter 23, this vol.). They must be kept in a mist propagator (25°C) for 7–10 d, and acclimatized by a gradual reduction in the relative humidity. In the absence of a mist propagator, plants should be covered with polyethylene bags for the first 10 d, and then exposed to the open-glasshouse environment for increasing periods of time. After 1 mo ex vitro, weaning is normally complete, and plants can be kept in the glasshouse or transplanted to the field.

3.5. Electro-Enhancement of Growth and Regeneration of Protoplast-Derived Cells of Woody Species

Protoplasts of fruit tree species, and those of woody species in general, tend to undergo a long lag phase in culture prior to the onset of division. It is, therefore, desirable to reduce this period. Protoplasts of several species, electroporated shortly after isolation, divide earlier in culture (*14*). Plant regeneration is also enhanced (*15*). The voltage and duration of the electric pulses required will vary with protoplast size and source; larger protoplasts are more sensitive to the electric pulse, as are, in general, cotyledon or leaf protoplasts. The following protocol applies specifically to protoplasts of woody species.

1. Protoplasts are suspended, at four times their final plating density, in 5 m*M* MES-based buffer solution with 6 m*M* $MgCl_2$ and 0.5–0.7*M* mannitol (pH 5.8).
2. Aliquots (400 µL) are given three successive exponential (DC) electric pulses, at 10-s intervals, in the chamber of an electroporator (DIALOG, G.m.b.H., 4 Dusseldorf 13, West Germany). Pulses of 250 V/cm

(for cotyledon or leaf protoplasts, or if the protoplasts are >30 μm in diameter) or up to 750 V/cm (for callus or cell suspension protoplasts, or those ≤ 20 μm in diameter) are given by discharging 30 to 10 nF capacitors, respectively.

3. Electropulsed protoplasts are diluted to their optimum plating density with culture medium (*see* Note 3).

3.6. Specific Tissue Culture Based Techniques for Top-Fruit Tree Breeding

3.6.1. Fusion of Fruit Tree Protoplasts and Somatic Hybridization

Fruit tree protoplasts can be readily fused to produce dividing heterokaryons, leading to callus and somatic hybrid plant regeneration. In this respect, an electroporation treatment of the separate parental protoplast populations prior to their fusion, using PEG/high pH (*see* Chapter 24, this vol.), not only increases the frequency of heterokaryon formation, but also enhances subsequent growth in culture. A model protocol is now available for the chemical fusion of woody species protoplasts; for example, leaf protoplasts of wild pear (3) with cell suspension protoplasts of Colt cherry (4).

1. Adjust the protoplast density to 1×10^6/mL (for wild pear) and 4×10^6/mL (for Colt cherry), and electroporate the two parental populations at 250 V/cm, as described in Section 3.5.
2. Readjust the density of both populations to 4×10^5/mL with K8P medium (*see* Appendix and ref. 17).
3. Protoplasts are then mixed and fused using PEG/high pH, as described in Chapter 24 (this vol.).
4. Protoplasts are washed twice in CPW 11M/Ca^{2+} solution, by resuspension and centrifugaion (100 x *g*, 5 min).
5. Fusion-treated protoplasts and the viability controls are adjusted to a density of 1×10^5/mL in semi-solid (0.625% agarose) K8P medium and dispensed into 3.5-cm Petri dishes. (In this example, protoplasts of both parental species do not divide, thereby providing a selection strategy for the recovery of somatic hybrids.)
6. All dishes are sealed with Nescofilm and maintained at 25°C, in the dark.
7. Reduce the osmotic pressure of the medium step-wise after 28 d.

8. Putative somatic hybrid microcalluses (produced after 8 wk) are transferred to MS medium with 2.0 mg/L NAA, 0.25 mg/L BAP, and 3% sucrose, with all cultures kept at 25°C with constant illumination (1000 lx, daylight fluorescent tubes). Subculturing is every 2–3 wk.
9. After approximately 20 wk, callus pieces (200 mg fresh wt) are transferred to MS medium with 1.0 mg/L Zeatin, for shoot bud differentiation.

3.6.2. Selection for Salt Tolerance in Vitro

For any improved (selected) trait to be useful, it must be stable and heritable. A recurrent selection strategy can be undertaken for woody species, in order to progressively eliminate physiologically adapted cells from the selected population while simultaneously enriching it with putatively tolerant lines. With fruit trees, and woody species in general, the gradual loss of regeneration capacity with time further complicates this situation, and therefore a direct, rather than stepwise selection strategy, must be adopted. The following protocol can be successfully used to select for salt tolerance in vitro using *Prunus* explant or protoplast-derived callus:

1. Callus tissue cultures are established in MSP1 medium, at 25°C and under a constant illumination of 1000 lx (daylight fluorescent tubes), and maintained under these conditions for at least six (3-wk) subculture periods.
2. Callus portions (50 mg fresh wt) are transferred to selection medium containing the appropriate salts (e.g., NaCl, KCl, or Na_2SO_4, alone or in combination [*13*], and at a concentration range of 25–200 m*M*).
3. Cultures are maintained on selection media for a further six subcultures and, once cell strains that show sustained growth over the entire six-subculture period are detected in selection media, the recurrent selection strategy is implemented.
4. Putatively tolerant cells are transferred to MSP1 medium without salts for two subcultures, and then returned (for two further subculture periods) to MSP1 medium with the specific salt and concentration as tolerated previously. The recurrent selection passages are repeated at least 3x.
5. Finally, calluses are transferred to regeneration medium, which is supplemented with salts at concentrations tolerated at the callus level. Differentiated buds are handled as described previously.
6. The stability of the acquired stress tolerance can be examined by reinitiating callus cultures using leaf or root explants from the first gen-

eration of regenerants, again using the selection media as tolerated at the callus level. In addition, protoplasts can be isolated from such regenerants and cultured on saline-protoplast media, in order to compare their response/tolerance to those of protoplasts from nontolerant plants.

7. Response to watering with saline solutions should also be assessed for the regenerated (tolerant) plants growing in the open glasshouse or field.

4. Notes

1. When preparing protoplasts, the composition of the enzyme solution will depend on the source tissue and genotype (Table 2). For most species, a mixture consisting of 1.0% (w/v) Cellulase Onozuka R-10, 1.0% (w/v) Hemicellulase, and 0.1% (w/v) Pectolyase Y-23 for leaf tissues, or 2/0% (w/v) Meicelase, 2.0% (w/v) Rhozyme HP-150, and 0.03% (w/v) Macerozyme R-10 for callus/cell suspension cultures, both in CPW salts with 13% mannitol (CPW 13M), will suffice. A supplement of 3–5 m*M* MES (2-*N*-morpholinoethane sulfonic acid) and 1.0% polyvinylpirrolidone (PVP-10) to the enzyme solution can improve protoplast yield and viability.
2. In general, protoplasts of fruit tree and woody species exhibit a marked decline in viability during culture, irrespective of the culture medium. In preliminary studies, and in order to optimize culture media requirements, it is recommended that a sample of protoplasts be stored at 4°C (in the dark and suspended in the osmoticum alone) in order to monitor the baseline decline in viability. This then provides a reference point for the optimization of culture media components for a given species, and places viability determinations on a comparative basis.

   In this respect, ammonium ions can have a negative effect on the survival of protoplasts from rosaceous species. A Murashige and Skoog (*16*, and *see* Appendix) based medium, lacking ammonium and supplemented with NAA and BAP, supports growth (e.g., *Pyrus* and some *Malus* genotypes). Similarly, ammonium ions are toxic to protoplasts of nonrosaceous fruit tree species (*6*). In some genera (e.g., *Pyrus*/*Malus*), differences are observed between scion varieties and rootstocks in terms of their medium organic component requirement. Rootstock protoplasts generally require media rich in organic constituents.

Table 3
Protoplast Culture Media

Species	Protoplast source	Basal medium,[a] reference	Growth regulators, mg/L						Additional components, mg/L
			2,4-D	NAA	IAA	BAP	Kin	Z	
Malus Xdomestica	Cell suspension	K8P (*17*)	0.8		0.5			0.2	Coconut milk (50); L-glutamine (250)
	Callus	K8P (*17*)	0.8		0.5	0.2			
		MS (*1*)	1.0	2.0			1.0		0.01*M* $CaCl_2$
Prunus avium	Leaf	MS		2.0		1.0		0.1	Casein hydrolysate (25)
P. avium x *pseudocerasus*	Leaf	MS (*4*)		2.0		0.5		0.5	
	Cell suspension	MS (*4*)		1.0		0.25		0.5	
P. cerasus	Leaf								
	CAB 4D	MS (*5*)						1.0	
	CAB 5H	MS (*5*)		1.0		0.5		1.0	
	CAB 11E	MS (*5*)		0.05				0.5	
	Cell suspension								
	CAB 11E	Binding (*5*)	0.1	1.0		0.2			
	Callus	MS		2.0		0.5		0.05	
P. dulcis	Cell suspension	MS + B5 vitamins (*4*)		0.5			0.1		
P. persica	Cell suspension	Nitsch-Nitsch (*4*)		2.0		0.2			Casein hydrolysate (1000); 5 m*M* glutamine
Pyrus communis	Callus	MS		2.0		0.5			(Ammonium-free)
	Leaf	MS (*10*)		1.0	1.0	0.8			Casein hydrolysate (50) (ammonium-free)
P. communis var *pyraster*	Leaf	MS + B5 vitamins (*3*)		1.0		0.4			Casein hydrolysate (50) (ammonium-free)

[a] *See* Appendix for media composition.

In contrast, protoplasts of many *Prunus* species are not affected by ammonium ions, and show no specific requirements for an enriched organic component. Cytokinins appear to be more important than auxins for sustained protoplast division, with zeatin a requisite in many cases. Exceptions to this include mesophyll protoplasts of sour cherry (*Prunus cerasus* L., clone CAB 4D), where auxins are not required for the initiation of division. Protoplast culture media for several top-fruit tree species are given in Table 3.

3. Protoplasts of fruit tree species handled in this way will typically enter division earlier than untreated material, with an 8-fold increase in the throughput of microcalluses. The frequency of regeneration, the number of shoot buds produced/callus, and the subsequent rooting capacity of the regenerated shoots are also enhanced.

References

1. James, D. J. (1987) Cell and tissue culture technology for the genetic manipulation of temperate fruit trees, in *Biotechnology and Genetic Engineering Reviews* vol. 5 (Intercept Ltd., Dorset, UK), 33–79.
2. Kobayashi, S., Uchimiya, H., and Ikeda, I. (1983) Plant regeneration from Trovita orange protoplasts. *Japan. J. Breed.* **33,** 119–122.
3. Ochatt, S. J. and Caso, O. H. (1986) Shoot regeneration from leaf mesophyll protoplasts of wild pear (*Pyrus communis* var *pyraster* L.). *J. Plant Physiol.* **122,** 243–249.
4. Ochatt, S. J., Cocking, E. C., and Power, J. B. (1987) Isolation, culture and plant regeneration of Colt cherry (*Prunus avium* x *pseudocerasus*) protoplasts. *Plant Sci.* **50,** 139–143.
5. Ochatt, S. J. and Power, J. B. (1988) An alternative approach to plant regeneration from protoplasts of sour cherry (*Prunus cerasus* L.). *Plant Sci.* **56,** 75–79.
6. Oka, S. and Ohyama, K. (1985) Plant regeneration from leaf mesophyll protoplasts of *Broussonetia kazinoki* (Sieb. paper mulberry). *J. Plant Physiol.* **119,** 455–460.
7. Schöpke, C. Müller, L. E., and Kohlenbach, H. W. (1987) Somatic embryogenesis and regeneration of plantlets in protoplast cultures form somatic embryos of coffee (*Coffea canephora* P. ex. Fr.). *Plant Cell Tissue Organ Culture* **8,** 243–248.
8. Vardi, A., Spiegel-Roy, P., and Galun, E. (1982) Plant regeneration from Citrus protoplasts: variability in methodological requirements among cultivars and species. *Theor. Appl. Genet.* **62,** 171–176.
9. Kouider, M., Hauptmann, R., Widholm, J. M., Skirvin, R. M., and Korban, S. S. (1984) Callus proliferation from *Malus Xdomestica* cv "Jonathan" protoplasts. *Plant Cell Rep.* **3,** 142–145.
10. Ochatt, S. J. and Power, J. B. (1988) Rhizogenesis in callus from Conference pear (*Pyrus communis* L.) protoplasts. *Plant Cell Tissue Organ Culture* **13,** 159–164.
11. Revilla, M. A., Ochatt, S. J., Doughty, S., and Power, J. B. (1987) A general strategy for the isolation of leaf mesophyll protoplasts from deciduous fruit and nut tree species. *Plant Sci.* **50,** 133–137.

12. Ochatt, S. J. (1987) Coltura di protoplasti come metodo per il miglioramento genetico nelle piante da frutto. *Frutticoltura* **49,** 58–60.
13. Ochatt, S. J. and Power, J. B. (1989) Selection of salt/drought tolerant Colt cherry plants (*Prunus avium* x *pseudocerasus*) from protoplasts and explant-derived tissue cultures. *Tree Physiol.* (in press).
14. Rech, E. L., Ochatt, S. J., Chand, P. K., Power, J. B., and Davey, M. R. (1987) Electro-enhancement of division of plant protoplast-derived cells. *Protoplasma* **141,** 169–176.
15. Ochatt, S. J., Chand, P. K., Rech, E. L., Davey, M. R., and Power, J. B. (1988) Electroporation-mediated improvement of plant regeneration from Colt cherry (*Prunus avium* x *pseudocerasus*) protoplasts. *Plant Sci.* **54,** 165–169.
16. Murashige, T. and Skoog, F. (1962) A revised medium for rapid growth and bio-assays with tobacco tissue cultures. *Physiol. Plant.* **15,** 473–497.
17. Kao, K. N. and Michayluk, M. (1975) Nutritional requirements for growth of *Vicia hajastana* cells and protoplasts at very low population density in liquid media. *Planta* **126,** 105–110.

Chapter 21

Fertilization In Vitro and Culture of Fertilized Ovules of Higher Plants

Gerard C. Douglas

1. Introduction

Sexual reproduction in higher plants consists of a series of complex processes, including pollination, double fertilization, embryo differentiation, and seed formation. Development of methods for pollination and fertilization under controlled conditions in vitro has greatly facilitated the study of these processes (*1*). In addition, manipulation of reproductive processes is possible by these methods. Thus, plants have been obtained following self-pollination in vitro with self-incompatible species, and hybrid plants have been obtained from interspecific and intergeneric crosses (*2,3*). Utilization of in vitro methods for wide hybridization has also provided a method for induction of parthenogenesis and development of haploid plants.

From: *Methods in Molecular Biology*, vol. 6, *Plant Cell and Tissue Culture*
Edited by Jeffrey W. Pollard and John M. Walker, ©1990 by The Humana Press

This chapter will describe methods used to achieve fertilization in vitro by culturing and pollinating whole florets of *Trifolium* spp. (4,7). Simple modification of the techniques allows application to *Nicotiana* spp. and *Petunia* spp. For culture of fertilized ovules, a method for obtaining interspecific hybrids of *Nicotiana rustica* x *N. tabacum* is described, which may be applied generally to other crosses in the genus *Nicotiana* (5).

2. Materials

2.1. *In Vitro Fertilization of* Trifolium repens

1. Seeds of *T. repens* or stolons collected from established plants in pastures.
2. Potting compost consisting of two parts fertilized peat with one part sand. Ensure compost is well moistened prior to seed sowing.
3. Glasshouse, temperature 10–20°C (supplementary lighting in the form of high-pressure sodium lamps is beneficial for growth in spring and autumn).
4. B5 (*see* Appendix) culture medium without growth regulators and modified by substitution of Sequestrene Fe 330 (Geigy) at 40.0 mg/L for Fe-EDTA (6). Add sucrose at 30 g/L and agar at 8 g/L and adjust pH to 5.8 prior to autoclaving.
5. A cold store (0–4°C) with supplementary lighting to supply a photon fluence rate of approximately 30 μmol/m^2/S.
6. A fresh solution of calcium hypochlorite prepared by placing 70 g of chemical in 1.0 L of distilled water, stirring for 15 min, followed by decanting and filtering through two layers of filter paper.
7. A culture room maintained at 22–25°C with a 16 h photoperiod provided by cool, white fluorescent tubes giving approximately 63 μmol/m^2/S at bench level.
8. Small weighted plastic baskets, binocular dissecting microscope, scalpels, tweezers, parafilm, and laminar airflow cabinet.

2.2. *Culture of Fertilized Ovules of* N. rustica *x* N. tabacum *and Other* Nicotiana *Species*

1. Seeds of *Nicotiana tabacum* L. and *N. rustica* L.
2. Seed compost, glasshouse (15–25°C) culture room, calcium hypochlorite, B5 medium, and other equipment as listed in points 2–8 above.

3. Methods

3.1. Method 1: In Vitro Fertilization of Trifolium repens

1. Seeds of *T. repens* consist of a variable percentage of "hard" seeds. For uniform germination, scarify seeds by placing them between two sheets of fine emery paper and gently rubbing them prior to sowing in 10-cm pots; cover pots with glass and opaque material, e.g., black plastic. Water pots from the base by placement in a deep saucer of water at a temperature of 8–15°C. Seeds will emerge within 2 wk, and may be transferred to 5-cm pots after 3 wk. Repot into 10- or 15-cm pots after 4 wk.
2. Grow plants in the greenhouse for a further 8–12 wk, with supplementary lighting to facilitate vigorous vegetative growth in early spring and autumn.
3. To induce flowering in plants of *T. repens* raised from seed in the spring or summer, a period of 8–12 wk of low temperatures (–1.0 to + 5.0°C) with short days (<8 h) is required. This is followed by transfer of plants to temperatures of 10–20°C with long days (>12 h), and flowering occurs after 8–12 wk. Alternatively, stolons collected in early spring from plants that were established in pasture do not require an additional period of vernalization to induce flowering. Stolons (5–10 cm in length) may be collected from established pasture and transferred to 1-cm pots, where rooting and establishment occurs. Flowering should occur after 12–16 wk of vegetative growth in the glasshouse.
4. Prepare modified B5 culture medium (*6*). Dispense medium (6 mL) into 50 x 20-mm Petri dishes.
5. Florets that are to provide pollen as the male partner of the cross are selected from the lower part of the flower head (Fig. 1A). Collect florets in which the keel petal is almost fully open. Dissect florets with aid of a clean scalpel, tweezers, and microscope by making an incision at the base of the floret in the sepals and through the petals (Fig. 1C). Open the floret to reveal the anthers connected near the base in a tube. This tube may be excised and removed with anthers attached. Anthers should have dehisced and show golden colored pollen.
6. *T. repens* is self-incompatible, and to achieve fertilization and seed formation, it is necessary to collect florets of the female partner and

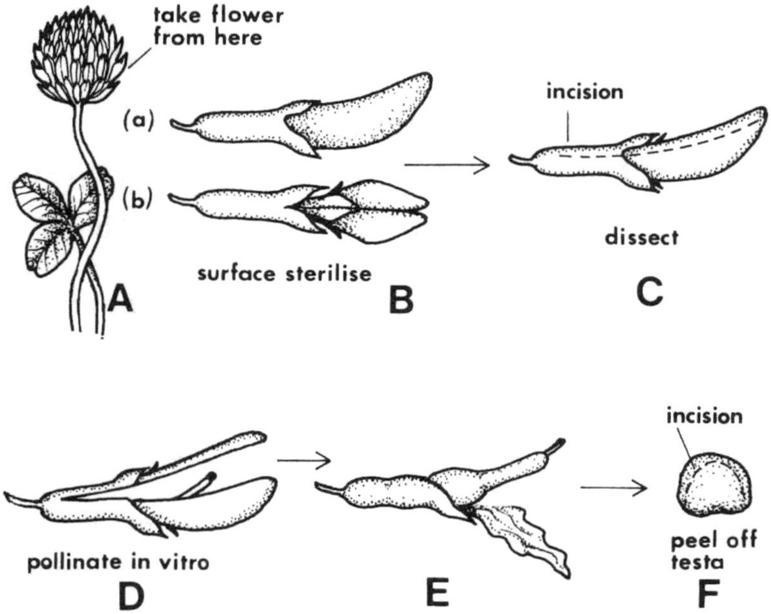

Fig. 1. Selection of florets of *Trifolium repens* (A,B), dissection of florets for in vitro pollination (C,D), and recovery of seed from fertilized ovules (E,F).

pollen of the male partner from different plants. Florets to be used as the female partner in the cross are selected at a more immature stage than those described (in step 5 above). Suitable florets in which the keel petal is just beginning to open are illustrated in a side (Fig. 1B,a) and top (Fig. 1B,b) view. In these florets, anther dehiscence will occur in the succeeding 24 h period (approximately). Take care to excise as much as possible of the floret stalks attached to each floret. Transfer florets to a small plastic histology basket for surface sterilization, performed by immersion for 20 min in a solution of calcium hypochlorite prepared by mixing 1 part of the 7% hypochlorite solution with 8 parts of distilled water. Agitate florets occasionally during this period. Then decant the hypochlorite solution and provide three rinses in sterile water (5 min/rinse). Asceptic conditions are provided during this and subsequent stages. Remove florets from the plastic baskets and transfer them to culture dishes, two per dish.

7. Pollination of each floret is performed separately. For pollination, transfer each floret to the lid of the Petri dish, and carefully make an incision, as illustrated, along the length of the floret (Fig. 1C). The

stigma and style may then be gently pushed out through the incision (Fig. 1D). It is not necessary to remove anthers, since *T. repens* is self-incompatible. The stigma should have a slightly moist appearance.

8. Transfer pollen to the stigma by rubbing anthers (collected as described in step 5 above) on the surface. Pollen should be applied to cover the stigmatic surface liberally.
9. Insert the stalk of the pollinated floret into the culture medium. Replace the lid on the culture dish, wrap with parafilm, and transfer to the culture room.
10. After 2 wk of culture, petals will have turned brown and withered, but sepals, flower stalk, and ovary will remain green and turgid (Fig. 1E). Transfer pollinated floret to the lid of the culture dish and asceptically dissect out the fertilized ovules. Care should be taken not to make deep incisions that will damage the immature seeds. Record the number of immature seeds present (usually 1–2/ovary), and carefully excise and remove the testa (outer covering) from each seed (Fig. 1F). This is essential to promote seed germination. Transfer these testaless seeds to fresh medium.
11. Germination of seeds begins after (approximately) 7 d. Culture the seedlings for a further 3–4 wk. Transfer the resultant plantlets to half strength medium for a further 3–4 wk prior to transfer to compost.
12. Transfer the plantlets to compost in a seed tray or pot. Maintain high humidity for 2 wk by covering the plantlets with plastic film, and by overhead shading. Thereafter, remove the plastic and transfer the plants to 5-cm pots.

3.2. Method 2: Culture of Fertilized Ovules from Hybridization of Nicotiana rustica x Nicotiana tabacum

1. Sow seeds in 10-cm pots in a glasshouse or phytotron at a temperature of 15–25°C, and cover with glass and opaque material, e.g., black plastic. Place pots in deep saucers, so that water rises by capillary action. Remove covers when seeds have germinated (10–15 d) and transplant seedlings into 5-cm pots. Transplant once or twice more.
2. Both *N. tabacum* and *N. rustica* are day neutral. They should be kept in a state of vigorous vegetative growth for at least 3–4 mo prior to flowering. This may require supplementary illumination during the

winter months. Flowering occurs after 3–4 mo growth for *N. tabacum,* and 4–5 m for *N. rustica.*

3. Prepare culture medium as described in Materials, but use sucrose at 40.0 g/L instead of 30.0 g/L.
4. Remove all immature seed capsules and opened flowers from *N. rustica* plants. Select flowers for emasculation. Flowers in which anther dehiscence and self-pollination would occur within 24 h are typified by a fully extended corolla (Fig. 2A). In such flowers, the distal end of the corolla shows widening prior to opening of the flower. It is best to select flowers just before this widening is evident. Emasculate flowers by slitting the corolla with a pointed forceps and removing all six anthers from their filaments (Fig. 2A). Pollinate 24 h later with pollen collected from *N. tabacum.*
5. For pollination, collect freshly ruptured anthers of *N. tabacum* and transfer pollen to the stigmas of emasculated flowers (Fig. 2A) of *N. rustica.*
6. Three days after pollination, excise flowers and flower stalk, carefully remove the sepals and corolla, and surface sterilize flower stalks with attached ovaries by immersion in 70% ethyl alcohol (1min) followed by immersion in 7% calcium hypochlorite for 20 min and by three rinses (5 min each) in sterile distilled water (Fig. 2B,C).
7. Holding the ovary at the flower stalk with tweezers, asceptically remove the ovary wall by making shallow longitudinal incisions along the edges of each of the two locules and around the base of the ovary (Fig. 2C). With a scalpel, excise the placentas bearing fertilized ovules from each of the locules, and transfer them to the culture medium (Fig. 2D).
8. Hybrid seedlings germinate on the placenta, and the number may be recorded after 3 wk of culture (Fig. 2D). A total of 25–50 seedlings may be expected for each flower pollinated (12–25/placenta) from interspecific crosses of *N. rustica* x *N. tabacum.*
9. Transfer seedlings to fresh medium containing 2% w/v sucrose to allow plantlet development. Transfer plantlets to the greenhouse when 2–3 leaves have expanded, and maintain high humidity by enclosing plantlets with a plastic bag or plastic film. Gradually acclimatize plants by making successively larger holes in the plastic over a 2-wk period.
10. Allow plants to flower, and observe morphological characters of flowers and leaves in hybrids derived from crosses.

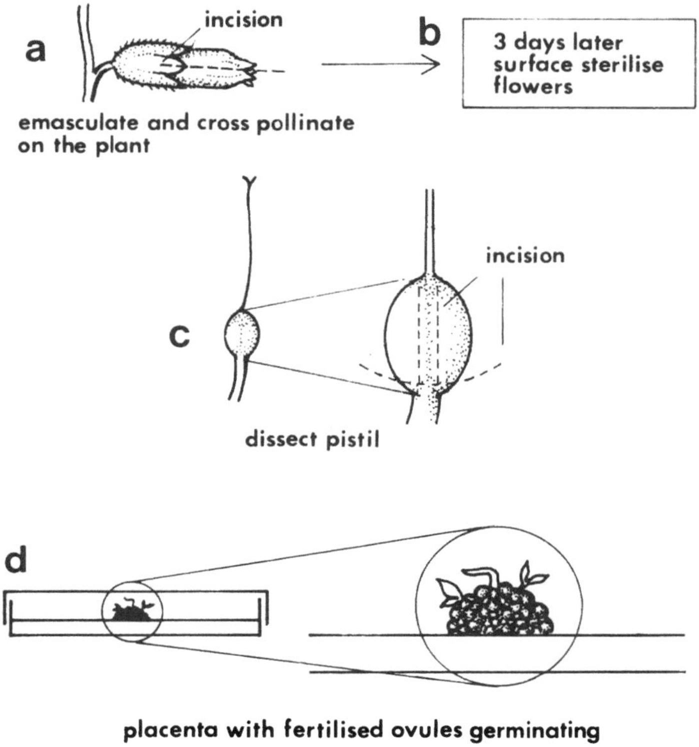

Fig. 2. Selection of flower of *Nicotiana rustica* for emasculation and pollination (A,B), dissection of ovary and culture of placentas bearing fertilized ovules (C,D).

4. Notes

4.1. Method 1

1. Florets selected as female parents of the cross should be excised with as much flower stalk as possible, and at the correct stage of development (Fig. 1B; Method 6.). Immature and older florets generally fail to set seeds. Florets with longer floret stalks may be obtained from flower heads that are already well opened.
2. The main precaution is to avoid injury to the ovary during handling and dissection procedures (Fig. 1C,D). Minor damage, such as a small incision in the ovary, will cause florets to die and fertilized ovules to abort.
3. The procedure described may also be successfully applied to other *Trifolium* species, e.g., *T. pratense*. With *T. pratense*, it is not necessary to provide a cold period to induce flowering (Method 3.). Plants may

be raised and grown continuously in the glasshouse where flowering occurs after approximately 3–4 mo. However, *T. pratense* has shorter floret-stalks than *T. repens* and each ovary contains fewer ovules; these factors may contribute to a lower yield of seedlings in crosses.

4. Although, *T. repens* is self-incompatible, progeny may be obtained and the self-incompatibility mechanism bypassed by following the procedure described. For self-pollination, both pollen and florets that are to be self-pollinated are collected from the same plant (7).
5. Growth of pollen tubes in the style and their entry to the ovule via the micropyle may be observed by fluoresence microscopy. From 6–20 h after pollination, collect florets, remove the corolla and calyx, and transfer the pistil (stigma style and ovary) to a solution of 1.0*M* NaOH for 30 min in order to soften tissues. Rinse 3x with distilled water, and transfer pistils to a drop of aniline blue solution on a slide. Aniline blue is prepared by dissolving 0.2 g of the stain in 100 mL of K_3PO_4. Place a coverslip on the specimen, squash gently, and examine with UV or blue light. Pollen tubes and plugs of callose are readily distinguishable by their bright fluoresence.

4.2. Method 2

1. In order to synchronize flowering in crosses of *N. rustica* x *N. tabacum*, sow seeds of *N. rustica* 3–4 wk prior to *N. tabacum*.
2. The method can be readily adapted for culturing self-fertilized ovules of *N. rustica*, *N. tabacum*, and *Petunia parodii*. For all of these species, it is important to remove all open flowers and immature seed capsules, and to select only 3–5 flowers for experimentation at any one time. Ensure that pollination occurs by observing dehiscence of anthers and by transfering pollen from anthers to the stigma, using a fine brush or wooden cocktail stick.
3. Fertilized ovules are larger than nonfertilized ovules and are distinguishable 3–5 d after pollination. Culture of placentas (with attached fertilized ovules) directly on the culture media is a convenient procedure (Fig. 2D). However, a doubling in the yield of seedlings may be obtained by removing the fertilized ovules from the placentas and placing them directly on the culture medium (5).

References

1. Zenkteler, M. (1980) Intraovarian and *in vitro* pollination, in *Perspectives in Plant Cell and Tissue Culture*, (Vasil, I. K., ed.), *Intern. Rev. Cytol.* **Suppl. 11B,** Academic Press, New York, pp. 137–156.

2. Zenkteler, M. and Slusarkiewicz-Jarzina, A. (1986) Sexual reproduction in plants by applying the method of test tube fertilization of ovules, in *Genetic Manipulation in Plant Breeding* (Horn, W., Jensen, C. J., Odenbach, W., and Schieder, O., eds.), Walter de Gruyter, Berlin, New York, pp. 415–423.
3. Stewart, J. McD. (1981) In vitro fertilization and embryo rescue. *Environ. Exp. Bot.* **21**, 301–315.
4. Richards, K. W. and Rupert, E. A. (1980) In vitro fertilization and seed development in *Trifolium*. *In Vitro* **16**, 925–931.
5. Douglas, G. C., Wetter, L. R., Keller, W. A., and Setterfield, G. (1983) Production of sexual hybrids of *Nicotiana rustica* x *N. tabacum* and *N. rustica* x *N. glutinosa* via *in vitro* culture of fertilized ovules. *Z. Pflanzenzuchtg* **90**, 116–129.
6. Gamborg, O. L., Miller, R. A., and Ojima, K. (1986) Nutrient requirements of suspension cultures of soybean root cells. *Exp. Cell. Res.* **50**, 151–158.
7. Douglas, G. C. and Connolly, V. (1989) Self fertilization and seed set in *Trifolium repens* L. by in situ and in vitro pollination. *Theor. Appl. Genet.* **77**, 71–75.

Chapter 22

Vegetative Propagation of Cacti and Other Succulents In Vitro

Jill Gratton and Michael F. Fay

1. Introduction

Maintenance of collections of succulent plants can be problematic, since many of these species are very susceptible to rots caused by bacteria and fungi. Rooting and establishment of cuttings can also be difficult. At Kew, methods for the micropropagation of cacti and other succulents have been developed over the last 10 yr. These have proved to be very useful for overcoming the problems mentioned above.

The techniques have been used most successfully with species in the *Asclepiadaceae*. For example, healthy plants have been produced from small pieces of unhealthy tissue in the genera *Caralluma, Huernia, Stapelia,* and *Ceropegia*. Some other genera, notably *Hoodia,* have proved intractable using the methods described below.

Cacti have proved to be more difficult. This is probably largely because of the structure of the plants. The meristematic area is hidden in the areole tissue. Removal of woody spines can damage this tissue or allow it to be damaged by the sterilant. Successes have, however, been achieved, and include species of *Opuntia* and *Mammillaria*.

From: *Methods in Molecular Biology*, vol. 6, *Plant Cell and Tissue Culture*
Edited by Jeffrey W. Pollard and John M. Walker, ©1990 by The Humana Press

Succulent *Euphorbia* spp. are also difficult, but several species have been micropropagated. Species with woody spines seem to be particularly problematic. Succulent species in *Aizoaceae* and *Crassulaceae* have also been propagated. The methods will not be described here, since they are based on those used for the *Asclepiadaceae*.

Micropropagation can be of particular value in the propagation of rare and endangered species, especially where viable seeds are not available. Bulk propagation and subsequent distribution of those species that are endangered by overcollection may alleviate pressure on the wild populations.

2. Materials

1. Media: Murashige and Skoog (MS) medium (*1* and *see* Appendix) is obtained as a ready-mixed powder, without sucrose, growth regulators, or agar. It is made up at the standard concentration with 30 g/L sucrose and a range of growth regulators, which are added before autoclaving. These are detailed in the Methods section.

   Adjust the pH of the medium to 5.6–5.8 using dilute NaOH or HCl as appropriate. Dissolve 9 g/L of Oxoid No. 1 agar in the medium using a microwave oven (1 min for each 100 mL of medium), and stir thoroughly. Dispense the medium into the required vessels and sterilize in the autoclave for 15 min at 121°C and 1.05 kg/cm^2. If the medium is not to be used immediately, store the vessels in plastic bags at 4°C.
2. Compost: Open, free-draining composts are used, since these have been shown to promote rooting and to decrease the losses from rotting off at the weaning stage. The most frequently used compost is composed of: 3 parts fine loam, 2 parts peat, 1 part sharp grit, 6 parts calcined montmorillonite (particle size up to 4 mm), and 6 parts Perlite. To this compost, slow-release fertilizer is added as required. Calcined montmorillonite possesses good absorption properties. This and Perlite make the compost considerably more open, thus aiding establishment.

3. Methods

The techniques to be described in this chapter have been developed at Kew in an attempt to make the best possible use of what little material is normally available. The basic principle on which the method is based is the induction of shoot formation from dormant meristems in the areole of cacti or, in succulents, from axillary buds, using media containing cytokinins, alone or in combination with low concentrations of auxins.

3.1. Initial Preparation of Tissue

1. Excise and discard dead or rotting material.
2. If the material is infested with pests, dip it in industrial methylated spirits for 10–30 s, and then wipe it with a soft tissue or paintbrush.
3. Remove any soil adhering to the plant material by washing under running water or agitating in water containing a few drops of Tween 80. A fine paintbrush can be useful for dislodging debris that is trapped by hairs or spines.
4. Remove any damaged tissue, taking care not to damage the underlying tissue.
5. Woody spines and hairs found in cacti and succulent *Euphorbia* spp. are difficult to surface sterilize effectively. They are therefore removed, using watchmaker's forceps, fine-pointed scalpel blades, or hypodermic needles (*see* Note 1). Great care must be taken not to damage the tissue underneath the spine, since the buds often lie very close to the base of the spines. Where the spines are less woody, they can be cut back as close as possible to the base, rather than being totally removed.

3.2. Surface Sterilization

1. The pieces of plant tissue should be left as large as possible for surface sterilization, to minimize damage to the healthy tissue by the sterilant. A range of sterilization times and sterilant concentrations are used where possible, but this depends on the amount of tissue available. Sodium hypochlorite (BDH, 10–14% available chlorine) is used as the sterilant, at concentrations ranging from 3–10% v/v in deionized water, with the addition of a few drops of Tween 80/L. Sterilization times range from 5–15 min. The most commonly used regime is 5% sodium hypochlorite plus Tween 80 for 10 or 15 min. Surface sterilization is improved if the beaker is put onto a magnetic stirrer set at a speed high enough to keep the plant tissue submerged, but not so high that it will damage the plant tissue.
2. Rinse the material 3x in sterile deionized water.

3.3. Preparation of the Surface Sterile Material for Culture

The size of the explant used for culture varies greatly, depending on the structure of the plant and the extent to which any rot present has advanced. The general procedures for each of the groups of plants are described below.

3.3.1. Asclepiadaceae

In asclepiads, axillary buds are found in the axils of the vestigial leaves. The buds are often visible to the naked eye. Remove these buds with some surrounding tissue, using a scalpel and forceps. The first cut should be made directly below the vestigial leaf and the second as far above the bud as the structure of the plant will allow. A vertical cut made behind these cuts will separate the required explant from the main body of the plant.

In species where the stems are very ridged and the rudimentary leaves are very close together, make vertical cuts between the ridges as far away from the buds as possible. In these species, we have found it better to use pieces of tissue with several rudimentary leaves for initiating cultures, rather than trying to separate out the individual buds.

In *Ceropegia* spp., nodal sections are taken. Where there are two buds present at the node, make a vertical cut through the node and culture the buds separately.

3.3.2. Cactaceae

Where the areoles are widely spaced, remove them with 3–5 mm of the surrounding tissue. In *Mammillaria* and other genera with tubercles, remove the areoles with as much of the tubercle tissue is possible. In cacti where the areoles are close together, make vertical incisions into the tissue, taking care not to cut right through, and excise small pieces consisting of several areoles.

3.3.3. Euphorbiaceae

These are cut up into explants similar to those from asclepiads, after spines have been removed where necessary.

1. Place the individual explants in boiling tubes containing 10–15 mL of medium, with the cut surface of the explant in contact with the medium, and the bud or areole facing upwards. Where sufficient tissue is available, explants are cultured on a range of semi-solid media

based on MS with added growth regulators. The most commonly used growth regulator levels used are 1 or 2 mg/L BAP (benzylaminopurine) ± 0.1 mg/L NAA (naphthylacetic acid) (*see* Note 2).
2. Maintain the cultures in a culture room at 22–25°C with a 16-h photoperiod (light intensity 1000–4200 lx) (*see* Note 3).
3. Examine the cultures after a few days for signs of contamination. Where no more material is readily available, it has proved possible in some cases to carry out a second surface sterilization if contamination has occurred. Solutions of 5 or 7.5% BDH sodium hypochlorite have been used for 5–30 min, depending on the type of material, but it is difficult to give more specific instructions.
4. After a variable length of time on the media described above, axillary buds on some of the explants will begin to grow out (*see* Note 4). When these have grown sufficiently (to 2–3 cm in asclepiads), cut them into sections, each with one or a few buds, and place them onto fresh medium in 1/2- or 1-lb honey jars. These sections and the stump on the original explant should continue to produce lateral shoots, thus establishing a proliferating system. Continue subculturing in the same way at 1–2 mo intervals until sufficient shoots have been obtained.
5. When this stage is reached, rooting in vitro is attempted. Place excised shoots onto a range of semi-solid media based on MS ± growth regulators. The most commonly used media for rooting are MS with: (a) no growth regulators; or (b) 0.01, 0.1, 0.25, 0.5, or 1.0 mg/L NAA, respectively. Alternative media are detailed in Note 5.
6. When the plantlets are well rooted, tease them out of the medium gently, and wash them carefully under running water to remove traces of agar.
7. Pot the plantlets in 2-in square pots with 1–4 plants/pot, using the compost described in the Materials (*see* Note 6). Place the pots on a bed of sand with under heating (if available) in a dry intermediate glasshouse (21°C day, 16°C night).
8. Where rotting off is a problem, fungicide drenches (benlate or captan) have proved useful. These plants can take up to 6 mo before they are fully established and begin to grow vigorously.

4. Notes

1. In cacti, the removal of spines can result in damage to the dormant buds located in the areole. In particularly sensitive species, steril-

ization with 0.1% mercuric chloride for 2–10 min has obviated the necessity of removing the spines prior to sterilization. As a result, the sensitive bud tissue is not exposed to damage by sodium hypochlorite.

Mauseth and Halperin (2) recommended singeing the spines on cacti instead of removing them. This has been tried with little success.

A major problem with removing spines can be that the newly exposed tissues are very susceptible to damage by the sterilant. Spines in some species of *Euphorbia* can be removed after sterilization. Where there is very little tissue available, the cut ends can be sealed with paraffin wax, thus minimizing the damage.

2. With some plants, it has proved necessary to use media with higher cytokinin concentrations to initiate proliferating cultures. The most frequently used concentrations are 5 or 10 mg/L BAP with 0.1 or 0.5 mg/L NAA.
3. During the initiation process with *Asclepiadaceae,* compact callus sometimes forms at the base of explants. Remove this at each subculture. If the callus threatens to take over the culture, transfer the tissue to a medium with lower growth regulator concentrations.
4. The pathway for cacti described in the Methods section is the ideal, but in certain species of cacti (e.g., some *Mammillaria* spp.), a callus is sometimes formed before shoots are regenerated.

   With some cacti, there is a tendency for vitrification to take place. Techniques for overcoming this that can be tried include:
   a. Incorporation of 1–3 g/L activated charcoal.
   b. Use of media of lower ionic strength (e.g., half-strength MS);
   c. Use of media with higher agar concentration (e.g., 1.4%).
   d. Allowing the medium to partially dehydrate before transferring the cultures.
5. Where rooting will not take place on any of the media listed above, the following variations have been used with some success:
   a. IAA (indoleacetic acid) can be used at the same concentrations in place of NAA.
   b. NAA can be used at higher concentrations (5–20 mg/L).
   c. Half-strength MS ± NAA or IAA can be used.
6. Where dehydration is a problem on weaning, steps must be taken to stop the plantlets dying. Keep the plants under a clear propagator lid until they have become acclimatized to the lower humidity. Gradually lower the humidity by opening the vents in the lid, and eventually remove the lid altogether.

With particularly difficult species, we have had some success using an intermediate stage between agar and pot culture. Sterilize the standard compost in honey jars, and then add sterile water or mineral salt solutions (e.g., half-strength MS without sucrose or agar). Then "pot" the plantlets in this and seal the vessels as normal. When the plantlets appear to have become established, transfer the jars to the glasshouse, gradually loosen the tops over a few days, and then pot the plantlets normally.

References

1. Murashige, T. and Skoog, F. (1962) A revised medium for rapid growth and bioassays with tobacco tissue cultures. *Physiol. Plant* **15,** 473–497.
2. Mauseth, J. D. and Halperin, W. (1975) Hormonal control of organogenesis in *Opuntia polyacantha* (Cactaceae). *Amer. J. Bot.* **62,** 869–877.

Chapter 23

The Preparation of Micropropagated Plantlets for Transfer to Soil Without Acclimatization

Andrew V. Roberts, Elaine F. Smith, and John Mottley

1. Introduction

Plantlets cultured in vitro on agar-based media in a water-saturated atmosphere wilt rapidly when transferred to normal greenhouse or field conditions. Water is rapidly lost from the leaves because stomata fail to respond to those stimuli that normally induce closure (*1–4*), and poor development of epicuticular wax results in loss of water through the cuticle (*5–7*). Uptake of water by the roots is limited by damage incurred during transplantation and by poor contact with the substrate. Problems of transplantation are accentuated in vitrified plantlets, which grow slowly and wilt rapidly. Reduced deposition of cellulose and lignin in these plantlets causes reduced cell wall pressure, leading to increased water uptake by the cells and a glassy turgescence of leaves and stems (*8,9*).

From: *Methods in Molecular Biology*, vol. 6, *Plant Cell and Tissue Culture*
Edited by Jeffrey W. Pollard and John M. Walker, ©1990 by The Humana Press

Sucrose is included in culture media to promote rapid growth, but leads to low levels of photosynthesis in the leaves (*10*). This has been attributed to low levels of chlorophyll and ribulose bisphosphate carboxylase activity (*10,11*). Leaves that were formed in vitro do not recover from this condition after transplantation, and continued growth of the plantlet depends on the formation of new leaves (*12,13*).

In order to maximize survival after transplantation, it is common practice to acclimatize the plantlets in a humid environment (*14*). Acclimatization over a period of 2–4 wk may be needed, depending on the rates at which new leaves and roots are produced. The cost of acclimatization in terms of capital items and labor is considerable, and must also be reckoned in terms of lost opportunities for distribution of plantlets in vitro to customers without such facilities. It is now possible to identify a number of factors that may be modified during in vitro culture to improve the fitness of plantlets for transfer to soil. Improved stomatal physiology and production of epicuticular wax occurs if the relative humidity of the culture vessel is reduced (*15–18*). A strengthening of shoots and roots has been reported in response to paclobutrazol and other growth retardants (*19,20,17*). In the case of paclobutrazol, an improvement in stomatal physiology, an increase in the deposition of epicuticular wax, an increase in chlorophyll/unit area of leaf, improved differentiation of the palisade mesophyll, and reduced wilting in response to water stress have also been reported (*17*). Damage to the root system can be reduced, and resistance to wilting significantly increase, by the use of cellulose plugs in which plantlets are rooted in vitro and transferred to the soil (*17,15*). Vitrification has been controlled in various species by reducing levels of one of the following: NH_4^+ (*21*), cytokinins (*8*), water potential of the substrate (*22,23*), relative humidity of the atmosphere (*23*), and production of ethylene by the plantlets (*24*).

Murashige (*25*) distinguished three stages of micropropagation:

Stage I: Establishment of the aseptic culture.

Stage II: Multiplication of propagula.

Stage III: Preparation for reestablishment of plants in soil (including rooting of shoots).

This is followed by acclimatization, which may be regarded as Stage IV (*26*). Debergh and Maene (*27*) proposed that rooting should be postponed until after transfer to soil, and subdivided Murashige's Stage III into:

Stage IIIa: The elongation of the buds into shoots and the preparation of uniform shoots for Stage IIIb.

Stage IIIb: The rooting and initial growth of the in vitro produced shoots under in vivo conditions.

Plantlets can be hardend at Stage IIIa by placing culture vessels on a cooled shelf, so that condensation is directed towards the bottom of the flask and the humidity of the culture vessel is reduced. This results in reduced water loss from the shoots and improved survival rates during acclimatization (*28*).

The method described in this chapter enables a reversion to Murashige's original scheme (*25*), but without the requirement for subsequent acclimatization. It is based on the use of a culture vessel that has a bacterial filter inserted into the lid to reduce humidity, and that also contains cellulose plugs (Sorbarods) in which paclobutrazol is incorporated. The method is ideally used in conjunction with bottom cooling (*28*), but is not dependent on this. Plantlets that are produced by the successful application of the method do not wilt after transplantation directly to soil under greenhouse conditions. Techniques for analysis, diagnosis and amelioration of potential problems are also presented.

2. Materials

1. It is assumed that facilities for in vitro culture of plantlets are available, including a laminar-air-flow cabinet, autoclave, culture room with lighting (16-h photoperiod, 25–100 µE/m^2/s), and air conditioning to maintain steady temperatures (± 0.5°C) within the range 23–28°C. It is also assumed that plantlets will have been multiplied in vitro before the recommended procedures are initiated, and that an appropriate rooting medium has been selected.
2. Trays and lids are supplied by Baumgartner Papiers SA, Lausanne, Switzerland, in twin packs. One pack contains four trays and the other pack contains five lids plus one tray (to stabilize the stack of lids). Both packs are sterilized by gamma radiation. Each tray contains 60 Sorbarods (cellulose transplantation plugs), held in place and sealed with a peel-off cover. The Sorbarods are cylindrical, measuring 20 mm in height and 18 mm in diameter, and consist of a cold-crimped cellulose paper wrapped with porous cellulose paper, on the model of a cigarette filter. The vessel is prepared by pressing a lid onto the tray. The lid remains firmly attached to the tray if lifted vertically, but can be removed by twisting from a corner. Paclobutrazol is incorporated in the Sorbarods at various levels during manufacture so that, when

300 mL (295 g) of rooting medium is added, paclobutrazol will reach a final concentration of 0.5, 1.0, or 2.0 mg/L (*see* Note 1).

3. Liquid culture media (no gelling agent) should be used. This permits the attractive option of filter sterilizing rather than autoclaving the medium. A wide range of bacterial filters are commercially available, the design and price of which depend upon the volume of medium that is to be prepared. The life of a bacterial filter (0.22-µm pores) will be extended if the medium is first passed through a 0.8-µm filter to remove larger particles.
4. Culture room shelving: The use of cooled shelves is ideal (*28*), but is not essential. Shelves used for cooling of beer or wine in restaurants and pubs are commercially available. Alternative systems can be custom-built by refrigeration engineers. If cooled shelves are not used, care should be taken to ensure that shelves are not significantly heated by lighting below. Suitable insulation can overcome this problem.
5. Acidic Potassium dichromate ($K_2Cr_2O_7$) for wax measurement: Add deionized water (20 mL) to powdered $K_2Cr_2O_7$ (10 g), then concentrated sulfuric acid (500 mL) and heat (below boiling and stirring vigorously) until a clear solution is obtained.

3. Methods

1. The rooting medium is prepared without the addition of a gelling agent, and may be sterilized either by autoclaving at 121°C or by the use of a bacterial filter. The period of autoclaving may depend on the volume used and the shape of the vessel. Fifteen minutes is usually adequate for 1 L of medium in a 2 L flask, but larger volumes or smaller vessels may require a longer period. Filter sterilized culture medium may be dispensed directly into the culture vessel or first stored in sterile screw-capped flasks. Measured quantities can be dispensed into the tray of the culture vessel (containing Sorbarods) on the basis of volume, or the tray (plus Sorbarods) may be placed on a balance and medium decanted on the basis of weight. After the medium has been dispensed, propagula can be inserted into the plugs and the vessel closed.
2. During culture maintenance, the culture medium may be allowed to evaporate, without detriment to the plantlets, to a point where it is retained only by the Sorbarod. Relative dryness of the vessel can be an advantage, because aeration promotes the formation of root hairs.

Table 1
Classification of Plantlets at Fixed Interval, e. g., 4 h, After Transfer from Culture Vessel, According to Degree of Wilting on a Scale of 0, Turgid, to 5, Completely Wilted

	% plantlets showing specified degree of wilting					
Example	0	1	2	3	4	5
1[a]	88	8	3	1	0	0
2[b]	59	15	3	2	5	16
3[c]	0	0	3	1	16	80

[a]Plantlets responsive to treatment.
[b]Variable response, most probably the result of irregular rooting response (*see* Table 2).
[c]Unsatisfactory response (*see* Table 2).

If plantlets are threatened by desiccation, sterile distilled water can be added to the tray in the laminar-air-flow cabinet.

3. During transfer from the culture vessel to compost, grasp the Sorbarod (not the plantlet) in order to minimize damage to the plantlet. If the plantlet is well rooted in the Sorbarod and the compost is adequately watered, wilting should not occur within the normal range of conditions encountered in a commercial greenhouse. Misting should not be used, because saturation of the soil impairs root growth and encourages fungal attack.

3.2. Assessment and Interpretation of Wilting (17)

In developing a protocol for a new species, wilting should be assessed. Ideally, this should be carried out under the most severe conditions to which plantlets are likely to be exposed. Plantlets may be classified according to the degree of wilting and the data interpreted as shown in Table 1. In the event of an unsatisfactory response in the wilting test, Table 2 provides a guide to the diagnosis of problems and suggests remedial action. Where transplantation has been satisfactory, but further refinement of culture conditions is desired, an assessment of the ability of leaves to withstand desiccation (stomatal response to water stress and formation of epicuticular wax) or an assessment of lignification may be needed, and methods for measuring this are described below. Analysis of variance can be used to determine whether or not variation between treatments significantly exceeds variation within treatments.

Table 2
Diagnosis and Treatment of Conditions that Lead to Wilting

Indicators	Interpretation	Possible remedies
Slow growth in culture vessel	Only leaves that were formed after transfer to culture vessel and exposed to reduced humidity and paclobutrazol would be expected to show benefit of treatment. This problem may arise if rooting medium excludes cytokinin	Adjust rooting medium to include a weak cytokinin, e.g., kinetin
Weak stems	This may arise if endogenous levels of gibberellins are supraoptimal	Check that rooting medium excludes gibberellic acids. Use higher levels of paclobutrazol
Stems succulent and/or translucent	Plantlets may be vitrified. This would probably have been initiated during multiplication, and would probably have been apparent then	One or more of the following modifications of the multiplication medium may ameliorate the problem: reduce level of NH_4^+, reduce level of cytokinin, increase concentration of agar
Absence of roots	Adjustment of rooting medium is required	May be solved by the use of two auxins (*8*) or reduction in level of NH_4^+ (*21*). In woody plants, rooting may be improved by reduction of pH in range 3–4 (*29*)
Irregular rooting	Explants from multiplication medium vary in size	Select cuttings more rigorously
	Roots are present, but short	Extend period of culture before transplantation
Weak roots easily detached from stem, associated with callusing at base of stem	Roots that originate in callus may develop poor vascular connections with the stem	Adjust cytokinin: auxin ratio to reduce callus formation

3.2.1. Stomatal Aperture (17)

1. Plantlets should be exposed to a dry atmosphere to stimulate closure of stomata. Impressions of stomata are then made by applying nail varnish to the abaxial surface of a mature leaf.
2. After the nail varnish has dried, it is removed by firmly applying sellotape (scotch tape) to the hardened film, peeling the strip away with the film intact, and reapplying to a glass slide.
3. Stomatal apertures are then measured using a microscope with micrometer eyepiece that has been calibrated against a slide graticule.

3.2.2. Measurement of Wax (30)

1. Excise the leaf, immerse it in 100% chloroform (10 mL) for 15 s, and remove. Estimate the leaf area by photocopying or drawing the leaf on graph paper and counting 1-mm squares.
2. Filter the chloroform solution to remove particles and evaporate to dryness at 100°C in a water bath.
3. Add acidic $K_2Cr_2O_7$ (5 mL) and heat at 100°C in a water bath for 30 min.
4. Cool, add deionized water (12 mL), allow 5 min for color development, and then measure the absorbance at 590 nm.
5. To prepare a calibration curve, a concentrated solution of wax is prepared by dipping several leaves in 50 mL of chloroform and filtering. Take 25 mL of the filtrate and, in a beaker, evaporate to dryness at 100°C. Accurately weigh the wax by difference. Dispense volumes of 0.1–10 mL into separate beakers, and calculate the weight of wax in each beaker. Evaporate to dryness, carry out steps 3 and 4 above, and prepare a calibration curve of absorbance against weight.
6. Convert absorbance to weight for test samples by reference to the calibration curve and estimate wax as μg/mm^2 (both surfaces). As an example of differences that might be achieved between treatments, leaves taken from chrysanthemum cultured on a medium containing 4 mg/L paclobutrazol contained 15.4 μg/mm^2 (SE $\pm$ 0.7) epicuticular wax, whereas leaves from untreated controls contained 5.5 μg/mm^2 (SE $\pm$ 0.7) (*17*).

3.3. *Indicators of Photosynthetic Potential*

If plantlets show satisfactory resistance to wilting, but grow slowly after transplantation, consideration should be given to the photosynthetic ability of in vitro and subsequently formed leaves. Poor photosynthesis has been attributed to low levels of chlorophyll and low activity of ribulose bisphosphate carboxylase (*see* Note 2). These problems may be ameliorated at the rooting stage by a reduction in the level of sucrose in the culture medium (*10*). Analysis of variance can be used to determine whether or not variation between treatments significantly exceeds variation within treatments.

3.4. Estimation of Chlorophyll (31)

1. Remove two leaf disks of 10–15 mm^2 with a cork borer.
2. Place in 3 mL of 100% methanol in stoppered tubes and incubate at 23°C for 2 h in darkness.
3. Measure the optical density of the extract at 650 nm and 665 nm.
4. Estimate the chlorophyll as µg/mL methanol as follows:

$$\text{Chlorophyll a} = 16.5\ A_{665} - 8.3\ A_{650} \quad (1)$$
$$\text{Chlorophyll b} = 33.8\ A_{650} - 12.5\ A_{665} \quad (2)$$
$$\text{Total chlorophyll} = 25.8\ A_{650} + 4.0\ A_{665} \quad (3)$$

where A_{650} and A_{665} are absorbances at 650 and 665 nm, respectively. Estimates of chlorophyll can then be converted to µg/mm^2 of the leaf area (*see* Note 4).

4. Notes

1. Paclobutrazol is active in a wide spectrum of plant species. At active levels, it can be expected to shorten and thicken stems, darken the color of the foliage, and shorten and thicken roots. The optimal level will be the highest level that does not cause unacceptable stunting of growth. For most species, 1 mg/L is likely to be suitable, and the optimum level will probably lie within the range 0.5–2.0 mg/L.
2. Estimation of ribulose bisphosphate carboxylase activity can be made spectrophotometrically from isolated chloroplasts or leaf extracts (*32*). Levels of activity in strawberry cultured without sucrose was more than double that with sucrose (*10*).
3. Measurement of lignin can be made spectrophotometrically (*33*) if poor development of vascular tissue is suspected. Low levels of lignin are often found in vitrified plantlets. The lignin content of vitrified carnation plantlets, for example, was found to be less than half that of nonvitrified plantlets (*21*).
4. Paclobutrazol can markedly affect levels of chlorophyll. Leaves from plantlets of chrysanthemum cultured on a medium containing 4 mg/

L paclobutrazol contained 0.56 μg/mm^2 (SE ± 0.018) of chlorophyll, whereas untreated controls contained only 0.37 μg/mm^2 (SE ± 0.018) (*17*).

References

1. Brainerd, K. E. and Fuchigami, L. H. (1982) Stomatal functioning of *in vitro* and greenhouse apple leaves in darkness, mannitol, ABA and CO_2. *J. Exp. Bot.* **33,** 388–392.
2. Wardle, K., Quinlan, A., and Simpkins, I. (1979) Abscisic acid and the regulation of water loss in plantlets of *Brassica oleracea*. L. var. *botrytis* regenerated through apical meristem culture. *Ann. Bot.* **43,** 745–752.
3. Wardle, K. and Short, K. C. (1983) Stomatal responses of *in vitro* cultured plantlets. 1. Responses in epidermal strips of chrysanthemum to environmental factors and growth regulators. *Biochem. Physiol. Pflanzen* **178,** 619–624.
4. Ziv, M., Schwartz, A., and Fleminger, D. (1987) Malfunctioning stomata in vitreous leaves of carnation (*Dianthus caryophyllus*) plants propagated *in vitro*; implications for hardening. *Plant Sci.* **52,** 127–134.
5. Fuchigami, L. H., Cheng, T. Y., and Soeldner, A. (1981) Abaxial transpiration and water loss in aseptically cultured plum. *J. Amer. Soc. Hort. Sci.* **106,** 519–522.
6. Grout, B. W. W. (1975) Wax development on leaf surfaces of *Brassica oleracea* var. *Currawong* regenerated from meristem culture. *Plant Sci. Lett.* **5,** 401–405.
7. Sutter, E. and Langhans, R. W. (1979) Epicuticular wax formation on carnation plantlets regenerated from shoot tip culture. *J. Amer. Soc. Hort. Sci.* **104,** 493–496.
8. Gaspar, Th., Kevers, C., Debergh, P., Maene, L., Paques, M., and Boxus, Ph. (1987) Vitrification: morphological, physiological and ecological aspects, in *Cell and Tissue Culture in Forestry,* (Bonga, J. M. and Durzan, D. J., eds.), vol. 1, Martinus Nijhoff, The Hague, Boston, London, pp. 152–166.
9. Kevers, C., Coumans, M., Coumans-Gilles, M. F., and Gaspar, Th. (1984) Physiological and biochemical events leading to vitrification of plants cultured *in vitro*. *Physiol. Plant.* **61,** 69–74.
10. Grout, B. W. W. and Price, F. (1987) The establishment of photosynthetic independence in strawberry cultures prior to transplanting, in *Plant Propagation in Horticultural Industries,* (Ducote, G., Jacobs, M., and Simeon, A., eds.), Belgian Plant Tissue Culture Group, Arlon, pp. 55–60.
11. Grout, B. W. W. and Donkin, M. E. (1987) Photosynthetic activity of cauliflower meristem cultures *in vitro* and at transplanting into soil. *Acta Hort.* **212,** 323–334.
12. Donnelly, D. J., Vidaver, W. E., and Colbow, K. (1984) Fixation of $^{14}CO_2$ in tissue cultured red raspberry prior to and after transfer to soil. *Plant Cell Tissue Organ Cult.* **3,** 313–317.
13. Grout, B. W. W. and Millam, S. (1985) Photosynthetic development of micropropagated strawberry plantlets following transplanting. *Ann. Bot.* **55,** 129–131.
14. Scott, M. A. (1986) Weaning of cultured plants, in *Micropropagation in Horticulture* (Alderson, P. G. and Dullforce, W. MN. eds.), Institute of Horticulture, London, pp. 173–182.
15. Short, K. C., Warburton, J., and Roberts, A. V. (1987) *In vitro* hardening of cultured cauliflower and chrysanthemum plantlets to humidity. *Acta Hort.* 212, 329–334.

16. Short, K. C., Wardle, K., Grout, B. W. W., and Simpkins, I. (1984) *In vitro* physiology and acclimatization of aseptically cultured plantlets in *Plant Tissue and Cell Culture Applications to Crop Improvement* (Novak, F. J., Havel, L., and Dolezel, J., eds.), *Czechoslovak Acad. Sci.*, pp. 475–486.
17. Smith, E. F., Roberts, A. V., and Mottley, J. The preparation *in vitro* of chrysanthemum for transplantation to soil. *Plant Cell Tissue Organ Cult.* (in press).
18. Wardle, K., Dobbs, E. B., and Short, K. C. (1983) *In vitro* acclimatization of aseptically cultured plantlets to humidity. *J. Amer. Soc. Hort. Sci.* **108,** 386–389.
19. Chin, C. K. (1982) Promotion of shoot and root formation in asparagus *in vitro* by ancymidol. *HortSci.* **17,** 590,591.
20. Khunachak, A., Chin, C-K., Le. T., and Gianfagna, T. (1987) Promotion of asparagus shoot and root growth by growth retardants. *Plant Cell Tissue Organ Cult.* **11,** 97–110.
21. Letouze, R. and Daguin, F. (1987) Control of vitrification and hypolignification process in *Salix babylonica* cultured *in vitro*. *Acta Hort.* **212,** 185–216.
22. Debergh, P. C., Harbaoui, Y., and Lemeur, R. (1981) Mass propagation of globe artichoke (*Cynara scolymus*): Evaluation of different hypotheses to overcome vitrification with special reference to water potential. *Physiol. Plant.* **53,** 181–187.
23. Ziv, M., Meir, G., and Halevy, A. H. (1983) Factors influencing the production of hardened glaucous carnation plantlets *in vitro*. *Plant Cell Tissue Organ Cult.* **2,** 55–65.
24. Kevers, C. and Gaspar, Th. (1985) Vitrification of carnation *in vitro*: changes in ethylene production, ACC level and capacity to convert ACC to ethylene. *Plant Cell Tissue Organ Cult.* **4,** 215–223.
25. Murashige, T. (1974) Plant propagation through tissue cultures. *Ann. Rev. Plant Physiol.* **25,** 135–166.
26. George, E. F. and Sherrington, P. D. (1984) *Plant Propagation by Tissue Culture.* Exergetics, LTD, Basingstoke, UK.
27. Debergh, P. C. and Maene, L. J. (1981) A scheme for commercial propagation of ornamental plants by tissue culture. *Sci. Hort.* **14,** 335–345.
28. Maene, L. J. and Debergh, P. C. (1986) Optimization of plant micropropagation. *Med. Fac. Landbouww. Rijksuniv. Gent.* **51/4**, 1479–1488.
29. Zatyko, J. and Molnar, J. (1987) Uj modszer a gyumolcsfajok gyokerkepzodesenek serkentesere a taptalaj H^+ koncentraciojanak novelesevel. *Kertgazdasag* **19,** 31–35.
30. Ebercon, A., Blum, A., and Jordan, W. R. (1977) A rapid colorimetric method for epicuticular wax content of sorghum leaves. *Crop. Sci.* **17,** 179,180.
31. Hipkins, M. F. and Baker, N. R. (1986) Spectroscopy, in *Photosynthesis Energy Transduction,* (Hipkins, M. F. and Baker, N. R. eds.), Oxford, Washington, IRL Press, pp. 51–101.
32. Lilley, R. McC. and Walker, D. A. (1974) An improved spectrophotometric assay for ribulose bisphosphate carboxylase. *Biochim. Biophys. Acta* **358,** 226–229.
33. Alibert, G. and Boudet, A. (1979) La liquification chez le peuplier. 1. Mise au point d'une method de dosage et d'analyse monomerique de lignine. *Physiol. Veg.* **17,** 67–71.

Chapter 24

Protoplasts of Higher and Lower Plants

Isolation, Culture, and Fusion

John B. Power and Michael R. Davey

1. Introduction

Plant protoplasts are cells from which the cell wall has been removed enzymatically. Thus, they retain all the normal cell organelles plus the nucleus; the latter is capable of expressing totipotency through the conversion of the protoplast to the regenerated plant using tissue culture technology. The fact that plant regeneration from protoplasts is possible for a large number of species, many of prime agronomic value, means that the protoplast is now seen as an ideal starting point for the gamut of genetic engineering technologies designed to bring about plant improvement. The protoplast is an invaluable tool for a variety of studies, including uptake of exogenously supplied materials, such as bacteria (*1*), algae (*2*), organelles (*3*), viruses (*4*), and macromolecules like DNA (*5*), as well as physiological investigations (*6*), transformation assessments (*7*), ultrastructural studies (*8*), and the isolation of subcellular components, including nuclei, chromosomes, and vacuoles (*9*).

From: *Methods in Molecular Biology*, vol. 6, *Plant Cell and Tissue Culture*
Edited by Jeffrey W. Pollard and John M. Walker, ©1990 by The Humana Press

It is in the context of genetic engineering that a clear role for plant protoplasts has emerged. In its simplest form, this could utilize the protoplast as the starting point for the generation of genetic (somaclonal) variation *(10)* following the mass regeneration of plants—a concept of potential and actual value *(11)*.

Fostering the generation of variation, protoplast fusion, leading to somatic *(12)* or gametosomatic *(13)* hybridization, is now well recognized as a powerful adjunct to conventional plant breeding methodologies. This section will concentrate on the basic techniques of isolation and culture, in principle essentially the same for all species and source tissues, and the more reproducible methods of chemical and electrofusion of protoplasts. Some of these techniques have been readily extended to lower plants, for example, mosses *(14)* and liverworts *(15)*, and as detailed in this section, lower vascular plants, such as the ferns.

The methodologies have been selected on the basis of reproducibility and general applicability, since irrespective of the use to which the protoplasts are to be put, the key considerations are good yields of highly viable protoplasts coupled with, in the context of genetic manipulations, an ability to induce sustained mitotic division of protoplasts leading to callus production and efficient plant regeneration.

2. Materials

2.1. Plant Material

A range of starting tissue provides adequate sources of protoplasts, and can be of whole plant or tissue culture origin. In terms of yields of protoplasts, leaves, in vitro grown shoots (*see* 3.1.1.), or callus/cell suspension cultures (*see* 3.1.2.) are best.

1. Greenhouse grown plants (leaf, stem): Plants should be nonflowering and grown under conditions giving "soft" growth (e.g., 20–25°C, high humidity, 16-h daylength of 10,000 lx provided by "daylight" fluorescent tubes). Fully expanded leaves should be used or, depending upon the species, those leaves whose lower epidermis can be removed by peeling. Pest control (red spider, white fly) is essential, and can be effected by employing a regular and sequential treatment of plants. Fumigants that are effective and nondamaging to protoplasts include DDT/Lindane (Murphy Chemicals Ltd., St. Albans, Herts, UK), Propoxur (Octavius Hunt Ltd., 5 Dove Lane, Redfield, Bristol, UK), or nicotine. Systemic insecticides should be avoided.

2. Seedling material (cotyledon, hypocotyl, root): Seeds are surface sterilized in 0.1% mercuric chloride with 0.1% sodium lauryl sulfate (10 min) followed by a minimum of six washes in sterile tap water. Alternatively, immersion in a 10% "Domestos" (or proprietary bleach) solution (30 min) followed by several rinses in sterile tap water will suffice. If necessary, steps 1 and 2 may be followed sequentially. Germinate seeds (in the dark), either on moistened sterile filter paper or on MS medium (*16* and *see* Appendix) with 0.8% agar and without phytohormones (pH 5.8), in jars or Petri dishes.
3. In vitro (micropropagated) shoots: This material offers the clear advantage of being contamination free, and the immediate environmental conditions can be fully regulated. Actively grown shoots can be used as a source of leaf or stem protoplasts.
4. Callus/cell suspension: Cells in their exponential phase of growth should be used (*see* section 1). Cell viability should be monitored (*see* F.D.A. staining, 3.2.2.) and should be 80% or higher, in order to give adequate yields of protoplasts.
5. Specialized plant parts:
   a. Pollen tetrad: Young buds can be identified as containing anthers that are at the tetrad stage by removing a single anther and squashing it in a drop of water. Remaining anthers can be assumed to be at the same developmental stage. Bud length can eventually be correlated with anther development, given a controlled plant environment.
   b. Petals: Freshly opened flowers are a suitable source of petal tissues. These can subsequently be handled in the same manner as leaf tissues.
   c. Root tips, coleoptiles, seed tissues (e.g., aleurone layer): Surface-sterilized seeds/seedlings can provide such tissues. Avoid roots that are necrotic.
6. Lower vascular plants:
   a. *Azolla* plants (frond tissue): *Azolla* plants are maintained on the medium of Watanabe et al. (*17*) in gravel trays (18 x 11 x 3 in) covered with propagator lids (height: 6 in), with a 16-h daylength (5000 lx, cool, white fluorescent tubes) (20–27°C; 80% humidity). Plant density is maintained at 50% of the liquid surface. Fronds, free of roots and necrotic regions, are surface sterilized in 0.12% sodium hypochlorite with 0.01% Triton X-100 (30 min) followed by six changes of sterile tap water.

b. *Pteridium* (spore/prothallus tissue): Prothallial tissues grown from spores are maintained on MS medium (*16*) with 3.0% (w/v) sucrose and 0.6% agar (pH 5.8) at 25°C with a constant illumination of 2,700 lx (daylight fluorescent tubes). Mutant (tumorous) prothalli that resemble higher plant callus can be produced by treating spores (e.g., 25 kR, X-rays) prior to germination on the culture medium.

2.2. Enzyme Mixtures/Washing Solutions

1. Enzymes: The choice of enzyme mixture will depend on the plant species and the source tissue for protoplast production. In Table 1, enzyme mixtures, plasmolytica, and incubation times are given that have been used for a variety of species and that are likely to be of general applicability. All enzyme solutions are filter sterilized and stored in the dark at –20°C until required.
2. Washing solutions: The compositions of the washing solutions (CPW media and *see* Appendix) are given in Table 1. Where floatation of protoplasts to remove cellular debris cannot be achieved with CPW21S medium (CPW salts with 21% sucrose), either a Percoll or Ficoll gradient will be required.
   a. Percoll: The protoplast suspension (in CPW13M medium or the appropriate osmoticum) (2 mL) is layered over a 30% Percoll solution (in the osmoticum) and centrifuged (100 x *g*; 10 min). Protoplasts are recovered at the Percoll/CPW13M interface.
   b. Ficoll: The method used is a modification of that reported by Attree and Sheffield (*18*) for the successful elimination of microbial contamination during the isolation of protoplasts from gametophytic cultures of *Pteridium*. If necessary, it can also be used for higher plant protoplasts.

      The protoplast pellet, obtained following centrifugation of the enzyme solution after incubation, is resuspended in 4 mL of CPW13M and mixed with an equal volume of 30% (w/v) Ficoll (400) in CPW13M. This suspension is transferred to a clean 16-mL centrifuge tube carefully, so as to prevent it from touching the sides of the tube. This is overlaid by 4 mL of 12% (w/v) Ficoll (CPW13M) followed by 1 mL of CPW13M. The tubes are centrifuged at 100–200 x *g* (10–12 min), during which time the protoplasts collect at the upper interface.

Table 1
Enzyme Mixture/Osmoticum/Incubation Conditions for Protoplast Production

Source tissue	Enzyme mixture (%, w/v) (pH 5.8): Mei-celase	Cellu-lase Rio	Cellu-lysin	Cellu-lase RS	Drise-lase	Hemi-cellu-lase	Macer-ozyme Rio	Rhozyme HP 150	Pecto-lyase Y23	Osmo-ticum	Incu-bation time[a] (h)	Anti-biotics[b]	Mode of incu-bation	Float-ation method[c]	Refer-ence	Special notes
Leaf (Dicot)	1.5	–	–	–	–	–	0.05	–	(0.1)	CPW13M	4–16	+	Static	CPW21S Percoll Ficoll	(22)	Leaf peeled or sliced. Pecto-lyase included for legume species
Leaf (Monocot)	–	–	2.0	–	–	0.5	0.2	–	–	CPW11M	2–16	+	Shake 40 cycles/min	CPW21S	(19)	1% Potassium dextran sulfate added to enzyme mixture
Stem	–	–	1.0	–	–	–	0.1	–	–	CPW13M	16	+	Static	CPW21S	(19)	No antibiotics are required if grown axenically
Cotyledon	–	–	1.0	–	–	–	0.1	–	–	CPW13M	4–16	+	Static	CPW21S	(23)	"
Hypocotyl	–	–	1.0	–	–	–	0.1	–	–	CPW13M	4–16	+	Static	CPW21S	(23)	"
Root tip	2.0	–	–	–	–	–	0.03	2.0	–	CPW13M	4–12	–	Static	CPW13M	(24)	"
In vitro cultured shoot	–	1.0	–	–	–	–	0.25	–	–	CPW9M	4–16	–	Shake 20 cycles/min	CPW21S	(25)	Leaves can be processed as above
Callus/cell suspension	2.0	–	–	–	–	–	0.03	2.0	–	CPW9M or CPW13M	1–16	–	Shake 20–60 cycles/min	CPW21S Percoll	(26)	Choice of osmoticum level depends on osmotic status of cells in culture
Petal	–	1.0	–	–	–	–	0.05	–	–	CPW13M	2–16	+	Static	CPW21S	(19)	Stem peeled or sliced.
Pollen tube	–	–	–	–	2.0	–	–	–	–	CPW13M	4	–	Shake 10 cycles/min	–	(19)	Pollen germinated (1–4 h) in 2% sucrose solution

(continued)

Table 1 *(continued)*
Enzyme Mixture/Osmoticum/Incubation Conditions for Protoplast Production

Source tissue	Enzyme Mixture (%, w/v) (pH 5.8)									Osmoticum	Incubation time[a] (h)	Antibiotics[b]	Mode of incubation	Floatation method[c]	Reference	Special notes
	Meicelase	Cellulase Rio	Cellulysin	Cellulase RS	Driselase	Hemicellulase	Macerozyme Rio	Rhozyme HP 150	Pectolyase Y23							
Pollen Tetrad	–	–	–	–	2.0	–	–	–	–	CPW9M	20 min 2 h	–	Shake 10 cycles/min	–	*(13)*	
Frond (e.g., *Azolla*)	–	–	–	2.0	–	–	0.1	–	–	CPW13M	12–16	+	Shake 40 cycles/min	Percoll ficoll	*(18)*	Use of Ficoll gradients
Prothallus (e.g., *Pteridium*)	–	–	2.0	–	–	–	0.1	–	–	CPW11M	2–3	–	Static	CPW21S CPW13M	*(27)*	

[a]Incubation: 23–30°C, dark or low light conditions.
[b]Antibiotics: Ampicillin, 400 mg/L; Gentamycin, 10 mg/L; Tetracycline, 10 mg/L.
[c]CPW Salts, mg/L (ref. *28*): KH_2PO_4, 27.2; KNO_3, 101.0; $CaCl_2 \cdot 2H_2O$, 1480.0; $MgSO_4 \cdot 7H_2O$, 246.0; KI, 0.16; $CuSO_4 \cdot 5H_2O$, 0.025; M = % mannitol; S = % sucrose; pH 5.8.

2.3. *Culture Media*

1. Protoplast media: The range of culture media is vast (*see* Appendix) and, of course, is determined by the species/protoplast source. However, three media are of general applicability, although may not be optimal for a given species. A medium based on MS (*16*) medium with 2.0 mg/L 1-naphthaleneacetic acid (NAA), 0.5 mg/L 6-benzylaminopurine (6-BAP) with mannitol (9–13%), 3% (w/v) sucrose, and a pH of 5.8 (MSP1 medium) (*19*) supports growth of many protoplast systems. Alternatively, a richer medium, KM8P (*20*) or K8P (*21*), may be used. The media counterpart, lacking an osmoticum, will also be required to facilitate the sustained division of regenerating protoplast systems (*see* 3.3.1.7.).
2. Plant regeneration media: These, like protoplast media, are very varied, primarily with respect to phytohormone content. Four regeneration media (*see* Appendix) are often suitable:
   a. MSP1—as MSP1 9*M*, but lacking mannitol, 0.8% (w/v) agar.
   b. MSZ—MS medium (*18*), but with zeatin (1.0 mg/L) as the sole growth regulator, 0.8% (w/v) agar.
   c. MSD3—MS medium (*18*) with IAA (2.0 mg/L), 6-BAP (1.0 mg/L), 0.8% (w/v) agar.
   d. MSD4—MS medium (*18*) with NAA (0.05 mg/L), 6-BAP (0.5 mg/L), 0.8% (w/v) agar.

   The pH is 5.8 in all cases.

2.4. *Fusion Solutions*

The most commonly used chemical fusogens are polyethylene glycol (PEG) and high pH/Ca^{2+} solutions, with their respective washing media. Such chemical fusogens can be used in large- and small-scale procedures. Normally, chemical fusogens and washing solutions contain high levels of Ca^{2+} ions. Electrofusion is a helpful technique, but necessitates specialized equipment.

Typical fusion solutions for the large-scale chemical procedure are:

1. PEG:
   30% w/v PEG 6000
   4% w/v sucrose
   0.15% w/v $CaCl_2 \cdot 2H_2O$
   Autoclave; store at 4°C in the dark until required.

2. High pH/Ca^{2+}:
   0.38% w/v glycine
   1.1% w/v $CaCl_2 \bullet 6H_2O$
   9% w/v mannitol
   pH 10.4 with NaOH
   Filter sterilize immediately prior to use. For the high pH/water fusion method (*see* Section 3.4.1.) the mannitol should be increased to 10% w/v.
3. CPW13M/Ca^{2+} washing solution. As CPW13M solution (Table 1), but with 0.74% w/v $CaCl_2 \bullet 2H_2O$ final concentration.

For small-scale fusion, the fusogen and washing solutions are as follows:

1. PEG:
   22.5% w/v PEG 6000
   1.8% w/v sucrose
   0.15% w/v $CaCl_2 \bullet 2H_2O$
   0.01% w/v KH_2PO_4
   pH 5.8 (adjusted with 1*N* KOH or HCl.)
2. Washing solution:
   0.74% w/v $CaCl_2 \bullet 2H_2O$
   0.38% w/v glycine
   11% w/v sucrose
   pH 5.8
3. A suitable electrofusion solution consists of:
   11% w/v mannitol
   0.03% w/v $CaCl_2 \bullet 2H_2O$
   pH 5.6
   Sterilize by autoclaving.

3. Methods

3.1. Isolation of Protoplasts

3.1.1. Leaf Protoplasts

1. Leaves are surface sterilized (7.5% Domestos, 30 min), their lower epidermis is removed by peeling using fine forceps, and then the leaves are floated on the surface of the enzyme solution (Table 1). Approximately 5 g fresh wt of peeled leaf tissue should be used with

20 mL of the enzyme mixture, maintained in a 14-cm Petri dish. Seal the dish with Nescofilm. Plasmolysis for 1 h (in CPW13M medium) prior to the addition of the enzyme solution can enhance protoplast yield and viability.

2. After incubation, and without disturbing the digested leaf pieces, remove the enzyme solution and squeeze the leaf pieces (using a Pasteur pipet against the edge of the dish) into 25 mL of CPW21S medium.
3. Transfer the protoplast suspension to centrifuge tubes and spin (100 x *g*; 10 min—swing out centrifuge). Protoplasts will collect at the surface. The Percoll or Ficoll procedures (*see* 2.2.; 2a or b) can be followed if a sucrose (CPW21S) floatation is unsuitable.
4. Transfer the protoplasts to a measured volume of washing (CPW13M) or culture medium (MSP1 9M, KM8P, KP8) for counting.

This basic procedure is also applicable for leaves of monocotyledons, stem, cotyledon, hypocotyl, root tip, in vitro cultured shoots, petal, and fern frond tissues, modified as shown in Table 1. A few protoplast systems (e.g., root tips) do not lend themselves to the standard washing procedures. They can be freed of cell debris and spent enzyme by repeated washing in CPW13M medium or by resuspension and centrifugation.

3.1.2. Cultured Cell Protoplasts (Callus / Cell Suspension)

1. For callus cultures, allow the cells to settle out of the medium and remove all of the culture medium.
2. Add approximately 30 mL of the enzyme solution (Table 1) to the cells (10–15 g fresh wt) in a 250-mL flask. Incubate on a slow shaker (Table 1).
3. Transfer, by pouring, the digested cells to centrifuge tubes. Top up the tubes with the appropriate washing solution (CPW13M) and spin (80 x *g*; 10 min).
4. Remove the supernatant (some protoplasts might float on this medium) and resuspend the protoplasts in CPW21S medium (5–10 mL).
5. Pour the contents onto a 64-µm nylon sieve, and collect the filtrate in a Petri dish. Small volumes (1–5 mL) of CPW21S can be added to the material on the sieve to facilitate, with gentle agitation, the release of protoplasts.
6. Transfer the protoplasts (in CPW21S medium) to centrifuge tubes and spin (100 x *g*; 10 min). Protoplasts collected at the surface can be transferred to the appropriate medium for counting, plating, or fusion treatment.

This basic method can be modified (Table 1) to produce protoplasts from stems, cotyledons, in vitro cultured shoots from prothallus material, and pollen tubes. In the last example, this will clearly be on a much reduced scale.

3.1.3. *Fern Frond Protoplasts* (Azolla)

1. Remove roots and necrotic areas (at stem base) from the *Azolla* fronds, and float fronds on the surface of the enzyme solution. Incubate on a slow shaker (40 cycles/min) at 23°C overnight (16 h).
2. Following incubation, pour the enzyme solution through a 64-μm sieve, and wash the protoplasts through with CPW13M (do not squeeze undigested frond tissues, since this releases large quantities of mucus).
3. Centrifuge (100 x *g*; 5 min) and resuspend the pellet in CPW13M. Protoplasts, free of undigested cells, may be recovered by the use of either Percoll or Ficoll. The former method results in the recovery of protoplasts with a bacterial contamination from the source tissue, whereas the latter method reduces this contamination.

3.2. Characterization of Protoplasts

3.2.1. *Counting Protoplasts*

For all subsequent manipulations, such as fusion and plant regeneration, the initial yield of protoplasts has to be determined, in order that the protoplast density (for culture) can be controlled. A double-chamber hemocytometer is used.

1. Transfer the protoplast suspension to a measured volume (e.g., 10 mL) of CPW13M medium and count using the hemocytometer.
2. Count the number of protoplasts in one triple-lined square (= n). The total yield = $n \times 5 \times 10$ (vol in mL) = $5n \times 10^4$. To calculate the volume that has to be added to (or in the case of low yields, subtracted from) the original 10 mL vol, divide the above figure by, for example, 2×10^5 (if a density of 2×10^5 is required).

3.2.2. *Determination of Viability*

Initial cell and protoplast viability is determined using fluorescein diacetate (FDA) (*see* Chapter 3, this vol.). FDA is nonfluorescent and freely permeable across the plasma membrane. The molecule is cleaved, by esterases in living cells, into fluroescein. It will, therefore, accumulate in viable cells and fluoresce yellow-green when excited by ultraviolet light.

1. Dilute 1 mL of FDA stock (5 mg/mL in acetone) with 10 mL of the appropriate culture medium, and then mix 1 mL of the medium plus FDA with 1 mL of the cell or protoplast suspension.
2. Examine the samples after 5 min, using an inverted microscope with fluorescence attachment.

3.2.3. Assessment of the Presence of a Cell Wall

1. Add one drop of a stock solution containing 0.1% (w/v) Calcofluor White in 0.4*M* Sorbitol to one drop of the protoplasts in culture medium. Leave for 1 min.
2. Add 0.5 mL of culture medium, allow protoplasts to settle, and remove supernatant with a Pasteur pipet.
3. Examine the preparation, using a microscope fitted with a mercury vapor lamp and the appropriate exciter and barrier filters. The presence of cell wall material is detected as an intense blue fluorescence.

3.3. Culture Methods

3.3.1. Protoplast Culture

Glass or plastic vessels can be used for protoplast culture. Plastic vessels now have excellent optical properties, and are preferred by most workers. Both small-scale and large-scale methods, some of which are summarized, are available for culturing protoplasts.

3.3.1.1. Liquid Culture

1. Suspend protoplasts at 5.0×10^4–2.5×10^5/mL in the required culture medium. The optimum density for protoplast division and the most suitable medium must be determined empirically.
2. Dispense 4.0 mL aliquots to 5 cm Petri dishes; 8.0 mL to 9 cm dishes. Seal dishes with Nescofilm.
3. Small-scale sitting drop cultures can be prepared by dispensing the culture medium, containing the protoplasts as 20–50 μL droplets, in the base of a Petri dish, using an air displacement pipet.
4. For hanging drop cultures, the Petri dish lid is removed, inverted, and the protoplast-containing droplets (20–50 μL) dispensed in the lid. A small volume of culture medium is placed in the base of the dish to maintain humidity, and the lid with its droplets carefully placed in position on the base of the dish.

Dishes are sealed with Nescofilm and placed in a moist environment, e.g., plastic sandwich boxes lined with damp tissue or containing small open jars of water.

3.3.1.2. Plating in Agar

1. Suspend protoplasts in liquid medium at double the required plating density.
2. Mix thoroughly with an equal volume of the same medium containing 1.2% w/v agar at 45°C. (The agar medium should be melted and maintained in a water bath until required.)
3. Dispense suitable aliquots into Petri dishes, and allow the agar to set at room temperature.
4. Cutting the agar into sectors with a scalpel (4–6 sectors/dish) and floating the sectors in the same liquid medium (8 mL for a 9 cm dish) frequently stimulates protoplast division (*29*). Seal dishes with Nescofilm and incubate in the culture room.
5. Small-scale agar droplets can be dispensed in the bottoms of Petri dishes, e.g., 90–120 50 µL droplets to a 9 cm dish. Surround the droplets by flooding the plate with 8.0 mL of liquid medium.

3.3.1.3. Liquid over Agar

1. Dispense a thin layer of agar medium into the bottom of a Petri dish (4.0 mL for a 5 cm dish; 8.0 mL for a 9 cm dish) and allow the medium to set.
2. Layer the same volume of liquid medium containing the protoplasts over the agar medium. The density of the protoplasts in the liquid layer should be twice the required density, since the overall plating density will be reduced by half.
3. Inclusion of a filter paper (Whatman 3MM) at the agar-liquid interface stimulates division in some protoplast systems (*30*), probably by encouraging gaseous exchange. Lay a sterile filter paper of suitable size over the agar. Spread the protoplast-containing liquid medium over the paper. Unfortunately, the filter paper prevents microscopic examination of the protoplasts during culture, but macroscopic colonies should be visible when the filter paper is viewed under a stereomicroscope.

3.3.1.4. Use of Agarose as a Gelling Agent. Cell division is stimulated in several protoplast systems by replacing agar with agarose, e.g., Seaplaque low melting temperature agarose or Sigma Type VII agarose.

1. Prepare 1.2% w/v agarose in distilled water. Autoclave. When required, melt in a steamer and keep at 40°C in a water bath.
2. Dilute the molten agarose with an equal volume of double-strength culture medium at 40°C.
3. Adjust the protoplasts in liquid medium to the required density. Pellet the protoplasts (120 x *g*; 5 min), remove the supernatant, and replace with the same volume of agarose medium. Dispense as layers or droplets in dishes.
4. If required, cut the agarose layers into sectors and float in liquid medium (*29*). Flood the agarose droplets with liquid medium (*31*).

3.3.1.5. Use of Conditioned Medium and Nurse Cultures. Dividing protoplasts can stimulate division of other protoplasts by releasing growth promoting compounds, e.g., amino acids, into the culture medium. Liquid medium "conditioned" in this way can be prepared by harvesting the medium from actively growing cultures during the first 7-d growth period. Conditioned medium can be filter sterilized (if required) and stored at 5°C.

Actively dividing protoplasts can be used as a "nurse" tissue.

1. Spread the nurse protoplasts at 5.0×10^4/mL in an agar or agarose layer, culture for 24–72 h, and plate the freshly isolated protoplasts of interest at a similar density in liquid medium over the nurse layer.
2. The freshly isolated protoplasts can also be spread in an agar or agarose layer, separating the latter from the nurse protoplasts with a filter paper.
3. Alternatively, morphologically distinct nurse protoplasts from an albino cell suspension can be used. Such nurse protoplasts are plated in liquid medium over agar or agarose, and the protoplasts of interest mixed with the nurse protoplasts.

Limited numbers of protoplasts, e.g., heterokaryons selected by micromanipulation, can be separated from nurse protoplasts using membrane chambers (*32*).

1. To prepare the chambers, cut "Sartorius" membrane filters (12 µm pore size) into 3 x 0.75 cm strips, form into cylinders, and seal the ends by applying acetone.
2. Autoclave and wash 3x with culture medium.
3. Dispense 3.0 mL of agarose medium in the bottom of a 5 cm Petri dish and stand the chamber vertically in the agarose in the center of the dish. Allow the agarose to set.

4. Plate nurse protoplasts at 5.0 x 104/mL in liquid medium over the agarose layer, outside the chamber.
5. Suspend the protoplasts of interest in 0.5 mL of the same liquid medium, and introduce into the chamber. Cover the dish and seal with Nescofilm.

3.3.1.6. Incubation of Protoplasts. Protoplasts should be cultured in the dark or under low intensity daylight fluorescent illumination (e.g., 0.6 Wm^{-2}) at 25–27°C. In some cases, it is beneficial to culture in the dark for the first 7 d before transferring to the light.

3.3.1.7. Protoplast Division and Reduction of the Osmoticum. The time required for protoplast division depends on the protoplast system, but in general, the first division in protoplasts isolated from cell suspensions can be expected within 24–72 h of culture. The osmotic pressure of the medium should be reduced, to accelerate division and to prevent secondary plasmolysis of newly formed cells.

For protoplasts spread in agar or agarose layers, sectors with dividing protoplasts are transferred to the surface of the same medium containing a lower level of plasmolyticum. Protoplasts on filter paper between liquid-agar layers can be moved *in situ* by transferring the filter paper to fresh agar medium.

It is readily possible (for liquid medium, liquid over agar or agarose, or agar or agarose sectors or droplets immersed in liquid medium), to dilute the liquid phase by adding measured volumes of fresh liquid medium of lower osmotic pressure: e.g., Petri dishes containing 8.0 mL of liquid medium with 9% mannitol can be diluted by addition of 2.0 mL aliquots of medium containing 6, 3, and 0% mannitol, respectively, at 5–7-d intervals. Finally, protoplast-derived cell colonies can be transferred to medium lacking mannitol. Increasing the number of dishes during the culture sequence may be necessary to accommodate the increasing volume of medium. Protoplasts mixed with nurse cells in liquid medium, or cultured in nurse chambers, are handled in a similar way by diluting the nurse culture.

3.3.2. Plant Regeneration from Protoplast-Derived Cells

Actively dividing protoplasts should produce macroscopic cell colonies within 4–8 wk of culture. At this stage, attempts can be made to induce regeneration through organogenesis or somatic embryogenesis, the pathway of regeneration and the medium required being dependent on the species. Often, protoplast-derived cell colonies of several species, particularly members of the *Solanaceae*, such as tobacco and petunia, can be induced to regenerate readily by transferring the cell colonies to the surface of MS based agar media with the appropriate combination of growth regulators. Protoplast-derived tissues may also regenerate on media prepared to other well-known formulations, e.g., those of wild *Glycine* species form shoots on B5 based media (*31* and *see* Appendix). Generally, high cytokinin to auxin ratios stimulate shoot formation; exposure to 2,4-D may induce embryogenesis. Somatic embryos may develop directly from compact cell colonies during dilution of the liquid protoplast medium, as in *Medicago* (*33*), whereas regeneration via callus in the same species necessitates a more involved media sequence (*30,34*). Exposure to 2,4-D-containing medium followed by transfer to hormone-free medium triggers embryogenesis with reproducible plant regeneration in rice (*35*). Overall, the optimal conditions for plant regeneration must be determined empirically, but prior knowledge of the conditions required to induce morphogenesis from explant-derived callus may facilitate regeneration from protoplast-derived tissues.

3.4. Fusion Methods

Chemical and physical methods are used for fusing protoplasts. Chemical methods are simple; electrofusion requires specialized equipment. In the large-scale chemical procedure, protoplasts are fused in 16 x 125 mm screw-capped tubes; in the small-scale method, protoplasts are fused on glass coverslips.

3.4.1. Procedure for Large-Scale Chemical Fusion

1. Adjust the densities of the parental protoplast suspensions to 2.0 x 10^5/mL, and dispense the protoplasts into seven centrifuge tubes as follows: Tube 1, 4 mL parent 1 (label P1 viability); Tube 2, 4 mL parent 2 (P2 viability); Tube 3, 8 mL parent 1 (P1 selfed); Tube 4, 8 mL parent 2 (P2 selfed); Tubes 5, 6, and 7, each with 4 mL parent 1 + 4 mL parent 2 (label fusion).
2. Centrifuge the tubes (100 x *g*; 5 min), except the viability controls, to pellet the protoplasts. Remove the supernatant and add the fusogen.
3. For PEG fusion, add 2.0 mL (one full Pasteur pipet) of PEG solution to each tube (except the viability controls). Leave for 10 min at 22°C. Dilute the PEG every 5 min by adding 0.5, 1.0, 2.0, 2.0, 3.0, and 4.0 mL of culture medium/tube. Resuspend protoplasts after each dilution.

Centrifuge (100 x *g*; 5 min), and remove the supernatant. Wash the protoplasts in culture medium by resuspension and centrifugation.

4. For high pH/Ca^{2+} fusion, add 4.0 mL of high pH/Ca^{2+} solution to each tube (except the viability controls), and incubate the tubes in a water bath at 30°C for 10 min. Centrifuge (100 x *g*; 5 min), remove the supernatant, and wash the protoplasts twice in CPW13M/Ca^{2+} solution by resuspension and centrifugation. Finally, replace the washing solution by culture medium.
5. After treatment, adjust the volume in the fusion tubes to 16 mL with culture medium and adjust the viability controls to 8 mL.
6. Prepare twelve 9 cm Petri dishes, each containing 8 mL of agar or agarose medium, for liquid over agar culture.
7. Dispense the protoplasts as follows: P1 viability control (8 mL; 1 dish); P2 viability control (8 mL; 1 dish); fusion treatment (8 mL in each of 6 dishes); P1 selfed (dispense 8 mL only to 1 dish); P2 selfed (dispense 8 mL only to 1 dish); P1 selfed + P2 selfed (postfusion mixture control—mix the remaining 8 mL of the two selfed tubes and dispense into 2 dishes).
8. Seal the dishes with Nescofilm and incubate at 25–27°C in the dark or under low-intensity fluorescent illumination (0.6 Wm^{-2}).

In some protoplast systems, a combination of PEG with high pH/Ca^{2+}, or a modification of the high pH/Ca^{2+} method (high pH/water fusion) is useful in optimizing fusion.

1. For PEG/high pH fusion, the PEG procedure should be followed. After 10 min in PEG, add 8 mL of high pH/Ca^{2+} solution to each tube, and incubate the mixture for 10 min. Subsequently, follow the general washing stages as for the high pH/Ca^{2+} method.
2. For high pH/water fusion, add 8 mL of the high pH/Ca^{2+} fusion solution (containing 10% w/v mannitol) to the pelleted protoplasts, and centrifuge gently (60 x *g*; 3 min) to pellet the protoplasts. Incubate the tubes at 30°C. After 12–17 min, add 2 mL of sterile water to the supernatant without disturbing the protoplasts. Incubate for a further 10 min, remove the supernatant, and wash the protoplasts in CPW13M/Ca^{2+} solution.

3.4.2. *Procedure for Small-Scale Chemical Fusion*

This method is ideal for small yields of protoplasts.

1. Suspend protoplasts of the two types in CPW13M at 2.5 x 10^5 protoplasts/mL, and mix equal volumes.

2. Place a drop of the protoplast mixture on a sterile coverslip in a 5 cm Petri dish, and allow the protoplasts to settle for 10–15 min.
3. Add two drops of the fusogen containing 22.5% w/v PEG to the opposite sides of the protoplast drop; allow the three drops to coalesce.
4. Incubate the protoplasts at 22°C for 20–25 min. Subsequently, add a drop of the Ca^{2+}–glycine–sucrose washing solution while withdrawing some of the fusion solution. Repeat this washing every 5 min for 20 min.
5. Finally, replace the washing solution with culture medium. The protoplasts are cultured *in situ* on the coverslip. Several sitting drops of medium should be added to the base of the dish to maintain humidity. The dish is sealed with Nescofilm prior to incubation.

3.4.3. *Electrofusion*

Electrofusion involves the alignment of protoplasts in an AC electric field, followed by a DC pulse to induce the coalescence of adhering plasma membranes. The first report of this technique employed a small-scale fusion chamber with two parallel wires (*36*). Since then, other workers have reported various chamber designs, including those with multiple wires (*37*). Some systems are available commercially, e.g., the apparatus developed by Krüss GmbH, D-2000 Hamburg 61, which utilizes frame, "meandering" and helical type chambers.

A simple apparatus has been described (*38*). The AC field is regulated by a function generator and the amplitude visualized with an oscilloscope. An electrophoresis power pack is used to charge a capacitor, and the DC pulse discharged through a resistor in parallel with the fusion electrodes. The electrode system, consisting of five copper plates (1.8 x 3.0 cm) separated by acrylic spacers (0.4 x 1.8 x 2.0 cm), fits the square wells of a 10 cm square, 5 x 5 compartment Sterilin plastic dish.

1. For electrofusion, the parental protoplasts are mixed in a 1:1 ratio in the electrofusion solution, the density adjusted to 1.0×10^5/mL, and 1.0 mL aliquots added to the square compartments of the dish.
2. The electrode is sterilized by immersion in 70% alcohol, and dried in the air stream of a laminar flow hood before being placed in a well containing the protoplasts. The latter are aligned into chains using an AC field of 500 KHz, 40 V/cm applied to the samples for 1–4 min.
3. For fusion, the AC field is reduced to zero, and a DC pulse of 500–800 V/cm for 0.2–2.0 ms applied to the aligned protoplasts. Subse-

quently, the electrode is transferred to the next well, and the procedure repeated.

4. 1.0 mL of culture medium is added to each well after electrofusion; the dishes are sealed with Nescofilm and incubated at 25–27°C in the dark. The protoplasts are transferred after 24 h to suitable Petri dishes (e.g., 5.0 cm) and cultured using the appropriate procedure.

4. Notes

4.1. Allied Information/Techniques

4.1.1. Determination of Chromosome Numbers

4.1.1.1. Callus. In general, good mitotic figures are difficult to find in callus cultures, and often much effort is needed to determine the optimal time after subculture that will yield mitotic figures. In most cases, the outermost parts of the callus (often most friable) have the greatest number of mitotic figures, whereas the innermost cells may have become necrotic and so are devoid of cell divisions.

Pretreat a small sample of callus in saturated α-bromonaphthalene (or other cytostatic agent) for 2–4 h at 4°C. Remove α-bromonaphthalene and wash the cells thoroughly in tap water, making sure the sample never dries out. Finally, fix and hydrolyze. Squash and stain a small sample (1–2 mm^2) on a glass slide, using acetocarmine or acetoorcein.

4.1.1.2. Suspension-Cultured Cells. Generally, higher frequencies of cell division are found in cell suspensions, but the optimum time after subculture for sampling must be determined experimentally. Remove, aseptically, approximately 20 mL of an actively growing cell suspension, and allow the cells to settle out of suspension before removing the supernatant. Always leave a small amount of liquid, so that the cells do not dry out. Pretreat, fix, hydrolyze, and squash as for callus.

4.1.1.3. Leaves and Shoots Regenerating From Callus. Mitotic figures can be found in the tips of very small leaves. The usual problem encountered is that of sufficiently clearing the tissues of chlorophyll to see the chromosomes distinctly. A number of fixatives can be used to overcome this problem.

Pretreat small leaves (approximately 1 mm in length) in 0.5% colchicine for 1 h at 10–12°C, and then wash the sample thoroughly in tap water. Fix in either:

a. Acetic alcohol (3:1) for 3–24 h;
b. Acetic alcohol (1:1) for 3–24 h; or

c. Carnoy's fixative for 3–24 h; or
d. Proprionic acid:chloroform:ethanol (1:3:6) for 6–12 h.

Change the fixative twice during the fixation period until, at the end of the treatment, it is no longer green. Finally, wash and hydrolyze the samples and squash/stain the tip of the leaf.

4.1.2. Characterization of Somatic Hybrids

Apart from chromosome number determinations, confirmation of hybridity usually requires some form of segregation analysis, assuming fertility in the hybrids. Characterization can be extended to the use of isoenzyme banding (*13*), fraction 1 protein analysis (*13*), or restriction analysis of DNA extracts (*39*).

4.1.3. Introduction of Selectable Markers into Plant Cells by Transformation

Several mutant cell lines have been reported, e.g., nuclear and cytoplasmic albinos, nitrate reductase deficient cell lines, streptomycin resistant lines, and a double mutant with nitrate reductase deficiency combined with streptomycin resistance (*40*), but such lines are naturally of infrequent occurrence or difficult to induce. Methods are now available for transforming plants, including the direct uptake of DNA into isolated plant protoplasts (*41,42*) and *Agrobacterium*-mediated vector delivery (*see* Chapters 28–30, this volume). The latter technique, involving disarmed vectors combined with regenerating leaf explants (*43*), is particularly useful in members of the *Solanaceae*. Several bacterial genes, including those conferring resistance to chloramphenicol, kanamycin, methotrexate, hygromycin, and bleomycin (*44*), will, under the correct promoter, express in transformed plants. Protoplasts, isolated directly from such transformed plants or from cell suspensions initiated from explants, are being employed in somatic hybridization since it is possible, with the use of antibiotic-containing media, to select hybrid and cybrid cells following protoplast fusion.

4.1.4. Use of Electromanipulation to Stimulate Protoplast Division and Plant Regeneration

Following the development of electroporation as a routine technique to transform animal cells (*45*), this approach has been applied successfully to several protoplast systems. This method, in which protoplasts are exposed to high-voltage electric pulses of short duration (micro- or millisecond), is believed to induce pores in the plasma membrane, through which exogenously supplied DNA (*46, see also* Chapter 30, this volume), RNA (*47*), or virus particles can enter (*48*).

An additional, important application of electroporation is that such electric fields also stimulate protoplast division and increase the throughput of protoplast-derived cell colonies (*49*). Significantly, such cell colonies from electropulsed protoplasts also show an increased shoot regeneration capability (*50*). The mechanisms involved in this stimulation are, at present, unknown. However, they may be similar in their effect to long-duration, low-voltage electrical impulses reported to stimulate shoot regeneration from callus cultures through their influence on auxin metabolism (*51,52*).

References

1. Davey, M. R. and Cocking, E. C. (1972) Uptake of bacteria by isolated higher plant protoplasts. *Nature (Lond.)* **239,** 455,456.
2. Davey, M. R. and Power, J. B. (1975) Polyethylene glycol-induced uptake of microorganisms into higher plant protoplasts: an ultrastructural study. *Plan. Sci. Lett.* **5,** 269–274.
3. Bonnett, H. T. and Eriksson, T. (1974) Transfer of algal chloroplasts into protoplasts of higher plants. *Planta* **120,** 71–79.
4. Otsuki, Y. and Takebe, I. (1973) Infection of tobacco mesophyll protoplasts by cucumber mosaic virus. *Virology* **52,** 433–438.
5. Ohgawara, T., Uchimiya, H., and Harada, H. (1983) Uptake of liposome-encapsulating plasmid DNA by plant protoplast and molecular fate of foreign DNA. *Protoplasma* **116,** 145–148.
6. Thom, M., Komor, E., and Maretzki, A. (1982) Vacuoles from sugarcane suspension cultures. II. Characterization of sugar uptake. *Plant Physiol.* **69,** 1320–1325.
7. Uchimiya, H., Fushimi, T., Hashimoto, H., Harada, H., Syono, K., and Sugawara, Y. (1986) Expression of a foreign gene in callus derived from DNA-treated protoplasts of rice (*Oryza sativa* L.). *Molec. Gen. Genet.* **206,** 204–207.
8. Fowke, L. C., Griffing, L. R., Mersey, B. G., and Van der Valk, P. (1983) Protoplasts for studies of the plasma membrane and associated cell organelles, in *Protoplasts 1983, Lecture Proceedings 6th Internatl. Protoplast Symposium, Basel. Experientia Supplementum* **46,** 101–110.
9. Bengochea, T. and Dodds, J. H. (1986) *Plant Protoplasts, a Biotechnological Tool for Plant Improvement* (Chapman and Hall, London, New York).
10. Evans, D. A., Sharp, W. R., and Medina-Filho, H. P. (1984) Somaclonal and gametoclonal variation. *Amer. J. Bot.* **71,** 759–774.
11. Brown, C., Lucas, J. A., Crute, I. R., Walkey, D. G. A., and Power, J. B. (1986) An assessment of genetic variability in somacloned lettuce plants (*Lactuca sativa*) and their offspring. *Ann. Appl. Biol.* **109,** 391–407.
12. Kinsara, A., Patnaik, S. N., Cocking, E. C., and Power, J. B. (1986) Somatic hybrid plants of *Lycoperisicon esculentum* Mill and *Lycopersicon peruvianum* Mill. *J. Plant Physiol.* **125,** 225–234.

13. Pirrie, A. and Power, J. B. (1986) The production of fertile, triploid somatic hybrid plants [*Nicotiana glutinosa* (n) + *N. tabacum* (2n)] via gametic:somatic protoplast fusion. *Theor. Appl. Genet.* **72,** 48–52.
14. Powell, A. J., Lloyd, C. W., Slabas, A. R., and Cove, D. J. (1980) Demonstration of the microtubular cytoskeleton of the moss, *Physcomitrella patens,* using antibodies against mammalian brain tubulin. *Plant Sci. Lett.* **18,** 401–404.
15. Ouo, K., Ohyama, K., and Gamborg, O. L. (1979) Regeneration of the liverwort *Marchantia polymorpha* L. from protoplasts isolated from cell suspension culture. *Plant. Sci. Lett.* **14,** 225–229.
16. Murashige, T. and Skoog, F. (1962) A revised medium for rapid growth and bioassays with tobacco tissue cultures. *Physiol. Plant.* **15,** 473–497.
17. Watanabe, I., Espinas, C. R., Berija, N. S., and Alimagno, B. V. (1977) The utilization of the Azolla–Anabaena complex as a nitrogen fertilizer for rice. *IRRI Res. Paper Ser.* **11,** 1–15.
18. Attree, S. M. and Sheffield, E. (1986) An evaluation of ficoll density gradient centrifugation as a method for eliminating microbial contamination and purifying plant protoplasts. *Plant Cell Rep.* **5,** 288–291.
19. Power, J. B. and Chapman, J. V. (1985) Isolation, culture and genetic manipulation of plant protoplasts in *Plant Cell Culture—A Practical Approach* (Dixon, R. A., ed.), IRL Press, pp. 37–66.
20. Kao, K. N. and Michayluk, M. R. (1975) Nutritional requirements for growth of *Vicia hajastana* cells and protoplasts at very low population density in liquid media. *Planta* **126,** 105–110.
21. Kao, K. N. (1977) Chromosomal behavior in somatic hybrids of soybean—*Nicotiana glauca. Mol. Gen. Genet.* **150,** 225–230.
22. Brown, C., Lucas, J. A., and Power, J. B. (1987) Plant regeneration from protoplasts of a wild lettuce species (*Lactuca saligna* L.). *Plant Cell Rep.* **6,** 180–182.
23. Bohorova, N. E., Cocking, E. C., and Power, J. B. (1986) Isolation, culture and callus regeneration of protoplasts of wild and cultivated *Helianthus* species. *Plant Cell Rep.* **5,** 256–258.
24. Power, J. B., Cummins, S. E., and Cocking, E. C. (1970) Fusion of isolated plant protoplasts. *Nature* **225,** 1016–1018.
25. Revilla, M. A., Ochatt, S. J., Doughty, S., and Power, J. B. (1987) A general strategy for the isolation of mesophyll protoplasts from deciduous fruit and nut tree species. *Plant Sci.* **50,** 133–137.
26. Yarrow, S. A., Cocking, E. C., and Power, J. B. (1987) Plant regeneration from cultured cell-derived protoplasts of *Pelargonium aridum, P.* x *hortorum* and *P. peltatum. Plant Cell Rep.* **6,** 102–104.
27. Partanen, C. R., Power, J. B., and Cocking, E. C. (1980) Isolation and division of protoplasts of *Pteridium aquilinum. Plant Sci. Lett.* **17,** 333–338.
28. Frearson, E. M., Power, J. B., and Cocking, E. C. (1973) The isolation, culture and regeneration of *Petunia* leaf protoplasts. *Dev. Biol.* **33,** 130–137.
29. Shillito, R. D., Paszkowski, J., and Potrykus, I. (1983) Agarose plating and a bead-type culture technique enable and stimulate development of protoplast-derived colonies in a number of plant species. *Plant Cell Rep.* **2,** 244–247.
30. dos Santos, A. V. P., Outka, D. E., Cocking, E. C., and Davey, M. R. (1980) Organogenesis and somatic embryogenesis in tissues derived from leaf protoplasts and leaf explants of *Medicago sativa. Z. Pflanzenphysiol.* **99,** 261–270.

31. Hammatt, N., Kim, H.-I., Davey, M. R., Nelson, R. S., and Cocking, E. C. (1987) Plant regeneration from cotyledon protoplasts of *Glycine canescens* and *G. clandestina*. *Plant Science* **48,** 129–135.
32. Gilmour, D. M., Davey, M. R., Cocking, E. C., and Pental, D. (1987) Culture of low number of forage legume protoplasts in membrane chambers. *J. Plant Physiol.* **126,** 457–465.
33. Lu, D. Y., Davey, M. R., and Cocking, E. C. (1983) A comparison of the cultured behaviour of protoplasts from leaves, cotyledons and roots of *Medicago sativa*. *Plant Sci. Lett.* **31,** 87–99.
34. Gilmor, D. M., Davey, M. R., and Cocking, E. C. (1987) Plant regeneration from cotyledon protoplasts of wild *Medicago* species. *Plant Science* **48,** 107–112.
35. Abdullah, R., Cocking, E. C., and Thompson, J. A. (1986) Efficient plant regeneration from rice protoplasts through somatic embryogenesis. *Biotechnology* **4,** 1087–1090.
36. Zimmerman, U. and Scheurich, P. (1981) High frequency fusion of plant protoplasts by electric fields. *Planta* **151,** 26–32.
37. Bates, G. W. (1985) Electrical fusion for optimal formation of protoplast heterkaryons in *Nicotiana*. *Planta* **165,** 217–224.
38. Watts, J. W. and King, J. M. (1984) A simple method for large scale electrofusion and culture of plant protoplasts. *BioScience Rep.* **4,** 335–342.
39. Pental, D., Hamill, J. D., Pirrie, A., and Cocking, E. C. (1986) Somatic hybridisation of *Nicotiana tabacum* and *Petunia hybrida*. Recovery of plants with *P. hybrida* nuclear genome and *N. tabacum* chloroplasts genome. *Mol. Gen. Genet.* **202,** 342–347.
40. Hamill, J. D., Pental, D., Cocking, E. C., and Muller, A. J. (1983) Production of a nitrate reductase deficient streptomycin resistant mutant of *Nicotiana tabacum* for somatic hybridisation studies. *Heredity* **50,** 197–200.
41. Paszkowski, J., Shillito, R. D., Saul, M., Mandak, V., Hohn, T., Hohn, B., and Potrykus, I. (1984) Direct gene transfer to plants. *EMBO J.* **3,** 2717–2722.
42. Shillito, R. D., Saul, M. W., Paszkowski, J., Muller, M., and Potrykus, I. (1985) High efficiency direct gene transfer to plants. *Biotechnology* **3,** 1099–1103.
43. Horsch, R. B., Fry, J. E., Hoffmann, N. L., Eichholtz, D., Rogers, S. G., and Fraley, R. T. (1985) A simple and general method for transferring genes to plants. *Science* **227,** 1229–1231.
44. Hille, J., Verheggen, F., Roelvink, P., Franissen, H., van Kammen, A., and Zabel, P. (1986) Bleomycin resistance: a new dominant selectable marker for plant cell transformation. *Plant Molec. Biol.* **7,** 171–176.
45. Neumann, E., Schaefer-Rider, M., Wang, Y., and Hofschneider, P. H. (1982) Gene transfer into mouse myeloma cells by electroporation in high electric fields. *EMBO J.* **1,** 841–845.
46. Fromm, M. E., Taylor, L. P., and Walbot, V. (1986) Stable transformation of maize after gene transfer by electroporation. *Nature* **319,** 791–793.
47. Nishigushi, M., Langridge, W. H. R., Szalay, A. A., and Zaitlin, M. (1986) Electroporation-mediated infection of tobacco leaf protoplasts with tobacco mosaic virus RNA and cucumber mosaic virus RNA. *Plant Cell Reports* **5,** 57–60.
48. Nishigushi, M., Sato, T., and Motoyoshi, F. (1987) An improved method for electroporation into plant protoplasts: infection of tobacco protoplasts by tobacco mosaic virus particles. *Plant Cell Reports* **6,** 90–93.
49. Rech, E. L., Ochatt, S. J., Chand, P. K., Power, J. B., and Davey, M. R. (1987) Electroenhancement of division of plant protoplast-derived cells. *Protoplasma* **141,** 169–176.

50. Ochatt, S. J., Chand, P. K., Rech, E. L., Davey, M. R., and Power, J. B. (1988) Electroporation-mediated improvement of plant regeneration from Colt cherry (*Prunus avium* x *pseudocerasus*) protoplasts. *Plant Sci.* **54,** 165–169.
51. Goldsworthy, A. and Rathore, K. S. (1985) The electrical control of growth in plant tissue cultures: the polar transport of auxin. *J. Exp. Bot.* **36,** 1134–1141.
52. Rathore, K. S. and Goldworthy, A. (1985) Electrical control of shoot regeneration in plant tissue cultures. *Biotechnology* **3,** 1107–1109.

Chapter 25

Plant Protoplast Enucleation by Density Gradient Centrifugation

Houri Kalilian Fakhrai, Nazmul Haq, and Peter K. Evans

1. Introduction

A number of different methods have been developed for the enucleation of cells, particularly mammalian cells. Centrifugation of cells adhering to a surface may lead to the nucleus being drawn out from the cytoplasm, resulting in the formation of an enucleated cell or cytoplast and a nucleus surrounded by a thin layer of cytoplasm bounded by a plasma membrane (mini-cell). In those cases where cells do not adhere sufficiently firmly to a surface, enucleation has been achieved by centrifugation in density gradients. This method takes advantage of the differing densities of the nucleus and the cytoplasm. When cells are centrifuged at high speed in a suitable density gradient, the nucleus and the cytoplasm tend to float in those regions of the gradient matching their own densities. If the distance between the regions with densities corresponding to that of the

From: *Methods in Molecular Biology*, vol. 6, *Plant Cell and Tissue Culture*
Edited by Jeffrey W. Pollard and John M. Walker, ©1990 by The Humana Press

cytoplasm and that of the nucleus is relatively long, considerable stretching forces are generated. Under optimum conditions, this force is usually sufficient to enucleate a high percentage of the cells. The use of density gradients and high-speed centrifugation for generating cytoplasts and mini-cells is much less explored in plant than in mammalian cells.

Discontinuous sucrose gradients (*1*) and mannitol/sucrose gradients (*2, 3*) have been used to produce enucleated plant protoplasts and mini-protoplasts, the equivalent to animal mini-cells. Percoll, a density gradient material consisting of colloidal silica coated with polyvinyl pyrrolidone (PVP) has also been used for cytoplast and mini-protoplast formation (*4*). Percoll consists of particles with high molecular weight and low charge, and solutions exhibit a low colloid osmotic pressure. Consequently, solutions differing in density, but with a constant osmotic pressure, can be prepared.

2. Materials

1. 0.21*M* $CaCl_2 \cdot 2H_2O$. Sterilize by autoclaving.
2. Mannitol solutions, differing in molarity as listed: 0.50*M*, 0.52*M*, 0.60*M*, 0.70*M*, 0.92*M*. Sterilize by autoclaving.
3. Percoll (Pharmacia Fine Chemicals, Uppsala, Sweden). This is a sterilized solution. Use under sterile conditions. Store in a refrigerator (0–4°C).
4. CPW5M Medium: KH_2PO_4, 27.2 mg/L; KNO_3, 101 mg/L; $CaCl_2 \cdot 2H_2O$, 1480 mg/L; $MgSO_4 \cdot 7H_2O$, 246 mg/L; $CuSO_4 \cdot 5H_2O$, 0.025 mg/L; KI, 0.16 mg/L; 5% Mannitol. Sterilize by autoclaving.
5. 1.5% cellulase "onozuka" R10 and 0.3% "macerozyme" R10 (both Kinki, Yakult Pharmaceutical Industry Co. Ltd, Nishinomiya, Japan). Enzymes are dissolved in CPW5M, pH adjusted to 5.8. Centrifuge at 700*g* for 10 min to remove debris. Filter sterilize the supernatant and store at –20°C. Do not freeze twice.
6. Fluorescein-diacetate (FDA), 5 mg/mL in acetone.
7. 4-6-diamidino-2-phenylindole (DAPI). DAPI is prepared (0.5 mg/mL) in a buffer solution containing 0.1*M* citric acid monohydrate and 0.2*M* anhydrous Na_2HPO_4 at pH 4.0 (*5*).
8. Fixative: 1% glutaraldehyde, 0.5*M* sorbitol in 50 m*M* Na-cacodylate buffer pH 7.2.
9. Rapidly growing cell suspension culture of winged bean. Subcultured 50:50 at 4-, 4-, 3-d interval prior to protoplast isolation.

10. Rapidly growing cell suspenion culture of crown gall of *Parthenocissus tricuspidata*, subcultured 50:50 at 4-, 4-, 3-d interval prior to protoplast isolation.

3. Method

1. The growth condition for the winged bean, and the methods for the development of cell suspensions and for the isolation and culture of protoplast have been described in Chapters 5 and 24 (*6*). Techniques for the maintenance of crown gall callus of *Partenocissus tricuspidata*, culture of cell suspension, and protoplast isolation have been described previously (*7*). However, the enzyme mixture used here consists of 1.5% onozuka and 0.3% macerozyme.
2. Iso-osmotic solutions of different densities are prepared using solutions of 0.21*M* $CaCl_2$, 0.50–0.92*M* mannitol and 5–50% Percoll (*see* Tables 1 and 2). These solutions are autoclaved separately. Volumes of each are added together under sterile conditions, in order to make up the required specific density and to achieve an osmolarity of 550 ± 15 mOsmol/kg H_2O.
3. Protoplasts, at a concentration of 2.5×10^5/mL in 2 mL of osmoticum containing 0.25*M* mannitol and 0.11*M* $CaCl_2$, are layered on to a discontinuous density gradient, prepared by successively layering solutions of decreasing density in 16-mL polypropylene tubes. Gradients PG1 and PG2, which differ in the combination of specific densities, are used for crown gall and winged bean protoplasts, respectively (*see* Notes 1–3).
4. Centrifuge at 60,000*g* at 12°C for 60 min (*see* Note 4). Centrifuge tubes are autoclaved before use. Centrifuge adaptors and adaptor tops are sterilized by immersion in 70% ethanol for 10 min, and air dried in a laminar flow cabinet. The sterilized tubes are loaded and balanced aseptically in a laminar flow cabinet.
5. After centrifugation, the fractionated material is removed under sterile conditions. Fractions F1, F2, and F3 are collected from successive interfaces, starting from the top of the gradient. To remove the Percoll, about 10 mL of CPW5M is added to each fraction, which is then centrifuged at 200*g* for 5 min. The pellet is resuspended in 2 mL of CPW5M.
6. To determine the proportion of cytoplasts, protoplasts, and miniprotoplasts in each fraction, samples from all three fractions are

Table 1
Composition, Osmolarity, and Specific Density of Discontinuous Iso-osmotic Percoll Gradient 1, PG1, Crown Gall

Osmolarity mOsmol/kg H_2O	Specific density g/mL	Total vol/ gradient mL	Components of gradient	Molarity of stock solution *M*	Vol added mL	Mixing ratio of component
550 ± 15	1.023	2	$CaCl_2$	0.21	1	1:1
			Mannitol	0.50	1	
550 ± 15	1.035	3	Mannitol	0.52	2.85	19:1
			Percoll	Undiluted	0.15	
550 ± 15	1.054	3	Mannitol	0.60	2.4	4:1
			Percoll	Undiluted	0.6	
550 ± 15	1.092	4	Mannitol	0.92	2	1:1
			Percoll	Undiluted	2	

Table 2
Composition, Osmolarity, and Specific Density of Discontinuous Iso-osmotic Percoll Gradient 2, PG2, Winged Bean

Osmolarity mOsmol/kg H_2O	Specific density g/mL	Total vol/ gradient mL	Components of gradient	Molarity of stock solution *M*	Vol added mL	Mixing ratio of component
550 ± 15	1.023	2	$CaCl_2$	0.21	1	1:1
			Mannitol	0.50	1	
550 ± 15	1.040	3	Mannitol	0.60	2.7	9:1
			Percoll	Undiluted	0.3	
550 ± 15	1.070	3	Mannitol	0.70	2.1	2.3:1
			Percoll	Undiluted	0.9	
550 ± 15	1.092	4	Mannitol	0.92	2	1:1
			Percoll	Undiluted	2	

fixed in 1% glutaraldehyde and 0.5*M* sorbitol in 50 m*M* Na-cacodylate buffer pH 7.2 for 30 min. Centrifuge at 100*g* for 5 min, and resuspend the pellet in CPW5M. A drop of DAPI is mixed with one drop of protoplast and subprotoplast fractions, and observed under the microscope.

7. Cell membrane integrity is monitored by staining with FDA. The stock solution of FDA (5 mg /mL) is diluted with CPW5M in the ratio of 1:24. One drop of diluted stain is added to one drop of subprotoplast fraction, and observed under the microscope.
8. All our observations are made with an "Olympus" microscope equipped with a mercury vapor lamp. Observations with DAPI are made with Neofluor Objectives, using a UV excitation filter in combination with a 470-nm suppression filter. In the case of FDA staining, a blue excitation filter is used in combination with a 510-nm suppression filter.
9. Fraction 1 (F1), which has the lowest density is always enriched with cytoplasts. Fraction 2 (F2), which has an intermediate density, is enriched with intact protoplasts (*see* Notes 5–9). Fraction 3 (F3), which has the highest density, is enriched with mini-protoplasts, multinucleated protoplasts, and isolated nuclei.

4. Notes

1. The combination of densities used in the gradient plays an important role in enucleation. Enucleation has been carried out on both protoplast systems using two different Percoll gradients (PG1 and PG2). The results showed the combination of higher densities in PG2 worked better for enucleation of winged bean protoplasts, whereas crown gall protoplasts were enucleated more effectively in PG1. An average 75 and 50% enucleation of winged bean protoplasts, and 50 and 85% enucleation of crown gall protoplasts, occurred with PG2 and PG1, respectively.
2. Sucrose-mannitol density gradients with or without cytochalasin B are not as effective for protoplast enucleation as Percoll density gradients.
3. Enucleation technique described here should be used on suspension culture derived protoplasts, and not on mesophyll protoplasts. Enucleation of mesophyll protoplasts is incomplete with both PG1 and PG2 gradients. With PG1, only one band in the interface between

Table 3
Influence of Time of Centrifugation and Gravity Forces on the Average Percentage Enucleation (Average PE) of Crown Gall Protoplasts

Gravity forces rotor	90-kg swing-out	60-kg swing-out	40-kg swing-out	20-kg swing-out	23-kg fixed angle SW41
30 min average PE	70 (± 5)	70 (± 8)	50 (± 7)	50 (± 9)	40 (± 3)
60 min average PE	95 (± 7)	85 (± 4)	65 (± 6)	62 (± 7)	45 (± 5)
90 min average PE	90 (± 9)	80 (± 9)	70 (± 4)	67 (± 3)	Not tested

Figures in brackets represent the standard deviation of the mean of samples of protoplasts with a total of 300 protoplasts counted for each treatment.

50–20% Percoll is obtained, whereas with PG2, two bands in the interfaces between 50–35% and 35–20% Percoll are obtained. No cytoplasts are obtained in any of these bands, instead there are quite a large number of protoplasts that appear to be budding. No chloroplasts are seen in the budded part of the protoplasts. Occasionally, where the band of cytoplasm connecting the two portions of the protoplasts becomes thin, the different portions of protoplasm pull apart into two separate structures.

4. The level of centrifugal force influences enucleation in both protoplast systems. The best results are obtained when protoplasts are centrifuged at 60,000*g* for 60 min in a swing-out rotor. The results obtained with crown gall protoplasts are shown in Table 3. A longer centrifugation time (90 min) compensates to some extent for a lower gravity force, but is not used routinely because protoplasts become elongated and acquire a long tube-like shape. Higher gravity forces (90,000*g*) in a swing-out rotor cause the protoplasts to disintegrate partially, and the enucleated protoplasts break up into many small fragments. This may cause a false reading on the number of cytoplasts. However, FDA staining test reveals that the plasma membranes of these fragments have been damaged.
5. Possibly the most important factor for a successful enucleation is the quality of the protoplasts used. Prior to enucleation, attempts should be made to optimize techniques for developing rapidly growing suspension cultures and for isolating, washing, and cleaning each protoplast system under investigation. When protoplasts with an average percentage viability of 95%, as measured by FDA staining, were used

Table 4
Influence of Protoplast Population Density and Average Percentage of Viability (Average PV) on Average Percentage of Enucleation (Average PE) of Crown Gall Protoplasts

	Protoplast/mL x10^5			
	1	2.5	5	7.5
Average PV = 95, average PE	50 (± 8)	85 (± 7)	75 (± 5)	55 (± 12)
Average PV = 80, average PE	40 (± 7)	64 (± 7)	60 (± 6)	40 (± 8)

Figures in brackets represent the standard deviation of the mean of samples of protoplasts with a total of 300 protoplasts counted for each treatment.

under optimum conditions, the average percentage of enucleation was 85%. When the average percentage viability was 85% under the same conditions, the average percentage enucleation was only 64% (*see* Table 4).

6. The population density of the protoplasts loaded on the top of the gradients before centrifugation influences the average percentage enucleation. Density of 2.5 x 10^5 protoplasts/mL over 12 mL of gradient worked best in our experiments (Table 4).
7. Protoplast storage conditions prior to enucleation affects their viability as measured by FDA staining test. Immediately after isolation, the protoplasts were 95% viable. These protoplasts were resuspended in CPW5M and incubated overnight, in the cold (0°C) or at room temperature. Protoplasts incubated at room temperature showed a higher viability than those incubated at 0°C (90% viability at room temperature, as compared to 75–80% at 0°C).
8. To check on the effect of high-speed centrifugation on the viability of protoplasts, fractions consisting of cytoplasts and mini-protoplasts were stained with FDA. The result shows no significant decrease in the viability of cytoplasts (85% viable). Viability of mini-protoplasts could not be detected by FDA staining test. This is not unexpected, since mini-protoplasts contain only a small amount of cytoplasm, and it is enzymes in the cytoplasm that convert FDA to fluorescein. The polar FDA passes through the plasma membrane into the cytoplasm, where it is cleaved by esterase activity, converting the FDA to fluorescein, which, being less polar, passes out through the plasma membrane less readily. Consequently, when the plasma membrane is

intact, the fluorescein molecules accumulate and cause the cell to fluoresce brightly.

9. Performing enucleation under aseptic conditions is necessary only if the intention is to culture the enucleated protoplasts subsequently.

References

1. Wallin, A., Glimelius, K., and Eriksson, T. (1978) Enucleation of plant protoplasts by Cytochalasin B. *Z. Pflanzenphysiol.* **87,** 333–340.
2. Bracher, M. and Sher, N. (1981) Fusion of enucleated protoplasts with nucleated mini protoplasts in onion (*Allium cepa* L.). *Plant Sci. Lett.* **23,** 95–101.
3. Lesney, M. S., Callow, P. C., and Sink, K. C. (1986) A technique for bulk production of cytoplasts and mini protoplasts from suspension culture derived protoplasts. *Plant Cell Rep.* **5,** 115–118.
4. Lorz, H., Paskowski, J., Dierks-Venthing, C. and Potrykus, I. (1981) Isolation and characterization of cytoplasts and mini protoplasts derived from protoplasts from cultured cells. *Physiol. Plant* **53,** 385–391.
5. Hull, H. M., Hoshaw, R. W., and Wang, J. C. (1982) Cytofluorometric determination of nuclear DNA in living and preserved algae. *Stain Tech.* **57,** 273–281.
6. Wilson, V. M., Haq, N., and Evans, P. K. (1985) Protoplast isolation, culture and plant regeneration in the winged bean *Psophocarpus tetragonolobus* (L) D. C. *Plant Sci.* **41,** 61–68.
7. Power, J. B., Frearson, E. M., Hayward, C., and Cocking, E. C. (1975) Some consequences of the fusion and selective culture of *Petunia* and *Parthenocissus* protoplasts. *Plant Sci. Lett.* **5,** 197–207.

Chapter 26

Isolation and Transplantation of Nuclei into Plant Protoplasts

Praveen K. Saxena and John King

1. Introduction

Plant protoplasts can internalize a variety of particles ranging from molecules (e.g., ferritin) to entire (algal) protoplasts, a size range of 0.09–23 μm (*1*). This capacity to take up foreign bodies provides a powerful tool with which to investigate the functions of cell organelles. Also, protoplasts are the material of choice from which to isolate organelles, because they are easy to rupture, and isolates are not contaminated with cell wall fragments.

We have developed efficient procedures for the isolation of nuclei and for their transplantation into protoplasts (*2,3,4*) as a means to transfer DNA between sexually incompatible species. The process of isolation involves a three-step destabilization and rupture of protoplasts to release all cell constituents, followed by the mechanical separation of nuclei using filters of appropriate sizes (Fig. 1). Pure preparations of nuclei that are morpho-

From: *Methods in Molecular Biology*, vol. 6, *Plant Cell and Tissue Culture*
Edited by Jeffrey W. Pollard and John M. Walker, ©1990 by The Humana Press

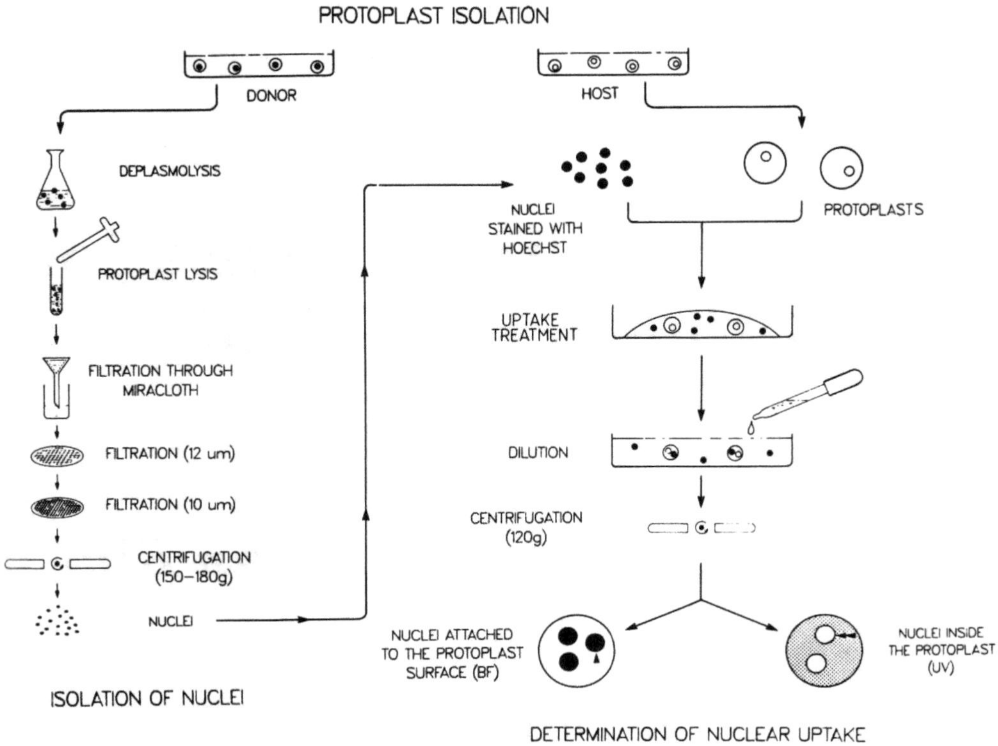

Fig. 1. A diagram showing various steps in the isolation and transplantation of nuclei. BF, bright light field; UV, ultraviolet light field. The nuclei attached to the surface of the protoplast appear blue (arrow) in BF, and those internalized by the protoplast fluoresce (arrows) in UV.

logically intact and biologically active can be obtained in high yield (70–90%) within 30 min after the isolation of the protoplasts from which the nuclei are to be obtained.

The uptake of nuclei into protoplasts is achieved by a modified protoplast fusion method, in which polyethylene glycol and Ca^{2+} are used. A common problem is distinguishing which nuclei have been taken up, and which are outside the cell attached to the plasmalemma. Recently, it has been possible to estimate accurately the frequency of nuclear uptake by employing a simple technique of differential staining (Fig. 1). This article describes these methods of the uptake of nuclei by protoplasts and the determination of uptake frequency.

2. Materials

2.1. Cells

1. Cell suspension cultures of a pantothenate-requiring auxotrophic variant of *Datura innoxia* P. Mill, hereafter referred to as Pn1.
2. Cell suspension cultures of *Brassica nigra*.

2.2. Isolation of Protoplasts

1. Enzyme solution A: 1% Cellulase Onozuka "R-10," 1% Cellulase "RS," 0.5% each of Macerozyme, Rhozyme "HP-150," Driselase, 5 m*M* $CaCl_2 \cdot 2H_2O$, and 0.5*M* mannitol.
2. Enzyme solution B: 1% Cellulase Onozuka "R-10," 0.5% each of Cellulase "RS," Driselase, Macerozyme, Rhozyme "HP-150," 5 m*M* $CaCl_2 \cdot 2H_2O$, and 0.6*M* mannitol.

   To prepare the enzyme solutions, dissolve twice the required quantity of the enzymes in 100 mL of distilled water. Centrifuge at 2500*g* for 15 min at 4°C, and then adjust the pH of the supernatant to 5.7. Filter sterilize the solution, dispense 10 mL aliquots into sterile tubes, and store at –20°C. Mix with an equal volume of autoclaved double-strength mannitol (1*M* for solution A and 1.2*M* for B) before use.
3. 0.6*M* sucrose, 5 m*M* $CaCl_2 \cdot 2H_2O$.
4. 0.6*M* and 0.5*M* mannitol, each supplemented with 5 m*M* $CaCl_2 \cdot 2H_2O$.
5. 85-μm screen.

2.3. Isolation of Nuclei

1. Nuclei Isolation Buffer (NIB): 10 mM MES, 0.2*M* sucrose, 0.02% Triton X-100, 2.5 m*M* EDTA, 2.5 m*M* dithiothreitol, 0.1 m*M* spermine, 10 m*M* NaCl, and 10 m*M* KCl (pH 5.3). Filter sterilize and keep at 4°C. Use within a week of preparation.
2. NIB-I: same as above, but lacking Triton X-100.
3. Miracloth.
4. Polycarbonate filters of 12 and 10 μm pore size (Nucleopore).

2.4. Uptake of Nuclei

1. 0.35*M* mannitol with $CaCl_2 \cdot 2H_2O$.
2. Hoechst "33258"; store in dark at 4°C.
3. 0.1% Evans' Blue prepared in 0.5*M* mannitol.

4. Uptake Inducing Solution (UIS): 30% polyethylene glycol, 0.1*M* $CaCl_2 \cdot 2H_2O$, 0.05*M* glycine, and 5% dimethyl sulfoxide (pH 6.8).

3. Methods

Perform all of the following manipulations in a sterile chamber.

3.1. Cell Maintenance

1. Maintain the cells in a preselected nutrient medium that supports the maximal proliferation of cells. Subculture the cells every week.
2. Cell suspension cultures should be maintained in diffuse light (10–15 μE/m/S) at 25–28°C on a gyrotory shaker (150 rpm).

3.2. Isolation of Protoplasts

1. Use actively growing cell suspension cultures, preferably 2–3 d old, or at a time when the cells are in exponential growth phase.
2. Collect the cells over Miracloth, using vacuum filtration. Transfer 2–2.5g of Pn1 cells to 10 mL of enzyme solution A dispensed into a 10 cm diameter Petri dish.
3. For the isolation of *B. nigra* protoplasts, digest 2–2.5g of cells in 10 mL of enzyme solution B. Prepare five Petri dishes.
4. Incubate the dishes at 25°C for 2–3 h on a horizontal shaker (50 rpm) in the dark. After complete digestion (monitored by microscopic examination), filter the suspensions through 85 μm nylon screens and centrifuge at 120*g* for 5 min. Remove the supernatant.
5. Dissolve the pellet in 5 mL of 0.6*M* sucrose, and layer 1 mL of mannitol solution on top of the sucrose solution. Use 0.6*M* mannitol to process the protoplasts to be used for nuclei isolation and 0.5*M* for those to be used as the host. Centrifuge at 120*g* for 5 min.
6. Withdraw the floating protoplasts from the interface between the sucrose and mannitol solutions with a Pasteur pipet. Dilute the protoplast suspension with 8–10 mL of 0.5 or 0.6*M* mannitol and centrifuge (120*g*, 5 min).
7. Suspend the pellet in mannitol solution.

3.3. Isolation of Nuclei

1. Pellet the protoplasts from which the nuclei are to be isolated, and resuspend the pellet in ice-cold NIB. Mix gently by tilting, and leave

the tubes on ice for 5–7 min. No more than 0.2 mL of pelleted protoplasts should be suspended in 15 mL of NIB.

2. Transfer the suspension, using a Pasteur pipet, to a glass homogenizer, and apply 10–15 gentle strokes. A syringe with an 18–26 gage needle can be used instead of the homogenizer. Five to ten passages through the needle will rupture the protoplasts.
3. Pass the resulting homogenate through two layers of Miracloth, presoaked in NIB, to remove large debris and partially ruptured protoplasts.
4. Filter the suspension through a polycarbonate filter (pore size 12 µm). A maximum of 15 mL of suspension should be filtered with one filter. The size of the filter should be determined by the size of the nuclei. For example, the diameter of *Brassica nigra* nuclei is approx. 8 µm, and the filters used for their isolation have pore sizes of 12 and 10 µm.
5. Repeat step 4 with another filter of lesser diameter (pore size 10 µm). Filtration should be done directly into centrifuge tubes. Centrifuge at 120–150*g*, depending upon the size of the nuclei. The speed of centrifugation required to pellet the nuclei may vary from one system to another, and should be determined empirically. Collect the supernatant in fresh tubes, and recentrifuge at a speed relatively higher than that used in the first centrifugation; this step improves the final yield, since not all nuclei will pellet in one cycle.
6. Pool the nuclear pellets from all tubes, mix with 5 mL NIB-I, and pellet the nuclei again by centrifugation. Repeat the step once more. An off-white pellet represents low contamination of starch grains; a white pellet indicates the presence of more starch than nuclei. Starch can be removed, to a certain extent, by repeated washing (2–4 x), but this step is not necessary if the nuclei are to be used for uptake. Alternatively, pellet the nuclei and remove the supernatant. Gently pour 200–300 µL of NIB-1 along the side of the tube and allow to stand for about a minute. Tilt the tube slightly, and collect the NIB-1 after 2 min, without disturbing the pellet. Repeat this step to collect more nuclei. The final yield of the nuclei by this method will be low, but with little starch.
7. Keep the nuclei at 4°C in a refrigerator, or leave the tubes in the ice bucket if the nuclei are to be used shortly thereafter.
8. The nuclear preparations obtained by this procedure are free of cytoplasmic contamination, and the nuclei are morphologically intact (Fig. 2A).

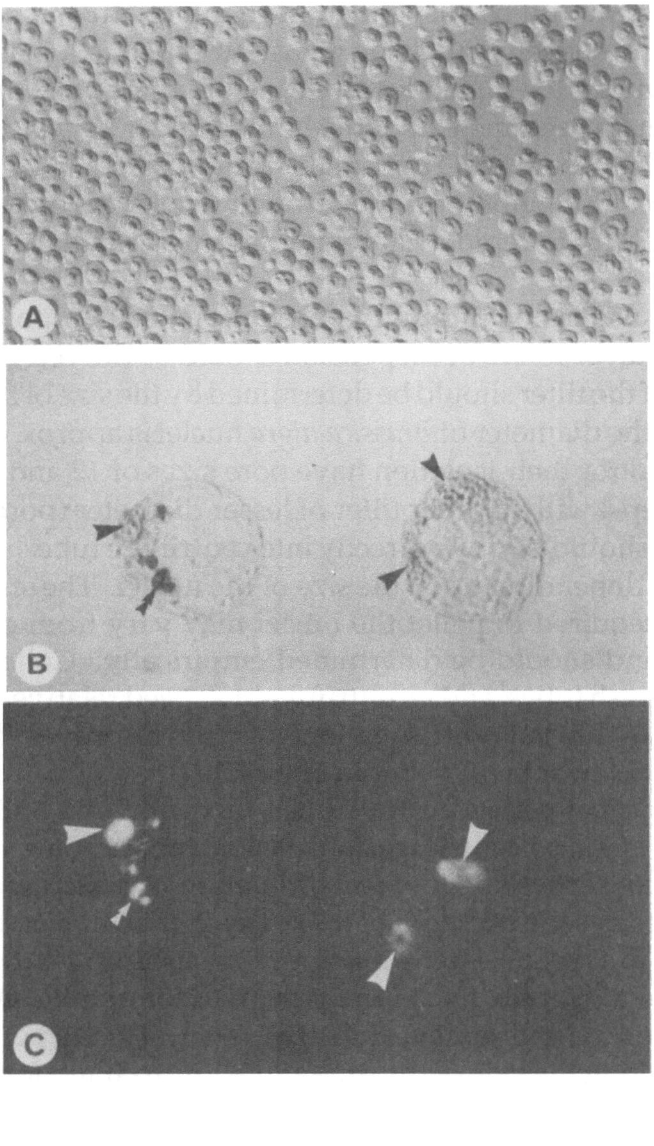

Fig. 2. The technique of differential staining to monitor the uptake of nuclei by protoplasts. The uptake is carried out using nuclei stained with Hoechst "33258," and the whole population is stained with Evans' Blue thereafter. Since the figures are in black and white, the blue color of the stain Evans' Blue and the fluorescence of Hoechst "33258" appear to be black and white, respectively. **A.** Nuclei isolated by the technique illustrated in Fig. 1. **B.** A protoplast (left) with a nucleus attached to its surface and, hence, stained blue with Evans' Blue (arrows). Note that the nucleus inside the same protoplast (arrow) and the nuclei inside the other protoplast (right, arrow) are not stained with Evans' Blue. **C.** The same protoplasts in UV field. Note that the nuclei inside the protoplasts do not show blue color, but do fluoresce (arrow). Figures **B** and **C** from ref. *4*.

3.4. Uptake of Nuclei by Protoplasts

1. Adjust the density of receptor protoplasts (Pn-1) to 2 x 10^6/mL and of the nuclei (*B. nigra*) to 10^8/mL.
2. Mix equal volumes of the protoplast and nuclear suspensions to obtain approximately a 1:50 ratio (protoplasts:nuclei). In the event of poor nuclear yield, the ratio can be as low as 1:20 without affecting the final results significantly.
3. Pipet 500 µL of the mixture obtained in step 2 into a Petri dish (35-mm diameter) in the form of a drop. Allow to settle for 5 min.
4. Add 100 µL of UIS at the edge of the drop. Repeat the step 4 x at 30-s intervals, using a total vol of 500 µL of the UIS. Incubate for 15–20 min.
5. Dilute the UIS gradually with 0.35*M* mannitol, using 2-mL aliquots every 5 min.
6. Transfer the suspension to centrifuge tubes, and centrifuge at 120*g* for 5 min.
7. Purify the protoplasts over 0.6*M* sucrose, as described earlier, and suspend in 0.5*M* mannitol.

3.5. Determination of Uptake Frequency

1. Stain the nuclei with 0.02% Hoechst "33258", for 5–10 min at 0°C. Remove the excess stain by washing with NIB-I, 3–4 x.
2. Carry out the uptake as described in the preceding section.
3. Mix 500–1000 µL of the protoplast suspension (from step 7) with an equal volume of 0.1% Evans' Blue solution (final concentration 0.05%) for 5 min. Wash the protoplasts with 0.5*M* mannitol to remove the excess stain.
4. Place a drop of the suspension on a slide, gently cover with a coverslip, and examine under the ultraviolet (UV) and bright light (BF) fields (Fig. 1) of a fluorescence microscope. Use excitation filters UG1 and BG38 in combination with a barrier filter K575.
5. The nuclei showing only the fluorescence of Hoechst without blue color are unambiguously inside the protoplast, and those with both the fluorescence and the blue color are outside the protoplasts (Fig. 2B,C). Score at least 500 protoplasts to estimate the uptake frequency.

4. Notes

The procedures described here were developed for the isolation and transplantation of nuclei from a cell suspension of *B. nigra* into the protoplasts of a pantothenate auxotroph of *D. innoxia*. However, the techniques *per se* have been used successfully in a number of different investigations of the isolation and transplantation of nuclei. The following factors should be paid special attention while applying these methods to other systems.

1. Optimize the process of protoplast isolation to recover maximal yields, as the uptake treatment requires a large number of nuclei. The enzyme composition of the isolation mixture should be effective in releasing the protoplasts within 2–3 h of incubation. Avoid overnight incubation, because the protoplasts released after longer exposure to enzymes produce nuclei that tend to stick together.
2. The pH of the NIB is very critical for the recovery of maximal yield and cleanliness of the nuclei, and thus, should be chosen after rigorous evaluation of a pH range from pH 4–7.
3. In certain cases, the starch content in nuclear preparations can be reduced by preculturing the cells in the dark in a low- to no-sucrose medium for a week or so.
4. For the experiment on the determination of uptake frequency, the nuclei, following staining with Hoechst, should be thoroughly washed to remove excess stain, which otherwise could stain the nuclei of the host protoplasts, creating an artifact in the estimation of uptake frequencies.
5. Only morphologically intact protoplasts should be scored to determine uptake frequency, because the nuclei of damaged protoplasts may have been stained with the Hoechst dye leached from pretreated, isolated nuclei.

References

1. Fowke, L. C. (1985) Ultrastructural cytology of cultured tissues, cells and protoplasts, in *Cell Cultures and Somatic Cell Genetics of Plants*, vol 3 (Academic, New York), pp. 323–339.
2. Saxena, P. K., Fowke, L. C., and King, J. (1985) An efficient procedure for isolation of nuclei from plant protoplasts. *Protoplasma*, **128**, 184–189.
3. Saxena, P. K., Mii, M., Crosby, W. L., Fowke, L. C., and King, J. (1986) Transplantation of isolated nuclei into plant protoplasts: a novel technique for introducing foreign DNA into plant cells. *Planta* **168**, 23–32.
4. Saxena, P. K., Lui, Y., and King, J. (1987) Nuclear transplantation into plant protoplasts: Optimal conditions for induction and determination of nuclear uptake. *J. Plant Physiol.* **128**, 451–460.

Chapter 27

Induction of Hairy Roots by *Agrobacterium Rhizogenes* and Growth of Hairy Roots In Vitro

Christopher S. Hunter and Steven J. Neill

1. Introduction

The bacterial genus *Agrobacterium* includes two species of considerable interest to plant physiologists and pathologists alike. Infection by virulent strains of *A. tumefaciens* induces the formation of tumors and infection by *A. rhizogenes* the proliferation of roots in a wide range of dicotyledenous plants: crown-gall and hairy root diseases are not naturally recorded in monocotyledenous plants (*1*). Invasion of plant tissues by these free-living soil bacteria usually occurs at a wound site—possibly caused by insect or mechanical damage; nonvirulent strains of the bacteria may invade the plant, but only virulent strains give rise to symptoms of disease.

From: *Methods in Molecular Biology*, vol. 6, *Plant Cell and Tissue Culture*
Edited by Jeffrey W. Pollard and John M. Walker, ©1990 by The Humana Press

Tumor and hairy root formation occur as a result of genetic transformation of host plant cells. *A. tumefaciens* and *A. rhizogenes* harbor large (> 200 kb) plasmids—the tumor-inducing (Ti) and root-inducing (Ri) plasmids, respectively—and the transformed plant phenotypes arise from the stable integration into the host cell of bacterial plasmid DNA.

Ti- and Ri-plasmids contain three regions of importance with regard to transformation (2). These include the transferred or T-DNA (the segment of DNA that becomes stably incorporated into the plant cell genome), the *vir* or virulence region (essential for successful transformation), and a region coding for the catabolism of opines (amino acid derivatives that can be utilized by the bacteria as a source of carbon and nitrogen). The T-DNA includes a gene coding for the synthesis of a particular opine and, in the case of Ti-plasmids, three oncogenes that code for enzymes involved in the synthesis of indoleacetic acid and cytokinins (3). Strains of *A. rhizogenes* are classified on the basis of the principal opine, such as agropine, mannopine, or cucumopine, produced by transformed cells (4).

The T-DNA of Ri-plasmids consists of one or two segments (T_L and T_R DNA), depending on the bacterial strain (5). Sequence and functional homology has been demonstrated between T_R-DNA and the *aux-1* and *aux-2* genes of Ti-plasmids (6,7), although increased synthesis of indoleacetic acid (IAA) by transformed tissues has not been demonstrated directly. There is no homology between the *aux* genes and the T-DNA of strains of *A. rhizogenes* that possess only a single T-region, but there is a strong homology between these single T-regions and T_L-DNA (4). Both T_L- and T_R-DNA have rhizogenic functions, but it has been suggested that T_L-DNA is more important in determining hairy root induction (8).

Plant molecular biologists have utilized the Ti-plasmid of *A. tumefaciens* as a vector for introducing "foreign" genes into plant cells (*1* and *see* Chapters 28 and 29, this volume). Foreign DNA is inserted into the T-region; this is then transferred to the plant cells during infection, and the recombinant DNA incorporated into the host cell genome. Transformed plant tissue can then be used as the starting material from which to regenerate whole plants, which will contain the inserted "foreign" gene.

The regeneration of whole plants from tumors is difficult, because expression of the oncogenes leads to disorganized growth. However, it is possible to "disarm" the bacterial vector by removal of these oncogenes; bacterial strains harboring such disarmed plasmids / vectors are then used to infect plant tissue (*1,9*). Usually, transformation is achieved by coincubation of leaf disks with a suspension of *A. tumefaciens* containing a recombinant Ti-plasmid. Transformed plants are then regenerated from these leaf disks. Regeneration of plants in this way may be difficult, but the regeneration of transformed plants from hairy roots is less so; consequently, there is considerable interest in this latter technique.

Whole plants can be regenerated either spontaneously or by appropriate modification of the tissue culture medium in which the transformed tissues are incubated: plants so regenerated include tobacco, oil-seed rape, potato, cabbage, cucumber, fennel, and sweet potato (*5*). Ri-plasmids have been used for genetic engineering: a gene coding for glyphosate resistance (*10*), a chimeric leghemoglobin gene (*11*), and various other genes (*4,5*) have been inserted into plant cells using Ri-vectors. Plants regenerated from hairy roots usually possess an abnormal phenotype (*12*), although it may be possible to construct semi-disarmed vectors to eliminate undesirable aspects of these hairy root phenotypes (*4*).

1.2. Use of In Vitro Hairy Root Cultures

Hairy root cultures have now been established from a large number and wide range of species (*8,13*). They may well prove useful for studies of obligate root parasites or symbionts that cannot be grown in isolation, and for investigations into the nature of root–microbe relationships (*5,13*). Untransformed roots in axenic culture normally require the inclusion of growth regulators in the culture medium to allow continued growth; transformed roots grow on simple media without growth regulators and exhibit high growth rates (Fig. 1, ref. *14*). In vivo, many secondary plant products (e.g., berberine, hyoscyamine, nicotine) are produced in roots (*14,15*).

Attempts to produce secondary products in vitro from callus and/or suspension cultures derived from roots or shoots of appropriate parent plants have generally been unsuccessful, and have led to only one commercial system (*16*). Most attempts to produce secondary products in vitro have failed, either because the cells have not produced the compound in sufficient quantity, or because yields have been unpredictably variable. These two factors have usually been considered to be associated with the use of undifferentiated cells (*17,18*). Secondary-product production by roots in vitro has been achieved, but usually only with media containing growth regulators (*19*). Various groups have investigated hairy root cul-

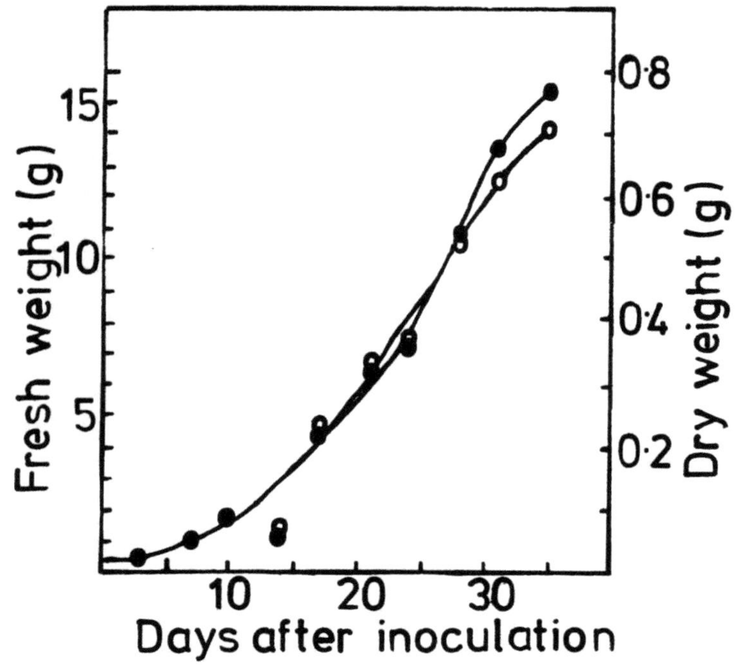

Fig. 1. Growth of hairy root suspension culture of *N. tabacum* cv. Wisconsin 38 (Hirons, unpublished); ● fresh wt, ○ dry wt.

tures for the synthesis of secondary metabolites (*8,14*); in such systems, transformed roots are grown in flasks, stirred tank bioreactors, or in immobilized systems where the culture medium is circulated over stationary roots held in a column (*15,20*). Hairy roots can produce the same spectrum of compounds as do nontransformed roots, at similar or increased concentrations (*21*).

For the production of some secondary metabolites, these systems do appear to offer considerable potential: perhaps even more potential will be realized when the polygenic, secondary-metabolite biosynthesis pathways are fully determined biochemically and can be "turned on" in transformed hairy roots. It may be possible to select for high-yielding cell lines (*22*) or enhance product synthesis and release through treatment with elicitors; for example, treatment of hairy roots of *Nicotiana tabacum* with elicitors derived from the cell walls of *Botrytis fabae* led to increased nicotine synthesis and release (Fillingham, unpublished). In addition, some hairy root cultures have been found to release a significant proportion of the secondary product into the culture medium—a bonus for the harvest of cell products. Secondary-metabolite production by transformed roots has been shown to be stable during many generations. The specific objectives of the following protocol are to inoculate an appropriate plant species with *A. rhizogenes*, to induce the formation of hairy roots, to clear the roots of bacteria, and to culture them in vitro.

2. Materials

1. Bacterium: The strain of *A. rhizogenes* we have used is LBA 9402, obtainable from Molbas Agrobacterium Collection, Leiden, Netherlands. This strain contains the plasmid pRi 1855, induces synthesis of the opine agropine, and has two regions of T-DNA, T_L- and T_R-DNA (*4*). It should be maintained on yeast mannitol agar (YMA) in Petri dishes at an ambient temperature of 18–25°C, and subcultured at regular intervals. However, it can be stored satisfactorily on a YMA slope in the dark at 5°C for several months between subcultures.
2. Yeast mannitol broth (YMB): 0.4 g/L yeast extract, 10g/L mannitol, 0.5 g/L K_2HPO_4, 0.2 g/L $MgSO_4 \cdot 2H_2O$, 0.1 g/L NaCl, pH 7.0. For yeast mannitol agar (YMA), add 15 g/L agar. Autoclave at 106 kPa, 121°C for 15 min.
3. *Beta vulgaris* cv. Boltardy (red beetroot), available from Suttons Seeds Ltd., Torquay, Devon, UK.
4. *Nicotiana tabacum*, available from Thompson and Morgan (Ipswich), Ltd., Ipswich, IP2 OBA, UK (*see* Note 1).
5. Tissue culture medium: Plants and excised hairy roots are grown in hormone-free Murashige and Skoog (MS) medium (*see* Appendix), containing 30 g/L sucrose, pH 5.7. For agar-based cultures, Oxoid L28 agar (Oxoid Ltd., Basingstoke, Berks, UK) is added at 6 g/L. Autoclave at 106 kPa, 121°C for 15 min.
6. Ampicillin: Stock solutions are prepared by dissolving a weighed quantity in a small volume of 1*M* NaOH, adjusting the pH to 7, and diluting to an appropriate volume. The stock solution is then sterilized by filtration through a sterile 0.22-µm Millipore filter and should be used fresh, because it is unstable. Aliquots of this stock solution are then added to MS medium that has cooled to at least 50°C, after autoclaving, to give the final desired concentration.
7. Culture vessels: Sterile 9-cm diameter Petri dishes are used for agar-

based hairy root cultures; 250-mL Erlenmeyer flasks, closed with two layers of aluminum foil, are used for growing seedlings and also for growing roots in suspension culture. Closed flasks are sterilized at 160°C for 90 min in a hot-air oven and then allowed to cool to room temperature before the addition of the sterile medium.

3. Methods

3.1. Preparation of Plants for Inoculation

1. Seeds are surface sterilized by immersing for 20 min in a closed, sterile flask containing sodium hypochlorite solution (5% available chlorine) containing 2–3 drops of Tween 20/100 mL (*see* Note 2). During this time, the flask is occasionally shaken.
2. The seeds are then rinsed several times in sterile distilled water. Seeds of other species may require sodium hypochlorite solutions of different concentrations.
3. Surface-sterilized seed is then distributed thinly over the surface of agar-based MS medium in a 250-mL closed flask. Seeds are left in lighted conditions at 25°C to germinate.
4. After 10 d, most seeds will have germinated: allow the seedlings to grow for a further period (2–3 wk) until the tobacco plants are ca. 6 cm tall or the beetroot plants are starting to show the third or fourth true leaves emerging.

3.2. Preparation of Bacterial Inoculum and Inoculation of Plants

Inoculations can be done using either *A. rhizogenes* taken directly from young YMA-grown colonies or a more concentrated bacterial inoculum.

1. To prepare the concentrated inoculum, aseptically remove several loopfuls of gelatinous bacterial colonies from the stock *A. rhizogenes* cultures and inoculate into 25 mL of YMB in a 100-mL sterile flask.
2. Incubate this on an orbital shaker (100 rpm, 25°C in light or darkened conditions) for 48 h.
3. Aseptically remove the culture, transfer it to a sterile, capped centrifuge tube, and spin down the bacteria at 1000 x *g* for 15 min. Remove and discard the supernatant with a sterile Pasteur pipet. Resuspend the soft pellet in as small a volume as possible of YMB.

4. Using a sterile scalpel or needle, scratch and stab the stem of the seedling.
5. Smear the bacterial suspension (from either YMA or concentrate in YMB) into the wound. When making the wound, do so at a height above the agar surface, to minimize the risk of the bacteria colonizing the agar.
6. Reclose the flasks and return them to lighted conditions at 25°C, so that hairy roots may develop from the wound sites: This usually occurs within 7–14 d.

3.3. Elimination of Bacteria and Subculture of Roots

The most effective way we have found to "clear" hairy roots of infecting *A. rhizogenes* is to grow root tips through several passages on MS medium containing 6 g/L agar and 500 µg/mL ampicillin.

1. Aseptically remove root tips (ca. 2 cm) from the infection sites and transfer to the MS + ampicillin plates.
2. Incubate the plates inverted and sealed with "Nescofilm" at 25°C.
3. After 2–3 wk, repeat the procedure, removing root tips (1–2 cm) and transferring to fresh MS + ampicillin plates.
4. Occasionally, small bacterial colonies are evident at the site of inoculation; care must be taken not to remove any of these when transferring root tips. After 2–3 passages, roots should be cleared of *A. rhizogenes* (Fig. 2), and tips can then be transferred to MS plates without ampicillin. If the roots are not bacteria-free, bacterial colonies will be evident within a few days, and passaging on MS + ampicillin media should be continued until the bacteria are eliminated. Roots can then be subcultured into liquid medium.
5. Transfer several (5–10) root tips (ca. 2 cm) to 60 mL of MS medium contained in a 250 mL Erlenmeyer flask. Incubate on an orbital shaker (100 rpm) at 25°C. After a lag phase, root growth is rapid (e.g., the doubling time for tobacco is 6.5 d). Hairy roots generally show extensive lateral branch formation and reduced geotropism (Fig. 3).
6. Repeated subcultures should be made by excising several root tips and transferring to fresh media at regular intervals (e.g., 3–4 wk). It is advisable to subculture the hairy roots while they are still growing—older cultures turn brown, and the root tips eventually become ne-

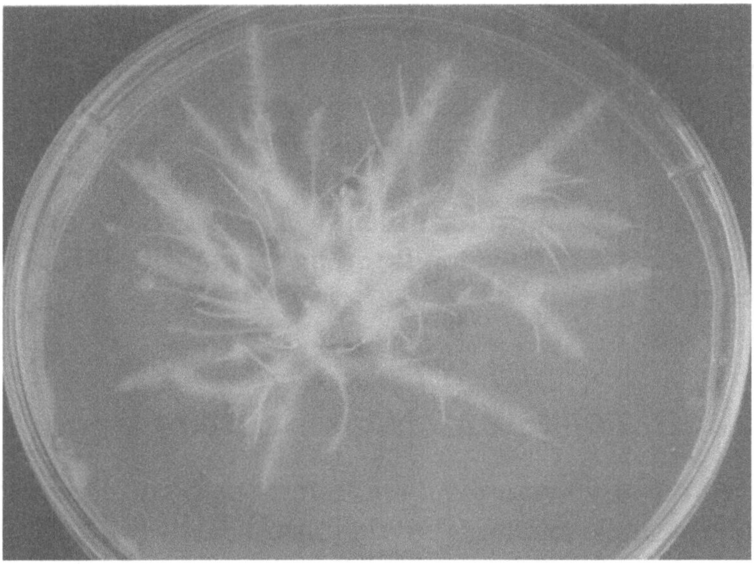

Fig. 2. Tobacco hairy root culture (21 d old) on MS + ampicillin.

crotic. Care is required during subculture: too large an inoculum and extensive damage to the roots can result in the formation of significant amounts of callus.

3.4. Confirmation of Transformation

The fact that hairy roots grow vigorously in simple media without any growth-regulator additions is a good indication of their transformed nature. Demonstration of opine synthesis by roots would provide firm evidence of transformation, but the expression of opine-synthesis genes is not stable (*23*). Direct confirmation of transformation requires identification of plasmid T-DNA in the hairy root cell genome by Southern blotting and hybridization to a labeled probe prepared from plasmid T-DNA (ref. *8* and Chapters 24 and 28, this vol.). Clearly, the details of such techniques are outside the scope of this chapter.

4. Notes

1. We use *N. tabacum* cv. Wisconsin 38, but most other cultivars are suitable for hairy root induction.
2. Sometimes a "dirty" seed lot will require a longer soak in sodium hypochlorite solution, e.g., 30 min.

Fig. 3. Tobacco hairy root suspension culture (21 d old).

References

1. Bryant, J. A. (1988) Putting genes into plants. *Plants Today* **1,** 23–28.
2. Nester, E. W., Gordon, M. P., Amasino, R. M., and Yanofsky, M. F. (1984) Crown gall: a molecular and physiological analysis. *Ann. Rev. Plant Physiol.* **35,** 387–413.
3. Weiler, E. W. and Schroder, J. (1987) Hormone genes and crown gall disease. *Trends in Biochemical Sciences* **12,** 271–275.
4. Birot, A. M., Bouchez, D., Casse-Delbart, F., Durand-Tardif, M., Jouanin, L., Pautot, V., Robaglia, C., Tepfer, D., Tepfer, M., Tourneur, J., and Vilaine, F. (1987) Studies and uses of the Ri plasmids of *Agrobacterium rhizogenes. Plant Physiol. Biochem.* **25(3),** 323–335.
5. Tepfer, M. and Casse-Delbart, F. (1987) *Agrobacterium rhizogenes* as a vector for transforming higher plants. *Microbiological Sciences* **4(1),** 24–28.
6. Huffman, G. A., White, F. F., Gordon, M. P., and Nester, E. W. (1984) Hairy root inducing plasmid: Physical map and homology to tumor-inducing plasmids. *J. Bacteriol.* **157,** 269–276.
7. Offringa, I. A., Melchers, L. S., Regensburg-Tuink, A. J. G., Costantino, P., Schilperoot, R. A., and Hooykaas, P. J. J. (1986) Complementation of *Agrobacterium tumefaciens* tumor-inducing *aux* mutants by genes from the T_R-region of the Ri plasmid of *Agrobacterium rhizogenes. Proc. Natl. Acad. Sci. USA* **83,** 6935–6939.
8. Hamill, J. D., Parr, A. J., Rhodes, M. J. C., Robins, R. J., and Walton, N. J. (1987) New routes to plant secondary products. *Biotechnology* **5,** 800–804.
9. Perani, L., Radke, S., Wilke-Douglas, M., and Bossert, M. (1986) Gene transfer methods for crop improvement: introduction of foreign DNA into plants. *Physiol. Plantarum* **68,** 566–570.
10. Comai, L., Facciotti, D., Hiatt, W. R., Thompson, G., Rose, R. E., and Stalker, D. M.

(1985) Expression in plants of a mutant *aro* gene from *Salmonella typhimurium* confers tolerance to glyphosate. *Nature* **317,** 741–744.

11. Jensen, J. S., Marcker, K. A., Otten, L., and Schell, J. (1986) Nodule-specific expression of a chimaeric soybean leghaemoglobin gene in transgenic *Lotus corniculatus*. *Nature* **321,** 699–674.
12. Sukhapinda, K., Trulson, A. J., Shahin, E. A., and Simpson, R. B. (1987) Phenotype of plants derived from hairy roots transformed with *Agrobacterium rhizogenes,* in *Molecular Biology of Plant Growth Control* (Fox, J. E. and Jacobs, M., eds.), Liss, New York, pp. 381–389.
13. Mugnier, J. (1988) Establishment of new axenic hairy root lines by inoculation with *Agrobacterium rhizogenes*. *Plant Cell Reports* **7,** 9–12.
14. Flores, H. E., Hoy, M. W., and Pickard, J. J. (1987) Secondary metabolites from root cultures. *Trends in Biotechnology* **5,** 64–69.
15. Payne, J., Hamill, J. D., Robins, R. J., and Rhodes, M. J. C. (1987) Production of hyoscyamine by "hairy-root" cultures of *Datura stramonium*. *Planta medica* **53(5),** 474–478.
16. Fujita, Y. and Tabata, M. (1987) Secondary metabolites from plant cells—pharmaceutical applications and progress in commercial production, in *Plant Tissue and Cell Culture* (Green, C. E., Somers, D. A., Hackett, W. P., and Biesboer, D. D., eds.), Liss, New York, pp. 169–185.
17. Charlwood, B., Brown, J. T., Moustou, C., and Charlwood, K. A. (1988) Pelargoniums: flavours, fragrances and the new technology. *Plants Today* **2,** 42–46.
18. Chandler, S. and Dodds, J. (1983) Solasodine production in rapidly proliferating tissue cultures of *Solanum laciniatum* Ait. *Plant Cell Reports* **2,** 69–72.
19. Anderson, L. A., Keene, A. T., and Phillipson, J. D. (1982) Alkaloid production by leaf organ, root organ and cell suspension cultures of *Cinchona ledgeriana*. *Planta medica* **46,** 25–27.
20. Jung, G. and Tepfer, D. (1987) Use of genetic transformation by the Ri T-DNA of *Agrobacterium rhizogenes* to stimulate biomass and tropane alkaloid production in *Atropa belladonna* and *Calystegia sepium* roots grown *in vitro*. *Plant Science* **50,** 145–151.
21. Parr, A. J. and Hamill, J. D. (1987) Relationship between *Agrobacterium rhizogenes* transformed hairy roots and intact uninfected *Nicotiana* plants. *Phytochemisty* **26,** 3241–3245.
22. Robins, R. J., Hamill, J. D., Parr, A. J., Smith, K., Walton, N. J., and Rhodes, M. J. C. (1987) Potential for use of nicotinic acid as a selective agent for isolation of high nicotine-producing lines of *Nicotiana rustica* hairy root culture. *Plant Cell Reports* **6,** 122–126.
23. Tepfer, D. (1984) Transformation of several species of higher plants by *Agrobacterium rhizogenes*: sexual transmission of the transformed genotype and phenotype. *Cell* **37,** 959–967.

Chapter 28

Agrobacterium Rhizogenes-Mediated Gene Transfer Using PRI 1855 and a Binary Vector

Houri Kalilian Fakhrai

1. Introduction

Agrobacterium rhizogenes causes a disease in susceptible dicotyledonous plants characterized by a proliferation of differentiated root tissue (hairy root) at the site of bacterial injection and following the transfer and integration of the T-DNA from the Ri-plasmid into the host plant genome (*1,2*). Recently, there has been increased interest in transformation of plants by *A. rhizogenes* (*3,4*). The ease with which Ri-TDNA-transformed plant cells of some plant species can be regenerated to whole plants suggests that an Ri-vector system might be a useful alternative to a Ti-vector system.

In our work, we use *A. rhizogenes* strain LBA 9402 (Rif^r), which contains the agropine pRi 1885, but which in addition has been transformed with BIN 19 the cloning component of the Ti binary vector system. BIN 19 contains a modified T-DNA consisting of left and right border repeats, a chimeric nopaline synthase (NOS)-neomycin phosphotransferase (NPT) gene conferring kanamycin resistance on transformed cells, and a polylinker

From: *Methods in Molecular Biology*, vol. 6, *Plant Cell and Tissue Culture*
Edited by Jeffrey W. Pollard and John M. Walker, ©1990 by The Humana Press

allowing insertion of any gene sequence. The strategy in using this combination of Ri and kan^r characteristics on separate replicons is to utilize the *trans* acting function of the *vir* region on the Ri-plasmid to cotransfer the Ri T-DNA as well as the BIN 19 T-DNA (containing additionally any desired genes) into the host plant. It is expected that the incidence of such cotransformation of plant cells with both Ri- and BIN 19 T-DNAs will be high and, thus, should produce abundant hairy roots exhibiting kanamycin resistance.

Having the novel gene on a separate, selectable, disarmed T-DNA will overcome the disadvantage of linkage to the Ri oncogenes (tms) that can give rise to aberrant phenotypes. Assuming independent integration of the two T-DNAs, at meiosis these will segregate independently; thus, kanamycin resistance and hairy root characteristics should also segregate independently in the seeds.

2. Materials

2.1. Tissue Culture

1. *Brassica napus* seeds cv. jet neuf.
2. *Agrobacterium rhizogenes* strain LBA 9402 containing pRi 1855 and P BIN 19 (*5*).
3. MS medium (*6* and *see* Appendix).
4. YMB medium: Mannitol 10 g/L, yeast extract 0.4 g/L, K_2HPO_4 0.5 g/L, $MgSO_4{\cdot}7H_2O$ 0.2 g/L, NaCl 0.1 g/L, agar 15 g/L pH 7.0.
5. Rifampicin: 50 mg/mL in methanol. Use fresh stock each time.
6. Kanamycin (50 mg/mL) and carbenicillin (50 mg/mL) in water sterilized by filtration through a 2 µm millipore filter and stored in aliquots at –20°C.
7. Cefotaxime (50 mg/mL) in water. Sterilize by filtration, and use fresh stock each time. Store the powder at 4°C in the dark.
8. 6, benzyl amino purine (BAP): Prepare by dissolving the powder in a few drops of 1*M* HCl, and then add H_2O to make up the volume. Store at 4°C. Use within 3 mo.
9. N^6–Δ^2 isopentenyl-adenine (IPA): Prepare by dissolving the powder in a few drops of 1*M* HCl. Then add H_2O to make up the volume. Store at 4°C. Use within 3 mo.
10. Gibberellic acid (GA_3): 10 mg are dissolved in water. Sterilize by filtration. Store at 4°C. Use within 3 mo.

11. 2-4-dichlorophenoxyacetic acid (2,4-D): Prepare by dissolving the powder in a few drops of 1*M* HCl, and then add H_2O to make up the volume. Store at 4°C. Use within 1 mo.
12. 1-Naphthylacetic acid (NAA): Prepare by dissolving the powder in a few drops of 1*M* NaOH, and then add H_2O to make up the volume. Store at 4°C. Use within 1 mo.
13. Thiamine-HCl (10 mg/mL), nicotinic acid (1 mg/mL), pyridoxin-HCl (1 mg/mL), and Myo-inositol (100 mg/mL) in water. Store at 4°C.
14. Macro- and micronutrients of RP_2B medium (*see* Table 1). Stock solution of each macronutrient salt should be made separately. Store at –20°C.
15. Bacto-agar (DIFCO).
16. Domestos. Commercial bleach containing NaOCl.
17. Mixture of sand and perlite (1:1), autoclaved.

    All media are sterilized by autoclaving. Antibiotics and GA_3 are added to the medium after autoclaving. The pH of all plant culture media is adjusted to 5.8 before the addition of agar.

2.2. Molecular Hybridization of DNA

1. Ethidium bromide (10 mg/mL)
2. Agarose beads: agarose 20 mg, glycerol 3.125 mL, Tris (50 m*M*, pH 8) 2 mL, EDTA (0.2*M*, pH8) 0.5 mL, H_2O 4.375 mL, bromophenol blue 10 mg. Autoclave without bromophenol blue and cool. Add bromophenol blue and squeeze several times through a 21-gage needle to form the beads.
3. Gel buffer (Alec's X10): Tris 96.8 g, EDTA 7.4 g, glacial acetic acid 32 mL. Make to 2 L with distilled water pH 7.7.
4. Denaturing solution: 0.5*M* NaOH, 1.5*M* NaCl, 1 m*M* EDTA.
5. Neutralizing solution: 1*M* Tris-HCl, pH 7.2, 1.5*M* NaCl, 1 m*M* EDTA.
6. 20 x SSC: Dissolve 175.3 g of NaCl and 88.2 g of Na-citrate in 800 mL of H_2O. Adjust pH to 7 with a few drops of 10*N* NaOH. Adjust the volume to 1 L. Dispense into aliquots. Sterilize by autoclaving.
7. Denhardt's solution X10: This contains the following ingredients in 100 mL: Ficoll 200 mg, PVP 200 mg, BSA 200 mg. Make up PVP in 100 mL of 3 x SSC. Add Ficoll after the PVP has dissolved. Dissolve BSA in a small amount of distilled water, and then add it to the solution. This is difficult to dissolve. Allow at least 1/2 h shaking to dissolve.
8. G-50 column buffer: 150 m*M* NaCl, 0.2% SDS in 100 mL of TE buffer (10 m*M* Tris-HCl, 1 m*M* EDTA, pH 8.0).

Table 1
Composition (mg/L) of Different Media used for Regeneration Procedure 2 (RP_2)

	RP_2A	RP_2B						RP_2C	RP_2D	RP_2E
		Macronutrients		Macronutrients		Inorganic supplements				
		NH_4NO_3	200	H_3BO_3	12.4	Inositol	100			
		KNO_3	1250	$MnSO_4 \cdot 4H_2O$	33.6	Nicotinic acid	1			
		$(NH_4)_2SO_4$	67	$ZnSO_4 \cdot 7H_2O$	21	Pyridoxine HCl	1			
Salts and		KH_2PO_4	35	KI	1.66	Thiamine	10			
vitamins	MS	$CaCl_2 \cdot 2H_2O$	525	$Na_2MoO_4 \cdot 2H_2O$	0.5			MS	MS	1/2 MS
		$MgSO_4 \cdot 7H_2O$	250	$CuSO_4 \cdot 5H_2O$	0.05					
		NaH_2PO_4	75	$CoCl_2 \cdot 6H_2O$	0.05					
				$FeSO_4 \cdot 7H_2O$	27.8					
				Na_2 EDTA	37.3					
Sugar										
Sucrose	30,000			20,000				10,000	10,000	–
Glucose	–			–				10,000	–	10,000
D-mannitol	–			–				10,000	–	–
Growth regulators										
BAP	2			–				0.5	0.5	–
IPA	–			–				0.5	–	–
NAA	0.2			–				1	0.1	–
2,4-D	–			3				–	–	–
Adenine sulfate	–			100				–	–	–
Antibiotics										
Carbenicillin	–			250				–	–	–
Cefotaxime	200			–				200	200	200
Kanamycin	100			100				100	100	50
Agar	1200			–				8000	8000	12,000

9. G-50 Sephadex column. Stir 1 g of G-50 (superfine) into 15 mL of the above buffer. Heat at 90°C for 1 h to hydrate the gel. Cool before poring the column.
10. TE buffer: 10 m*M* Tris-HCl, 1 m*M* EDTA, pH 8.0.
11. Salmon sperm DNA: 100 µg/mL in TE buffer.
12. [^{32}P]-dCTP random primed DNA labeling kit.
13. EcoRI, Sac II and Hind III used as recommended by supplier.

3. Methods

3.1. Transformation Through the Origin of the Wound: Regeneration Procedure 1 (RPI)

1. Single colonies of *A. rhizogenes* strain LBA 9402, containing PRI 1855 and P BIN 19 (5), are selected on YMB agar containing 100 mg/L rifampicin and 50 mg/L kanamycin. Inoculate colonies in 10 mL of YMB liquid culture containing 50 mg/L kanamycin and grow for 2 d at 28°C prior to infection.
2. Surface sterilize seeds in 10% "Domestos" for 1 h on a shaker (100 rpm), wash 4x with sterile distilled water (SDW), transfer to the surface of a wet filter paper in a Petri dish, and incubate at 0°C in the dark for 2 d.
3. Wash again in SDW and transfer (10 seeds/jar) to the surface of agar in 300-mL cylindrical glass jars containing 40 mL of MS medium, supplemented with sucrose (2%) and agar (1.5%). Cover the jar with plastic Petri dish lids, seal with Nescofilm, and incubate at 180 µE/m^2/s light intensity with a 16-h photoperiod. After 3 wk, well-developed seedlings can be used for transformation (*see* Note 1).
4. Seedlings with their attached agar are pulled halfway up the jars and laid horizontally on a surface of sterile tile. Then each seedling is injured with a hypodermic needle on the epicotyl and a 5 µL drop of bacterial suspension applied to the wound (*see* Notes 2 and 3). Seedlings and their attached agar are then transferred to their original position and incubated at 80 µE/m^2/s light intensity with a 16-h photoperiod for a further 2 wk.
5. Hairy roots 1–2 cm in length with the origin of the wound (1–2 mm thick section of epicotyl) are excised from seedlings and cultured horizontally on the surface of agar in medium RPIA.

6. Green shoots that developed from the origin of the wound attached to hairy roots are excised and cultured in RPIB medium.
7. After 2 wk, shoots are cut from the surface of callus and transferred to rooting media (RIPC) for a further 2 wk.
8. Rooted transgenic plants are transferred to the mixture of perlite and soil, covered by a polybag, and incubated for a further 2 wk at 180 μE/m²/s light intensity with a 16-h photoperiod (*see* Table 2).

3.2. Transformation Through the Hairy Roots: Regeneration Procedure 2 (RP2)

1–3. The procedure is the same as for RPI.

4. Hairy roots (1–2 cm) are excised above the inoculation site, cultured horizontally on the surface of agar in medium RP2A, and incubated at 80 μE/m²/s light intensity with a 16-h photoperiod of 2–3 wk. Hairy roots grow as closely interwoven masses over the surface of culture media and up the sides of the culture vessels.
5. Fragments of 0.5 cm of the most plagiotrophic hairy roots are excised and cultured in liquid medium RP2B. Culture approximately 20 fragments from individual clones in 150 mL conical flasks containing 40 mL of the above medium, and incubate at 150 rpm on a rotary shaker for 2 wk. Root fragments are thickened, and some form callus during this period.
6. Root fragments are then cultured individually on medium RP2C and incubated at 180 μE/m²/s light intensity for 4 wk. These root fragments exhibit various forms of development, and eventually give rise to cultures that are a mixture of hairy roots and callus.
7. Calluses devoid of roots should be subcultured in the same medium for a further 4 wk, during which time certain callus lines develop dark green areas, and others form more hairy roots.
8. Shoot buds develop on the surface of green callus in 0.7% of the cultured root fragments, after 2–3 subcultures in medium RP2C.
9. Transfer small shoots (1 cm tall) and their attached callus to RP2D medium, for leaf expansion.
10. After approximately 4 wk, well-developed shoots are cut above the callus and transferred to rooting medium RP2E for a further 2 wk.
11. Rooted transgenic plants are transferred to the mixture of sand and perlite, covered by a polybag, and incubated for a further 2 wk at 180 μE/m²/s light intensity with a 16-h photoperiod.

Table 2
Composition (mg/L) of Different Media
Used for Regeneration Procedure 1 (RPI)

Salts and vitamins	RPIA MS+, 2 g/L $CaCl_2$	RPIB MS+, 2 g/L $CaCl_2$	RPIC 1/2 MS
Sugar			
Sucrose	30,000	30,000	15,000
Growth regulators			
BAP	2	2	–
NAA	0.2	–	–
GA_3	–	10	–
Antibiotics			
Cefotaxime (*see* Note 4)	200	200	200
Kanamycin	100	50	
Agar	12,000	8000	12,000

12. Vernalization is carried out on transgenic shoots from both RPI and RP2. These shoots are incubated in the cold (0°C) for 6 wk. They are then transferred to the growth room with 25°C and 8 h of high light with 210 μE/m^2/s light intensity, 8 h of low light with 60 μE/m^2/s light intensity, and 8 h of dark. Flowering commences after 3–6 wk in the growth room.

    The transgenic state of plants from both RPI and RP2 is confirmed by their ability to grow in the presence of kanamycin and by Southern blot hybridization of leaf genomic DNA, with NPT II gene probe (*see* Notes 5 and 6).

3.3. Molecular Hybridization of DNA from Transformed Plant Tissue

1. Approximately 10 μg of plant DNA sample is digested with EcoRI and SacII under conditions recommended by the supplier (*see* Note 7).
2. The digested DNA is then fractionated by electrophoresis on a 0.8% agarose gel.
3. Gels are prepared for Southern blotting by washing in 1% HCl (20 min), 0.5*M* NaOH, 1.5 *M* NaCl (3 x 10 min), 1.0*M* Tris-HCl pH 7.2, 1.5*M* NaCl (3 x 20 min), 20 x SSC (0.3*M* Na-citrate, and 3*M* NaCl pH 7.2 (1 x 20 min).

4. Gels are then blotted onto nitrocellulose filters, and the filters baked under vacuum (80°C, 30 min).
5. Hybridization is carried out as described by Maniatis et al. (1982) (7, and *see* vol. 2 of this series). Filters are prehybridized in 3 x SSC, 10x Denhardt's solution and 100 µg/mL salmon sperm DNA for 5 h at 65°C, and then placed into hybridization solution containing the same ingredients as for prehybridization, but with added denatured [^{32}P]-labeled probe. Hybridization is carried out at 65°C for 16–20 h. Radioactive filters are washed in 3 x SSC (4 x 15 min), 1 x SSC (2 x 15 min) and 0.1 SSC (2 x 15 min), and then exposed to preflashed X-ray film (X-Omat S) with an intensifying screen (Cronex lighting plus) at –80°C for 7 d.

3.4. Isolation of DNA Fragments for Use as Radioactive Probes

1. DNA fragments for use as radioactive probes are isolated by digesting PABDI with Hind III and running on a 0.8% agarose gel. Bands containing the relevant fragments are cut from the gel and added to a mixture of 0.9 mL of H_2O, 0.1 mL 3*M* Na-acetate, and 50 µL of 0.2*M* EDTA, incubated in the dark for 15 min, blotted dry, and stored at –80°C for 30 min. Liquid is squeezed from the gel while it remains frozen, and then extracted with phenol.
2. The DNA is precipitated with a mixture of 1 mL of 100% ethanol, 20 µL 3*M* Na-acetate, and 5 µL of 1*M* $MgCl_2$ at –20°C for 30 min. The DNA pellet is collected by centrifugation, vacuum dried, and resuspended in 10 µL of water. Approximately 1 µg is labeled with [^{32}P] dCTP by random primed DNA labeling kit. A specific activity of about 2.5×10^8 cpm/µg should be obtained.

4. Notes

1. For a successful transformation, a vigorously growing seedling developing in vitro is a prerequisite. Seeds should be undamaged and intact, and allowed to be germinated in a glass cylindrical jar. Population density of the seedlings is also important, since the seedling growth is impaired if they are growing too far from or too close to each other.
2. Extra care should be taken when wounding the epicotyl and applying *Agrobacterium* to the in vitro growing seedling. Too much wounding

may cause extensive damage to the cell and, consequently, will prevent integration of T-DNA into plant genome.

3. Application of excessive amounts of *Agrobacterium* (in excess of 5 μL) should be avoided, since this may cause bacterial infections of the agar, and will eventually infect the seedlings.
4. Cefotaxime should be used only in agar cultures, not in liquid cultures. The antibacterial properties of cefotaxime last for 3 wk in agar cultures; therefore cultures containing cefotaxime should be subcultured every 3 wk.
5. The two procedures for RI and BIN 19 mediated gene transfer reported here are different in many ways. Each system has certain technical advantages and disadvantages over the other. These are briefly discussed below.
   a. In regeneration procedure 1, the efficiency of shoot regeneration from hairy root + origin of the wound (HR + O) cultures is very high (36%) in medium containing 50 mg/L kanamycin. The time required to produce a transgenic plant is short (we conclude that the minimum time required to obtain a fully established plant is 10 wk: 2 wk of incubation after inoculation, 2 wk for the culture of HR + O clones, 2 wk for shoot elongation, 2 wk for rooting shoots, and 2 wk to establish in soil. Regeneration procedure 1 through HR + O cultures is relatively simpler than HR culture; therefore, transgenic plants could be established with much less effort. However, the disadvantage of this system is that many escape shoots will develop through tissue culture. Molecular hybridization of leaf genomic DNA from transgenic plants with radiolabeled NPT II probe showed a low frequency (6%) of cotransfer of RI and BIN 19 T-DNA.
   b. In regeneration procedure 2, the efficiency of shoot regeneration from hairy roots is low (0.7% of all single hairy root fragments regnerate), and it occasionally takes two or three subcultures in RP2C medium for shoot bud development. These shoots form root in medium supplemented with kanamycin. The time required to establish transgenic plants is longer than in procedure 1, and it needs a more complex subculturing regime to induce shoot organogenesis from single hairy root tips. However, the advantages of this regeneration system over system 1 are that escape shoots are not developed and the frequency of foreign gene integration in the plant genomic DNA is very

high. On the basis of results from molecular hybridization of seven independently transformed shoot lines with radiolabeled NPT II probe, we have shown that all lines possess homologous sequences to the probe, i.e., 100% cotransformation of RI and BIN 19 T-DNA is achieved.

6. When transgenic shoots developed through regeneration procedure 2 were rooted in RP2E containing 50 mg/L kanamycin, they all rooted readily, the apical bud remained green, and the roots grew vigorously, but when transgenic shoots were rooted in the same medium with 100 mg/L kanamycin, the response of the shoots varied, approximately 50% remaining green, and others showing varying degrees of bleaching, from streaking of the leaves to complete loss of color. Although their root growth remained for the most part vigorous, when these bleached shoots were transferred to soil plus perlite (50:50), a few died, but most produced green leaves at their apices and grew normally. It would appear, therefore, that a kanamycin concentration of 100 mg/L is too stringent a selection medium for all but the most robust shoots.
7. EcoRI and Sac II-restricted DNA of all the hairy roots and shoot lines derived from hairy roots, transformed with PRI 1855 and P BIN 19 contain fragments that hybridize to the NPT II gene probe. Sac II has two restriction sites on BIN 19 T-DNA, one on the right and the other on the left hand side of the NPT II gene, within the right and left border repeats. Using a ^{32}P labeled Hind III fragment from plasmid ABDI (which lies internal to the T-DNA and includes NPT II gene for kanamycin resistance), we have observed that hybridizing fragments of 800 bp (NPT II) are common to all the lines of hairy roots—hairy roots growing from a single hairy root tip with or without selection pressure, hairy roots growing on an inoculated plant, shoots derived from hairy root tips (TS1, TS2, TS3, TS4, TS6, TS7, TS8), and shoots derived from HR + O clones 2.1.1. DNA from roots and shoots transformed with LBA 9402 (containing PRI 1855) showed no homology to this probe. Transformed tissues contained between one and five NPT II genes.

Acknowledgment

This work was supported by The Department of Trade and Industry UK as a part of the "Plant Gene Tool Kit programme."

References

1. Chilton, M. D., Tepfer, D. A., Petit, A., David, C., Casse-Delbart, F., and Tempe, J. (1982) *Agrobacterium rhizogenes* inserts T-DNA into the genomes of the host plant root cells. *Nature (London)* **295,** 432–434.
2. Guerche, P., Jouanin, L., Tepfer, D., and Pelletier, G. (1987) Genetic transformation of oilseed rape (*Brassica napus*) by the Ri-T-DNA of *Agrobacterium rhizogenes* and analysis of inheritance of the transformed phenotype. *Mol. Gen. Genet.* **206,** 382–386.
3. Simpson, R. B., Spielmann, A., Margossian, L., and McKnight, T. D. (1986) A disarmed binary vector from *Agrobacterium tumefaciens* functions in *Agrobacterium rhizogenes*. *Pl. Mol. Biol.* **6,** 403–415.
4. Ooms, G., Bains, A., Burrell, M., Karp, A., Twell, D., and Wilcox, E. (1985) Genetic manipulation in cultivars of oilseed rape (*Brassica napus*) using *Agrobacterium*. *Theor. Appl. Genet.* **71,** 325–329.
5. Bevan, M. (1984) Binary *Agrobacterium* vectors for plant transformation. *Nucleic Acid Res.* **12,** 8711–8721.
6. Murashige, T. and Skoog, F. (1962) A revised medium for rapid growth and bioassays with tobacco tissue cultures. *Physiol. Pl.* **15,** 473–497.
7. Maniatis, T., Fritsch, E. F., and Sambrook, J. (1982) *Molecular Cloning. A Laboratory Manual* (Cold Spring Harbor Laboratory, Cold Spring Harbor, NY).

Chapter 29

Transformation of Rape with *Agrobacterium tumefaciens*-Based Vectors

Houri Kalilian Fakhrai

1. Introduction

Agrobacterium tumefaciens, a soil bacterium that causes crown gall disease on a wide range of dicotyledonous plants, has been used most widely as a vehicle for gene transfer into plants (*1,2*). The transferred genes are stably integrated into the genomes of the transgenic plants and are transmitted to progeny as dominant mendelian traits (*3,4*). Since the production of auxins and cytokinins coded by oncogenes on T-DNA is often incompatible with normal plant development, it is often desirable to "disarm" the plasmids by removing the oncogenes and, therefore, allow the transformed plant cells to differentiate in a normal manner, rather than grow in a tumorous fashion.

To fulfill the above requirement, we have used a disarmed nopaline Ti plasmid pGV 3850 (2) in *A. tumefaciens* strain C58. This plasmid contains the T-DNA border regions and a genetic marker to monitor the presence

From: *Methods in Molecular Biology*, vol. 6, *Plant Cell and Tissue Culture*
Edited by Jeffrey W. Pollard and John M. Walker, ©1990 by The Humana Press

of the T-DNA, but no genes that prevent normal differentiation of transformed plant cells. In addition, pGV 3850 contains the widely used cloning vehicle pBR 322 between the borders of the mutated T-DNA. This makes this Ti plasmid a versatile acceptor for introduction of many foreign genes. In addition to pGV 3850, we have used the binary vector pJIT 73. This vector contains both the neomycin phosphotransferase (NPT II) and the *E. coli* β-glucuronidase (GUS) coding sequences driven by the cauliflower mosaic virus 35S promoter. Inoculated explants produced transgenic shoots in selective media containing kanamycin. The transgenic plants express the 35S CAMV-GUS mRNA, which confers β-glucuronidase activity to the plants.

2. Materials

2.1. Tissue Culture

1. *Brassica napus* cv. Raphal.
2. *Agrobacterium tumefaciens* C58 strain carrying the disarmed nopaline plasmid pGV 3850, and the binary vector pJIT 73 containing both the "NPT II" and "GUS" coding sequences driven by the CAMV-35S promotor.
3. YEB medium: yeast extract (Bacto) 1 g/L, beef extract 5 g/L, peptone (Bacto) 5 g/L, sucrose 5 g/L, $MgSO_4 \cdot 7H_2O$ 493 mg/L, agar 15 g/L, pH 7.2. (Omit the agar for liquid medium).
4. Kanamycin: 50 mg/mL in water. Sterilize by filtration through a 2 µm millipore filter and store in aliquots at –20°C.
5. Carbenicillin: 50mg/mL in water. Sterilize by filtration through a 2 µm millipore filter and store in aliquots at –20°C.
6. Cefotaxime: 50mg/mL in water. Sterilize by filtration through a 2 µm millipore filter. Use fresh stock each time.
7. 6, benzyl amino purine (BAP). Prepare by dissolving the powder in a few drops of 1*M* HCl, and then add H_2O to make up the volume. Store at 4°C. Use within 3 mo.
8. 1-Naphthyl acetic acid (NAA). Prepare by dissolving the powder in a few drops of 1*M* NaOH, and then add H_2O to make up the volume. Store at 4°C. Use within 1 mo.
9. RP3A medium: MS (*5* and *see* Appendix) salts and vitamins, 30 g/L sucrose, 2 mg/L BAP, 0.2 mg/L NAA, 12 g/L agar, pH 5.8.
10. RP3B medium: MS salts and vitamins, 10 g/L sucrose, 10 g/L glucose, 2 mg/L BAP, 0.2 mg/L NAA, 500 mg/L carbenicillin, 50 mg/L kanamycin, 8 g/L agar, pH 5.8.

11. RP3C medium: MS salts and vitamins, 10 g/L sucrose, 0.5 mg/L BAP, 0.1 mg/L NAA, 250 mg/L cefotaxime, 50 mg/L kanamycin, 8 g/L agar, pH 5.8.
12. RP3D medium: MS salts and vitamins, 10 g/L glucose, 0.01 mg/L NAA, 250 mg/L cefotaxime, 50 mg/L kanamycin, 12g/L agar, pH 5.8.
13. Bacto-agar (DIFCO).
14. Domestos. Commercial bleach containing NaOCl.

All media are sterilized by autoclaving. Antibiotics are added to the medium after autoclaving.

2.2. *β-Glucuronidase (GUS) Assay*

1. 4-methyl umbelliferyl glucuronide (MUG) (Sigma M9130) substrate. Dissolve 1 mg MUG in 0.27 mL MUG-minus buffer. This is 10 m*M* MUG.
2. MUG-minus buffer containing the following ingredients: 50 m*M* sodium phosphate buffer, pH 7.0, 10 m*M* EDTA, 0.01% v/v Triton-X100, 10 m*M* mercaptoethanol.
3. Stop solution: 0.2*M* sodium carbonate.
4. Acid-washed sand.

3. Method

3.1. *Stem Explant Transformation*

1. Seeds of *Brassica napus* cv. Raphal are grown in 4-in pots in the greenhouse with a day/night temperature of 15/10°C, relative humidity of 75%, 16-h photoperiod, and a light intensity of 80 $\mu E/m^2/s$ (*see* Note 1).
2. Internodes from plants just prior to bolting, or in the process of bolting but before flowering (6) are removed and washed for 1–2 h in running tap water, then surface sterilized in 10% "Domestos" for 30 min, and rinsed 3x in sterile distilled water.
3. Stem explants (approximately 5 mm long) are excised from internodes. Longitudinal stem sections are used for inoculation (7).
4. *A. tumefaciens* strain C58 containing PGV 3850 and PJIT 73 is grown at 28°C in YEB + kanamycin plates. Single colonies are transferred to 5 mL YEB + kanamycin liquid cultures and grown for 48 h at 28°C. Longitudinal stem segments of *Brassica napus* are inoculated with approximately 5 µL of bacterial suspension, then transferred to a filter paper placed on top of RP3A medium, and incubated at 50 $\mu E/m^2/s$ light intensity.

5. After 2 d, the stem tissues are transferred to RP3B medium without the filter paper and incubated at 180 μE/m²/s light intensity (*see* Note 2).
6. After 4 wk, shoot buds develop in about 30% of the explants. Shoots regenerate from the stem explant at first. Excision of these shoots and subculture of the explants in PR3B medium for the second time results in the growth of callus and regeneration of more shoots through callus—occasionally more than 10 shoot buds develop from one stem callus (*see* Notes 3–5).
7. Regenerated shoots with attaching callus are transferred to RP3C medium (*see* Note 6) for further shoot development and leaf expansion at 180 μE/m²/s light intensity.
8. Well-developed shoots are cut above the surface of callus and transferred to RP3D medium for rooting at 180 μE/m²/s light intensity (*see* Note 7).
9. Rooted plantlets are transferred to the mixture of perlite and soil (50:50), covered by a polybag, and incubated for a further 2 wk at 180 μE/m²/s light intensity and 16-h photoperiod, before weaning out.

3.2. *Assay for Transformation* (*GUS*) (See *Notes 8 and 9*)

Wear gloves at all stages.

1. Remove plant material and wash in 80–70% ETOH, and then rinse in sterile water.
2. Place in a sterile Eppendorf tube, add MUG-minus buffer (0.1 mL) and sand, grind the tissue, and centrifuge for 5 min at high speed in a microfuge at 0°C.
3. Remove the supernatant (approximately 90 μL) to a sterile Eppendorf tube, add 0.21 mL MUG-minus buffer, vortex for 5 s, and then place in a 37°C water bath.
4. Add MUG substrate (35 μL/Eppendorf), vortex, and incubate for 1 h at 37°C.
5. Remove 0.1 mL of the reaction mixture, add to a sterile Eppendorf tube containing 0.9 mL of stop solution, and vortex.
6. For qualitative analysis, view on a transilluminator (*see* Note 10).
7. For quantitative analysis, set up fluorimeter as follows:
   a. Excitation 365 nm.
   b. Emission 455 nm.
   c. Scale Factor 1.
   d. Slit width excitation 10 nm.

e. Slit width detection 2.5 nm.
f. Blank on stop solution.

8. Add 0.1 mL of stopped reaction mixture to 2.9 mL of stop solution. Measure the fluorescence:

$$100 \text{ U of fluorescence} = 730 \text{ pM MU/min/mg protein} \tag{1}$$

The fluorimeter should be calibrated with freshly prepared 4-methyl umbelliferyl (MU) standards of 100 n*M* and 1 µ*M* MU in the same buffers. Fluorescence is linear, from nearly as low as 1 n*M* to 5–10 µ*M* MU.

4. Notes

1. The growing condition of the mother plant and its physiological age play an important role for success of transformation/regeneration.
2. Longitudinal and horizontal cut stem sections of intact stem explants have been tested for their ability to regenerate shoots in RP3B medium without antibiotics. Results showed that shoot regeneration in longitudinal cut sections is 3x as high as in intact stem explants, and twice as high as in horizontal cut stem explants.
3. The kanamycin concentration used for selection of transgenic shoots is the lowest concentration at which the control stem tissues are killed. Shoot regeneration was inhibited by 50 mg/L kanamycin in non-inoculated explants.
4. With kanamycin as a selectable marker, transgenic plants are obtained at a frequency of about 30%. When these transgenic plants were tested for β-glucuronidase, 30% expressed GUS activity.
5. Shoots initially regenerate directly on the stem explant, probably from the cambium layer, in the first 4 wk after inoculation in 30% of the explants. The growth of these shoots is not accompanied by the growth of callus. In the rest of the inoculated explants, yellow and friable callus grows around the two cut edges of the explants, and after 6–8 wk, adventitious shoots develop through the formation of *de novo* meristems in certain callus lines. It has been observed that transformation efficiency is higher among these shoots; therefore, it is likely that cell dedifferentiation is required for transformation.
6. When using cefotaxime, subculture to a fresh medium every 3 wk, since it looses its antibacterial properties beyond 3 wk in culture.
7. A number of stem explant callus lines produced green, club-like struc-

tures without a meristem. These structures usually would not go on to produce a plant with a normal phenotype.

8. In this work, the *E. coli* β-glucuronidase gene (GUS) has been used as a gene marker for analysis of gene expression in transformed plants. Control plants tested lack β-glucuronidase activity. Expression of GUS is directed by the cauliflower mosaic virus (CAMV) 35S promotor in transformed plants. Activity of GUS can be measured using fluorometric analysis of very small amounts of transformed tissue. GUS is very stable, and tissue extracts continue to show high levels of GUS activity after prolonged storage at –70°C. Storage of extracts at –20°C should be avoided, since it seems to inactivate the enzyme.
9. Expression of GUS genes might be influenced by the physiological differences in transgenic plants; therefore, care must be taken to choose uniform plant material for the GUS assay. Also, β-glucuronidase activity in transgenic plants can be measured very easily by fluorimetry; therefore, a large number of plants should be screened to compensate for these differences in small populations of transgenic plants.
10. A simple and sensitive qualitative GUS assay can be done by placing the tubes containing the assay mixture on a long-wave UV light box and observing the blue fluorescence. This assay can be scaled down to measure very small volumes (reaction volume 50 μL) terminated with 25 μL 1*M* Na_2CO_3 in microtiter dishes or Eppendorf tubes.

Acknowledgment

This work was supported by the Department of Trade and Industry, UK as a part of the "Plant Gene Tool Programme" in Durham.

References

1. Fraley, R. T., Rogers, S. G., Horsch, R. B., Eichholtz, D. A., Flick, J. S., Fink, C. L., Hoffman, N. L., and Sanders, P. R. (1985) The SEV system; a new disarmed Ti plasmid vector system for plant transformation. *Biotechnology* **3**, 629–635.
2. Zambryski, P., Joos, H., Genetello, C., Leemans, J., van Montagu, M., and Schell, J. (1983) Ti plasmid vector for the introduction of DNA into plant cells without alteration of their normal regeneration capacity. *EMBO J.* **2**, 2143–2150.
3. De Block, M., Herrera-Estrella, L., van Montagu, M., Schell, J., and Zambryski, P. (1984) Expression of foreign genes in regenerated plants and their progeny. *EMBO J.* **3**, 1681–1684.

4. Horsch, R. B., Fraley, R. T., Rogers, S. G., Sanders, P. R., Lloyd, A., and Hoffmann, N. (1984) Inheritance of functional foreign genes in plants. *Science* **223,** 496–498.
5. Murashige, T. and Skoog, F. (1962) A revised medium for rapid growth and bio-assays with tobacco tissue cultures. *Physiol. Plant* **15,** 413–497.
6. Fry, J., Barnson, A., and Horsch, R. (1987) Transformation of *Brassica Napus* with *Agrobacterium tumefaciens* based vectors. *Plant Cell Rep.* **6,** 321–325.
7. Pau, E. C., Palta, A. M., Nagy, F., and Chua, N. H. (1987) Transgenic plants of *Brassica napus L. Biotechnology* **5,** 815–817.

Chapter 30

Transient and Stable Expression of Foreign DNA Introduced into Plant Protoplasts by Electroporation

George W. Bates, Dietmar Rabussay, and William Piastuch

1. Introduction

Electroporation utilizes high-voltage electric fields for cell permeabilization. This technique has been used for promoting the cellular uptake of exogenous molecules and macromolecules, including nucleotides, dyes, RNA, DNA, and even small proteins (*1–7*). Electroporation's useful attributes are its simplicity and its general effectiveness with a wide range of cell types. Because of its general efficacy, electroporation is becoming a valuable technique for the introduction of DNA into cell types that are resistant to transformation by other procedures (*6–8*). In this chapter, we will outline procedures for introducing foreign DNAs into plant protoplasts by

From: *Methods in Molecular Biology*, vol. 6, *Plant Cell and Tissue Culture*
Edited by Jeffrey W. Pollard and John M. Walker, ©1990 by The Humana Press

electroporation. DNA uptake is verified by the transient expression of chloramphenicol acetyltransferase (CAT), and by the recovery of plants stably transformed to kanamycin resistance through the introduction of the neomycin phosphotransferase gene.

2. Materials

1. Instrumentation: Electroporation equipment consists of two components, a power source capable of delivering short, high-voltage, direct-current pulses and a chamber to hold the protoplasts during electroporation (*see* Notes 1–3). Two different types of electric-field pulses have been used for electroporation. The approach originally described (*9*) was designed for cell fusion and utilized very-high-voltage (2000–4000 V/cm), short (10–100 µs) square-wave pulses. The equipment required to provide such short, high-voltage square waves is complex and expensive, and usually requires that the protoplasts be suspended in solutions of low conductivity (for example, 0.5 *M* mannitol without added salts). Subsequently, electroporation systems have been described (*6,7,10*) that electroporate cells through the action of lower-voltage (500–750 V/cm), longer 1–50 ms) pulses delivered by means of capacitive discharges (Fig. 1). Such systems are more easily built, cost less, and allow the cells to be electroporated in media containing much higher levels (100–200 m*M*) of salts. A lower-voltage, long-pulse, capacitive-discharge system is used in the experiments described in this chapter. Electroporation by means of short, high-voltage, square-wave pulses is also used. Although both systems work, in our hands the low-voltage, long-pulse system gives more efficient transient expression.

   Complete and relatively inexpensive electroporation systems (including electronics and chambers) are available from several commercial sources (for example, Bethesda Research Laboratories, Gaithersburg, MD, USA; Bio-Rad Laboratories, Richmond, CA, USA; Biotechnologies & Experimental Research Inc., San Diego, CA, USA; Hoefer Scientific Instruments, San Francisco, CA, USA) for prices ranging from about $1500–3500 (in 1989 dollars).

   Homemade capacitive-discharge systems have also been described (*10*). Homemade systems use an electrophoresis power supply to charge a capacitor or capacitor bank. A switching circuit (for discharging the capacitor in a controllable fashion) and an electropora-

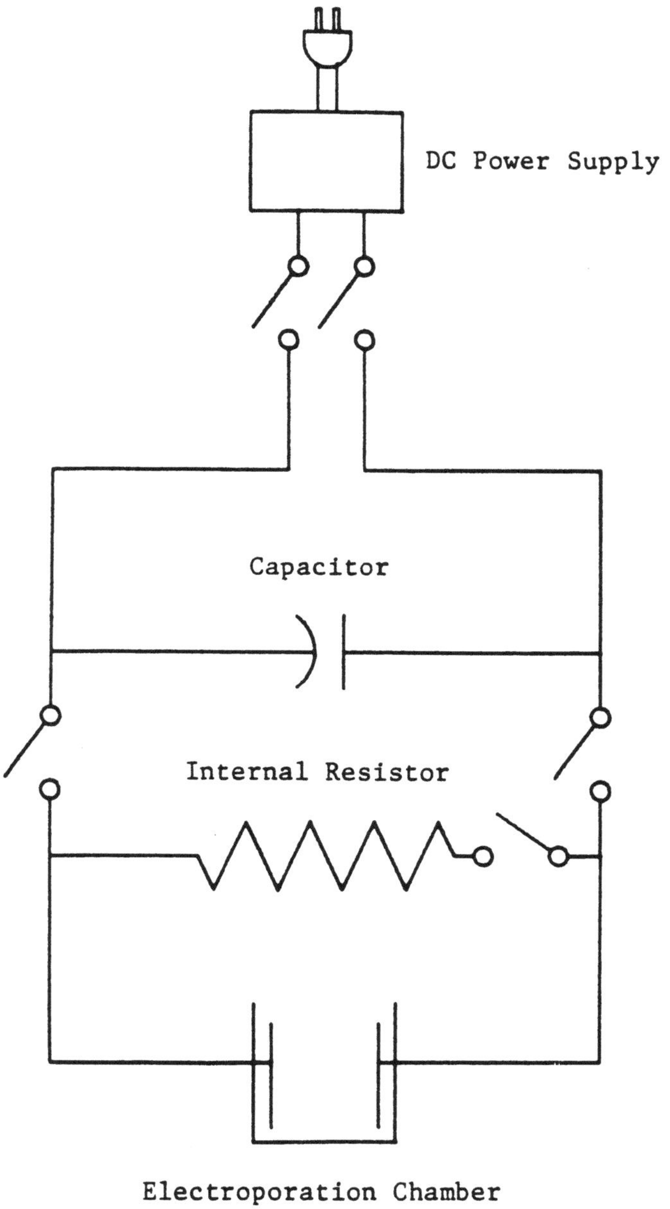

Fig. 1. Schematic diagram of a capacitive-discharge electroporation system. The switch between the power supply and the capacitor is used to prevent recharging of the capacitor during pulse discharge. The internal resistor can be used to short the capacitor to ground (if one decides not to apply a pulse), or by selecting a resistor of an appropriate size, this resistor can be used to modulate pulse length.

tion chamber must also be built. The homemade electroporation chambers usually consist of a disposable spectrophotometer cuvet (0.4 cm wide) into which two aluminum (or platinum) plates are glued to serve as electrodes. These chambers can be sterilized before use by immersion in 70% ethanol.

In most of our work, and in the experiments described here, we have utilized the Bethesda Research Laboratories Cell-Porator™ Electroporation System (*11*). Compared with homemade equipment, commercial equipment offers a higher degree of operator safety (from high-voltage shocks) and reliability, and is ready to use without requiring any development time by the experimenter. Electrical equipment should be selected that offers a range of pulse lengths (1–100 ms when a high-salt medium, such as that described in this chapter, is used) and voltages (100–400 V or 250–1000 V/cm), because the optimal settings for electroporation may vary for different species or cell types.

2. Protoplasts: Protoplasts may be isolated from leaves or suspension cultures by conventional procedures (*see* Chapter 24, this vol.). We have used protoplasts from a nonregenerable carrot suspension culture for transient expression work, and tobacco mesophyll protoplasts for stable integration experiments.
3. DNA: For transient expression of CAT, the plasmid pUC8CaMVCATΔN was obtained from Fromm and Walbot (Stanford University, Stanford, CA, USA). This plasmid contains the CAT coding region under the control of the cauliflower mosaic virus 35S promoter and the nopaline synthase polyadenylation signals. For the recovery of stable transformants, the plasmid pMON200 was obtained from Fraley, Horsch, and Rogers (Monsanto Co., St. Louis, MO, USA). This plasmid carries the neomycin phosphotransferase coding region under the control of the nopaline synthase promoter and polyadenylation signals. Introduction of pMON200 into plant cells confers resistance to the antibiotics kanamycin and G418. Plasmids were maintained in *E. coli* and isolated by conventional procedures. Purified plasmids should be free of RNA and protein.
4. Poration medium: HBS (HEPES buffered saline): 150 m*M* KCl, 4 m*M* $CaCl_2$, 10 m*M* HEPES (pH 7.2), and mannitol. The amount of mannitol must be adjusted to match the tonicity of the protoplasts. For carrot suspension-cell protoplasts, we used 0.11*M* mannitol, whereas 0.21*M* mannitol was required to stabilize the tobacco mesophyll protoplasts.

5. Culture media and plasticware:
   a. Carrot protoplasts: Errikson's salts (*see* Appendix) and vitamins plus 40 g/L sucrose, 1 mg/L a-naphthaleneacetic acid, 0.02 mg/L kinetin, and 0.2*M* mannitol.
   b. Tobacco mesophyll protoplasts: K3 salts (*see* Appendix), vitamins, and hormones (*12*) containing 0.4*M* glucose as both the carbon source and the osmoticum.
   c. Tobacco callus: Murashige and Skoog salts (*see* Appendix) plus 1 mg/L thiamine, 100 mg/L inositol, 3% sucrose, 1 mg/L benzyladenine, and 1 mg/L α-naphthaleneacetic acid (plus 0.8% agar if the medium is to be solid.
   d. Tobacco regeneration: This medium is the same as the solidified tobacco callus medium, except that the α-naphthaleneacetic acid is reduced to 0.1 mg/L.
6. Sonicator for cell disruption.
7. Tris-HCl (1*M*) pH 7.5.
8. Phenylmethylsulfonyl fluoride (PMSF): stock solution; 100 m*M* in isopropanol, stable at room temperature for several mo.
9. ^{14}C-chloramphenicol (53 Ci/mol).
10. Acetyl coenzyme A:10 mg dissolved in 0.3 mL of 100 m*M* Tris-HCl, pH 7.5. This stock can be stored for 2–3 wk at –20°C.
11. Ethyl acetate.
12. Thin layer chromatography system: Bakerflex Silica Gel Plates 1B-F (J. T. Baker Chem. Co., Phillipsburg, NJ, USA) and chloroform:methanol (95:5 v/v).
13. X-ray film is required for visualizing radioactive spots on the TLC plates, and a liquid scintillation counter is needed for quantifying CAT activity.
14. A 0.1% solution of Evans' Blue dissolved in the appropriate protoplast culture medium.
15. SeaPlaque agarose.
16. Kanamycin sulfate.

3. Methods

3.1. Electroporation

All operations are carried out at room temperature with sterile solutions in a laminar-flow hood.

1. Wash the protoplasts (*see* Note 4) twice with HBS + mannitol and resuspend them in HBS + mannitol at a density of 4×10^6 protoplasts/mL.
2. Mix an equal volume of protoplast suspension with a solution of supercoiled plasmid DNA (40 µg/mL dissolved in HBS + mannitol); leave to stand for 5 min.
3. Resuspend the protoplasts by gentle agitation, and introduce 0.5 mL of the protoplast/DNA mixture into a sterile electroporation chamber.
4. Apply a single 8–10-ms (RC time constant) pulse of 300 V. (Using the BRL system and 0.5 mL of protoplast/HBS/DNA suspension, a 325 µF capacitor will deliver an 8–10-ms pulse.) The voltage indicated here (300 V) assumes an interelectrode spacing of 0.4 cm in the electroporation chamber, to yield an electrical field of 750 V/cm. If a different electrode spacing is used, the voltage should be adjusted accordingly (*see* Note 5).
5. Let the protoplasts stand for 2 or 3 min, and then dilute them 1:10 (final vol 5 mL) with culture medium. Centrifuge (100 g, 5 min) and resuspend in 2 mL of culture medium. Culture the protoplasts at 28°C (*see* Note 6).
6. Controls should include protoplasts mixed with DNA and cultured without being electroporated.

3.2. Viability Test

After 24–48 h of culture, remove one drop of protoplast suspension, and mix it with one drop of Evans' Blue solution. Allow 2–3 min for dye uptake. Only the dead cells and dead protoplasts will take up the dye and stain blue. It is best to observe the cells with an inverted microscope, since addition of a coverslip to the preparation would crush some of the protoplasts.

3.3. Transient Expression Assay of CAT

The following CAT assay (*see* Note 7) is adapted from Gorman et al. (*13* and Chapter 44 of vol. 5, this series).

1. Electroporate the protoplasts in the presence of a plasmid carrying a CAT reporter gene construct.
2. Twenty-four to 48 h after electroporation (*see* Note 8), collect the cells by centrifugation (100*g*, 5 min). Discard the supernatant, and re-

suspend the cells in 100 μL of ice-cold 0.25*M* Tris-HCl, pH 7.5 + 0.1 m*M* PMSF. Disrupt the cells by sonication and centrifuge (10,000 rpm, 5 min). The supernatant (cell extract) is saved for the CAT assay and may be stored frozen (–20°C).

3. In a 1.5-mL microcentrifuge tube, mix 50 μL of the cell extract with 25 μL of 1*M* Tris-HCl, pH 7.5, 1.5 μL (0.15 μCi) ^{14}C-chloramphenicol, 1.5 μL 40 m*M* acetyl coenzyme A, and 72 μL of distilled water. Vortex, and allow the reaction to proceed for 2 h at 37°C.
4. Stop the reaction by extracting the reaction mixture twice with 150 μL of ice-cold ethyl acetate. Combine the solvent (upper) phases of the two extractions and evaporate to dryness (60°C for 1–2 h).
5. Redissolve the extracts with 5–10 μL of ethyl acetate, and spot on the TLC plate. Rinse the tubes with 5–10 μL more ethyl acetate, and combine this rinse with the first aliquots on the TLC plate. Chromatograph using chloroform:methanol (95:5) until the solvent front has migrated about 10 cm. Dry the plate and autoradiograph it overnight at room temperature.
6. Radioactive spots on the chromatogram (located by means of the autoradiogram; *see* Fig. 2) are cut out and counted in a liquid scintillation counter. CAT activities are usually expressed as the percent of ^{14}C-chloramphenicol converted to acetylated forms (i.e., cpm of acetylated chloramphenicol x 100/total cpm) or as moles of chloramphenicol converted to acetylated forms/min/10^6 cells (*see* Note 9).

3.4. Recovery of Stable Transformants

1. Electroporate tobacco protoplasts in the presence of a plasmid containing a chimeric neomycin phosphotransferase construct.
2. After electroporating the samples and diluting them with tobacco protoplast culture medium, split each 2 mL sample into two subsamples, and culture them in 30 x 15 mm Petri plates. One subsample will be subjected to selection on kanamycin; the other will serve as an unselected control. Controls should also contain samples incubated with DNA, but not electroporated.
3. Day 4: Add 0.5 mL of tobacco callus medium to each sample.
4. Day 10: Add another 0.5 mL of tobacco callus medium and transfer culture to a 60 x 15 mm Petri plate.
5. Day 12: To the samples that are to undergo selection, add enough filter-sterilized kanamycin sulfate to make the final kanamycin concentration 100 μg/mL (*see* Note 10). Leave the other half of the

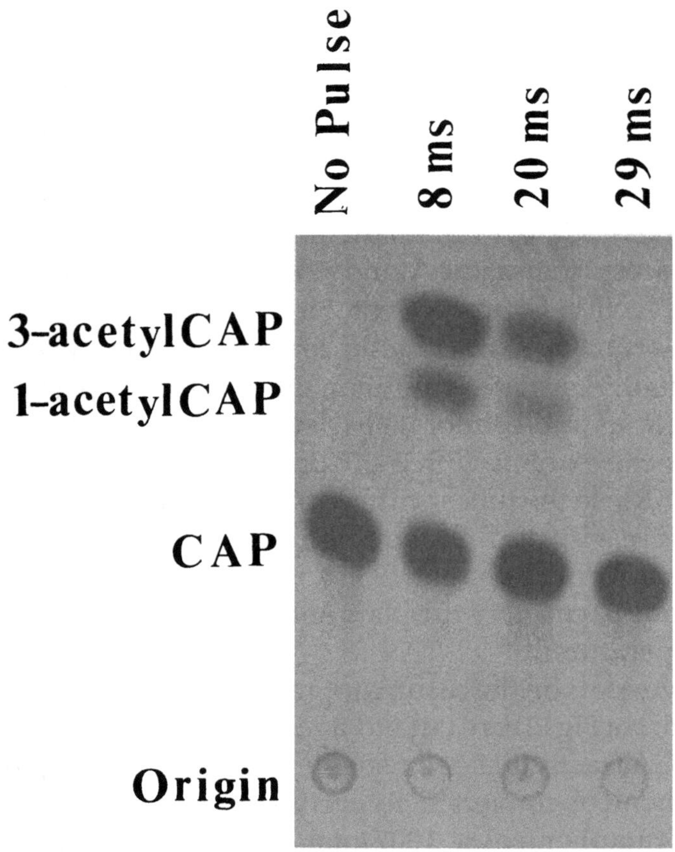

Fig. 2. Example of a CAT assay autoradiogram. Carrot protoplasts were electroporated in the presence of pUC8CaMVCAT and were assayed for CAT activity 24 h later. The electroporation voltage was 300 V (750 V/cm); the pulse duration was varied from 0–29 ms. Very little CAT activity is seen in the sample receiving the 29–ms pulse, because nearly all the protoplasts were killed by heat generated by this treatment (*see* Fig. 3). CAP, chloramphenicol.

samples without kanamycin. Prepare two batches of tobacco callus medium; make each of them 1.8% in SeaPlaque agarose. After autoclaving, when the media have cooled enough to be handled, make one batch of medium 100 μg/mL in kanamycin. When the agarose-containing media have cooled to just above their gelling temperature, add 2 mL of the appropriate medium (plus or minus kanamycin) to each of the cultures. Mix thoroughly by swirling, and place the cultures on ice or in the refrigerator to harden. Cut the hardened cultures into wedges, and transfer them to 100 x 15-mm Petri plates. Add 5 mL of liquid callus medium (plus or minus kanamycin) to each plate.

6. Day 14: Add another 5 mL of liquid callus medium (plus or minus kanamycin) to each plate.
7. Replace half of the liquid callus medium (5 mL) weekly with fresh medium. Kanamycin-resistant colonies should be apparent after 3–4 wk.
8. After 4–6 wk, resistant calluses should be large enough to be transferred to solid callus medium (containing kanamycin; *see* Note 11). Estimate the transformation frequency by comparing the numbers of colonies recovered in selected (kanamycin-treated) and unselected cultures derived from the same original sample.
9. Six to ten wk after electroporation, calluses can be transferred to regeneration medium (plus or minus kanamycin) for shoot induction. Induce rooting of regenerated shoots by transferring them to media lacking phytohormones but containing kanamycin (100 µg/mL).

4. Notes

1. Pulse lengths and voltages should be verified by oscilloscope recordings. This is especially important when working with homemade electroporation equipment, but even commercial equipment should be checked periodically. Although a storage oscilloscope is ideal for this purpose, a relatively inexpensive substitute can be obtained by adapting a personal computer. For example, a hardware/software package for upgrading an IBM PC to a digital storage oscilloscope can be purchased from Cyber Research, Inc., New Haven, CT, USA, for about $1000.
2. Caution should be exercised when working with reusable electroporation chambers. Aluminum electrodes will degrade with use. Also, residual DNA is usually not destroyed by the methods commonly used for resterilizing the chambers (i.e., ethanol). Treatment with agents that destroy DNA (chlorine, acids) can also attack the electrodes and plastic chambers.
3. Table 1 summarizes the relevant characteristics of some of the more commonly used commercial electroporation apparatuses.
4. High-quality, clean protoplast preparations are essential. Optimal electroporation conditions usually kill about half of the original protoplasts (Fig. 3). Cell death is often unacceptably high in low-quality protoplast preparations.
5. Optimal field strengths and pulse lengths are often different for different species and cell types. Thus, initial experiments should explore

Table 1
Features of Some Commercial Electroporations

Feature	Suppliers		
	A[a]	B[b]	C[c]
Type of pulse generator[d]	Capacitive discharge	Capacitive discharge	Electronic switch
Voltage range[e]	0–1000 V/cm	0–5500 V/cm	0–3600 V/cm
Pulse length range[f]	Adequate	Adequate	Adequate
Compatible with low ionic strength buffers	Yes	Yes	Yes
Compatible with high ionic strength buffers[g]	Yes	Yes	No
Dedicated oscilloscope jack	Yes	No	No
Disposable electroporation chambers	Yes	Yes	Yes
Disposable electrodes[h]	Yes	Yes	No
Assurance of sterile handling of the sample[i]	Good	Good	Fair
Temperature control during electroporation	Yes	No	No
Overall safety of the system[j]	Very good	Very good	Fair

[a]Bethesda Research Laboratories.

[b]Bio Rad Laboratories.

[c]Hoefer Scientific Instruments, progenetor.

[d]A and B generate exponentially decaying pulses, whereas C generates an irregularly shaped pulse.

[e]A and B provide homogeneous electric-field pulses. C uses ring electrodes, which produce a nonhomogenous field.

[f]Although the three systems differ in their pulse lengths, all provide an adequate selection of pulse lengths.

[g]C can be used only with media of low conductivity (e.g., mannitol solution).

[h]A and B use built-in electrodes. C uses a reusable electrode that has to be rinsed between individual electroporations.

[i]A and B provide sterile, individually wrapped electroporation chambers with tightly closing lids. C uses 24-well tissue culture plates with no lids during electroporation.

[j]With systems A and B, no conductive components can be touched during pulse delivery. System C uses an unshielded electrode.

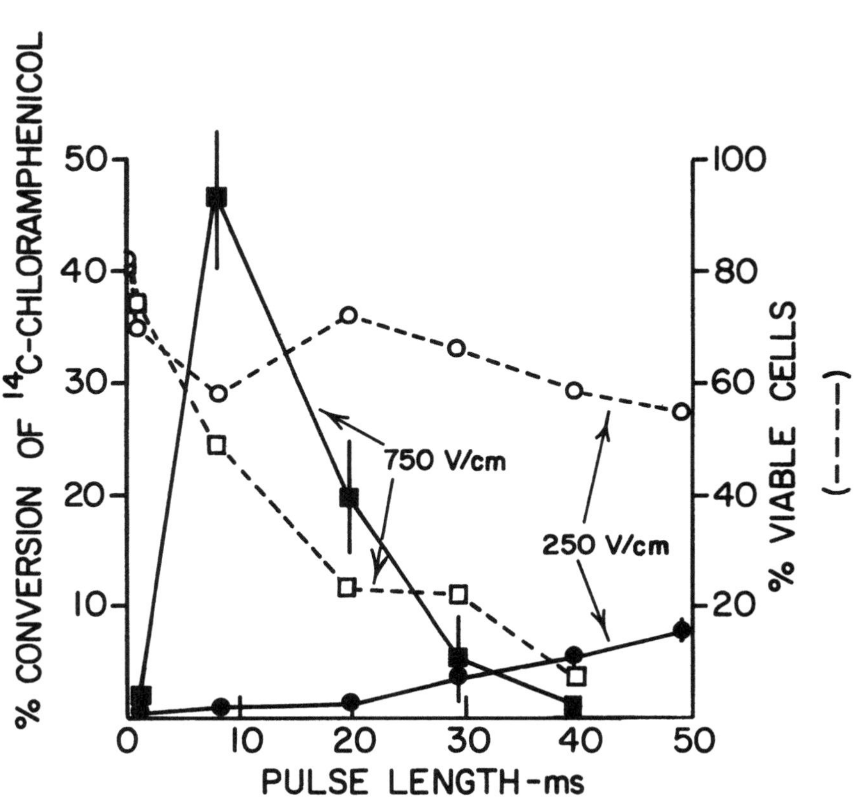

Fig. 3. The effect of varying pulse length and voltage on CAT expression and the viability of carrot protoplasts. Maximum CAT activity is obtained for samples receiving an 8–ms, 300–V (750 V/cm) pulse. These conditions kill approximately half of the protoplasts. Increasing the pulse length at this voltage results in decreased CAT expression because of increased cell death. Pulses of 100 V (250 V/cm) kill fewer protoplasts, but also give much less CAT expression.

the effects of different pulse length and voltage settings (as in Fig. 3). Because field-strength and pulse-length optima are sometimes rather narrow, it is recommended that field strength be changed in 100 V/cm increments (and pulse length in 10-ms increments) in the range where maximal electroporation efficiency is expected. To a limited extent, shorter pulses can be compensated for by higher field strength, and longer pulses can partially compensate for lower field strength.

6. Other factors important for successful electroporation include:
   a. Ionic strength of the medium. Changing the ionic strength alters pulse length.
   b. Temperature. Maximal transient expression is obtained for cells electroporated at room temperature. However, long pulses of high field strength can result in enough heating of the medium to cause significant heat-induced cell death (*see*, for example, the 40 ms pulses of 750 V/cm in Fig. 3). The temperature of the electroporation medium also affects pulse length.
   c. DNA concentration and conformation. Efficient transformation is often observed with DNA concentrations as low as several µg/mL. However, the efficiency of transient gene expression increases with increasing DNA concentration. Supercoiled and linear DNA yield similar transient expression, and the addition of carrier DNA (sonicated, heat-denatured, salmon-sperm DNA) to the plasmid DNA is also beneficial.
   d. Time after electroporation before the addition of culture medium. Maximum transient gene expression is observed when the protoplasts are left in the DNA-containing electroporation medium for 1–3 min after electroporation. In our experience, longer incubation times actually result in a reduction of transient expression.
7. Luciferase (*14*) and neomycin phosphotransferase can also be used as reporter genes in electroporation experiments. Strong constitutive promoters give the best results (developmentally regulated and light-sensitive promoter elements usually function poorly in protoplasts).
8. Maximum expression is usually observed 24–36 h after electroporation.
9. The CAT assay is nonlinear beyond 50–60% conversion of chloramphenicol to acetylated forms. The amount of cell extract can be reduced or the incubation time can be shortened in order to keep the assay in its linear range.
10. The timing of dilution of the cultures and of initiation of selection should be adapted to the rate of growth of the cultures. The first addition of callus medium should be given when the cultures are at the microcolony stage (4–10 cells/colony). If this first addition of medium results in a great deal of cell death, add the medium instead

in two aliquots spaced 2 d apart. Selection may be started at 7–10 d; delaying the start of selection much beyond 2 wk may permit some untransformed colonies to escape selection.

11. Kanamycin-resistant lines should be kept under selection at least until plants have rooted. Two simple phenotypic tests can be used to verify the recovery of transformed plants. First, untransformed shoots will not form roots when exposed to media containing kanamycin, whereas transformed shoots will root. Second, leaf sections taken from untransformed plants will not form callus in the presence of kanamycin. Ultimate proof of transformation requires Southern blotting to demonstrate the presence of the transforming DNA, and/or a genetic analysis of the inheritance of this DNA.

References

1. Nishiguchi, M., Langridge, W. H. R., Szalay, A. A., and Zaitlin, M. (1986) Electroporation-mediated infection of tobacco leaf protoplasts with tobacco mosaic virus RNA and cucumber mosaic virus RNA. *Plant Cell Reps.* **5,** 57–60.
2. Okada, K., Nagata, T., and Takebe, I. (1986) Introduction of functional RNA into plant protoplasts by electroporation. *Plant Cell Physiol.* **27,** 619–626.
3. Sokoloski, J. A., Jastreboff, M. M., Bertino, J. R., Sartorelli, A. C., and Narayanan, R. (1986) Introduction of deoxyribonucleoside triphosphates into intact cells by electroporation. *Anal. Biochem.* **158,** 272–277.
4. Mehrle, W., Zimmerman, U., and Hampp, R. (1985) Evidence of asymmetrical uptake of fluorescent dyes through electropermeabilized membranes of *Avena* mesophyll protoplasts. *FEBS Lett.* **185,** 89–94.
5. Sowers, A. E. and Lieber, M. R. (1986) Electropore diameters, lifetimes, numbers, and locations in individual erythrocyte ghosts. *FEBS Lett.* **205,** 179–184.
6. Potter, H., Weir, L., and Leder, P. (1984) Enhancer-dependent expression of human κ immunoglobulin genes introduced into mouse pre-B lymphocytes by electroporation. *Proc. Natl. Acad. Sci. USA* **81,** 7161–7165.
7. Fromm, M. E., Taylor, L. P., and Walbot, V. (1986) Stable transformation of maize after gene transfer by electroporation. *Nature* **319,** 791–793.
8. Ou-Lee, T. M., Turgeon, R., and Wu, R. (1986) Expression of a foreign gene linked to either a plant-virus or a *Drosophila* promoter, after electroporation of protoplasts of rice, wheat and sorghum. *Proc. Natl. Acad. Sci. USA* **83,** 6815–6819.
9. Zimmerman, U. (1982) Electric field-mediated fusion and related electrical phenomena. *Biochim. Biophys. Acta* **694,** 227–277.
10. Fromm, M., Callis, J., Taylor, L. P., and Walbot, V. (1987) Electroporation of DNA and RNA into plant protoplasts. *Methods in Enzymol.* pp. 351–366.
11. Rabussay, D., Uher, L., Bates, G., and Piastuch, W. (1987) Electroporation of mammalian and plant cells. *Focus* **9(3),** 1–5.
12. Nagy, J. I. and Maliga, P. (1976) Callus induction and plant regeneration from proto-

plasts of *Nicotiana sylvestris*. *Zeit. Pflanzenphysiol.* **78,** 453–455.

13. Gorman, C. M., Moffat, L. F., and Howard, B. H. (1982) Recombinant genomes which express chloramphenicol acetyltransferase in mammalian cells. *Mol. Cell. Biol.* **2,** 1044–1051.
14. Ow, D. W., Wood, K. V., DeLuca, M., DeWet, J. R., Helinski, D. R., and Howell, S. H. (1986) Transient and stable expression of the firefly luciferase gene in plant cells and transgenic plants. *Science* **234,** 856–859.

Chapter 31

Protoplast Microinjection Using Agarose Microdrops

Wendy Ann Harwood and David Roy Davies

1. Introduction

A number of plant species have now been transformed, but there remains considerable uncertainty regarding the optimal procedure to be adopted. *Agrobacterium* species have been used as vectors for the DNA in some instances, whereas in others direct uptake has been achieved following chemical, physical, or electrical treatment of protoplasts or even of cells (*see* Chapters 27–30 of this volume). Viruses are another vehicle being considered in this context. One of the most effective techniques of transforming animal cells is that of microinjection, and this, together with the recent claim of very high rates of transformation of plant cells by microinjection (*1*), has heightened interest in the more general exploitation of this method of introducing DNA to plants. The technique is now sufficiently well developed to enable it to be more generally adopted, and this chapter will describe the method that we have developed. An important feature of the method is the use of low melting point (LMP) agarose, both for holding protoplasts during microinjection and for their subsequent culture. The method will be described in detail for tobacco mesophyll protoplasts, and then variations of the method for rape hypocotyl and sugar beet suspension culture protoplasts will be listed.

From: *Methods in Molecular Biology*, vol. 6, *Plant Cell and Tissue Culture*
Edited by Jeffrey W. Pollard and John M. Walker, ©1990 by The Humana Press

2. Materials

2.1. Protoplast Preparation and Culture

All solutions except those used for sterilizing plant material are autoclaved, unless otherwise specified. Analar-grade reagents and Analar water are used for all solutions and culture media. Hormone stock solutions are kept at 4°C and discarded after 1 mo. Autoclaved culture media are not kept for more than 3 mo.

2.1.1. Tobacco Mesophyll Protoplasts

1. 10% Domestos (household bleach).
2. 0.5*M* mannitol.
3. NT medium (*see* Appendix and ref. 2).
4. Enzyme solution: 0.3 strength NT medium in 0.5*M* mannitol containing 0.05% Macerozyme R10 and 0.25% Onozuka R10 cellulase (Yakult Honsha Co. Ltd.), freshly made up for each experiment and filter sterilized.
5. 0.8*M* sucrose.
6. NAA (1-Naphthalenacetic acid): 0.1 mg/mL and 0.01 mg/mL stocks in water, filter sterilized.
7. BAP (6-Benzylaminopurine): 0.1 mg/mL stock in water, filter sterilized.
8. NT media with 0.6, 0.9, and 1.4% LMP agarose.
9. MS medium (*see* Appendix and ref. 3) with 0.7% LMP agarose, 0.25*M* mannitol, and 0.7% agarose.

2.1.2. Rape Hypocotyl Protoplasts

1. 90% ethanol.
2. Domestos or other proprietary bleach.
3. 0.3*M* Sorbitol with 0.05*M* $CaCl_2$.
4. K3 medium (*see* Appendix and ref. 4).
5. Enzyme solution: 0.5% cellulysin (Calbiochem) and 0.25% Macerozyme R10 in K3 medium, freshly made up and filter sterilized.
6. BE Salts solution (*see* Appendix and ref. 5) with 16% sucrose.
7. W5 solution (*see* Appendix and ref. 6).
8. A-medium (*see* Appendix and ref. 7), with 0.03*M* sucrose and 0.39*M* mannitol, but no glucose.
9. A-medium as above with 0.6, 0.9, and 1.4% LMP agarose.
10. MS medium with 0.7% agar.

11. MS salts (macronutrients only) solidified with 0.7% agar.
12. A-medium as above, but with half the quantity of mannitol and 0.7% agar.

2.1.3. Sugar Beet Suspension Culture Protoplasts

1. MS medium.
2. BAP 0.1 mg mL and 0.01 mg mL stocks in water, filter sterilized.
3. MS medium containing half the concentration of macronutrients and 0.7% agar (1/2 MS).
4. Enzyme solution containing 2% cellulysin, 0.5% macerozyme R10, 0.8*M* mannitol, 0.3 m*M* MES buffer, 0.64 m*M* NaH_2PO_4, and 6.8 m*M* $CaCl_2$, freshly made up and filter sterilized.
5. BE salts with 16% sucrose.
6. W5 solution.
7. A-medium as described for rape protoplasts, but with the standard hormones replaced with 0.1 mg/L NAA and 0.5 mg/L BAP.
8. A-medium as above with 0.6, 0.9, and 1.4% LMP agarose.
9. A-medium (*see* ref. 7) with half the quantity of mannitol and 0.7% agarose.

2.2. Microinjection

1. Micromanipulator, such as that manufactured by Leitz, with fixed stage microscope inside a laminar flow cabinet.
2. Syringe system to control injection, e.g., Beaudouin syringe (Micro-instruments, Oxford). Alternatively, injection may be controlled electronically, e.g., PLI-100 picoinjector (Medical Systems Corp., NY).
3. Sigmacote or other siliconizing solution.
4. Needle or electrode puller.
5. Heavy paraffin oil (specific gravity 0.875–0.885) for use with Beaudouin syringe system, or compressed oxygen or nitrogen supply for picoinjector.
6. Borosilicate glass tubing with outer diameter to fit needle holder.
7. Nylon rings, 9-mm inner diameter and 1.5 mm deep (Micro Plastics Int. Ltd.), autoclaved.
8. Sterile charcoal powder.
9. Sterile filter paper circles to fit 9-cm Petri dishes.
10. Solution for injection, e.g., DNA 50–100 µg/mL in 10 m*M* Tris 1 m*M* EDTA. This is spun in an Eppendorf centrifuge for 10 min before use.

3. Methods

3.1. Microinjection of Tobacco Mesophyll Protoplasts

3.1.1. Preparation of Protoplasts for Microinjection

All operations are carried out in a laminar-flow cabinet, and sterile techniques are used throughout.

1. Mesophyll protoplasts are prepared from leaves that are approximately 75% expanded. These are first sterilized in 10% Domestos for 10 min, and then washed at least 4x in sterile distilled water.
2. The lower epidermis is stripped from leaf sections, using curved forceps.
3. The sections are floated, mesophyll down, on 0.5*M* mannitol for 30 min to preplasmolyze the cells.
4. The mannitol is removed from under the leaf sections and replaced with the enzyme solution. Dishes containing leaf sections are incubated at 25°C in the dark for 16 h or overnight.
5. Protoplasts released from the leaves are filtered through a sterile 100-µm nylon mesh to separate leaf debris. They are collected by centrifugation at 600 rpm for 4 min and resuspended in 0.8*M* sucrose.
6. After centrifugation for a further 4 min, the band of floating protoplasts is collected and washed twice in 0.5*M* mannitol. Protoplasts are then resuspended in NT medium containing 3 mg/L NAA and 1 mg/L BAP. After 1 h, protoplasts can be transferred to NT medium with 1 mg/L BAP, but only 0.1 mg/L NAA. This medium is used for all subsequent culture unless otherwise stated (*see* Notes 1 and 2).
7. The unit developed for protoplast culture and injection consists of a sterile coverslip, on which is placed a nylon ring. The ring is filled with NT medium containing 0.9% LMP agarose, so that the top agarose surface is flat (*8*).
8. A sample of the prepared protoplasts is diluted to approximately 2×10^4/mL in NT medium. This is then mixed 1:1 with NT medium that has been cooled to 35°C, and contains 0.6% LMP agarose and a small amount of sterile charcoal (*see* Note 3). The charcoal enables protoplasts to be located for injection, and the agarose holds the protoplasts while the needle is being inserted.
9. One µL of this mixture is placed in the center of the agarose-filled

ring. About 12 of these units, referred to as microdrops, should be prepared at any one time, and can be held for up to 2 h in 9-cm Petri dishes lined with moist filter paper. The number of protoplasts plated in each ring varies from 5 to 15 (*see* Note 4).

3.1.2. Microinjection

1. Glass tubing for the injection needles is cut to appropriate lengths, siliconized by dipping in Sigmacote, and then dried at 60°C. There is no need to sterilize glass or needles, but they must be prepared on the day they are to be used.
2. The needle or electrode puller is set to pull needles with open tips of 1 μm or less in diameter. Once prepared, needles are stored in a closed, dust-free container.
3. Assuming that the oil-filled injection system is used, needles are back filled with oil from a long spinal needle and syringe.
4. A needle is clamped firmly into the needle holder, taking care not to introduce air into the oil-filled system. The oil should run easily to the very tip of the needle.
5. The injection solution is introduced to the tip of the needle by lowering the needle into a drop of the solution on a sterile slide. The flow of injection solution into the needle is controlled by suction/pressure from the syringe system, and is stopped when the injection solution/oil meniscus is still in the field of view.
6. One of the prepared microdrops is transferred from its moist environment onto a sterile slide, and the needle is positioned so that it and the protoplasts are in focus. Once a protoplast is penetrated with the needle, pressure is applied from the syringe system to force the injection solution from the needle. This can be monitored by watching movement of the meniscus of the oil/injection solution interface. The needle should be removed quickly from the protoplast and moved into position for another injection. The volume injected is usually between 10^{-8} and 10^{-9} mL.
7. The needle can be refilled, as necessary, by moving back to a drop of injection solution on the slide and repeating the procedure.
8. After being refilled once or twice, the needle may become blocked, and must be replaced. Sometimes a needle will fail after only one or two injections (*see* Note 5).

3.1.3. Culture of Microinjected Protoplasts

1. Immediately after injection, microdrops are replaced in the moist Petri dishes. Within 20 min, the microdrops are covered with NT

medium containing 0.9% LMP agarose that has been cooled to 35°C. Sufficient agarose is added to form a meniscus above the ring.

2. Once the agarose has set, the microdrops containing injected protoplasts are transferred to a "feeder dish" developed to allow culture of low density protoplasts. This exploits the "bead" culture system described by Shillito et al. (*9*).
3. To prepare feeder "beads," a sample of protoplasts at 1 x 10^5/mL in NT medium is mixed with an equal volume of NT medium containing 1.4% LMP agarose, and 5-mL aliquots are set in 5-cm Petri dishes. Once set, the agarose is cut into eight sections or "beads," and two of these are placed in a 5-cm Petri dish containing 3.5 mL NT medium. This constitutes the "feeder dish." The protoplasts at high density "feed" the few protoplasts in the microdrop via the liquid medium. Protoplasts used to prepare feeder beads may be up to 3 d old.
4. Feeder dishes are incubated at 25°C with gentle shaking (40 rpm). After 1 wk, the microdrop is transferred to a freshly prepared feeder dish. The nylon ring containing the agarose will probably have floated off the coverslip by this time, but can still be easily transferred.
5. After a further 2 wk, the calluses derived from protoplasts in the microdrop are usually visible to the naked eye. The nylon ring can be lifted off, and the agarose drop placed on MS medium with 0.25*M* mannitol and 0.7% LMP agarose. At this stage, the agarose surface is cut to increase aeration of the calluses.
6. If calluses are not large enough after 3 wk in feeder dishes, then a third feeder dish can be used for a further week before transfer to MS medium.
7. Once calluses can be picked up individually, they are transferred to MS medium with 0.7% LMP, but without mannitol. On average, 40% of injected protoplasts give rise to calluses, but the figure may be up to 71% in individual experiments.

3.2. Microinjection of Rape Hypocotyl Protoplasts

3.2.1. Preparation of Protoplasts for Microinjection

1. Seeds of *Brassica napus* are surface sterilized for 4 min in 90% ethanol, and then for 10 min in Domestos.
2. Seeds are then washed at least 5 x in sterile distilled water and left on moist filter paper for 30 min before sowing.
3. Approximately 20 seeds are sown per 9-cm Petri dish containing MS

salts, and the dishes are incubated in the dark at 25°C for at least 3 d.

4. Protoplast isolation is essentially according to Glimelius (*10*). The hypocotyls are excised, and each is cut once longitudinally before being preplasmolyzed in 0.3*M* sorbitol/0.05*M* $CaCl_2$ for at least 30 min. The sorbitol solution is then replaced with the enzyme solution, and the dishes are sealed, wrapped in foil, and incubated overnight on a slow shaker at 25°C.
5. The released protoplasts are filtered through a sterile 100 µm nylon filter, and BE salts with 16% sucrose are added to make up half the total volume.
6. After centrifugation at 600 rpm, the bands of floating protoplasts are removed and washed twice in W5 solution. Protoplasts are finally resuspended in liquid A-medium.
7. Microdrops for microinjection are prepared exactly as for tobacco protoplasts (steps 7–9, inclusive), but with A-medium substituted for NT medium.

3.2.2. Microinjection

The procedure is exactly as described for tobacco protoplasts. However, with rape hypocotyl protoplasts, the nucleus is easily visible, enabling nuclear injections to be made.

3.2.3. Culture of Microinjected Protoplasts

1. Sections 1–3, inclusive, from the tobacco protoplast method apply, but A-medium should be substituted for NT medium. Also, the feeder beads contain protoplasts at a final concentration of 2×10^4/mL, not 6×10^4/mL.
2. It may be more difficult to obtain large numbers of hypocotyl protoplasts than mesophyll protoplasts. If this is the case, the feeder dish can be miniaturized, with 3-cm Petri dishes being used both to form the feeder beads and as the feeder dish. A 3-cm Petri dish needs only 1.5 mL of protoplasts in agarose from which to cut the beads, and 1 mL of liquid medium to make contact between the microdrop and feeder beads.
3. Microdrops must be transferred to a freshly prepared feeder dish after 1 wk. Additionally, the liquid in the second feeder dish must be changed each week until the calluses are large enough to remove.
4. Microdrops containing calluses that are easily visible are transferred to A-medium with half the quantity of mannitol and 0.7% agar.
5. After 1 wk, they are transferred to MS medium with 0.7% agar. The

plating efficiency of injected hypocotyl protoplasts is approximately 23%; this is lower than that of tobacco mesophyll protoplasts. The small calluses derived from injected hypocotyl protoplasts are difficult to culture unless their feeder protoplasts are kept in an optimum condition.

3.3. Microinjection of Sugar Beet Suspension Culture Protoplasts

3.3.1. Preparation of Protoplasts for Microinjection

1. Sugar beet protoplasts are prepared from established suspension cultures that are subcultured every 3 d into fresh MS medium with 0.1 mg/L BAP.
2. The enzyme solution is added to a similar volume of suspension culture, and the mixture is shaken gently for 16 h at 25°C.
3. The released protoplasts are filtered through a 100-µm sterile nylon mesh and then pipeted onto a cushion of BE salts with 16% sucrose.
4. After centrifugation at 600 rpm for 7 min, the band of protoplasts is removed, washed twice in W5 solution, and then resuspended in A-medium.
5. Microdrops are prepared as described in steps 7–9 of the method for tobacco protoplasts, but A-medium is substituted for NT medium (*see* Note 7).

3.3.2. Microinjection

The procedure is again as described for tobacco protoplasts.

3.3.3. Culture of Microinjected Protoplasts

1. After injection, microdrops containing sugar beet protoplasts are placed in a moist environment, but are not covered with agarose medium as in the previous methods. Aeration is particularly important for these protoplasts, so the feeding system is modified accordingly.
2. Feeder dishes consist of a layer of protoplasts at 5×10^4/mL embedded in A-medium with 0.7% agarose (5 mL/5-cm Petri dish).
3. Microdrops are transferred directly to these feeder dishes by sliding them from coverslips onto the feeder-dish surface. Thus, there is no liquid medium involved in this feeding system.
4. Dishes are incubated at 25°C, and after 1 wk, microdrops are lifted from the feeder dish surface and placed on a freshly prepared feeder dish.

5. When calluses derived from injected protoplasts are 1 mm in diameter, they are removed from the feeder dishes and placed on A-medium with half the quantity of mannitol and 0.7% agarose.
6. After 1 wk, the calluses are transferred to 1/2 MS medium with 1 mg/L BAP. On this medium, the callus multiplies rapidly. Approximately 16% of injected sugar beet protoplasts give rise to large calluses. The plating efficiency of sugar beet protoplasts is lower than that of tobacco protoplasts, hence the lower percentage that give rise to callus. With sugar beet, losses of injected protoplasts are more likely to occur at an early stage of culture, rather than when small calluses have formed.

4. Notes

1. Tobacco mesophyll protoplasts contain a large vacuole, and injections of DNA into this vacuole are wasted. It is possible to use evacuolate protoplasts for microinjection to avoid this problem. The method for evacuolation is that of Griesbach and Sink (*11*), except that centrifugation is for 1 h at 30,000 rpm using a SW50.1 rotor (Beckman). Washed evacuolate protoplasts are then treated exactly as normal vacuolate ones.
2. Protoplasts are useful for microinjection only on the day of their preparation, and possibly on the following day. This is because cell wall reformation after this time makes penetration with the needle difficult. The time available for microinjection can be extended by holding protoplasts at a lower temperature to delay wall reformation. Keeping tobacco and rape protoplasts at 10°C, and sugar beet protoplasts at 4°C, allows one further day for injections to be made.
3. 0.3% LMP agarose is sufficient to hold protoplasts in most cases. However, if a partial cell wall has reformed, the concentration of agarose for holding can be increased. Alternative methods of holding protoplasts include poly-L-lysine (*12*) and a holding pipet (*13*).
4. The method of preparing protoplasts for microinjection gives from 5–15 protoplasts per microdrop. Sometimes it is not possible to inject this number before the microdrop must be returned to a moist environment. In order to avoid culturing noninjected protoplasts, these can be destroyed by bursting them with the injection needle. The average plating efficiencies of injected protoplasts are: tobacco, 40%; rape, 23%; and sugar beet, 16%.

5. One of the main problems with the microinjection technique is that of blocked needles. The glass used to prepare the needles must be clean, and the siliconizing agent filtered to ensure that there are no particles that might later block the needles. Similarly, the injection solution should be centrifuged before use to remove any contaminating particles. However, the main problem is caused by fragments of cell membrane adhering to the needle after a few injections. The problem of needles becoming blocked may be partly overcome if a more sophisticated injection system is available. A system based on compressed gas (e.g., Picoinjecter) rather than oil allows greater pressures to be generated, and a high pressure applied for a very short time may clear a blocked needle. Such a system also allows finer control of the injection process.
6. To check that the injection solution enters a protoplast, it is useful to inject fluorescent dyes. 4'6-Diamidino-2-phenylindole (DAPI) causes the nucleus to fluoresce brightly under a UV light source when a solution of 100 µg/mL is injected.
7. The plating efficiency of sugar beet protoplasts in microdrops can be improved if the agarose microdrops are preconditioned for 1 d before the protoplasts are added. Agarose microdrops are prepared as previously described, but without the addition of the 1 µL drop containing the protoplasts. These microdrops are placed on sugar beet feeder dishes and kept at 25°C for 1 d before the protoplasts to be injected are added. For injection, the microdrop can then be lifted directly from the feeder dish, and returned immediately after the injection is complete. It is possible to use sugar beet suspension culture cells to feed protoplasts, but the plating efficiency is reduced slightly compared to protoplast feeders.
8. A variety of macromolecules can be introduced in protoplasts by microinjection. However, the injection of DNA in order to obtain transformants is the most common application. Transformation frequencies can be as high as 26% (*12*) if intranuclear injections are made, and as high as 6% for cytoplasmic injections (*13*). Injections into microspore-derived embryoids of rape result in even higher transformation frequencies than the protoplast injections, 27–51% (*1*). With transformation rates at this level, selectable markers are not essential. Problems can arise with the use of selectable markers in the method described, because the addition of an antibiotics, such as kanamycin, adversely affects the feeder protoplasts, and therefore reduces the chance of transformants surviving the initial stages of culture.

References

1. Neuhaus, G., Spangenberg, G., Mittlestein-Scheid, O., and Schweiger, H.-G. (1987) Transgenic rapeseed plants obtained by the microinjection of DNA into microspore-derived embryoids. *Theor. Appl. Genet.* **75,** 30–36.
2. Nagata, T. and Takebe, I. (1971) Plating of isolated tobacco mesophyll protoplasts on agar medium. *Planta* **99,** 12–20.
3. Murashige, T. and Skoog, F. (1962) A revised medium for rapid growth and bioassays with tobacco tissue cultures. *Physiol. Plant.* **15,** 473–497.
4. Nagy, J. I. and Maliga, P. (1976) Callus induction and plant regeneration from mesophyll protoplasts of *Nicotiana sylvestris*. *Z. Pflanzenphysiol.* **78,** 453–455.
5. Banks, M. S. and Evans, P. K. (1976) A comparison of the isolation and culture of mesophyll protoplasts from several *Nicotiana* species and their hybrids. *Plant Science Letters* **7,** 409–416.
6. Menczel, L., Nagy, F., Kiss, Zs. R., and Maliga, P. (1981) Streptomycin resistant and sensitive somatic hybrids of *Nicotiana tabacum* and *Nicotiana knightiana*: Correlation of resistance to *N. tabacum* plastids. *Theor. Appl. Genet.* **59,** 191–195.
7. Kao, K. N. and Michayluk, M. R. (1981) Embryoid formation in Alfalfa cell suspension cultures from different plants. *In Vitro* **17,** 645–648.
8. Lawrence, W. A. and Davies, D. R. (1985) A method for the microinjection and culture of protoplasts of very low densities. *Plant Cell Reports* **4,** 33–35.
9. Shillito, R. D., Paszkowski, J., and Potrykus, I. (1983) Agarose plating and a bead type culture technique enable and stimulate development of protoplast-derived colonies in a number of plant species. *Plant Cell Reports* **2,** 244–247.
10. Glimelius, K. (1984) High growth rate and regeneration capacity of hypocotyl protoplasts in some *Brassicaceae*. *Physiol. Plant.* **61,** 38–44.
11. Griesbach, R. J. and Sink, K. C. (1983) Evacuolation of mesophyll protoplasts. *Plant Science Letters* **30,** 297–301.
12. Reich, T. J., Iyer, U. N., and Miki, B. L. (1986) Efficient transformation of Alfalfa protoplasts by the intranuclear microinjection of Ti plasmids. *Biotechnology* **4,** 1001–1004.
13. Crossway, A., Oakes, J. V., Irvine, J. M., Ward, B., Knauf, V. C., and Shewmaker, C. K. (1986) Integration of foreign DNA following microinjection of tobacco mesophyll protoplasts. *Mol. Gen. Genet.* **202,** 179–185.

Chapter 32

Transformation of Plants Via the Shoot Apex

Roberta H. Smith, Eugenio Ulian, and Jean H. Gould

1. Introduction

The terms "meristem" and "shoot tip" culture have often been indiscriminately interchanged. According to Cutter (*1*), the apical meristem refers to only that portion of the shoot apex lying distal to the youngest leaf primordium. The shoot apex, or shoot tip, consists of the apical meristem and one to three subjacent leaf primordia. True apical meristem culture of higher plants was first demonstrated by Smith and Murashige (2). The isolated tissues of the apical domes develop directly into plants, demonstrating the developmental automony of the angiosperm shoot apical meristem.

True meristem culture has been used to eliminate virus infestation in higher plants. However, because of the size of this explant (80–100 μm in height), virus eradication is usually achieved using the shoot apex, which includes the meristem and two to three primordial leaves. This technique has been widely used with both monocot and dicot species. A few crop species that have been freed of viruses by the shoot-tip method include: citrus, garlic, pineapple, horseradish, asparagus, cauliflower, taro, strawberry, soybean, hop, sweet potato, rye grass, apple, cassava, banana,

From: *Methods in Molecular Biology*, vol. 6, *Plant Cell and Tissue Culture*
Edited by Jeffrey W. Pollard and John M. Walker, ©1990 by The Humana Press

rhubarb, gooseberry, raspberry, sugar cane, potato, and ginger (*3*). This list is certainly not inclusive; however, it indicates that published methodology for the isolation and regeneration of plants from shoot apex explants of a wide variety of species is available. Furthermore, there are no reports of variant plant types resulting from this explant in culture. It is believed that true genetic type as well as agronomic and phenotypic character is maintained because the plants are derived from a preexisting shoot meristem.

Asexual reproduction using shoot tip or axillary bud explants gives rise to plants that are genetically identical to the stock plant. The propagation of many foliage ornamentals, orchids, flowers, shrubs, fruit and nut crops, vegetable, condiment, oil seed, beverage and latex-producing crops, medicinal, and forest tree crop species from this explant source is well-documented, and true clonal characteristics have been maintained (*4*). In contrast, plants arising from callus or suspension cultured cells, where the somatic embryo or shoot meristem develops adventitiously, exhibit a variety of culture-induced changes termed somaclonal variation. Somaclonal variation (*5*) is the genetic variability observed in plants derived from a callus intermediate.

To date, the genetic transformation of plants with the *Agrobacterium* vector has utilized plant regeneration methods that involve adventitious shoot development from callus or primary explants of leaf (*6*), hypocotyl (*7*), epidermal peel (*8*), or protoplasts (*9*). Significant limitations of these regeneration procedures are: (1) the lack of wide-spread adaptability to the vast majority of crop species and the diverse varieties within species; and (2) the increased risk of somaclonal variation because the plants arise via adventitious organogenesis.

By far, the most limiting factor is the difficulty in adapting these regeneration methods for use with the majority of crop species. Adventitious shoot development in vitro from leaf or epidermal peel explants is rare in most agriculturally important crop species. Additionally, regeneration from callus, suspension, or protoplast cultures is extremely limited. The limitations of the existing cell culture systems have been recognized as a major obstacle to the application of biotechnology and genetic engineering to the improvement of crop plants.

In retrospect, it would appear that the most universal explant for plant regeneration and true clonal propagation is the tissue of the shoot apex. The literature concerning the in vitro establishment of plants from the shoot meristem and shoot apex spans nearly 30 yr, beginning with the work of Morel (*10,11*). The shoot apex explant is widely recognized as the method of choice in the generation of virus-free germplasm, and apparently can be used with all plant species (*12*). Therefore, the adaptation of this classic procedure to transformation studies can eliminate many of the present restrictions as to species and variety that result from the lack of an appropriate in vitro regeneration method. In addition, somaclonal variation can be minimized in plants regenerated from shoot apex explants. The importance of this aspect becomes obvious when the insertion of specific genes into an otherwise desirable cultivar and retention of genetic fidelity are of paramount concern.

Transformed plants can be obtained using shoot apex explants in cocultivation with *Agrobacterium tumefaciens*. The report of Ulian et al. (*13*) was the first to demonstrate that in vitro shoot apex transformation was feasible, and that the rates of transformation were comparable to those of the leaf disk explant system. The removal of one barrier often exposes another. In this case, once the ability to regenerate almost any desired plant in vitro is in hand, the understanding and manipulation of *Agrobacterium* genetic transfer processes will become even more critical.

2. Materials

1. Incubator or light and temperature controlled culture shelves, 16:8-h photoperiod, 27 ± 3°C, 200 ft-c, 28–30 $\mu Em^{-2} s^{-1}$.
2. Suspension of *Agrobacterium tumefaciens*, disarmed, containing kanamycin resistance gene and desired reporter genes. Materials for assay of chosen reporter gene or GUS assay (*see* Chapters 29 and 30 for details; *see also* ref. *14*).
3. Seeds, Petunia "Rose Flash" (Geo. Ball, Co., Chicago, IL), #10 & #11 scalpel blades & #7 scalpel handle, forceps.
4. Commercial bleach (5.25% Sodium hypochlorite).
5. Wetting agent: Tween-20 or liquid detergent.
6. Murashige and Skoog inorganic salt (MS) formulation (*15* and *see* Appendix).
7. N^6-benzyladenine (BA) stock (10 mg/100 mL). Weigh out 0.010 g of crystals, and dissolve in a beaker with a few drops of 1*N* HCl and a few drops of water, with gentle heating to dissolve crystals. Add the remaining water rapidly to bring up to volume in a 100-mL volumetric flask. Store at 4°C.
8. Antibiotics: (a) Carbenicillin. Weigh out 0.5 g and dissolve in 5 mL of double-distilled water, and (b) Kanamycin. Weigh out 0.2 g and dissolve in 5 mL of double-distilled water.

9. Seed germination medium: MS salts plus 30 g/L sucrose, 7 g/L agar.
10. Primary explant culture medium: MS salts plus 0.1 mg/L BA, 30 g/L sucrose, 2 g/L gelrite (KC Biological, Kansas City, MO), pH 5.7.
11. Antibiotic-containing selection media: MS salts, 0.1 mg/L BA, 30 g/L sucrose, 2 g/L gelrite, 200 mg/L kanamycin, 500 mg/L carbenicillin. The antibiotics are filter sterilized using a 5 mL syringe and disposable 0.2 µm membrane filter units and added to the autoclaved, cooled medium.
12. Rooting/selection medium: MS salts, 30 g/L sucrose, 100 mg/L kanamycin, 500 mg/L carbenicillin.
13. ABS medium for *Agrobacterium*: the *A. tumefaciens* LBA4404 containing desired reporter gene(s) is cultured in a medium prepared as follows: (a) 100 mL salt solution is prepared comprising 3.9 g dibasic potassium phosphate •$3H_2O$, 1 g sodium monobasic phosphate, 1 g NH_4Cl and 0.15 g potassium chloride. The salt solution is then autoclaved at 121°C and 18 psi for 15 min. (b) A separate 900 mL media solution is prepared containing: 0.59 g/L sucrose; 13 mg/L calcium chloride, 0.5 g/L magnesium sulfate, 10 µL ferric sulfate stock (250 mg/mL $FeSO_4$•$7H_2O$). Stock cultures of *Agrobacterium* are maintained on the same medium as above, but to which 15 g of agar has been added. For suspension cultures, agar is omitted. Medium (b) is autoclaved for 25 min, cooled, and combined with medium (a); plus 50 mg of kanamycin in 5 mL ddH_2O. *A. tumefaciens* LBA4404 is grown on 3 mL of medium for 2 d. The medium and cultured *A. tumefaciens* are then used to inoculate the shoot apices in the manner described below.

3. Methods

3.1. Transformation

1. Surface sterilize petunia seeds with 20% (v/v) Clorox™, to which a wetting agent (one or two drops/100 mL) has been added, for 30 min.
2. Rinse 5x with sterile water, and then transfer the sterilized seeds to seed germination medium.
3. Incubate for 1 wk at 25°C with a day/night cycle of 16/8 h.
4. After 1 wk of incubation, seedlings will be ready for shoot apex excision. With the aid of a dissection microscope (use 8, 12, and 20x magnification settings), isolate the shoot apex consisting of the apical meristem and two primordial leaves (approximately 0.3 x 0.6 cm).

5. Culture shoot tips on primary explant medium for 2 d.
6. After 2 d of culture, innoculate each explant with a 5-μL drop of *A. tumefaciens* suspension. Leave the Petri dishes open in the laminar-flow transfer hood until the drops of bacterial suspension have dried.
7. Transfer the explants to fresh medium, seal the plates with Parafilm, and continue to culture for 2 d at 25°C, with a day/night cycle of 16/ 8 h.
8. After 2 d of incubation, transfer the explants to selection medium containing kanamycin and carbenicillin, and incubate for 3 wk. Shoot apices will enlarge, and the two primary leaves will expand, but the leaves will bleach in response to kanamycin. The transformed tissues will appear as green regions at the base of the pale or white primary leaves.
9. Remove the bleached leaves and reculture the green shoot onto medium with kanamycin reduced to 100 mg/L. Transformed shoots will elongate within 1 wk.
10. After 1 wk, transfer the elongated shoots to rooting medium. All of the transformed shoots will root in the presence of kanamycin.

3.2. GUS Assay

Rooted plants are assayed for the presence of the GUS gene as follows:

1. Approximately 50 mg of plant tissue is homogenized with a pestle in an Eppendorf tube containing 200 μL of 50 m*M* $NaPO_4$(sodium phosphate buffer), pH 7.0, 10 m*M* EDTA, 0.1% Triton X-100, 0.1% Sarkosyl, 10 m*M* β-mercaptoethanol.
2. One hundred μL of the extract is added to 100 μL of 2 m*M* 4-methylumbelliferyl glucuronide dissolved in the same buffer.
3. The reaction is incubated at 37°C for 5 h, and stopped with 1 mL of 0.2 M Na_2CO_3. The production of methylumbelliferone is quantified with a spectrofluorimeter at hourly intervals with excitation set at 365 nm and emission at 455 nm (*see* Chapter 29, this vol. for details).

References

1. Cutter, E. C. (1965) Recent experimental studies of the shoot apex and shoot morphogenesis. *Bot. Rev.* **31,** 7–113.
2. Smith, R. H. and Murashige, T. (1970) *In vitro* development of the isolated shoot apical meristem of angiosperms. *Amer. J. Bot.* **57,** 562–568.
3. Bhojwani, S. S. and Razdan, M. K. (1983) Clonal Propagation, in *Plant Tissue Culture:*

Theory and Practice (Bhojwani, S. S. and Razdan, M. K., eds.), Elsevier, Amsterdam, Oxford, New York, and Tokyo, pp. 313–372.

4. Bhojwani, S. S. and Razdan, M. K. (1983) Production of Pathogen-free Plants, in *Plant Tissue Culture: Theory and Practice* (Bhojwani, S. S. and Razdan, M. K., eds.), Elsevier, Amsterdam, Oxford, New York, and Tokyo, pp. 287–312.
5. Larkin, P. J. and Scowcroft, W. R. (1981) Somaclonal variation—a novel source of variability from cell cultures for plant improvement. *Theor. Appl. Genet.* **60,** 197–214.
6. Horsch, R. B., Fry, J. E., Hoffmann, N. L., Wallroth, M., Eichholtz, D. Z., Rogers, S. G., and Fraley, R. T. (1985) A simple and general method for transferring genes into plants. *Science* **227,** 1229–1231.
7. Firoozabady, E., DeBoer, D. L, Merlo, J., Halk, E. L., Amerson, L. N., Radhka, K. E., and Murry, E. E. (1987) Transformation of Cotton (*Gossypium hirsutum* L.) by *Agrobacterium tumefaciens* and regeneration of transgenic plants. *Pl. Mol. Bio.* **10,** 105–116.
8. Trinh, T. H., Mante, S., Pua, E. C., and Chua, N. H. (1987) Rapid production of transgenic flowering shoots and F1 progeny from *Nictoiana plumbaginifolia* epidermal peels. *Biotechnology* **5,** 1081–1084.
9. Marton, L., Wullems, G., Molendijk, L., and Schilperoort, R. A. (1979) *In vitro* transformation of cultured cells from *Nicotiana tabacum* by *Agrobacterium tumefaciens. Nature* **279,** 129–131.
10. Morel, G. (1972) Morphogenesis of stem apical meristem cultivated *in vitro:* Applications to clonal propagation. *Phytomorphyology* **22,** 265–277.
11. Morel, G. M. and Martin, G. (1952) Guerison de Dehlias atteints d'une maladie a virus. *C. R. Hebd. Seanc. Acad. Sci., Paris* **235,** 1324, 1325.
12. Murashige, T. (1974) Plant propagation through tissue cultures. *Ann. Rev. Plant Physiol.* **25,** 135–166.
13. Ulian, E. C., Smith, R. H., Gould, J. H., and McKnight, T. D. (1988) Transformation of plants via the shoot apex. *In Vitro Cell and Dev. Bio.* **24,** 951–954.
14. Jefferson, R. A., Burgess, S. M., and Hirsch, D. (1986) Betaglucurondiase from *Escherichia coli* as a gene-fusion marker. *Proc. Nat. Acad. Sci. USA* **83,** 8447–8451.
15. Murashige, T. and Skoog, F. (1962) A revised medium for rapid growth and bioassay with tobacco tissue cultures. *Physiol. Plant* **15,** 473–497.

Chapter 33

Methods of Gene Transfer and Analysis in Higher Plants

Timothy J. Golds, Michael R. Davey, Elibio L. Rech, and John B. Power

1. Introduction

The rapid development of tissue culture and recombinant-DNA technology in recent years has enabled plants of many species to be regenerated from cultured cells and their genetic information to be manipulated in various ways. Gene transfer has already resulted in the production of a wide range of transgenic dicotyledons, particularly in the *Solanaceae*. Although success with monocotyledons is more limited, transgenic rye (*1*), maize (*2*), and rice (*3*) plants have been reported.

The techniques for inserting genes into dicotyledons have been based primarily on the tumor-inducing (Ti) plasmid of *Agrobacterium tumefaciens* and the root-inducing (Ri) plasmid of *A. rhizogenes* (*4*) (*see* Chapters 27–29, this volume). The main procedures involve inoculation of seedlings or explants, particularly leaf disks (*5*), or cocultivation of protoplast-derived cells (*6*) with the bacteria. Exposure of imbibed seeds to the bacteria has recently been used as a transformation method (*7*). Transformation of plant cells by wild-type strains of *Agrobacterium* involves the transfer of part of the Ti- or Ri-plasmid (the T-DNA) to recipient cells, followed by its integration into the plant genome, with subsequent expression. Foreign genes inserted between the T-DNA borders can also be transferred to plant cells. Thus, cointegrate-type Ti- and Ri-vectors have been constructed in which a gene of interest, cloned on an

From: *Methods in Molecular Biology*, vol. 6, *Plant Cell and Tissue Culture*
Edited by Jeffrey W. Pollard and John M. Walker, ©1990 by The Humana Press

E. coli plasmid, is inserted into the T-DNA by recombination at a region of common homology (*8*). Since oncogenicity genes on the Ti T-DNA incite a tumorous response in plant cells, these genes are generally deleted and replaced by an antibiotic resistance gene that can be used as a marker to select transformed cells (*9*). Such disarmed vectors still retain T-DNA border sequences to permit integration into the plant genome.

Binary vector systems, in which the *Agrobacterium* strain carries two plasmids (*see* Chapter 28, this volume), are also used extensively. One is a small plasmid that has a selectable marker and cloning sites between T-DNA borders, and that can replicate in both *E. coli* and *Agrobacterium*. The other is a Ti- or Ri-plasmid that acts *in trans*, supplying virulence functions to enable the transfer to plant cells of genes between the T-DNA borders of the smaller vector. The Ti or Ri "helper" plasmid may be intact (*10,11*), in which case its T-DNA may also be transferred to plant cells during transformation. More frequently, the resident Ti-plasmid is deleted of its T-DNA (*12*).

The general inability of *Agrobacterium* to induce a response in monocotyledons has resulted in the development of methods for direct DNA delivery into cells, overcoming host range specificity (*see* Chapters 30–32, this vol.). Such techniques include DNA uptake into protoplasts stimulated by chemical treatment or electroporation (*13*), injection of DNA into developing inflorescences (*1*), and the use of pollen as a DNA transfer agent (*14*). Microinjection (*15*) and ballistic methods (*16*) also hold promise for DNA transfer into monocotyledons as well as dicotyledons (*17*). Vectors for direct gene transfer are based on small plasmids that can be amplified in *E. coli* and that can be isolated in large (mg) quantities by standard procedures. They can be used both in transient studies, in which gene expression is assayed in the recipient cells a short time (usually hours) after DNA uptake, and for the production of stably transformed tissues and transgenic plants. Genes that are readily assayed in transient systems include those for chloramphenicol acetyltransferase (CAT) (*18*), β-glucuronidase (GUS) (*19*), and the bioluminescent luciferase gene (*20*). Those used for stable transformation include genes for resistance to antibiotics, such as kanamycin (*21*), bleomycin (*22*), and hygromycin (*23*), and resistance to herbicides (*24*), and viruses (*25*). Transient expression genes can also be assayed in stably transformed tissues, provided they are linked to a dominant marker for initial selection of transformed cells.

This chapter summarizes some of the main procedures used in the Plant Genetic Manipulation Group at Nottingham to transfer genes to plant cells and methods for analyzing their expression in transformed tissues and transgenic plants. Because of the number of cellular and molecular techniques currently available, the following protocols are intended to serve only as a general guide.

2. Materials

2.1. *Transformation of* E. coli *with Calcium Chloride*

1. LB medium: 1.0% w/v Difco Bacto Tryptone, 0.5% w/v Difco Bacto yeast extract, 0.5% w/v NaCl pH 7.0.
2. Cell resuspension buffer: 50 m*M* $CaCl_2$, 10 m*M* Tris-HCl pH 8.0.
3. TE buffer: 1 m*M* EDTA, 10 m*M* Tris-HCl pH 8.0.
4. Ampicillin. (10 mg/mL stock in water).

2.2. *Conjugation of Recombinant Plasmids into* Agrobacterium *by Triparental Mating*

1. LB medium (*see above*).
2. APM medium: 0.5% w/v yeast extract, 0.05% w/v casamino acids, 0.8% w/v mannitol, 0.2% w/v $(NH_4)_2SO_4$, 0.5% w/v NaCl.
3. Nutrient agar: 1.0% w/v Lab Lemco beef extract, 1.0% w/v peptone, 0.5% w/v NaCl, 1.5% w/v Bacto Difco agar, pH 7.0.
4. Minimal agar: 3 g/L K_2HPO_4, 1 g/L NaH_2PO_4, 1 g/L NH_4Cl, 0.3 g/L $MgSO_4 \bullet 7H_2O$, 0.15 g/L KCl, 0.01 g/L $CaCl_2$, 2.5 mg/L $FeSO_4 \bullet 7H_2O$, 0.5% w/v glucose, 1.5% w/v Bacto agar.

2.3. *Small-Scale Plasmid Preparation from* E. coli

1. LB medium (*see above*).
2. Plasmid extraction buffer: 8% w/v sucrose, 0.5% w/v Triton X-100, 50 m*M* EDTA, 10 m*M* Tris-HCl, pH 8.0.
3. Lysozyme: 10 mg/mL in 10 m*M* Tris-HCl pH 8.0. Prepare fresh as required.
4. 2.5*M* sodium acetate.
5. Isopropanol.
6. TE buffer pH 8.0 (*see* 2.1.).
7. DNase-free RNase. Prepare a stock solution (10 mg/L) in 15 m*M* NaCl, 10 m*M* Tris-HCl pH 7.5, and boil for 15 min with slow cooling to room temperature before use.

2.4. Large-Scale Plasmid Preparation from E. coli Using Alkaline Lysis

1. LB medium (*see above*).
2. Chloramphenicol: 30 mg/mL stock solution in ethanol.
3. STE buffer: 100 m*M* NaCl, 1.0 m*M* EDTA, 10 m*M* Tris-HCl pH 8.0.
4. Glucose/buffer solution: 50 m*M* glucose, 10 m*M* EDTA, 25 m*M* Tris-HCl, pH 8.0.
5. NaOH/SDS solution: 0.2*M* NaOH, 1.0% w/v SDS.
6. 5*M* potassium acetate, pH 6.2. Add 11.5 mL of glacial acetic acid to 60 mL of 5*M* potassium acetate, and make up to 100 mL with water.
7. Isopropanol.
8. TE buffer pH 8.0 (*see* 2.1. *above*).
9. Cesium chloride.
10. Ethidium bromide: 10 mg/mL in water.

2.5. Preparation of Total Nucleic Acid from Agrobacterium

1. APM medium (*see* 2.2.).
2. TE buffer (*see* 2.1.).
3. 25% w/v SDS.
4. Pronase: 5 mg/mL. Remove contaminating nucleases by predigesting at 27°C for 2 h.
5. Equilibrated phenol: melt the phenol at 68°C, and then extract with an equal volume of 1.0*M* Tris-HCl, pH 8.0, followed by several extractions with 0.1*M* Tris-HCl, pH 8.0, 0.2% β-mercaptoethanol, until the pH of the aqueous phase exceeds 7.6.
6. Chloroform/isoamyl alcohol: 24:1, v/v.
7. Phenol/chloroform: phenol:chloroform, 1:1. (The chloroform component is 24:1 v/v chloroform:isoamyl alcohol.)

2.6. Agrobacterium-Based Gene Transfer

1. Commercially available bleach, e.g., Domestos (Lever Bros. London).
2. MS-based medium lacking growth regulators (MSO). *See* Appendix.
3. Cefotaxime (Claforan; Roussel Laboratories, Wembley, UK).

2.7. Polyethylene Glycol-Induced DNA Uptake into Protoplasts

1. Polyethylene glycol (PEG): 40% w/v PEG 6000 (Koch Light Ltd.).

2. F solution: 8.12 g/L NaCl, 0.27 g/L KCl, 0.11 g/L Na_2HPO_4, 0.9 g/L glucose, and 18.36 g/L $CaCl_2 \cdot 2H_2O$, pH 7.0 (*26*).

2.8. Electroporation

1. Commercially available electroporator (DIA-LOG, GmbH, 4 Dusseldorf 13, F.R.G.).
2. Electroporation medium (modified from ref. *27*): 0.8 g/L NaCl, 0.02 g/L KCl, 0.02 g/L KH_2PO_4, 0.115 g/L Na_2HPO_4, 100 g/L glucose pH 7.1.
3. Calf thymus DNA.

2.9. Detection of Octopine, Nopaline, Agropine, or Mannopine

1. MSO medium (*see* Appendix) containing 100 m*M* L-arginine HCl.
2. MN 214 chromatography paper (Macherey Nagel, 516 Duren, Germany).
3. Stock solutions of arginine, octopine, and nopaline, each at 0.5 mg/mL, and agropine and mannopine (1.0 mg/mL).
4. Electrophoresis buffer: formic acid:acetic acid:water, 5:15:80, v/v/v.
5. Phenanthrenequinone stain (for octopine and nopaline): Dissolve 2 mg of phenanthrenequinone in 10 mL of absolute ethanol. Add this solution to 10 mL of 60% v/v ethanol containing 1.0 g of NaOH. Prepare fresh as required.
6. Stain for agropine or mannopine: prepare two solutions, A and B. Solution A consists of 0.625 g of silver nitrate in 0.25 mL of water. Add to 50 mL of acetone; a silver precipitate forms. Add water dropwise (approximately 50 drops from a Pasteur pipet) until the silver precipitate dissolves. Solution B consists of 10 mL of 20% w/v NaOH mixed with 90 mL of methanol.
7. 5% w/v sodium thiosulfate.

2.10. Neomycin Phosphotransferase Assay

1. Acrylamide/bisacrylamide stock solution: Dissolve 30 g of acrylamide and 0.8 g of bisacrylamide in 100 mL distilled water. Pass through a millipore filter (0.2 μm) and store at 4°C.
2. 1.5% w/v ammonium persulfate (prepare fresh).
3. 1.0*M* Tris-HCl pH 6.8.
4. TEMED.
5. Tris-glycine buffer stock (5x): 139.6 g glycine plus 15.2 g of Tris base made up to 1 L, pH 8.3.

6. Extraction buffer: 10% glycerol, 40 m*M* EDTA, 150 m*M* NaCl, 100 m*M* NH_4Cl, 3.0 mg/mL dithiothreitol, 0.2 mg/mL leupeptin, 0.2 mg/mL trypsin inhibitor, 10 m*M* Tris-HCl pH 5.0.
7. Bovine serum albumin: 10 mg/mL stock solution.
8. Phenylmethylsulfonyl fluoride (7.0 mg/mL in isopropanol).
9. Reaction buffer: dissolve 8.0 g of Tris base, 8.5 g of $MgCl_2 \cdot 6H_2O$ and 21.4 g of NH_4Cl in 500 mL of distilled water. Titrate to pH 7.1 with solid maleic acid and make up to 1 L.
10. Low gelling temperature SeaPlaque agarose.
11. [γ^{32}P] ATP.
12. Kanamycin sulfate solution (25 mg/mL stock).
13. Whatman P81 paper.
14. 0.2*M* phosphate buffer, pH 7.7: Mix 13 mL of 0.2*M* $NaH_2PO_4 \cdot 2H_2O$ (31.2 g/L) with 87 mL of 0.2*M* $Na_2HPO_4 \cdot 12H_2O$ (71.7 g/L) and make up to 1 L with distilled water.

2.11. Plant DNA Isolation

1. Extraction buffer: 200 m*M* Tris-HCl pH 7.5, 250 m*M* NaCl, 25 m*M* EDTA pH 8.0, and 0.5% w/v SDS.
2. Equilibrated phenol (*see* 2.5.).
3. DNase-free RNase (*see* 2.3.). 10 mg/mL stock.
4. Proteinase K (10 mg/mL aqueous stock).
5. TE buffer: 10 m*M* Tris-HCl pH 7.5, 1.0 m*M* EDTA.

2.12. Southern Blotting

1. 2.5*M* sodium acetate.
2. Gel loading buffer: 0.25% w/v bromophenol blue, 0.25% w/v xylene cyanol, 25% w/v Ficoll (Type 400) in water.
3. TBE buffer (5x stock): 54 g of Tris base, 27.5 g of boric acid, 20 mL of 0.5*M* EDTA, pH 8.0 in 1 L of distilled water.
4. Ethidium bromide solution (10 mg/mL stock).
5. 0.25*M* HCl.
6. Denaturing solution: 0.5*M* NaOH, 1.5*M* NaCl.
7. 0.5*M* Tris-HCl pH 7.5, 1.5*M* NaCl.
8. SSPE Buffer (20 x stock): Dissolve 210 g NaCl, 27.6 g $NaH_2PO_4 \cdot H_2O$ and 7.4 g EDTA in 800 mL of water. Adjust the pH to 7.4 with 10*M* NaOH solution, and then make up to 1 L with distilled water.
9. Prehybridization solution: 100 mL of prehybridization solution consists of 50 mL of deionized formamide, 5 mL of SDS (10% w/v stock solution), 10 mL of dextran sulfate (50% w/v stock solution), 15 mL

of 20 x SSPE and 10 mL of 50 x Denhardt's solution. Deionized formamide is prepared by stirring 10 g of Dowex MR3 resin with 100 mL of formamide (30 min, dark, 22°C) and filtering. The Denhardt's solution consists of 1% w/v of Ficol, 1% w/v of polyvinylpyrrolidone, and 1% w/v BSA in 100 mL of distilled water. Sheared, freshly boiled, and chilled salmon sperm DNA is added to a final concentration of 100 µg/mL from a stock solution (10 mg/mL in water).

2.13. Dot Blotting

1. Denaturing solution: 0.5*M* NaOH, 1.5*M* NaCl.
2. Neutralizing solution: 1.5*M* NaCl, 0.5*M* Tris-HCl, pH 7.5, 1 m*M* EDTA.

2.14. Chloramphenicol Acetyltransferase (CAT) Assay

1. CAT extraction buffer: 0.2*M* Tris HCl pH 7.8.
2. 10 m*M* acetyl coenzyme A.
3. ^{14}C chloramphenicol (Amersham, CFA 754 at 25 µCi/mL).
4. Purified CAT enzyme.

2.15. β-Glucuronidase (GUS) Assay

1. Extraction buffer: 50 m*M* sodium phosphate buffer, pH 7.0, 10 m*M* EDTA, 10 m*M* β-mercaptoethanol and 0.1% w/v Triton X-100.
2. 25-gage needle.
3. 4-methyl umbelliferyl glucuronide (1 m*M* stock solution).
4. 0.2*M* Na_2CO_3.

3. Methods

3.1. Vectors for Gene Transfer

Although protocols for gene cloning through restricting and ligating DNA molecules are beyond the scope of this chapter, this section summarizes some of the standard procedures in which vectors carrying a gene of interest can be transferred to a suitable host strain of *E. coli*. Such vectors can be multiplied in *E. coli* prior to large-scale isolation and used in direct gene uptake [e.g., pABD1 (*28*), pCaMVNEO (*27*)], or where their molecular configuration permits, transferred to *Agrobacterium* by triparental mating. The verification of vector structure is crucial prior to its use in experiments in gene transfer, through small-scale plasmid preparations from *E. coli* followed by agarose gel electrophoresis, and total nucleic acid preparation from *Agrobacterium* strains and subsequent Southern blot analysis.

3.1.1. *Transformation of* E. coli *by a Calcium Chloride Procedure (29)*

This method is based on the observation that competent bacterial cells of a suitable strain of *E. coli*, e.g., HB101, are capable of high-frequency transformation when treated with intact plasmid in the presence of calcium chloride. Selection of transformants is based on their ability to express an antibiotic resistance gene carried on the plasmid (e.g., ampicillin resistance in pCaMVNEO).

1. Remove a single bacterial colony of HB101; transfer to 5 mL of LB medium. Grow at 37°C (overnight) with vigorous shaking.
2. Transfer 1 mL of the overnight culture to 100 mL of LB medium. Grow at 37°C with vigorous shaking to OD_{550} = 0.5.
3. Chill the culture on ice (10 min). Centrifuge (4,000*g*, 5 min, 4°C) in a Beckman J2-21 centrifuge or equivalent.
4. Remove the supernatant; resuspend the cells in 50 mL of ice-cold sterile cell resuspension buffer.
5. Place on ice for 15 min; recentrifuge as in step 3.
6. Remove the supernatant and resuspend the cells in 7 mL of ice-cold cell resuspension buffer.
7. Transfer 200 µL aliquots of cells into small, precooled glass tubes.
8. Add 25 ng of plasmid DNA in 10 µL of sterile TE buffer pH 8.0, mix, and transfer to ice for 30 min.
9. Heat shock in a water bath at 42°C for 2 min.
10. Add 1 mL of LB to each tube; incubate for 1 h at 37°C (stationary). This step allows the bacteria to express the antibiotic resistance prior to plating.
11. Spread an appropriate quantity of cells (prepared by serial dilution) onto LB plates with 1.5% w/v Bacto agar and ampicillin (50 µg/mL) or other selective agent, depending on the vector. After air drying, incubate the plates (inverted) at 37°C for 16 h.
12. Count the number of colonies/plate and calculate the transformation frequency (number of transformants/µg plasmid DNA).

3.1.2. *Conjugation of Recombinant Plasmids into* Agrobacterium *by Triparental Mating*

Triparental mating is the most convenient way of transferring a plasmid from *E. coli* to *Agrobacterium*. The method detailed here describes the transfer of a vector, containing a ColE1 replicon, to *Agrobacterium*, to produce a cointegrate vector. The mating procedure relies on the use of a helper strain of *E. coli* (GJ23) containing pGJ28 and R64drd11. These plasmids are transferred to the *E. coli* host strain containing the vector of interest. Subsequently, all three plasmids are transferred to *Agrobacterium*, where the vector of interest becomes integrated at a low frequency into the host Ti- or Ri-plasmid through recombination at a common site of homology (the vector contains a cloned fragment of *Agrobacterium* T-DNA).

1. Grow *E. coli* strains in LB medium with selection at 37°C for 16 h, and *Agrobacterium* in APM medium at 28°C for 16–24 h.
2. Pipet 100 µL of each of the three overnight cultures onto a nutrient agar plate, mix by spreading, and incubate (inverted) at 28°C for 16 h. Provide suitable control plates containing:
   a. *Agrobacterium* + helper strain
   b. *Agrobacterium* + vector strain
   c. Helper + vector strains and
   d. Each strain plated alone.
3. Remove a loopful of cells and resuspend in 250–500 µL of LB medium. Streak onto minimal agar plates containing suitable antibiotics and incubate for 48 h at 28°C. Repeat this step for each of the bacterial controls.
4. Transconjugants should appear only where the vector for transfer has stably integrated into the *Agrobacterium* Ri- or Ti-plasmid and expressed the antibiotic resistance gene. Since the ColEl replicon is unable to function in *Agrobacterium*, cells taking up the vector without integration are not able to multiply; *E. coli* should show little growth on minimal agar.
5. Restreak (X3-X4) putative transconjugants onto fresh minimal agar (with selection) and incubate as in step 3.
6. Perform Southern blot analysis (*see below*), on a total nucleic acid sample isolated from selected putative transconjugants, to confirm the molecular structure of the cointegrate.

Cloning vectors based on the wide host range RK2 origin of replication are mobilized into *Agrobacterium* using a helper *E. coli* strain containing pRK2013, allowing stable maintenance in *Agrobacterium* under the correct selection, without the need for homologous recombination. Triparental matings of this type give rise to a high frequency of transformants of the binary vector type.

3.1.3. *Small-Scale Plasmid Preparation from* E. Coli *(30)*

The following rapid method can be used to produce samples of sufficiently pure plasmid DNA to permit digestion with restriction enzymes and subsequent confirmation of vector structure by agarose gel electrophoresis, ethidium bromide staining, and visualization under UV illumination.

1. Grow 1.0-mL cultures of *E. coli* in LB medium at 37°C for 16 h, with shaking.
2. Transfer each aliquot to an Eppendorf tube and centrifuge at high speed in a microcentrifuge for 1 min.
3. Remove the supernatant; resuspend the pellet in 350 µL of plasmid extraction buffer. Mix by vortexing (2–3 s).
4. Add 25 µL of a freshly prepared solution of lysozyme. Mix by vortexing (2–3 s).
5. Transfer the tubes to a boiling water bath for 40 s. Spin in a microcentrifuge (4°C, 10 min).
6. Remove a known volume of supernatant, and add 1/10 the volume of 2.5*M* sodium acetate and an equal volume of isopropanol. Mix, and place at –70°C for 15 min.
7. Spin down the precipitated DNA in a microcentrifuge for 15 min at 4°C.
8. Remove the supernatant and dry the pellet in a vacuum desiccator (5–10 min). Resuspend in 50 µL of TE buffer (pH 8.0). Contaminating RNA can be removed by digesting each DNA sample with 2 µL of a 10 mg/L stock solution of DNase-free RNase for 2 min.
9. Remove 10 µL of the DNA solution and digest with restriction enzyme(s). The digestion is performed in a reaction volume of 20 µL, comprising 10 µL of DNA solution, 2 µL of 10x appropriate digestion buffer (according to the manufacturer's instructions) and 1 µL of restriction enzyme (10 U/µL). It is best to use this generous amount of enzyme in each reaction to allow for any impurities in the DNA. Digest for 1 h, usually at 37°C.

3.1.4. *Large-Scale Plasmid Preparation from* E. Coli *Using Alkaline Lysis (31)*

Vectors carrying a Co1E1 replicon can be amplified by treating the bacterium with chloramphenicol to yield milligram quantities of plasmid. The protocol described has been optimized for filling four Beckman quick seal ultracentrifuge tubes.

1. Transfer a single *E. coli* colony to 10 mL of LB medium with the appropriate antibiotic(s) for selection. Incubate for 16 h at 37°C with shaking.
2. Inoculate four flasks, each containing 300 mL of warmed (37°C) LB medium with selection, with 2.5-mL aliquots of the overnight culture. Incubate at 37°C, 150 rpm, until $OD_{600} = 0.7$.
3. Add chloramphenicol to a final concentration of 170 μg/mL from the stock solution. Incubate for 16 h at 37°C using 150 rpm.
4. Centrifuge the cells in four large screw-capped bottles (400-mL capacity; Nalgene) in a JA-10 rotor at 5000 rpm (4420 x *g*) for 15 min in a Beckman J2-21 centrifuge or equivalent.
5. Resuspend the pellets in 20 mL of STE buffer. Combine the contents into two bottles and respin as in step 4.
6. Resuspend each bacterial pellet in 10 mL of glucose/buffer solution containing lysozyme at 5 mg/mL. Swirl gently and allow to stand at 22°C for 5 min.
7. Cool on ice (5 min) and transfer the contents of the bottles to a sterile 250 mL Erlenmeyer flask.
8. Lyse the cells by adding 40 mL of freshly prepared NaOH/sodium dodecyl sulfate (SDS) solution. Seal with Nescofilm (Bando Chemical Ind. Ltd. Japan), mix by gentle inversion. Return to ice for 10 min.
9. Precipitate chromosomal DNA and cellular debris by adding 30 mL of 5*M* potassium acetate (pH 6.2). Mix by inversion and place on ice for 15 min.
10. Decant into 50-mL capacity Nalgene tubes and spin at 18,000 rpm (39,200 x *g*) for 30 min at 4°C in a Beckman JA-20 rotor.
11. Remove a measured volume of the supernatant to a clean, sterile Erlenmeyer flask and add 0.6 vol of isopropanol. Mix by inversion, and incubate at 22°C for 15 min. Plasmid DNA can be observed by the formation of a milky white precipitate.
12. Transfer the solution to Corex tubes, and pellet the DNA by centrifugation (10,000 rpm, 12,100 x *g*; 30 min) at room temperature to avoid salt precipitation.
13. Remove the supernatants; wash the pellets 3x in 5 mL of 70% v/v ethanol, centrifuging for 10 min between each wash.
14. After vacuum desiccation, dissolve the pellets in exactly 16 mL of TE buffer, pH 8.0.
15. Add 16.8 g of cesium chloride and 800 μL of ethidium bromide solution. Mix well to dissolve the salt and transfer to four Beckman

quick-seal ultracentrifuge tubes. Top up with mineral oil if required, heat seal, and centrifuge (20 h at 40,000 rpm, 20°C) in a VTi65 rotor in a Beckman L5-65B ultracentrifuge or equivalent.

16. After centrifugation, view tubes under UV illumination, and remove the lower plasmid band by side puncture of the tubes with a syringe and 19-gage needle.
17. Remove the ethidium bromide from the plasmid by extracting with an equal volume of isopropanol saturated with sodium chloride. (Prepare a saturated solution of sodium chloride, add to an equal volume of isopropanol, mix, and allow to settle; use the upper phase.) Repeat until the pink coloration is removed from the lower phase. This is best performed by centrifuging in a bench centrifuge (80 x *g*) to separate the phases between extractions.
18. Dialyze against 1 L of TE buffer at 4°C for 5 h; repeat this procedure 4x.
19. Reprecipitate the plasmid DNA by adding 1/10 the vol of 2.5*M* sodium acetate and 3 vol of ethanol; place at –70°C for 1 h. Centrifuge at 10,000 rpm (12,100 x *g*) in a JA-20 rotor, and wash the pellet 3x in 70% v/v ethanol. Vacuum desiccate, and redissolve in a known volume of sterile TE buffer.
20. Determine the concentration of plasmid DNA by spectrophotometry. At 260 nm, an OD of 1.0 corresponds to approximately 50 µg/mL of double-stranded DNA.

3.1.5. Preparation of Total Nucleic Acid from Agrobacterium *(32)*

This method can be used to obtain samples of total DNA from *Agrobacterium* species suitable for restriction enzyme digestion and Southern blotting.

1. Pick a single colony of *Agrobacterium* and grow overnight in 10 mL of APM medium for 16–20 h, with vigorous shaking (28°C). Pellet 1.5-mL aliquots of the overnight culture in Eppendorf tubes (microcentrifuge, 5 min).
2. Resuspend each pellet in 400 µL of TE buffer, pH 8.0. Add 20 µL of 25% w/v SDS and 100 µL of pronase.
3. Incubate at 37°C for 30 min.
4. Add 50 µL of 5*M* sodium chloride solution; incubate at 68°C for 30 min.
5. Add an equal volume of equilibrated phenol and spin in a microcentrifuge for 2 min to separate the phases. Reextract the upper phase with an equal volume of phenol/chloroform (the chloroform consists of 24:1 chloroform:isoamyl alcohol). Finally, reextract the

upper phase with an equal volume of chloroform/isoamyl alcohol (24:1).

6. Precipitate the DNA by adding 2 vol of ethanol, and place at –70°C for 1 h. Pellet the DNA in a microcentrifuge (10 min).
7. Wash the DNA pellet 3x in 70% v/v ethanol, vacuum desiccate, and resuspend in 50–100 µL of sterile TE, pH 8.0.

3.2. Plant Systems for Agrobacterium-Based Gene Transfer

Intact seedlings, plant explants, and protoplast-derived cells can be inoculated with *Agrobacterium*. The chosen method for transformation depends on the plant species and availability of suitable tissues.

3.2.1. Sterilization of Plant Material

3.2.1.1. Seeds

1. Surface sterilize seeds in a dilute solution of commercially available bleach, e.g., 10% v/v "Domestos," for 10–30 min. In some cases, a prerinse in 70% v/v ethanol for 10–20 s, or immersion in 0.1% w/v mercuric chloride with 0.1% w/v sodium lauryl sulfate for 10 min, may be beneficial for difficult-to-sterilize seeds.
2. Wash the seeds 3x in sterile tap water.
3. Transfer the seeds to the surface of agar-solidified (0.8% w/v Sigma agar) MS-based medium (*33*, and *see* Appendix) lacking growth regulators (MSO). Incubate at 27°C (1000 lx).

3.2.1.2. Explants of Glasshouse Grown Material

1. Excise young leaves and stems from rapidly growing plants, and surface sterilize as described above.
2. Leaves can be used for the isolation of protoplasts or the preparation of leaf disks (*5*), both for subsequent inoculation with *Agrobacteria*. Sterile stems should be cut into 3–5 cm lengths, discarding any damaged regions, and inserted into agar-solidified MSO medium in the normal orientation.

3.2.2. Inoculation of Plant Material with Agrobacterium tumefaciens and A. rhizogenes

3.2.2.1. Stem and Seedling Explants. Using the following method, sterilized stem explants can be infected, and will give crown galls or hairy roots.

1. Fill a 1.0-mL syringe, connected to a 23-gage needle, with an overnight culture of *Agrobacterium*.

2. Remove the prepared explants from MSO agar medium, place on a sterile surface (ceramic tile or Petri dish), and wound the explant by puncturing the distal region with the needle. Return the explant to MSO medium. Alternatively, inoculate explants or decapitated seedlings by applying a loopful of bacterial culture to the recently cut surface. This procedure can be performed without removing explants from the culture vessels.
3. Incubate at 25°C under low light (700 lx; daylight fluorescent tubes) until a response is observed at the inoculation site (10 d to 6 wk).
4. Excise proliferating tumors induced by *A. tumefaciens*, and transfer to liquid MSO medium containing 500 µg/mL of cefotaxime (4.0 mL in a 5-cm Petri dish). Incubate statically for 5 d under the conditions described in step 3. Transfer the tumors to MSO agar medium containing 500 µg/mL cefotaxime. Excise individual hairy roots, and float on the surface of liquid MSO medium containing 500 µg/mL cefotaxime (4.0 mL in a 5-cm Petri dish). Incubate statically. Hairy roots can either be maintained in MSO liquid medium with regular subculture every 2–3 wk or transferred to MSO agar medium as described for tumors.
5. Tumors and hairy roots should be checked periodically for sterility by incubating a small sample of tissue in nutrient broth or LB medium at 28°C overnight. The antibiotic can be omitted from the medium when the cultures are sterile.
6. Hairy roots of several plant species, particularly members of the *Solanaceae* (e.g., *Nicotiana tabacum, Solanum nigrum*) produce plants spontaneously. These can be excised and grown to maturity following transfer to the glasshouse. However, plant regeneration in some species necessitates an intervening callus phase followed by culture on a suitable regeneration medium (the one normally used to obtain plants from nontransformed callus) to give plants by shoot production (e.g., *Solanum tuberosum*) or somatic embryogenesis (e.g., *Medicago sativa*).
7. Tumors induced by wild-type nopaline strains of *A. tumefaciens* frequently produce morphologically abnormal shoots (teratomas) that can be excised and cultured on MSO agar medium. Such shoots can be grown to maturity, but must be grafted to nontransformed stock plants, since they do not form their own roots.

3.2.2.2. Leaf Disks (*34*). Leaf explants of some species can be induced to undergo direct shoot regeneration. This procedure, combined with the use of disarmed strains of *A. tumefaciens* (oncogenic region of the

T-DNA deleted and replaced with a dominant selectable marker gene, such as that for kanamycin resistance), can be employed to produce transgenic plants that can be grown on their own roots.

1. Cut disks or squares (0.5–1.0 cm in size) from sterile leaves, using a cork borer or scalpel.
2. Immerse the explants in a dilute *Agrobacterium* suspension (an overnight culture diluted approximately 1:20 with fresh bacterial medium) for 30 min.
3. Blot the explants on sterile filter paper; transfer to the surface of agar solidified medium normally used for regeneration of the plant species under investigation (e.g., for *N. tabacum*, use agar-solidified MS-based medium with 0.05 mg/L NAA and 0.5 mg/L BAP). Incubate at 25°C (700 lx) for 2 d. Suitable controls (leaf disks not incubated with *Agrobacterium*) should also be set up.
4. Transfer the leaf discs (*Agrobacterium*-treated and control) to the same regeneration medium containing 250–500 µg/mL of cefotaxime, both with and without the appropriate antibiotic for selection, e.g., kanamycin at 50–100 µg/mL. Incubate as in step 3 until callus or shoots appear.
5. Excise developing shoots that continue to grow in the presence of antibiotic selection, and transfer to MSO medium containing 250 µg/mL cefotaxime with antibiotic selection. Only those transgenic shoots arising from leaf explants infected with *Agrobacterium* should root and undergo further development. This method can also be applied to explants, such as those of seedlings, hypocotyls, and cotyledons that undergo organogenesis.

3.2.2.3. Cocultivation of Protoplast-Derived Cells with Agrobacterium. Protoplasts that have regenerated new cell walls and that are undergoing division are susceptible to *Agrobacterium*-induced transformation. This method has been used to transform mesophyll protoplasts of *Nicotiana tabacum* (*35*) and those of *Solanum nigrum* with *A. rhizogenes* (*36*), but is only suitable for very vigorous protoplast systems.

1. Isolate mesophyll protoplasts of *N. tabacum* using a standard procedure.
2. Culture the protoplasts in 3.0 mL of liquid MS-based medium containing 9% w/v mannitol, 2.0 mg/L NAA, and 0.5 mg/L BAP (MSP19M) plated over 12 mL of the same agar-solidified (0.8% w/v) medium in 9.0-cm Petri dishes. The protoplast density in the liquid layer should be 3.0×10^6. Incubate at 25°C (700 lx).

3. Harvest the protoplasts after 7 d and replate as in step 2, but in MS-based medium containing 7% w/v mannitol (MSP17M).
4. After 8–14 d from the beginning of culture, inoculate each plate with 400 µL of an overnight culture of *Agrobacterium*. The resulting mixture should contain approximately 5.0 x 10^7 bacteria/mL of liquid medium and 0.8 x 10^2 bacteria/protoplast-derived cell. Incubate for 36 h at 25°C in the dark (stationary).
5. After the cocultivation period, disrupt the plant cell-bacterial aggregates by gently pipeting with a Pasteur pipet. Harvest the plant cells by centrifugation at 100 x *g* for 5 min. Resuspend the cells in MSP17M liquid medium containing 500 µg/mL cefotaxime and recentrifuge. Repeat the washing procedure twice.
6. Replate the colonies as in step 3 in MSP17M, adding 500 µg/mL of cefotaxime to the liquid and agar layers. After 7 d, replate the cells in MSP1 with 3% w/v mannitol (MSP13M).
7. Select transformants by transferring individual cell colonies to agar selection medium (MSO medium for cells transformed by agrobacteria carrying oncogenic Ti- or Ri-plasmids; MSP1 medium with appropriate antibiotic selection for engineered Ti- or Ri-plasmids). Select at least 1000 colonies/treatment (50 colonies/9 cm dish; 15–20 mL of medium/dish).

3.3. Direct Gene Transfer to Plant Protoplasts

Direct gene transfer is useful in those systems, particularly monocotyledons, which do not respond readily to *Agrobacterium* infection.

3.3.1. Polyethylene Glycol-Induced DNA Uptake into Protoplasts

Polyethylene glycol (PEG) has been used extensively to induce DNA uptake into protoplasts. The following method has been used to transform sugarcane cell suspension protoplasts (*37*) with the vector pABD1 carrying a kanamycin resistance gene.

1. Isolate protoplasts and resuspend in their normal culture medium at 2.0 x 10^6/mL.
2. Transfer 1.0-mL aliquots of protoplast suspension to 16 x 110 mm screw-capped polycarbonate tubes.
3. Heat shock by incubating the tubes at 45°C for 5 min; cool by plunging into ice for 10 s.
4. Add 10 µg of linearized plasmid DNA (cut with a restriction enzyme at a single site) and 50 µg of sheared salmon sperm DNA in a total volume of 75 µL of sterile distilled water. The plasmid and

carrier DNA should be sterilized by ethanol precipitation before use. Set up suitable controls with carrier DNA only. Incubate for 5 min at 25°C.

5. Add 500 µL of a filter sterilized (0.2-µm pore size) solution of 40% w/v PEG 6000 (Koch Light Ltd.) in F solution.
6. Incubate for 30 min at 25°C.
7. Dilute the PEG solution stepwise by adding 2.0 mL vol of F solution (with the glucose increased to 20 g/L) to the incubation tubes every 5 min. Gently rock the tubes after each addition to mix the contents.
8. Pellet the protoplasts at 100 x *g* for 5 min, remove the supernatant, and resuspend in fresh culture medium. Count the intact protoplasts.
9. Respin the preparation as in step 8; remove the supernatant. Plate the protoplasts in agarose droplets or sectors (*see* Chapter 24, this volume) at the plating density normally used for the protoplast system.
10. After 7–14 d, select transformed colonies by replacing the liquid medium bathing the droplets or sectors, with fresh medium containing the selective agent (e.g., kanamycin) and with a reduced level of osmoticum. The selective agent should be used at a concentration already known to inhibit the growth of nontransformed protoplast-derived cells of the same age. Subsequently, remove individual putative transformants, and maintain on culture medium with selection.

Although this method employs linearized plasmid and carrier DNA, transformation of some protoplast systems can also be achieved using supercoiled plasmid in the absence of carrier DNA.

3.3.2. *Electrical Stimulation of DNA Uptake into Protoplasts*

Electroporation, involving exposure of cells to electric field impulses, increases the permeability of the plasma membrane, probably by the formation of pore-like openings. The technique is used routinely to introduce foreign genes into plant protoplasts (*3, 38–40* and *see* Chapter 30, this vol.). In addition, it also stimulates DNA synthesis in protoplasts (*41*) and promotes plant regeneration from protoplast-derived cells (*42,43*). The following method has been used to transform rice protoplasts (*13*) with a vector carrying a kanamycin resistance gene (pCaMVNEO).

1. Resuspend protoplasts at 2.0–5.0 x 10^6/mL in electroporation medium.
2. Mix 1.0 mL aliquots of protoplast suspension with 10 µg of super-

coiled plasmid and 50 µg of sheared calf thymus carrier DNA. In the electroporation chamber, give 400 µL samples 3 pulses at 10 s intervals, at voltages in the range of 55–2500 V with 20–50 nF capacitance.

3. Incubate protoplasts on ice for 10 min after electroporation. Plate protoplasts as for PEG-induced plasmid uptake, including suitable controls.

To screen the theoretical optimum conditions for plasmid uptake, protoplasts should be electroporated at a range of voltages and capacitances, utilizing a readily assayable transient gene expression system, e.g., CAT or GUS (*see* section 3.5). Although commercially available electroporators are expensive, a simple instrument can be constructed from the following circuit diagram (Fig. 1).

Electroporation chambers can be constructed by lining disposable plastic spectrophotometer cuvets (0.4- or 1.0-cm distance between the internal walls) with platinum or aluminum foil, attached to the internal walls with Vaseline™. Two important parameters that should be measured are the strength of the electrical field and the pulse duration to which the protoplasts are exposed. The electric field (EF) strength is

$$EF = V.d^{-1} \qquad (1)$$

where V is the voltage applied and d is the distance between the electrodes. The pulse duration (the exponential pulse decay constant), τ, is defined as the time taken for the pulse to decay to 63% of the initial voltage, and is expressed as

$$\tau = R.C \qquad (2)$$

where R is the resistance of the sample and C the capacitance of the discharge capacitor. R can be estimated by placing an aliquot of electroporation buffer (the same volume as used for electroporation of protoplasts) into the electroporation chamber and measuring the resistance with a suitable meter.

Electroporators are potentially lethal. The circuit should be checked by an electrician and extreme care exercised in their use.

3.4. Analysis of Transformed Plant Material

This section summarizes the methods that can be used to confirm the transformed nature of plant tissues. *Agrobacterium*-transformed cells usually synthesize novel amino acids termed opines, encoded by T-DNA genes. Ti-encoded opines include octopine and nopaline; agropine and mannopine are readily detected Ri opines. In cases where tissues are

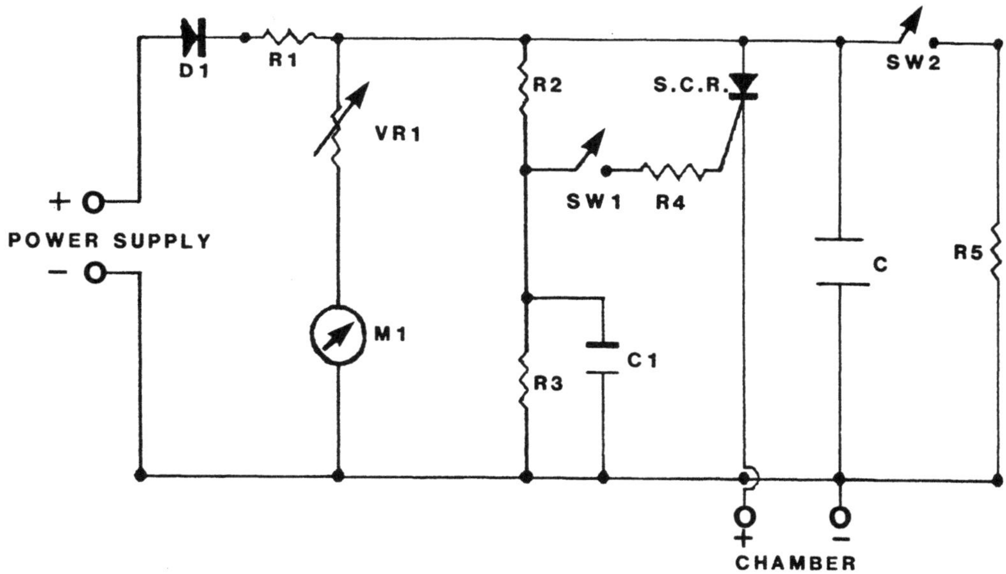

Fig. 1. Circuit diagram for the construction of an electroporator. Circuit reference: D1 = 1N4007; R1 = 22KΩ/10 W; R2,R3 = 10 MΩ/0.5 W; R4 = 0.47KΩ/5 W; R5 = 1 KΩ/10 W; VR1 = 1 MΩ/2 W; M1 = 3 mA; C1 = 0.1μF/630 V; S.C.R. = TIC116M, 600 V; SW1 = Push button switch; SW2 = Toggle switch; C = 2 x 10, 2 x 22, 2 x 47, 4 x 100, and 2 x 1000μF capacitors. An electrophoresis power supply can be used as the input. The circuit should not be charged above 450 V.

transformed by engineered Ti- or Ri-plasmids carrying a foreign gene, assay methods are used to detect expression of the inserted gene. Opine synthesis may also be used where these T-DNA genes remain in the engineered vector. Confirmation of the presence of foreign DNA in the plant genome necessitates subsequent Southern or dot blot hybridization of plant DNA to suitable probes.

3.4.1. Detection of Octopine or Nopaline in Transformed Tissues (44)

1. Incubate 200 mg of tissue in 2.0 mL of MSO liquid medium containing 100 m*M* L-arginine•HCl at 25°C for 24–48 h.
2. Rinse with sterile distilled water; blot dry with absorbent tissue. Transfer to a microcentrifuge tube and homogenize with a plastic rod.
3. Centrifuge at 12,000 rpm for 5 min.
4. Cut a 20 x 21 cm sheet of MN 214 chromatography paper. Make a pencil line 4.0 cm in from one of the longer sides. Mark this base line (anode) at 1.5-cm intervals.

5. Lightly fold the paper 3.5 cm from the longer sides.
6. Spot references (2 µL of arginine + 2 µL of octopine or nopaline) in the two outer lanes. Spot 5 µL samples of the suspected transformed tissues and nontransformed tissue extracts on the paper. Also spot 5 µL of tissue extracts mixed with 2 µL of octopine or nopaline. Dry using warm air.
7. Wet the paper uniformly with cotton wool soaked in electrophoresis buffer, taking care that the buffer wets the spot uniformly, but does not saturate the paper.
8. Fill each side of a Shandon flat bed tank with equal volumes of buffer to a depth of 3 cm. Place the electrophoresis paper in the tank, ensuring that the base line of the paper corresponds to the anode of the tank. Electrophorese for 35–45 min at 400 V and 30 mA.
9. Dry the paper with warm air for 30 min.
10. Prepare phenanthrenequinone stain.
11. Dip the paper in the stain solution, passing the anodal end through the reagent first.
12. Dry with cool air for 10 min.
13. Observe under long wave UV illumination (366 nm). Octopine and nopaline have a yellowish green fluorescence that soon fades, although electropherograms can be stored in airtight plastic bags in the dark at –20°C. Photograph electropherograms on Ilford Pan F or FP4 film using a yellow filter.

3.4.2. Detection of Agropine and Mannopine in Transformed Tissues (45)

1. Homogenize 200 mg of material with 200 µL of 0.1*N* HCl. Centrifuge (microcentrifuge; 10 min).
2. Prepare the electrophoresis paper as described for the detection of octopine and nopaline.
3. Spot 2 µL of agropine or mannopine standard stock solution in the two outer lanes.
4. Spot 10-µL samples of plant extracts. Also spot 10 µL of samples mixed with 2 µL of standards. Dry using warm air.
5. Wet the paper uniformly with buffer (formic acid:acetic acid:water [5:15:80]). Electrophorese for 1 h at 300 V.
6. Air dry.
7. Prepare stain solutions A and B as described in Section 2.9.
8. Dip the electropherogram in solution A, passing the end furthest from the baseline through the solution first. Air dry for 20 min.
9. Dip in solution B. Dry for 30 min.

10. Fix by immersion in 5% w/v sodium thiosulfate solution.
11. Wash for 1–2 h in cold running tap water.
12. Air dry. Agropine and mannopine appear as black spots.

3.4.3. Neomycin Phosphotransferase II (NPTII) Assay (46)

This method involves preparation of a crude protein extract from plant tissues transformed by a kanamycin resistance gene, followed by detection of endogenous NPTII enzyme activity. The functional enzyme is separated from contaminating ATPases by nondenaturing polyacrylamide gel electrophoresis, and enzyme activity is detected by pouring a layer of low gelling temperature agarose, containing the necessary reaction components, over the gel. Subsequently, phosphorylated kanamycin is transferred to a sheet of phosphocellulose P81 paper. The radiolabeled phosphorylated kanamycin is visualized by autoradiography.

1. In a 250-mL Buchner flask and in the following order, prepare a mixture consisting of 13.4 mL of an acrylamide/bisacrylamide stock solution, 15 mL of 1*M* Tris-HCl pH 8.8, and 7.0 mL of distilled water. On a vacuum line, degas for 1 min, followed by addition of 20 µL of N,N,N^1,N^1-tetramethylene diamine and 2.0 mL of a freshly prepared solution of 1.5% w/v ammonium persulfate.
2. Pour the acrylamide solution between clean glass plates assembled with the appropriate spacers (1.5 mm thick gel), leaving enough room for a 3.0-cm stacking gel.
3. Place 2.0 mL of a mixture of isopropanol/water (1:1, v/v) over the gel using a Pasteur pipet; rock gently to level the gel.
4. Allow to polymerize (up to 1 h), remove the isopropanol/water, and rinse the gel surface with distilled water, removing any excess liquid with filter paper.
5. In a Buchner flask, mix 2.5 mL of acrylamide/bisacrylamide stock solution, 2.5 mL of 1.0*M* Tris-HCl, pH 6.8, and 13.8 mL of distilled water. After degassing for 1 min, add 5 µL of TEMED and 1.0 mL of freshly prepared 1.5% w/v ammonium persulfate. Pour between the glass plates over the resolving gel. Insert the gel comb; allow to polymerize, and remove the comb as soon as polymerization has occurred (approximately 15 min).
6. Rinse the wells with 1 x Tris-glycine reservoir buffer prepared from the 5 x stock.
7. Fill the bottom reservoir of the gel apparatus with 1 x reservoir buffer, and remove the lower spacer from between the gel plates. Lower the gel plate assembly into the reservoir, removing air

trapped beneath the gel. Secure the glass plates to the gel kit with clips. Fill the top reservoir with 1 x reservoir buffer; transfer the whole apparatus to the cold room.

8. Prepare plant protein samples by homogenizing 200 mg of tissue on ice with 50 µL of extraction buffer in an Eppendorf tube with a disposable plastic rod. During homogenization, add 5 µL of bovine serum albumin solution and 2 µL of phenylmethylsulfonyl fluoride solution. Macerate thoroughly.
9. Remove cell debris by centrifugation in a microcentrifuge for 3 min. Transfer the supernatant to a clean Eppendorf tube. Remove a small aliquot, and determine the protein concentration spectrophotometrically, using a Bio-Rad protein assay kit/dye reagent concentrate and making an adjustment for the bovine serum albumin added. Prepare samples of approximately 20 µg of total protein, diluting with 0.1 x extraction buffer as required, and adding bromophenol blue (5 mg/mL stock) to a final concentration of 1.0%. It is also important to prepare samples from nontransformed plant tissues and a bacterial extract by sonicating (on ice) an aliquot of an overnight culture of an *E. coli* strain containing a plasmid harboring the NPTII gene, e.g., pKm2.
10. Load samples onto the polyacrylamide gel in the cold room. Electrophorese for 16 h at 60–70 V until the dye front is 2 cm from the bottom of the gel.
11. Remove the gel by separating the plates; remove the stacking gel.
12. Transfer the gel to a plastic sandwich box and rinse twice with distilled water at 4°C. Soak in 100 mL of reaction buffer.
13. Transfer the gel to a glass plate of the same size as the gel, and surround the plate and gel with electrical tape, so that the tape protrudes above the gel to a height of 5 mm.
14. Prepare a reaction gel overlay by mixing 15 mL of reaction buffer, with 15 mL of molten 2.0% w/v low gelling temperature SeaPlaque agarose (both at 37°C), 100 µCi of [γ-^{32}P]ATP, and 40 µL of kanamycin sulfate solution. Pour over the surface of the polyacrylamide gel; allow to set at room temperature. This mixture is sufficient to cover a 200-cm^2 gel to a depth of 1.5 mm, and the volume should be adjusted according to the size of the gel.
15. Place a sheet of Whatman P81 paper onto the surface of the reaction gel, followed by a stack of 12 sheets of Whatman 3MM chromatography paper. Place a 1–2 kg weight resting on a glass plate on top of the stack. Leave for 3 h at 22°C.

16. Remove the P81 paper and wash in 50 mL of a solution of 1.0% w/v SDS containing 1.0 mg/mL of trypsin for 30 min at 60°C.
17. Wash 5x (5 min each wash) in 50 mL aliquots of 0.2*M* phosphate buffer, pH 7.7 at 75°C. Prepare the phosphate buffer by mixing 13 mL of 0.2*M* $NaH_2PO_4 \cdot 2H_2O$ (31.2 g/L) with 87 mL of 0.2*M* $Na_2HPO_4 \cdot 12H_2O$ (71.7 g/L) and making up to 1 L with distilled water.
18. Air dry; cover with Saran Wrap™ (Dow Chemical Co) and expose to X-ray film (Amersham Hyperfilm-MP) in a cassette with an intensifying screen, at –70°C. A strong overnight signal is usually obtained when 20 µg of total protein is loaded into each well of the gel.

3.4.4. Plant DNA Isolation (47)

There are numerous methods for the small-scale isolation of plant DNA. The one described here has been used to prepare DNA from several plant species pure enough to be cut by a range of restriction endonucleases.

1. Grind 200 mg of fresh tissue with liquid nitrogen in a mortar and pestle. If required, plant material can be harvested and stored at –70°C prior to DNA extraction. Thorough grinding is essential to maximize DNA recovery.
2. Transfer the ground tissue to a sterile Eppendorf tube and add 500 µL of extraction buffer. Mix thoroughly with a plastic rod.
3. Add 0.7 vol of equilibrated phenol and 0.3 vol of a 24:1 chloroform/isoamyl alcohol mixture. Emulsify; centrifuge (minicentrifuge, 10 min, 22°C).
4. Transfer the supernatant to a clean Eppendorf tube and repeat the phenol:chloroform extraction.
5. Transfer the supernatant to a fresh tube. Add 1 vol of chloroform/isoamyl alcohol (24:1). Emulsify and respin for 10 min at 22°C.
6. Remove contaminating RNA by addition of RNase (DNase-free) to a final concentration of 50 µg/mL from the stock solution. Incubate at 37°C for 30 min.
7. Add proteinase K (Boehringer) to a final concentration of 50 µg/mL. Incubate at 37°C for 30 min.
8. Repeat the phenol/chloroform extraction.
9. Extract with chloroform.
10. Add 0.6 vol of cold (–20°C) isopropanol to the supernatant to precipitate the DNA.

11. Pellet the DNA by centrifugation (1 min). Wash the pellet twice with 70% v/v ethanol. Centrifuge briefly between washes.
12. Vacuum desiccate (5–10 min) and resuspend the DNA in 50–100 µL of TE buffer. Store the DNA at 4°C until required.

3.4.5. Detection of Foreign DNA in Transformed Plant Tissues

Foreign genes can be detected in plant DNA by Southern (*48*) and dot (*49*) hybridization. The hybridization protocol is based on a procedure reported for Ri T-DNA (*50*). The conditions work well for the detection of large Ti and Ri T-DNA sequences, and also detection of individual genes introduced directly into protoplasts by chemical treatment or electroporation, e.g., the NPTII gene.

3.4.5.1. Southern Blotting and Hybridization

1. Restrict 10 µg of each plant DNA sample with an appropriate restriction endonuclease. This is best performed in 100 µL reaction volumes containing 10 µL of suitable restriction buffer (10 x concentration), 10 µL of spermidine (0.1*M* stock), 1 µL of restriction enzyme (> 40 U/µL), and the DNA in sterile water. Digest overnight (usually at 37°C for most of the common restriction enzymes).
2. Precipitate DNA, after digestion, by the addition of 1/10 vol of 2.5*M* sodium acetate and 2.5 vol of absolute ethanol. Place at –20°C (30 min). Centrifuge (10 min). Wash the pellet twice in 70% v/v ethanol, vacuum desiccate, and resuspend in 40 µL of sterile distilled water.
3. Add 8 µL of gel loading buffer.
4. Prepare 100 mL of agarose (Sigma Type I) in 1 x TBE buffer by diluting the 5x buffer stock. Cast a 11 x 14 cm gel.
5. Immerse the gel in 1 x TBE buffer; load the plant DNA samples. In addition, DNA samples that are homologous to the probe should be loaded at a concentration suitable for balancing the hybridization signals that occur in the plant DNA (i.e., of comparable copy number). This provides a reference for foreign gene signals seen in the plant DNA. In practice, 100 ng of total DNA from *Agrobacterium tumefaciens* or *A. rhizogenes* or 10 pg of purified plasmid (both cut with the same restriction enzymes as used for the plant DNA) are adequate. Add one fifth volume of gel loading buffer. Usually, it is also beneficial to include molecular weight markers on the gel, e.g.,

λ DNA cut with EcoR1 or HindIII. The relative positions of these fragments can be observed during UV photography of the gel (*see* step 7 *below*).

6. Electrophorese for 14–18 h at 25 V until the dye front is 1 cm from the end of the gel.
7. Stain the gel in ethidium bromide solution (0.5 μg/mL from a 10 mg/mL stock) for 30 min. Photograph under UV illumination on a transilluminator.
8. Depurinate the DNA by soaking the gel for 30 min in 200 mL of 0.25*M* HCl.
9. Rinse the gel 3x in water. Denature the DNA by gently rocking the gel in denaturing solution (2 x 15 min).
10. Rinse the gel in water and neutralize by immersing in a solution of 0.5*M* Tris HCl pH 7.5 and 1.5*M* NaCl with gentle rocking (2 x 15 min).
11. Soak the gel for 30 min in 20 x SSPE buffer.
12. Transfer the DNA from the gel to nitrocellulose or nylon membrane (Amersham Hybond) by overnight blotting, using the arrangement shown in Fig. 2. Nitrocellulose must be soaked thoroughly in distilled water before use; Hybond membrane is used dry. Care must be taken to avoid air bubbles being trapped between the membrane and the gel surface.
13. Mark the orientation of the filter (remove one corner) and wash briefly in 2 x SSPE solution. Bind the DNA to the membrane by baking (80°C, 2 h) under vacuum, or by UV irradiation (Hybond) on a transilluminator according to the manufacturer's instructions.
14. Incubate the filter at 37°C (16 h) in 20 mL (11 x 14 cm filter size) of prehybridization solution contained in a heat-sealed plastic bag.
15. Remove the prehybridization buffer and replace with 8.0 mL of fresh buffer. Add the denatured (boil for 2 min and chill on ice) radiolabeled probe. Remove as much air as possible from the bag. Heat seal and incubate at 37°C in a water bath (24 h, stationary). Circular plasmid, ^{32}P-labeled by nick translation (BRL nick translation kit), or DNA fragments isolated from low gelling temperature agarose gels and ^{32}P-labeled using an oligolabeling kit (Amersham), can be used as homologous probes. Labeling reactions are performed according to the manufacturer's instructions.
16. Remove the filter and wash, using a shaking water bath, in 2 x SSPE with 0.1% SDS for 10 min at 22°C, and then in 2 x SSPE with 0.1% SDS for 30 min at 65°C twice, followed by two washes in 0.3x SSPE with

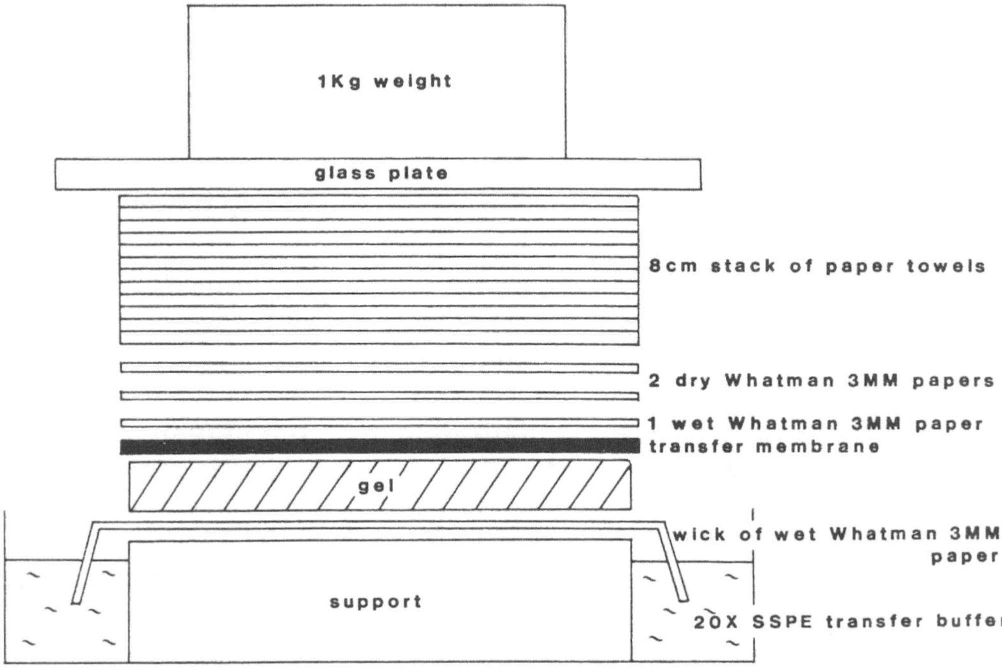

Fig. 2. Arrangement for the transfer of DNA restriction fragments from an agarose gel to a nitrocellulose or Hybond nylon membrane.

0.1% SDS for 30 min at 65°C. Air dry the filter, enclose in Saran Wrap™ and expose to X-ray film in a cassette with an intensifying screen, at –70°C for 2–14 d.

3.4.5.2. *Dot Blotting*

1. Denature 5–10-µg DNA samples in approximately 20 µL vol by heating to 95°C for 5 min. Chill on ice for 2 min.
2. Spot the samples onto nylon Hybond membrane in 2 µL aliquots, using an air displacement pipet. Dry the aliquots using cool air.
3. Wet the membrane in a denaturing solution for 1 min.
4. Transfer the membrane to neutralizing solution for 1 min.
5. Place the membrane on a piece of Whatman paper; allow to air dry.
6. Fix the DNA by wrapping the Hybond membrane in Saran Wrap™ and irradiating on a UV transilluminator for the time recommended by the manufacturer. The side of the membrane with the DNA should be in contact with the platen of the transilluminator.
7. Hybridize as in the Southern procedure (steps 14–16).

3.5. Transient Gene Expression Systems

Studies of gene uptake and promoter activity are facilitated by linking the promoter sequence to a gene that encodes an enzyme whose activity is readily detectable. The chloramphenicol acetyl transferase (CAT) and β-glucuronidase (GUS) genes are used extensively in transient gene expression studies.

3.5.1. Chloramphenicol Acetyltransferase (CAT) Assay

1. Electroporate protoplasts mixed with a plasmid carrying the CAT gene (e.g., pNOSCAT; pCaMVCAT [39]) at a range of voltages and capacitances, as described in section 3.3. Also electroporate protoplasts in the absence of plasmid.
2. Culture the protoplasts for 2–48 h in the liquid medium normally employed for the protoplast system under test.
3. Harvest the protoplasts (microcentrifuge, slow speed setting for 2 min). Discard the supernatant and add 200 µL of ice-cold CAT extraction buffer.
4. Lyse the protoplasts by passing 5x through a 25-gage needle, or by sonication on ice (3 x 20 s bursts). Centrifuge in a microcentrifuge for 8 min at 4°C.
5. Remove the supernatant and transfer to a new Eppendorf tube. Add 10 µL of a freshly prepared solution of 10 m*M* acetyl coenzyme A and 1 µL of (^{14}C)chloramphenicol. Prepare suitable positive control samples with known amounts of purified CAT enzyme, e.g., 1, 2, 5, and 10 U; (Sigma).
6. Mix by vortexing; incubate at 37°C for 2 h.
7. Add 66 µL of ethyl acetate and mix by vortexing. Centrifuge for 2 min, and transfer the upper phase to a clean Eppendorf tube.
8. Remove the ethyl acetate slowly under vacuum in a glass desiccator (this step may take 2–3 h).
9. Redissolve the radiolabeled and acetylated chloramphenicol by adding 20 µL of ethyl acetate; vortex thoroughly for 30 s.
10. Spot 10 µL of the samples onto a 20 x 20 cm silica thin layer chromatography plate (Sigma), 1 cm from the edge of the plate and 1.5 cm apart. Air dry using warm air.
11. Carry out ascending chromatography in a TLC tank containing a 5-mm depth of solvent (chloroform/methanol; 95:5, v/v). Allow the solvent front to ascend for 5–15 cm. The TLC tank should be lined with filter paper and equilibrated with the solvent prior to use.
12. Remove the plate from the tank and air dry. Wrap in Saran Wrap™ and expose to X-ray film for 12–36 h to quantitate and localize the

products. Reaction products may then be scraped off the TLC and quantitated directly in a liquid scintillation counter.

CAT inactivates chloramphenicol by acetylation. Three forms of radiolabeled acetylated chloramphenicol can be observed on the autoradiograph, these being (in ascending order) 1-acetylchloramphenicol, 3-acetylchloramphenicol and, under optimum conditions, 1,3-diacetylchloramphenicol.

3.5.2. b-Glucuronidase (GUS) Assay

1. Electroporate protoplasts mixed with a plasmid carrying the GUS gene (e.g., pBI221 [*19*]), as in section 3.3. Also electroporate protoplasts in the absence of plasmid. This control is important, since many protoplast systems have high endogenous levels of GUS activity.
2. Culture the protoplasts for 12–36 h in the liquid medium normally used for the protoplast system under test.
3. Pellet the protoplasts (microcentrifuge, slow speed setting; 2 min). Discard the supernatant and add 300 µL of ice-cold extraction buffer. Disrupt the cells by passage 5x through a 25-gage needle. Maintain on ice.
4. Transfer 190 µL aliquots of the samples to the wells of a microtiter plate standing on ice.
5. Add 10 µL of 4-methyl umbelliferyl glucuronide (4-MUG) stock solution to each well. Incubate at 37°C.
6. Remove 50 µL aliquots, at 0, 1, and 2 h; transfer to a new microtiter plate. At these times, add 50 µL of 0.2*M* Na_2CO_3 to terminate the reaction.
7. Carry out a qualitative assay by observing the microtiter plate under UV illumination on a transilluminator. In the presence of functional GUS enzyme, MUG is converted to 4-methyl umbelliferone (4-MU), which fluoresces blue under UV illumination when ionized. Na_2CO_3 has the dual function of terminating the reaction and maximizing the fluorescence of 4-MU. Photograph on Ilford FP4 film using a Kodak™ Wratten gelatin filter (No. 2E; Cat. No. 1760875).
8. A quantitative assay can also be carried out using a fluorimeter. Calibrate the machine by preparing 10 n*M*, 100 n*M*, and 1 µ*M* solutions of 4-MU sodium salt by diluting a 1 m*M* aqueous stock solution with extraction buffer.
9. Zero the machine using 0.2*M* Na_2CO_3. Take 100 µl aliquots of the standards, add 2.9 mL of 0.2*M* Na_2CO_3, and measure the fluores-

cence (excitation at 365 nm; emission at 455 nm). Prepare a calibration curve.

10. Mix 100 μL of the reaction mixtures from step 6 above with 2.9 mL of 0.2*M* Na_2CO_3. Measure the fluorescence of the sample and compare to the calibration curve.

References

1. de la Pena, A., Lörz, H., and Schell, J. (1987) Transgenic rye plants obtained by injecting DNA into young floral tillers. *Nature* **325,** 274–276.
2. Rhodes, C. A., Pierce, D. A., Mettler, I. J., Mascarenhas, D., and Detmer, J. J. (1988) Genetically transformed maize plants from protoplasts. *Science* **240,** 204–207.
3. Zhang, H. M., Yang, H., Rech, E. L., Golds, T. J., Davis, A. S., Mulligan, B. J., Cocking, E. C., and Davey, M. R. (1988) Transgenic rice plants produced by electroporation-mediated plasmid uptake into protoplasts. *Plant Cell Rep.* **7,** 379–384.
4. Davey, M. R., Gartland, K. M. A., and Mulligan, B. J. (1986) Transformation of the genomic expression of plant cells, in *Plasticity in Plants. Symposium XXXX of the Society for Experimental Biology* (Jennings, D. H. and Trewavas, A. J., eds.), The Company of Biologists, Department of Zoology, University of Cambridge, pp. 85–120.
5. Horsch, R. B., Fry, J. E., Hoffmann, N. L., Eichholtz, D., Rogers, S. G., and Fraley, R. T. A simple and general method for transferring genes into plants. *Science* **227,** 1229–1231.
6. Márton, L., Wullems, G. J., Molendijk, K. L., and Schilperoort, R. A. (1979) *In vitro* transformation of cultured cells from *Nicotiana tabacum* by *Agrobacterium tumefaciens*. *Nature* **227,** 129–131.
7. Feldmann, K. A. and Marks, M. D. (1987) *Agrobacterium*-mediated transformation of germinating seeds of *Arabidopsis thaliana*: A non-tissue culture approach. *Mol. Gen. Genet.* **208,** 1–9.
8. Morgan, A. J., Cox, P. N., Turner, D. A., Peel, E., Davey, M. R., Gartland, K. M. A., and Mulligan, B. J. (1987) Transformation of tomato using an Ri plasmid vector. *Plant Sci.* **49,** 37–49.
9. Zambryski, P., Joos, H., Genetello, C., Leemans, J., Van Montagu, M., and Schell, J. (1983) Ti plasmid vector for the introduction of DNA into plant cells without alteration of their normal regeneration capacity. *EMBO J.* **2,** 2143–2150.
10. An, G., Watson, B. D., Stachel, S., Gordon, M. P., and Nester, E. W. (1985) New cloning vehicles for transformation of higher plants. *EMBO J.* **4,** 277–284.
11. Simpson, R. B., Spielmann, A., Margossian, L., and McKnight, T. D. (1986) A disarmed binary vector from *Agrobacterium tumefaciens* functions in *Agrobacterium rhizogenes*: Frequent cotransformation of two distinct T-DNAs. *Plant Mol. Biol.* **6,** 403–416.
12. Bevan, M. (1984) Binary *Agrobacterium* vectors for plant transformation. *Nucleic Acids Res.* **12,** 8711–8721.
13. Yang, H., Zhang, H. M., Davey, M. R., Mulligan, B. J., and Cocking, E. C. (1988) Production of kanamycin resistant rice tissues following DNA uptake into protoplasts. *Plant Cell Rep.* **7,** 421–425.
14. Ou-Lee, T. M., Turgeon, R., and Wu, R. (1986) Expression of a foreign gene linked

to either a plant-virus or a *Drosophila* promoter, after electroporation of protoplasts of rice, wheat, and sorghum. *Proc. Natl. Acad. Sci.* **83,** 6815–6819.

15. Crossway, A., Hauptli, H., Houck, C. M., Irvine, J. M., Oakes, J. V., and Perani, L. A. (1986) Micromanipulation techniques in plant biotechnology. *Biotechniques* **4,** 320–334.
16. Klein, T. M., Wolf, E. D., Wu, R., and Sanford, J. C. (1987) High-velocity microprojectiles for delivering nucleic acids into living cells. *Nature* **327,** 70–73.
17. McCabe, D. E., Swain, W. F., Martinell, B. J., and Christou, P. (1988) Stable transformation of soybean (*Glycine max*) by particle acceleration. *Biotechnology* **6,** 923–926.
18. Herrera-Estrella, L., Depicker, A., Van Montagu, M., and Schell, J. (1983) Expression of chimeric genes transferred into plant cells using a Ti plasmid-derived vector. *Nature* **303,** 209–213.
19. Jefferson, R. A. (1987) Assaying chimeric genes in plants: the GUS gene fusion system. *Plant Mol. Biol. Rep.* **5,** 387–405.
20. Ow, D. W., Deluca, M., De Wet, J. R., Helsinki, D. R., and Howell, S. H. (1986) Transient and stable expression of the firefly luciferase gene in plant cells and transgenic plants. *Science* **234,** 856–859.
21. Bevan, M., Flavell, R., and Chilton, M.-D. (1983) A chimeric antibiotic resistance gene as a selectable marker for plant cell transformation. *Nature* **304,** 184–187.
22. Hille, J., Vetheggen, F., Roelvink, P., Franssen, H., van Kammen, A., and Zabel, P. (1986) Bleomycin resistance: a new dominant selectable marker for plant cell transformation. *Plant Molec. Biol.* **7,** 171–176.
23. Van Den Elzen, P. J. M., Townsend, J., Lee, K. L., and Bedbrook, J. R. (1985) A chimeric hygromycin resistance gene as a selectable marker in plant cells. *Plant Mol. Biol.* **5,** 299–302.
24. Comai, L., Facciotti, D., Hiatt, W. R., Thompson, G., Rose, R. E., and Stalker, D. M. (1985) Expression in plants of a mutant *aroA* gene from *Salmonella typhimurium* confers tolerance to glyphosate. *Nature* **317,** 741–744.
25. Nelson, R. S., McCormick, S. M., Delannay, X., Dube, P., Layton, J., Anderson, E. J., Kaniewska, M., Proksch, R. K., Horsch, R. B., Rogers, S. G., Fraley, R. T., and Beachy, R. N. (1988) Virus tolerance, plant growth, and field performance of transgenic tomato plants expressing coat protein from tobacco mosaic virus. *Biotechnology* **6,** 403–409.
26. Krens, F. A., Molendijk, L., Wullems, G. J., and Schilperoort, R. A. (1982) *In vitro* transformation of plant protoplasts with Ti-plasmid DNA. *Nature* **296,** 72–74.
27. Fromm, M. E., Taylor, L. P., and Walbot, V. (1986) Stable transformation of maize after gene transfer by electroporation. *Nature,* **319,** 791–793.
28. Paszkowski, J., Shillito, R. D., Saul, M., Mandak, V., Hohn, T., Hohn, B., and Potrykus, I. (1984) Direct gene transfer to plants. *EMBO J.* **3,** 2717–2722.
29. Mandel, M. and Higa, A. (1970) Calcium dependent bacteriophage DNA infection. *J. Mol. Biol.* **53,** 154–162.
30. Holmes, D. S. and Quigley, M. (1981) A rapid boiling method for the preparation of bacterial plasmids. *Anal. Biochem.* **114,** 193–197.
31. Birnboim, H. C. and Doly, J. (1979) A rapid alkaline extraction procedure for screening recombinant plasmid DNA. *Nucleic Acids Res.* **7,** 1513–1523.
32. Garfinkel, D. J., Simpson, R. B., Ream, L. W., White, F. F., Gordon, M. P., and Nester, E. W. (1981) Genetic analysis of crown gall: fine structure map of the T-DNA by site directed mutagenesis. *Cell* **27,** 143–153.
33. Murashige, T. and Skoog, F. (1962) A revised medium for rapid growth and bio-

assays with tobacco tissue cultures. *Physiol. Plant.* **15,** 473–497.
34. McCormick, S., Niedermeyer, J., Fry, J., Barnason, A., Horsch, R., and Fraley, R. (1986) Leaf disc transformation of cultivated tomato (*L. esculentum*) using *Agrobacterium tumefaciens. Plant Cell Rep.* **5,** 81–84.
35. Davey, M. R., Cocking, E. C., Freeman, J., Draper, J., Pearce, N., Tudor, T., Hernalsteens, J.-P., De Beuchleer, M., Van Montagu, M., and Schell, J. (1979) The use of plant protoplasts for transformation by *Agrobacterium* and isolated plasmids, in *Advances in Protoplast Research* (Ferenczy, L. and Farkas, G. L., eds.), Akademiai Kiado, Budapest, pp. 425–430.
36. Wei, Z.-M., Kamada, H., and Harada, H. (1986) Transformation of *Solanum nigrum* L. protoplasts by *Agrobacterium rhizogenes. Plant Cell Rep.* **5,** 93–96.
37. Chen, W. H., Gartland, K. M. A., Davey, M. R., Sotak, R., Gartland, J. S., Mulligan, B. J., and Power, J. B. (1987) Transformation of sugarcane protoplasts by direct uptake of a selectable chimaeric gene. *Plant Cell Rep.* **6,** 297–301.
38. Shillito, R. D., Saul, M. W., Paszkowski, J., Muller, M., and Potrykus, I. (1985) High efficiency direct gene transfer to plants. *Biotechnology* **3,** 1099–1103.
39. Fromm, M., Taylor, L. P., and Walbot, V. (1985) Expression of genes transferred into monocot and dicot plant cells by electroporation. *Proc. Natl. Acad. Sci. USA* **82,** 5824–5828.
40. Langridge, W. H. R., Li, B. J., and Szalay, A. A. (1985) Electric field mediated stable transformation of carrot protoplasts with naked DNA. *Plant Cell Rep.* **4,** 355–359.
41. Rech, E. L., Ochatt, S. J., Chand, P. K., Davey, M. R., Mulligan, B. J., and Power, J. B. Electroporation increases DNA synthesis in cultured plant protoplasts. *Bio/Technology* **6,** 1091–1093.
42. Ochatt, S. J., Chand, P. K., Rech, E. L., Davey, M. R., and Power, J. B. (1988) Electroporation-mediated improvement of plant regeneration from colt cherry (*Prunus avium X pseudocerasus*) protoplasts. *Plant Sci.* **54,** 165–169.
43. Chand, P. K., Ochatt, S. J., Rech, E. L., Power, J. B., and Davey, M. R. Electroporation stimulates plant regeneration from protoplasts of the woody medicinal species *Solanum dulcamara* L. *J. Exper. Bot.* **39,** 1267–1274.
44. Aerts, M., Jacobs, M., Hernalsateens, J.-P., Van Montagu, M., and Schell, J. (1979) Induction and *in vitro* culture of *Arabidopsis thaliana* crown gall tumors. *Plant Sci. Lett.* **17,** 43–50.
45. Petit, A., David, C. C., Dahl, G. A., Ellis, J. G., Guyon, P., Casse-Delbart, F., and Tempé, J. (1983) Further extension of the opine concept: plasmids in *Agrobacterium rhizogenes* cooperate for opine degradation. *Mol. Gen. Genet.* **190,** 204–214.
46. Reiss, B., Sprengel, R., Will, H., and Schaller, H. (1984) A new sensitive method for qualitative and quantitative assay of neomycin phosphotransferase in crude cell extracts. *Gene* **30,** 211–218.
47. Rech, E. L., Golds, T. J., Hammatt, N., Mulligan, B. J., and Davey, M. R. *Agrobacterium rhizogenes* mediated transformation of the wild soybeans *Glycine canescens* and *G. clandestina*: production of transgenic plants of *G. canescens. J. Exp. Bot.* **39,** 1275–1285.
48. Southern, E. M. (1975) Detection of specific sequences among DNA fragments separated by gel electrophoresis. *J. Mol. Biol.* **98,** 503–517.
49. Khandjian, E. W. (1987) Optimized hybridization of DNA blotted and fixed to nitrocellulose and nylon membranes. *Bio/Technology* **5,** 165–167.
50. White, F. F., Taylor, B. H., Huffman, G. A., Gordon, M. P., and Nester, E. W. (1985) Molecular and genetic analysis of the transferred DNA regions of the root-inducing plasmid of *Agrobacterium rhizogenes. J. Bacteriol.* **164,** 33–44.

Chapter 34

Electrofusion of Plant Cells

Anne Donovan, Susan Isaac, and Hamish A. Collin

1. Introduction

Isolated plant protoplasts can be induced to fuse with protoplasts from different species and, therefore, provide an ideal system for genetic modification and for use in plant breeding. Techniques for electrofusion of plant protoplasts have been developed relatively recently, and specialized apparatus is required, although this is now becoming more widely available. Electrofusion offers definite advantages over more commonly used protocols that make use of chemical stimulation to induce fusion. The most valuable aspects of electrofusion techniques are the high fusion frequencies attained, often tenfold higher than analogous chemical systems (*1*). The use of potentially toxic chemical stimulants is avoided, and the zones of membrane disturbance are limited to regions of membrane contact alone, all of which tends to preserve protoplast viability. Additionally, fusion events can be monitored microscopically, allowing the precise determination of the effects of various electrical parameters and, therefore, the use of optimum electrical values. Such control, together with manipulation of electrodes, allows more precise definition of parentage (2) than previously was possible with other methods.

From: *Methods in Molecular Biology*, vol. 6, *Plant Cell and Tissue Culture*
Edited by Jeffrey W. Pollard and John M. Walker, ©1990 by The Humana Press

The protocol described here was developed in connection with a program designed to introduce higher levels of disease resistance into a crop species by various means. The technique could be easily manipulated to suit other source material, using the same basic approach.

All varieties of celery, and particularly the self-blanching varieties now popular for cultivation, are susceptible to the disease late blight, caused by the fungus *Septoria apiicola*. The fungus infects by direct penetration of leaf surfaces, proliferates between cells, and eventually forms pycnidia, which break through the leaf surface, producing the characteristic brown/black spots of late blight disease. There is evidence that resistance is correlated with flavor compound levels in host tissues. Traditional plant breeding methods have failed to increase resistance levels in celery varieties. This electrofusion protocol (Fig. 1) is now being used to introduce disease resistance from a related and morphologically similar, but late blight resistant, species *Levisticum officinale* (lovage).

2. Materials

1. Solution for seed surface sterilization: 20% Domestos (or any commercial bleach).
2. Germination agar (GA): MS medium (*3*, and *see* Appendix) supplemented with 3% sucrose and 1% agar.
3. Callus induction medium (CIM): MS salts supplemented with 3% sucrose; 0.5 mg/L 2,4-dichlorophenoxyacetic acid (2,4-D); 0.6 mg/L kinetin and 1% agar.
4. Cell suspension medium (CSM): MS salts supplemented with 3% sucrose; 0.5 mg/L 2,4-D and 0.6 mg/L kinetin.
5. Solution for plasmolysis treatment: 0.6*M* mannitol.
6. Lytic mixture: Enzymes as appropriate, in 0.6*M* mannitol. Make up fresh mixture each time, filter sterilize and adjust pH to 5.6 if necessary.
   a. Celery: Pectolyase Y23 (R. W. Unwin & Co. Ltd.) 0.1% and Cellulysin (Cambridge Bioscience) 2%.
   b. Lovage: Pectolyase Y23 0.1% and Meicellase (R. W. Unwin & Co. Ltd.) 3%.
7. Protoplast washing solution: 0.6*M* mannitol containing 2 m*M* calcium chloride.
8. Medium for culture of electrofused protoplasts (EPM): NTK medium (*4*) lacking ammonium salts, but supplemented with 100 m*M* $CaCl_2$, 0.5 mg/L 2,4-D and 0.6 mg/L kinetin.

ELECTROFUSION OF PLANT PROTOPLASTS

ISOLATED PROTOPLASTS

↓

Celery / Lovage protoplasts mixed 1 : 1
in 0.6M mannitol plus 2 mM $CaCl_2$
at final density of 10^5 protoplasts/mL

ZIMMERMANN CELL FUSION ™ SYSTEM

↓

Helical Fusion Chamber
(200µl capacity)

ALIGNMENT

↓

alignment frequency	:	2.0 MHz
alignment voltage	:	5 $V.cm^{-1}$
alignment field	:	45 secs duration

(assess protoplast membrane associations microscopically)

FUSION

↓

fusion voltage	:	30V
number of pulses	:	2 at 1 sec intervals
pulse duration	:	90µsecs

(determine fusion events microscopically)

INCUBATION

Fig. 1. Summary protocol for plant protoplast electrofusion.

3. Method

3.1. Growth of Sterile Seedlings

1. Surface sterilize seeds in 20% Domestos for 12 min.
2. Wash 3x in sterile distilled water.
3. Place seeds, aseptically, onto germination agar (GA) in Universal vials.
4. Incubate at 4°C in dark for 5 d followed by 22°C in light for 8 wk (*Apium graveolens*—celery var. Celebrity) or 12–16 wk (*Levisticum officinale*—lovage), as appropriate. Conditions of high light and high humidity are most favorable.
5. Use young leaf material for protoplast isolation.

3.2. Growth of Cell Suspensions

1. Place petiole explants (1 cm) from laboratory-grown seedlings onto callus induction medium (CIM).
2. Incubate at 22°C under 12 h light/dark regime for 8 wk.
3. Subculture and incubate for 4 wk.
4. Transfer callus tissue to 50 mL cell suspension medium (CSM) in 250- mL flasks. Incubate at 22°C with shaking (100 rpm) under 12 h light/dark regime.
5. Subculture at intervals of 4 wk.
6. Use 7-day-old suspension culture material for protoplast isolation.

3.3. Isolation of Protoplasts

1. Finely chop young leaf material (0.5 g fresh weight) and plasmolyze in 0.6*M* mannitol for 10 min.
2. Alternatively, use cell suspension cultures (7 d) as the source material. Allow suspension cultures to settle for 10 min, and use cells from 10 mL of suspension.
3. Transfer to lytic mixture (10 mL) as appropriate (*see* Note 1). Incubate at 22°C, with gentle shaking, for 3 h (celery mesophyll tissue); 5 h (lovage mesophyll tissue), or overnight (celery cell suspension material).
4. Separate protoplasts from lytic mixture and remove cell debris by filtration through 70 µm nylon net, followed by centrifugation at 100*g* for 2 min.
5. Resuspend the resulting pellet in protoplast washing solution (PWS). Centrifuge and wash twice more in PWS.

6. Resuspend in PWS and determine protoplast number. A yield of about 5 x 10^6 protoplasts/g mesophyll tissue will result.

3.4. Protoplast Fusion

A Zimmermann Cell Fusion™ System (GCA Corporation, Chicago, Illinois, USA) as described by Zimmermann and Scheurich (*5*) was used to generate electric fields. The protocol given here was developed for a flat fusion chamber similar to that described by Pilwat et al. (*6*), consisting of two fine, wired electrodes arranged in parallel on a polyethylene slide, and having a 17 µm capacity. The protocol is equally useful for larger capacity 200 µL helical chambers (GCA Corporation).

1. Isolate and resuspend protoplasts in washing solution (PWS). Mix equal quantities of protoplasts from those species to be fused, to give a final concentration of 10^5 cells/mL.
2. Fill the fusion chamber with the protoplast mixture.
3. Apply alignment frequency (2 MHz) and alignment voltage (5 V). Protoplasts align at electrodes on exposure to a heterogeneous (AC) electrical field, and form "pearl chains" of variable length.
4. Determine the percentage of protoplasts aligning and forming pearl chains. Observe the degree of membrane/membrane contact.
5. Apply fusion voltage (30 V) in two pulses of 90 µs duration at 1-s intervals. Protoplasts between the electrodes align and fuse when exposed to short DC pulses. After fusion, protoplasts round up over a period of several min (Fig. 2 and Notes 2–5).
6. Transfer to culture medium.

3.5. Culture of Electrofused Protoplasts

1. Wash the protoplasts into Universal vials using culture medium (EPM) in a ratio of 1:5.
2. Culture the protoplasts as sitting drops (0.1–0.2 mL) in 6-cm diameter Petri dishes (7 drops/dish) sealed with Labfilm at 22°C.
3. Culture protoplast samples not treated in electrofusion apparatus as controls.
4. Examine drops microscopically after 3 d to detect any cell division. Evans' Blue vital dye may be used to assess protoplast viability.
5. Transfer dividing cells to callus induction medium (CIM) for further cultivation.

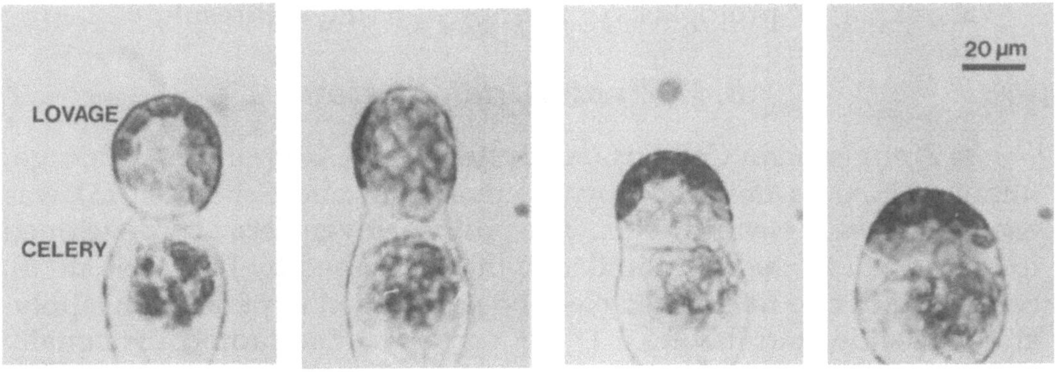

Fig. 2. Electrofusion between a celery cell suspension protoplast and a lovage mesophyll protoplast.

4. Notes

1. In order to preserve maximum protoplast viability, it is important to minimize lytic incubation times and to use low enzyme concentrations, as far as is practical. Excessive or prolonged incubation in lytic mixtures may result in some protoplast loss from lysis. Protoplasts of good quality are most likely to regenerate well. This method reliably provides large, spherical protoplasts and preparations generally lacking cellular debris.
2. Identification of fused (celery/lovage) products can be aided by the use of various source materials. Lovage mesophyll protoplasts contain more chloroplasts and, therefore, appear "greener" than celery mesophyll protoplasts. Alternatively, celery cell suspension protoplasts do not contain any chloroplasts and are readily identifiable.
3. On electrofusion of celery/lovage protoplast mixtures, a fusion frequency of 40% results, and 20% of fused products are heterokaryons.
4. Small protoplasts require a higher field strength and frequency to form pearl chains. Larger protoplasts tend to align at the electrodes first. Additionally, the density of protoplasts in the fusion chamber influences the number of protoplasts in each pearl chain. At a final density of over 10^5 protoplasts/mL, there is an increase in pearl chain length. Ideally, protoplast pairing at electrodes is most useful for one-to-one fusion. In lovage/celery protoplast mixtures where pairing occurred, 50% of pearl chains contained both protoplast types.

5. Small protoplasts require a higher DC voltage to fuse. If there is a large size discrepancy between protoplast types to be fused, then the larger protoplasts may be stabilized and encouraged to fuse by treatment with Pronase E (Pronase, Type XXV, Sigma Co. Ltd.). Incubation at 30°C for 30 min–1 h with 1 mg/mL Pronase E increased fusion frequency with recalcitrant material.

References

1. Zachrisson, A. and Bornman, C. H. (1986) Application of electric field fusion in plant tissue culture. *Physiologia Plantarum* **61,** 314–320.
2. Koop, H. U. and Schweiger, H. G. (1985) Regeneration of plants after electrofusion of selected pairs of protoplasts. *European Journal of Cell Biology* **39,** 46–49.
3. Murashige, T. and Skoog, F. (1962) A revised medium for rapid growth and bioassays with tobacco tissue culture. *Physiologia Plantarum* **15,** 473–487.
4. Scowcroft, W. R. and Larkin, P. J. (1980) Isolation, culture and plant regeneration from protoplasts from *Nicotiana debneyi. Australian Journal of Plant Physiology* **7,** 635– 664.
5. Zimmermann, U. and Scheurich, P. (1981) High frequency fusion of plant protoplasts by electrical fields. *Planta* **151,** 26–32.
6. Pilwat, G., Zimmermann, U., and Richter, H. P. (1981) Giant culture cells by electric field induced fusion. *FEBS Letters* **133,** 169–174.

Chapter 35

Production of Cybrids in Rapeseed (*Brassica napus*)

Stephen Yarrow

1. Introduction

Protoplast fusion produces cells that contain a mixture of the DNA-containing organelles (nuclei, chloroplasts, and mitochondria) from both fusion parents. This chapter describes the production of rapeseed (*B. napus*) fusion-product cells where one of the nuclei is eliminated. In the absence of selection pressure, the remaining chloroplast/mitochondrial mixture segregates randomly as the cell divides, eventually creating different nuclear/chloroplast/mitochondrial combinations. These novel nucleocytoplasmic combinations are called cytoplasmic hybrids, or "cybrids."

As defined here, cybrids can include cases in which the chloroplast/mitochondrial combination is novel, as well as alloplasmic substitutions, in which the cytoplasm of one parent is substituted for that of the other. The existence of new types of mitochondria or chloroplasts derived from the recombination of organelle DNA also constitutes a cybrid.

With few exceptions, novel chloroplast/mitochondrial combinations are unobtainable with conventional techniques for plant hybridization, since the male gamete rarely contributes functional cytoplasm during fertilization. The identification of agronomically important traits that are

From: *Methods in Molecular Biology*, vol. 6, *Plant Cell and Tissue Culture*
Edited by Jeffrey W. Pollard and John M. Walker, ©1990 by The Humana Press

conferred by the organelles of certain major crop species, like cytoplasmic male sterility (CMS) and cytoplasmic tolerance to the triazine herbicides (CTT), as found in rapeseed, has made the production of cybrids commercially important.

CMS is a valuable trait in breeding crops like rapeseed, since, by preventing self-fertilization, it can allow the economic production of single-cross F_1 hybrids. In addition, incorporating CTT into rapeseed hybrids can allow the crop to be grown in weed-infested areas. The weeds are eliminated with applications of triazine (*1*).

In rapeseed, CTT is conferred by the chloroplast genome (2), and CMS, by the chondriome (*3–6*). For commercial production of single-cross hybrid CTT canola, as proposed by Beversdorf et al. (*7*), both CMS and CTT traits must be present in the female parent. However, since these are generally only transmitted through the female, it is not possible to combine these traits by conventional breeding approaches. Protoplast fusion, however, allows the biparental transmission of both CMS and CTT traits (*3,5,6*).

A variety of methods of producing and selecting for cybrids has been reported. For a detailed review on cybrid research, *see* Kumar and Cocking (*8*). This chapter describes the production of rapeseed (*B. napus*) cybrids combining CMS and CTT traits, using a "donor–recipient" aided selection scheme first described for *Nicotiana* by Zelcer et al. (*9*). The nuclei of the "donor" parent are inactivated prior to fusion, so that effectively it donates only cytoplasm to the other, "recipient," nucleus-bearing parent. Nuclear inactivation can be achieved by gamma or X-irradiation (*6,10–12*). The selection is complemented by the treatment of the "recipient" parent with an inhibitor, iodoacetic acid (IOA), that kills the cells by metabolic inactivation of the cytoplasm (*6,11,12*). Only fusion products survive because of complementation. This selection is only for fusion products, and not for specific cybrids; these are formed after organelle segregation during cell division and plant regeneration (Fig. 1).

In the procedures described below, mesophyll protoplasts of the CTT rapeseed "donor" parent are fused with "recipient" CMS rapeseed hypocotyl protoplasts. Fusion is achieved by a modified combination of polyethylene glycol (PEG) (*13,14*) and Ca^{2+} ions at high pH (*15*). Protoplasts are cultured, following a protocol that was devised for rapeseed (*16*), with the addition of tobacco nurse cells (*5*) to augment fusion-product survival (*see* Notes 1 and 2). Although these protocols have been tried and tested only for rapeseed (*B. napus*) (*5,6,17*), they are applicable to other species, provided that plant regeneration is possible.

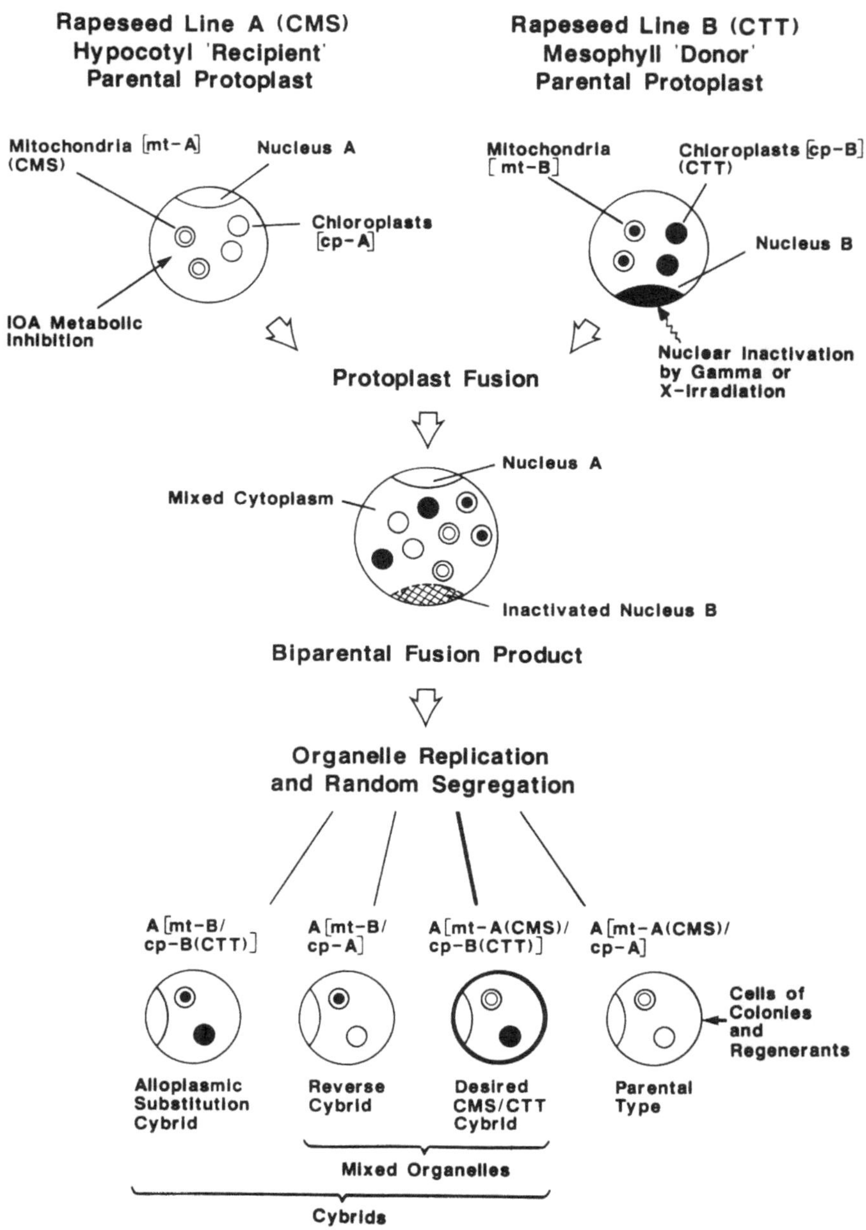

Fig. 1. Protoplast fusion and possible organelle segregation.

2. Materials

2.1. Protoplast Source Material

1. Mesophyll: CTT plants are grown from seed and established in "Metro-mix" potting compost, fertilized daily with "Peters" 20-10-20 fertilizer (both from W. R. Grace and Co., Canada), and maintained within growth chambers under a regimen of a 12-h photoperiod of 160 $\mu E/m^2/s^1$ of photosynthetically active radiation (provided by 40 W cool white fluorescent bulbs) at a constant temperature (23°C). Typically, 4–6 CTT plants are needed for each individual fusion operation.
2. Hypocotyl: CMS seedlings are grown from surface sterilized seed.
3. 10% (v/v) solution of 6% commercial preparation of hypochlorite.
4. Paper towels, wrapped in aluminum foil and autoclaved.
5. MSS medium: hormone-free MS basal media (*18* and *see* Appendix) containing 10 g/L sucrose, solidified with 4 g/L agarose (Sigma Type I), pH adjusted to 5.7, poured into 10-cm Petri plates (70 mL/plate).
6. Sieves, pore size <0.5 mm, autoclaved.

2.2. Protoplast Isolation

1. Soft nylon paintbrush, sterilized by immersion in 70% ethanol for 5 min and dried on sterile paper towels.
2. Scalpel with #11 blade, sterilized by flaming.
3. Sterile filtering funnels (double-layer cheesecloth lining a glass funnel).
4. Porcelain tile, wrapped in foil and autoclaved.
5. Soak solution (for 1 L):

KNO_3	475 mg
$CaCl_2 \cdot 2H_2O$	110 mg
$MgSO_4 \cdot 7H_2O$	92.5 mg
KH_2PO_4	42.5 mg
Thiamine•HCl	0.125 mg
Glycine	0.5 mg
Nicotinic acid	1.25 mg
Pyridoxine•HCl	0.125 mg
Folic acid	0.125 mg
Biotin	0.0125 mg
Casein Hydrolysate	25.0 mg
6-benzylamino purine	0.5 mg

2,4-dichlorophenoxyacetic acid	1.0 mg

pH 5.6, autoclaved.

The soak solution can be stored at 4°C for several weeks.

6. Enzyme solution for 100 mL:

Macerozyme R-10*	0.1 g
Cellulase R-10*	1.0 g
2 (*N*-morpholino) ethanesulphonic acid (MES)	0.1 g
Polyvinylpyrrolidone	1.0 g
Sucrose	12.0 g

pH 5.6, filter sterilizedthrough 0.2 µm Nalgene filter.

*Yakult Honsha Co., Ltd., Tokyo, Japan.

The enzyme solution can be stored frozen for up to 6 mo.

Mesophyll protoplast isolation requires a 1/10 strength enzyme solution, prepared by dilution with rinse solution (*see below*). For the isolation of hypocotyl protoplasts, full-strength enzyme solution is used.

7. Rinse solution (for 1 L):

Sucrose	120 g
MES	0.5 g

pH 5.6, autoclaved. Stored at room temperature in sealed bottles for several months.

2.3. Protoplast Inactivation

1. Iodoacetic Acid (IOA) solution: The IOA solution for protoplast inactivation is prepared from a 30 m*M* stock, diluted with rinse solution, pH adjusted to 5.6, and filter sterilized with a 2-µm Nalgene filter. The final concentration of IOA necessary for protoplast inactivation is determined prior to fusion.

   30 m*M* stock: 0.279 g IOA in 50 mL of H_2O. This stock solution can be stored at 4°C in the dark for up to 3–4 mo, with no noticeable loss of activity.

2.4. Protoplast Fusion

1. PEG solution for 100 mL:

Polyethylene glycol 8000	25.0 g
$CaCl_2 \cdot 2H_2O$	0.147 g
Sucrose	4.0 g

Autoclaved in foil-wrapped vials. Can be stored at 4°C for up to 6–8 wk.

2. Ca^{2+} solution (for 1L):

$CaCl_2 \cdot 6H_2O$	0.735 g
Glycine	0.375 g
Sucrose	12.0 g

pH 10.5, filter sterilized with 0.2-μm Nalgene filter. Solution must be prepared on the day of fusion.

2.5. Protoplast Culture

1. Preparation of media (Table 1):
   a. PM medium: MS basal medium (*18* and *see* Appendix) containing 10 g/L sucrose, 1.0 mg/L 2,4-D, 0.1 mg/L kinetin, 4.0 g/L agarose (Sigma Type I), pH adjusted to 5.7.
   b. RM1 medium: MS basal medium (*18* and *see* Appendix) containing 2.0 g/L sucrose, 2.0 mg/L zeatin, 2.0 mg/L kinetin, 0.1 mg/L IAA*, 4.0 g/L agarose (Sigma Type I), pH adjusted to 5.7.
   c. RM2 medium: same as RM1 medium, except the sucrose is reduced to 1.0 g/L, and the zeatin and IAA are each increased to 3.0 mg/L.
   d. SP medium: Gamborg's basal medium (*20* and see Appendix) containing 2.0 g/L sucrose, 0.03 mg/L GA_3 and 4.0 g/L agarose (Sigma Type I), pH adjusted to 5.7.

   All media are adjusted to pH 5.6, and sterilized by autoclaving.
2. Preparation of R plates: Sterile, deep Petri plates (100 x 25 mm) are filled with 70 mL of molten R medium. Upon setting, an X is cut into the medium, reaching to the edges of the plate. Two diametrically opposite quadrants are then removed with a sterile spatula to complete the R-plate preparation. Twenty or so R plates should be prepared in advance. Unused plates can be stored at 4°C in the dark for up to 4 wk.
3. Preparation of RM1, RM2, and SP media plates: RM1, RM2, and SP media are dispensed into sterile deep Petri plates (100 x 25 mm), 70 mL/plate. Plates can be stored at 4°C in the dark for several weeks.

*Abbreviations: IAA—3-indoleacetic acid; GA_3—gibberelic acid.

Table 1

Medium	Cl[a]	R[a]	DM[a]
	Major elements, mg/L		
KNO_3	7600	1900	1900
$CaCl_2 \cdot 2H_2O$	1760	440	440
$MgSO_4 \cdot 7H_2O$	1480	370	370
KH_2PO_4	680	170	170
	Iron & minor elements, mg/L		
$Na_2 \cdot EDTA$	18.5	18.5	18.5
$FeSO_4 \cdot 7H_2O$	13.9	18.5	18.5
H_3BO_3	3.1	3.1	3.1
$MnCl_2 \cdot 4H_2O$	9.9	9.9	9.9
$ZnSO_4 \cdot 7H_2O$	4.6	4.6	4.6
KI	0.42	0.42	0.42
$Na_2MoO_4 \cdot 2H_2O$	0.13	0.13	0.13
$CuSO_4 \cdot 5H_2O$	0.013	0.013	0.013
$CoSO_4 \cdot 7H_2O$	0.015	0.015	0.015
	Organics, mg/L		
Thiamine•HCl	0.5	0.5	0.5
Glycine	2.0	2.0	2.0
Nicotinic acid	5.0	5.0	5.0
Pyridoxine•HCl	0.5	0.5	0.5
Folic acid	0.5	0.5	0.5
Biotin	0.05	0.05	0.05
Casein hydrolysate	50.0	100	100
	Osmoticum, g/L		
Sucrose	85.6	34.25	34.25
Myo-Inositol	5.7	0.1	0.1
D-Mannitol	5.7	18.25	18.25
Sorbitol	5.7	–	–
Xylitol	4.7	–	–
	Hormones, mg/L		
NAA[b]	0.1	0.1	–
BAP[b]	0.4	–	–
2,4-D[b]	1.0	1.0	1.0
Kinetin	–	–	0.1
	Agarose, g/L		
Type VII (Sigma)	0.6	4.5	0.6

[a]From Shepard (*19*), modified table reprinted by permission from The University of Minnesota Press.

[b]Abbreviations: NAA—1-napthaleneacetic acid; BAP—6-benzylaminopurine; 2,4-D—2,4-dichlorophenoxyacetic acid.

2.6. Triazine Tolerance Test

Triazine paste: A paste of the triazine herbicide Atrazine WP (80%; Ciba Giegy) is prepared by mixing 1.47 g of Atrazine in 100 mL water.

3. Methods

All procedures must be performed aseptically within a laminar flow sterile air cabinet.

3.1. Production of Etiolated Hypocotyls

1. Seed of CMS-rapeseed are surface sterilized by immersion in 10% (v/v) commercial hypochlorite solution for 10 min and then rinsed for 5 min in sterile water. These procedures are best performed by placing the seed in presterilized sieves, for ease of handling, and immersing into the different solutions.
2. For some batches of seed, this procedure may be ineffective. In these cases, seeds are surface sterilized by immersion in a solution of 0.05% mercuric chloride (taking particular care in the handling of this substance) and 0.1% sodium dodecyl sulfate for 10 min, and then rinsed 6 x in sterile water.
3. Seeds are transferred to paper towels to remove excess moisture, and then onto MSS medium (approximately 20 seeds/Petri plate).
4. Plates are incubated in the dark at constant temperature (25°C) for germination. Typically, 20–30 plates of CMS-rapeseed seed are required/fusion operation.

3.2. Isolation of Protoplasts

3.2.1. Mesophyll Protoplasts

1. Leaves from 3-wk-old CTT plants are surface sterilized by immersing in 10% (v/v) commercial hypochlorite solution for 3 min, rinsing once in sterile water and once for 30 s, in 70% ethanol.
2. The leaves are placed onto paper towels to remove excess moisture, care being taken not to dry the leaves to the point of wilting.
3. The lower epidermis of the leaves is abraded by gently stroking with the paintbrush until bristle marks are just visible. Excessive brushing should be avoided to prevent damage to the subepidermal cells, and consequent possible reduction of protoplast yields.
4. Leaves are cut into 1-cm wide strips and transferred to the soak solu-

tion. Approximately 3.0 g of tissue (around 6–9 leaves) are immersed in each flask of soak (100 mL of soak/flask). Typically, two flasks of material are prepared for each individual fusion operation.

5. The flasks of leaf material are then incubated for 2–4 h in the dark at 4°C. Longer incubations, of up to 12 h, are possible without undue harm.
6. Following cold incubation, the soak solution is carefully poured off and replaced with 1/10 strength enzyme solution. Typically, leaf material is prepared early in the afternoon, and the soak substituted with enzyme in the early evening.
7. Flasks are incubated for 15–18 h (i.e., overnight) on a rotary shaker, set to give a very gentle agitation. Successful enzyme digestion is evident when the leaf pieces are floating and have a translucent appearance. However, leaves that have sunk and/or appear undigested can still yield protoplasts, but the numbers are generally reduced.

From this stage, all manipulations must be performed particularly carefully to avoid damaging the fragile protoplasts.

8. The contents of each flask are poured through a filtering funnel into two Babcock bottles. A retort stand is used to hold the funnel. Gentle agitation with a Pasteur pipet will assist the liberation of protoplasts.
9. The Babcock bottles are then topped up with rinse solution and recapped with sterile foil.
10. Protoplasts will float to the top of the bottle following centrifugation at 500 rpm (350*g*) for 10 min (in a bench model centrifuge with Babcock bottle-size swing-out buckets).
11. Protoplasts are "rinsed" free of enzyme solution by transferring them with a Pasteur pipet to a Babcock bottle previously half filled with rinse solution, thereby combining into one bottle the protoplasts from two Babcocks.
12. The bottle is then filled completely with rinse solution, resuspending the protoplasts, recapped, and centrifuged once again, as described before.

3.2.2. Hypocotyl Protoplasts

1. Hypocotyls excised from etiolated CMS-rapeseed seedlings (3–5 d after sowing) are chopped on a porcelain tile into 2–5 mm transverse segments. These are immediately placed into full-strength enzyme solution. Generally, 200–300 chopped hypocotyls are added to each flask filled with 100 mL of enzyme solution.

2. The hypocotyl material is incubated overnight and protoplasts isolated as described for mesophyll CTT material (*see* Note 3).

3.3. Prefusion Treatment of Protoplasts

3.3.1. Donor Parent

Nuclear inactivation of mesophyll CTT protoplasts can be accomplished by either γ-ray or X-ray irradiation (*see* Note 4). Whichever is used, it is necessary to establish the dosage and check the protoplast response in control cultures. Irradiation should produce protoplasts that do not divide, but remain "dormant" for several days, under normal culture conditions, before eventually dying. However, a low frequency of first divisions is acceptable, provided there is no colony formation. Irradiation is performed immediately after protoplast isolation, in the Babcock bottle.

3.3.2. Recipient Parent

The effective dosage level of IOA must be established prior to fusion experiments. The IOA concentration should be sufficient to kill the protoplasts, under normal culture conditions, within 24 h. For *B. napus* rapeseed, 3–4 m*M* IOA is typical.

1. Following isolation, CMS protoplasts are carefully transferred to a sterile centrifuge tube of 10–25 mL capacity.
2. The tube is then filled with the diluted IOA solution, resuspending the protoplasts.
3. Following 10 min incubation at room temperature, the tube is centrifuged at 500 rpm for 10 min.
4. The IOA solution is removed by two consecutive centrifugations, each time transferring the floating protoplasts to separate Babcock bottles filled with rinse solution. It is typical for about 25% of the protoplasts to be lost as a result of the centrifugation steps.

3.4. Protoplast Fusion

1. Protoplasts of both donor and recipient parents are carefully combined in a sterile round-bottom tube (13 x 100 mm). Actual numbers of protoplasts from each parent will vary widely with each fusion, because of yield variations of different tissue samples, and through varying degrees of cell death caused by centrifugation. Ratios of the CTT and CMS parents should be in the 1:1–1:4 range. Some adjustments may be necessary.

2. Prior to initiating the fusion process, approximately 1/5 of the mixed protoplast suspension is removed and transferred to a separate tube. This will constitute the "unfused mixed control" (UFMC) to be cultured separately (*see below*).
3. The volume of the remaining protoplast suspension should be adjusted to 0.5–1.0 mL, by the addition of extra rinse solution, or by centrifuging the tube and removing excess rinse solution from underneath the floating protoplasts.
4. To this is added an equal volume (0.5–1.0 mL) of PEG solution, followed by the gentle resuspension of the protoplasts by careful rotations of the tube for 30 s or so.
5. After 10 min incubation at room temperature, 2 mL of Ca^{2+} solution is gradually added, further resuspending the protoplasts.
6. The tube is then filled with further volumes of Ca^{2+} solution, and centrifuged at 400 rpm for 10 min.
7. Floating protoplasts are carefully removed and resuspended in rinse solution contained in a Babcock bottle.
8. After a final centrifugation at 500 rpm for 10 min, the fusion process is complete.

3.5. Protoplast Culture

Fusion protoplasts and the UFMC are plated in nurse culture plates prepared 24 h in advance.

3.5.1. Nurse Culture Preparation

1. *Nicotiana tabacum* (tobacco) plants are grown from seed under the same conditions as described on page 391. Conditions for the preparation of the leaf material and the isolation of protoplasts are the same as described on page 392. Protoplasts are mitotically inactivated by irradiation as described in Section 3.3.
2. Irradiated tobacco protoplasts are then transferred from the rinse solution to several mL of CL medium and resuspended.
3. With a hemocytometer, the number of cells in the suspension is calculated from a small sample.
4. Using further volumes of CL medium, the density of the protoplast suspension is adjusted to 10^5 protoplasts/mL.
5. Next, 4 mL of the adjusted suspension is transferred to each empty quadrant in the R medium plates. Generally, 10–12 tobacco-filled nurse plates should be sufficient for each fusion and its corresponding UFMC.

6. Completed nurse plates are incubated at 25°C in the dark.

3.5.2. Culture of Fused and UFMC Protoplasts

1. Protoplasts from fusion and the UFMC sample are separately resuspended in a few mL of CL medium.
2. As with the nurse protoplasts above, the cell density of each suspension is estimated using a hemocytometer, and adjusted to 10^5 protoplasts/mL.
3. One mL of the postfusion suspension is then dispensed into each nurse quadrant, giving a final *Brassica* protoplast density of 2×10^4 protoplasts/mL. The UFMC suspension is dispensed similarly.
4. Fusion and UFMC plates are then sealed with parafilm and incubated at 25°C in the dark.
5. One week to 10 d after plating, suspensions are transferred to regular Petri plates (one quadrant/100 x 15 mm plate) and diluted with an equal volume of molten DM medium. The combined suspension and DM medium should be spread evenly over the bottom of the Petri plates, to form a thin layer for maximum colony formation.

Subsequent culture conditions are changed to 25°C with a 16-h photoperiod of 150–200 $\mu E/m^2/s^1$ of photosynthetically active radiation.

3.6. Callus Proliferation and Plant Regeneration

From 10–14 d after dilution, colonies in the fusion plates will have grown to approximately 0.5–1.0 mm in size. Ideally, there should be no colony development in the UFMC plates, since the irradiation and IOA treatments should have inactivated or killed the unfused protoplasts. However, it is likely that some colonies will develop from the irradiated population, since the irradiation treatment is often not 100% effective. If the number of colonies in the UFMC plates approaches the colony count in the fusion plates, then the fusion must be discarded and the inactivation treatment checked and reestablished.

1. Assuming that there is no colony formation in the UFMC plates, or at least considerably less than in the fusion plates, colonies of 0.5–1.0 mm size from the fusion plates are individually transferred to the surface of PM medium, 35–40 colonies/plate, for further proliferation.
2. Colonies that are 1.0–2.0 mm in diameter (usually after 10–14 d, at which time they should be colorless or very pale green) are transferred to RM1 medium, 20–25 colonies/plate, to induce differentiation.

3. Within 2–3 wk, it is likely that some of the colonies will have regenerated small shoots. These colonies must then be transferred to SP medium to induce shoot proliferation and grow out.
4. The remaining colonies are transferred to RM2 medium, to further induce differentiation.
5. Any shoots developing on the RM2 medium should also be transferred to SP medium. It is unlikely that more than 1–5% of the colonies will respond on the differentiation medium, since the various treatments and fusion processes typically reduce the regeneration response considerably.
6. Shoots that have clearly differentiated meristems are transferred to peat pellets ("Jiffy 7," Jiffy Products [N.B.] Ltd., Shippegan, Canada; sterilized by autoclaving) to induce root formation.
7. Pellets are placed in sterile glass jars, to maintain a high humidity. Peat pellets should be moistened with filter sterilized fertilizer (*see* Materials, section 2.1., item 1). To prevent premature flowering, the photoperiod is reduced to 10 h.
8. When several roots are protruding from the sides of the peat pellets, the plantlets are transferred to potting compost for further plant development.
9. Regenerated plants are maintained at 23°C with a 16-h photoperiod of 160 $\mu E/m^2/s^1$ of photosynthetically active radiation to induce flowering.

3.7. Identification of Cybrids

1. Upon flowering, the flower morphology of the regenerants are examined for male sterility (absence of pollen production), which indicates the presence of CMS-mitochondria.
2. Plants are then tested for their sensitivity to triazine herbicides. Two to three mature leaves, still attached to each plant, are painted with the atrazine paste. Those plants carrying CTT-chloroplasts should remain undamaged. The other plants will suffer yellowing of the leaves, and maybe death, within 7–14 d.

Those plants that are male sterile and are tolerant to atrazine are the desired cybrids. Male fertile, atrazine susceptible plants are also cybrids, but with the opposite organelle combination, and so are called reverse-cybrids. In addition, male fertile, atrazine tolerant plants can also be considered cybrids, since by alloplasmic substitution, they possess the nucleus of the original CMS parent. However, if the original parents had identical nuclei, e.g., if they were both versions of the same rapeseed variety, then these plants cannot be considered as cybrids (*see* Fig. 1, and Note 5).

Male sterility sometimes appears in regenerants that have aneuploid chromosome complements, independent of the mitochondria. The genotype of the mitochondrial population can be distinguished by "restriction fragment length polymorphism" (RFLP) analysis of the mitochondrial DNA. Plastomes can be similarly identified. The techniques involved are beyond the scope of this chapter, and the reader is referred to Kemble (*21*).

4. Notes

1. The described procedure can produce cybrids in 8–12 mo, from sowing the seed of the protoplast source material to identifying the cybrids in the whole plant regenerants. This time period predominantly depends on the rapidity of shoot differentiation. This, in turn, depends on the careful timing of each transfer onto the different culture media.
2. These procedures can be utilized with different parents to produce rapeseed cybrids with other desired organelle combinations. For example, similar techniques have transferred the CMS trait from spring-planted rapeseed to winter forms, producing cybrids that can be useful to a single-cross hybrid winter rapeseed production program (*17*). In this case, conventional crossing also could have produced CMS winter lines, but this would have taken up to 3 yr. Fusion accomplished the same goal in just 1 yr, a significant saving of time.
3. In some circumstances, it may be necessary to use hypocotyl protoplasts as the "donor" parent when attempting to produce a particular rapeseed cybrid. Hypocotyl protoplasts will require higher dosages of irradiation for inactivation, since their nuclei are more resistant to radiation damage than their mesophyll counterparts.
4. When irradiating donor parent protoplasts prior to fusion, the exposure period should not be longer than 1–2 h, to prevent undue cell wall regeneration that can inhibit fusion. If the exposure period has to be longer because of the nature of the irradiation equipment, then the problem can be circumvented by directly irradiating the leaves instead, prior to preparations for protoplast isolation. Indeed, this may be more convenient for inactivating both the donor parent and the nurse material.
5. Figure 1 suggests that one-quarter of the regenerants will be the desired cybrids, which indeed would be true if there were random

segregation of the organelles following cell divisions of the fusion-product cells. However, the proportions of the different segregants is, in practice, usually not equal, owing to differences in protoplast tissue sources, differential responses in culture, and adverse effects of the cell inactivation processes.

References

1. Beversdorf, W. D., Weiss-Lerman, J., Erickson, L. R., and Souza-Machado, Z. (1980) Transfer of cytoplasmic inherited triazine resistance from bird's rape to cultivated oilseed rape (*Brassica campestris* and *B. napus*). *Can. J. Genet. Cytol.* **22,** 167–172.
2. Reith, M. and Straus, N. A. (1987) Nucleotide sequence of the chloroplast gene responsible for triazine resistance in canola. *Theor. Appl. Genet.* **73,** 357–363.
3. Pelletier, G., Primard, C., Vedel, F., Chetrit, P., Remy, R., Rouselle, P., and Renard, M. (1983) Intergeneric cytoplasmic hybridization in *Cruciferae* by protoplast fusion. *Molec. Gen. Genet.* **191,** 244–250.
4. Chetrit, P., Mathieu, C., Vedel, F., Pelletier, G., and Primard, C. (1985) Mitochondrial DNA polymorphism induced by protoplast fusion in *Cruciferae. Theor. Appl. Genet.* **69,** 361–366.
5. Yarrow, S. A., Wu, S. C., Barsby, T. L., Kemble, R. J., and Shepard, J. F. (1986) The introduction of CMS mitochondria to triazine tolerant *Brassica napus* L., var. "Regent," by micromanipulation of individual heterokaryons. *Plant Cell Reports* **5,** 415–418.
6. Barsby, T. L., Chuong, P. V., Yarrow, S. A., Wu, S. C., Coumans, M., Kemble, R. J., Powell, A. D., Beversdorf, W. D., and Pauls, K. P. (1987) The combination of Polima cms and cytoplasmic triazine resistance in *Brassica napus. Theor. Appl. Genet.* **73,** 809–814.
7. Beversdorf, W. D., Erickson, L. R., and Grant, I. (1985) Hybridization process utilizing a combination of cytoplasmic male sterility and herbicide tolerance. U.S. Patent No. *4517763*.
8. Kumar, A. and Cocking, E. C. (1987) Protoplast fusion: A novel approach to organelle genetics in higher plants. *Amer. J. Bot.* **74(8),** 1289–1303.
9. Zelcer, A., Aviv, D., and Galun, E. (1978) Interspecific transfer of cytoplasmic male sterility by fusion between protoplasts of normal *Nicotiana sylvestris* and X-ray irradiated protoplasts of male sterile *N. tabacum*. *Z. Pflanzenphysiol.* **90,** 397–407.
10. Aviv, D., Arzee-Gonen, P., Bleichman, S., and Galun, E. (1984) Novel alloplasmic plants by "donor-recipient" protoplast fusion: cybrids having *N. tabacum* or *N. sylvestris* nuclear genomes and either or both plastomes and chondriomes from alien species. *Mol. Gen. Genet.* **196,** 244–253.
11. Sidorov, V. A., Menczel, L., Nagy, F., and Maliga, P. (1981) Chloroplast transfer in *Nicotiana* based on metabolic complementation between irradiated and iodoacetate treated protoplasts. *Planta* **152,** 341–345.
12. Ichikawa, H., Tanno-Suenaga, L., and Imamura, J. (1987) Selection of *Daucus* cybrids based on metabolic complementation between X-irradiated *D. capillifolius* and iodoacetamide-treated *D. carota* by somatic cell fusion. *Theor. Appl. Genet.* **74,** 746–752.
13. Kao, K. N. and Michayluk, M. R. (1974) A method for high frequency intergeneric fusion of plant protoplasts. *Planta* **115,** 355–367.

14. Wallin, A. K., Glimelius, K., and Eriksson, T. (1974) The induction of aggregation and fusion of *Daucus carota* protoplasts by polyethylene glycol. *Z. Pflanzenphysiol.* **74,** 64–80.
15. Keller, W. A. and Melchers, G. (1973) The effect of high pH and calcium on Tobacco leaf protoplast fusion. *Z. Naturforsch.* **28c,** 737–741.
16. Barsby, T. L., Yarrow, S. A., and Shepard, J. F. (1986) A rapid and efficient alternative procedure for the regeneration of plants from hypocotyl protoplasts of *Brassica napus*. *Plant Cell Reports* **5,** 101–103.
17. Barsby, T. L., Yarrow, S. A., Kemble, R. J., and Grant, I. (1987) The transfer of cytoplasmic male sterility to winter-type oilseed rape (*Brassica napus* L.) by protoplast fusion. *Plant Science* **53,** 243–248.
18. Murashige, T. and Skoog, F. (1962) A revised medium for rapid growth and bioassays with tobacco tissue cultures. *Physiol. Plant.* **15,** 473–497.
19. Shepard, J. F. (1980) Mutant selection and plant regeneration from Potato mesophyll protoplasts, in, *Genetic Improvement of Crops: Emergent Techniques* (Rubenstein, I., Gengenbach, B., Philips, R. L., and Green, C. E., eds.), University of Minnesota Press, Minnesota, pp. 185–219.
20. Gamborg, O. L., Miller, R. A., and Ojima, K. (1968) Nutrient requirements of suspension cultures of soybean root cells. *Exptl. Cell Res.* **50,** 151–158.
21. Kemble, R. J. (1987) A rapid, single leaf, nucleic acid assay for determining the cytoplasmic organelle complement of rapeseed and related *Brassica* species. *Theor. Appl. Genet.* **73,** 364–370.

Chapter 36

Selection of Somatic Hybrids by Resistance Complementation

Randal M. Hauptmann and Jack M. Widholm

1. Introduction

Protoplast fusion provides a nonsexual system for the transfer of genetic information between cell types. This transfer can be between species, genera, families, or kingdoms, thereby allowing unique opportunities to study somatic cell genetics in plants. Individual chromosomes (*1*) or pieces of chromosomes (*2*) have been transferred between species in unstable hybrids, but the technology also allows the transfer and recombination of organelles (*3*). The major obstacle in most protoplast fusion experiments has been developing strategies for the selection of somatic hybrids. This is mainly because of the fact that the fusion process itself is random. The isolation of heterokaryons from homokaryons and unfused protoplasts requires the use of a selection method. There are many selection strategies available (*4*), but one of the most efficient methods employs the use of resistance complementation (*5*).

Carrot hybrids have been isolated between a cell strain selected as doubly resistant to DL-5-methyltryptophan (5MT) and azetidine-2-carboxylate (A2C) and a cell strain resistant to S-2 (aminoethyl)-L-cysteine (AEC) (*6*). *Nicotiana sylvestris* hybrids have been selected using 5MT and

From: *Methods in Molecular Biology*, vol. 6, *Plant Cell and Tissue Culture*
Edited by Jeffrey W. Pollard and John M. Walker, ©1990 by The Humana Press

A2C resistant cell strains (*7*). Carrot and tobacco hybrids have been isolated using an AEC resistant carrot and a 5MT resistant tobacco (*8*), and using a 5MT and A2C resistant carrot and an AEC resistant tobacco cell strain (*9*). In these cases, the amino acid analog resistance has been expressed in a dominant or semi-dominant fashion in the somatic hybrids.

The possibility of extending biochemical selection methods to other types of resistances could be beneficial, not only by allowing the use of a selectable marker, but also by transferring a desirable trait, such as herbicide resistance. This has been demonstrated using a carrot cell strain selected as tolerant to the herbicide *N*-(phosphonomethyl) glycine (glyphosphate) (*10*). This strain was 52 x more tolerant to glyphosate than the wild type, and did not show any reduction in glyphosate uptake. The mechanism of resistance has been found to be the result of amplification of EPSP synthase and has been shown to act as a dominant selectable marker in protoplast fusion experiments (*11,12*).

The various selection methods used to obtain plant cell mutants have been described in detail by Duncan and Widholm in Chapter 40. In addition, a number of plant mutants are currently available, including strains that can act as universal hybridizers (*12*).

The selection of somatic hybrids involves three steps: (a) the isolation of protoplasts, (b) the fusion process, and (c) the application of selection pressure for heterokaryons. The selection of heterokaryons in this chapter will be illustrated in carrot using the amino acid analog azetidine-2-carboxylate (A2C) and the herbicide *N* (phosphonomethyl)-glycine (glyphosate) by dominant resistance complementation (*11*).

2. Materials

All materials listed should be sterilized by autoclaving except for reagents, which must be filter sterilized where noted. All reagents are prepared with double distilled or distilled deionized water. All work must be conducted under aseptic conditions.

2.1. Cell Suspension Culture and Protoplast Isolation

1. Plant cell mutant strains C123 (A2C resistant), PR (glyphosate resistant), or other mutant isolated as described in Chapter 40.
2. Culture media: Murashige and Skoog (*13*, and *see* Appendix) powdered medium containing organic nutrients and sucrose with 0.4 mg/L 2,4-dichlorophenoxyacetic acid (2,4-D) (MS). Dispense 50

mL in 125-mL flasks for liquid medium, and for agar solidified medium, add 0.8% Difco Bacto-Agar and autoclave.

3. Protoplast wash solution: 10% mannitol and 0.1% $CaCl_2 \cdot 2H_2O$, pH 5.7.
4. Enzyme mixture: Dissolve 2% Cellulase "Onozuka" R-10, 0.1% Macerozyme R-10 (Yakult Honsha Co., Medicine Department Enzyme Division, 1-1-19, Higashi-shinbashi, Minatoku, Tokyo, 105, Japan) in protoplast wash solution and adjust pH to 5.7. Centrifuge 10 min at 10,000*g* and filter sterilize the supernatant through 0.45-μm filter.
5. Protoplast purification solution: 20% (w/v) sucrose.
6. Special equipment: Stainless-steel mesh (105 μm) or Miracloth (Calbiochem) lined funnel mounted atop a flask; Babcock bottles; hemocytometer counting chamber, 0.2 mm deep.

2.2. Protoplast Fusion

1. Fusion plates: 100 x 15 mm "X Plate" Petri dish (Falcon).
2. Dextran solution: Prepare an 8% (w/v) NaCl solution and add dextran (500,000 mol wt) to give a final concentration of 15% (w/v). The dextran will usually go into solution during autoclaving, but may require further mixing while hot.
3. Fusion solution: 0.1*N* NaOH containing 8% (w/v) NaCl.
4. Fusion wash: 5% mannitol and 2% $CaCl_2 \cdot 2H_2O$, pH 5.5.
5. Glucose stock: 50% glucose.

2.3. Protoplast Culture and Selection of Heterokaryons

1. Protoplast culture media: Add 1/10 vol of 50% glucose stock to MS suspension culture media to give a final concentration of 5% glucose.
2. Selection media: Prepare MS suspension culture media with 0.8% Bacto-Agar and autoclave. Cool to 45°C, and add 1/10 vol of 50% glucose stock. Add appropriate selection agents at double the desired final concentration (500 μ*M* A2C and 30 m*M* glyphosate).

3. Methods

3.1. Protoplast Isolation

1. Grow the carrot suspension cultures in 50 mL of MS culture media in 125-mL flasks on gyrotory shakers at 130 rpm at 27–28°C. The

cells should be subcultured every 4 d, using a 10-mL inoculum, to maintain them in logarithmic growth. Cells used for protoplast isolation should be harvested on day 3 or 4 of subculture.

2. Place 4 mL of packed cell volume of suspension cultured cells in 20 mL of the enzyme mixture (*see* Note 1). Incubate the mixture for approximately 12 h in a 100 x 25 mm Petri dish at 27–28°C. Gentle agitation on a gyrotory shaker at 30 rpm can aid in protoplast production for cultures containing large cell aggregates.
3. Filter the protoplasts through a 105-μm stainless-steel mesh or a miracloth-lined filter attached to a funnel. The filtration removes undigested cell aggregates and debris.
4. Remove the enzyme mixture by centrifugation (100 x *g* for 5 min). Resuspend the protoplasts in protoplast wash solution.
5. Wash twice by centrifugation (100 x *g*) for 5 min in the protoplast wash solution to remove residual enzyme.
6. Suspend the protoplasts in a 20% sucrose solution and centrifuge at 100 x *g* for 10 min in Babcock bottles or 15-mL tubes. The purified protoplasts will float to the surface and form a band. With a Pasteur pipet, gently remove the band and resuspend the protoplasts in the washing solution. The Babcock bottle has a constricted neck that allows for easy recovery of the purified protoplasts (*see* Note 2).
7. Remove an aliquot and determine the density of protoplasts using a hemocytometer. Adjust the volume of protoplasts with wash solution to obtain a density of 1 x 10^6 protoplasts/mL. Total cell wall digestion results in perfectly spherical protoplasts. This is essential for successful protoplast fusion.

3.2. *Dextran Fusion* (See *Note 3*)

1. Mix purified protoplasts from the two resistant strains, C123 and PR, in a 1:1 ratio. Reserve some of the mixture of protoplasts for use as unfused controls.
2. Place 1 mL of the mixture in the center of one well of the 100 x 15 mm "X Plate" Petri dish.
3. Add 4 mL of the dextran fusion solution around the edges of the protoplast mixture, and then over the surface. Incubate for 10–15 min at room temperature.
4. Add one to three drops of a solution containing 8% NaCl in 0.1*N* NaOH, and incubate for 10–15 min. The amount added varies, since increased amounts will increase fusion frequency, but can become toxic at higher concentrations depending on the sensitivity of the plant protoplasts used.

5. Add 5 mL of the fusion wash solution gently to the surface of the mixture. Incubate for 1 h at room temperature, and then place at 2°C overnight. Fusion of the protoplasts should be visible microscopically.
6. Pipet the mixture into a 50-mL centrifuge tube, rinsing the protoplasts from the plate with the fusion wash solution, bringing the final vol to 50 mL. Centrifuge at 100 x *g* for 10 min.
7. Resuspend the protoplasts in protoplast culture medium and adjust the density to 7 x 10^4 protoplasts/mL. Pipet the mixture in 60-mm plastic Petri dishes with 2.5 mL/plate. Gently agitate the mixture to completely cover the bottom of the dish, and seal with Parafilm.

3.3. Selection of Somatic Hybrids

1. Culture the protoplasts for approximately 10 d until small colonies are formed.
2. Add an equal vol (2.5 mL) of liquified agar selection media (2x concentration of A2C, 500 µ*M*, and glyphosate, 30 m*M*), which has been cooled to 45°C. Double-resistant heterokaryons will form 1–5 mm colonies in 3–4 wk. Controls, consisting of a mixture of unfused protoplasts, should be plated under identical conditions and should show complete growth inhibition (*see* Notes 4–6).

4. Notes

1. The isolation of protoplasts involves the use of enzyme mixtures that are largely undefined with respect to specific enzymatic activities. Empirical methods are still employed to determine the appropriate mixture for each tissue type or source (*14*). Many commercial preparations are available using mixtures of hemicellulases, pectinases, and cellulases, and will be successful for the isolation of large numbers of protoplasts.
2. The most critical factor for protoplast fusion, regardless of the method used, is the quality and purity of the protoplast preparation. Purification of the protoplasts by floating them on a 20% sucrose solution works well for this purpose. The concentration of sucrose may have to be adjusted for different systems, since the density of the purified protoplasts may vary.
3. We have previously used three fusion methods in our laboratory: the dextran (*15*), polyethylene glycol (*16*), and polyvinyl alcohol (*17*) methods. They have been found to be similar in their relative effectiveness in producing hybrids based on amino acid analog resis-

tance complementation frequencies (*8*). We generally use the dextran fusion method, since the dextran is relatively nontoxic and large volumes of protoplasts can be treated.

4. When using different selection agents with other plant species, it is essential that the levels of complete growth inhibition are precisely determined. This is done by plating protoplasts from the individual species in various concentrations of the selection agents, using the same conditions that will be employed for selection of heterokaryons. The concentration of inhibitor that allows good growth for the resistant strain and completely inhibits growth of the sensitive strain should be used for the selection of the somatic hybrids.
5. Using dominant selectable markers for protoplast fusion experiments generally gives unambiguous selection for heterokaryons when the selection agents are applied to the cultured, fused protoplasts after colony formation is observed. If a large number of colonies are observed in control mixtures of unfused protoplasts or no growth is observed in fused protoplasts plated under selection conditions, then the addition time of the selection agent should be varied.
6. The isolation of somatic hybrids should be confirmed by additional techniques. This can be through chromosome and isozyme analysis or through the use of molecular probes. The choice of the technique will differ depending on the individual characteristics of the species used in the fusion experiments.

Acknowledgments

We would like to thank Dr. Toshiaki Kameya, who developed the dextran fusion method, for instruction in the use of the technique. We would also like to thank Dr. Rob Ferl for review of the manuscript.

References

1. Kao, K. N. (1977) Chromosomal behavior in somatic hybrids of soybean-*Nicotiana glauca*. *Molec. Gen. Genet.* **150,** 225–230.
2. Dudits, D., Fejer, O., Hadlaczky, G., Koncz, C., Lazar, G. B., and Horvath, G. (1980) Intergeneric gene transfer mediated by plant protoplast fusion. *Molec. Gen. Genet.* **179,** 283–288.
3. Kumar, A. and Cocking, E. C. (1987) Protoplast fusion: A novel approach to organelle genetics in higher plants. *Amer. J. Bot.* **74,** 1289–1303.
4. Widholm, J. M. (1982) Selection of protoplast fusion hybrids, in *Plant Tissue Culture 1982* (Fujiwara, A., ed.), Maruzen, Tokyo, pp. 609–612.

5. Widholm, J. M., Hauptmann, R. M., and Dudits, D. (1984) Carrot protoplast fusion. *Plant Molec. Bio. Reporter* **2,** 26–31.
6. Harms, C. T., Potrykus, I., and Widholm, J. M. (1981) Complementation and dominant expression of amino acid analogue resistance markers in somatic hybrid clones from *Daucus carota* after protoplast fusion. *Z. Pflanzenphysiol.* **101,** 377–390.
7. White, D. W. R. and Vasil, I. K. (1979) Use of amino acid analogue-resistant cell lines for selection of *Nicotiana sylvestris* somatic cell hybrids. *Theor. Appl. Genet.* **55,** 107–112.
8. Hauptmann, R. M., Kumar, P., and Widholm, J. M. (1983) Carrot + tobacco somatic cell hybrids selected by amino acid analog resistance complementation, in *Protoplasts 1983: Poster Proceedings* (Potrykus, I., Harms, C. T., Hinnin, A., Hutter, R., King, P. J., and Shillito, R. D., eds.), Birkhauser, Basel, pp. 92,93.
9. Harms, C. T. and Oertli, J. J. (1982) Complementation and expression of amino acid analog resistance studied by intraspecific and interfamily protoplast fusion, in *Plant Tissue Culture 1982* (Fujiwara, A., ed.), Maruzen, Tokyo, pp. 467,468.
10. Nafziger, E. D., Widholm, J. M., Steinrucken, H. C., and Killmer, J. L. (1984) The selection and characterization of a carrot cell line tolerant to glyphosate. *Plant Physiol.* **76,** 571–574.
11. Hauptmann, R. M., della-Cioppa, G., Smith, A. G., Kishore, G. M., and Widholm, J. M. (1988) Expression of glyphosate resistance in carrot somatic hybrid cells through the transfer of an amplified 5-enolpyruvylshikimic acid-3-phosphate synthase gene. *Molec. Gen. Genet.* **211,** 357–363.
12. Ye, J., Hauptmann, R. M., Smith, A. G., and Widholm, J. M. (1987) Selection of a *Nicotiana plumbaginifolia* universal hybridizer and its use in intergeneric somatic hybrid formation. *Molec. Gen. Genet.* **208,** 474–480.
13. Murashige, T. and Skoog, F. (1962) A revised medium for rapid growth and bioassays with tobacco tissue cultures. *Physiol. Plant* **15,** 473–497.
14. Davey, M. R. (1983) Recent developments in the culture and regeneration of plant protoplasts, in *Protoplasts 1983: Lecture Proceedings* (Potrykus, I., Harms, C. T., Hinnin, A., Hutter, R., King, P. J., and Shillito, R. D., eds.), Birkhauser, Basel, pp. 19–29.
15. Kameya, T., Horn, M. E., and Widholm, J. M. (1981) Hybrid shoot formation from fused *Daucus carota* and *D. capillifolius* protoplasts. *Z. Pflanzenphysiol.* **104,** 459–466.
16. Kao, K. N. and Michayluk, M. R. (1974) A method for high-frequency intergeneric fusion of plant protoplasts. *Planta* **115,** 355–367.
17. Nagata, T. (1978) A novel cell-fusion method of protoplasts by polyvinyl alcohol. *Naturwissenschaften* **65,** 263.

Chapter 37

Dual Fungal and Plant Cell Culture

Anne Donovan, Susan Isaac, and Hamish A. Collin

1. Introduction

Plant tissue cultures are now well-recognized as valuable experimental systems for use in the study of host–pathogen interactions. These techniques have obvious major advantages for the examination of obligately biotrophic fungi and also those with a necrotrophic life style, and it is in these areas that much research effort has been concentrated (*1*). Success with combined fungal–plant cultures has been variable, especially in terms of establishing cultures that may be maintained in a balanced state for prolonged periods, but there is no doubt that such systems are useful places for the study of cell–cell interactions.

Callus cultures provide host tissue that is comparable to that of the intact plant, but that can be maintained in an undifferentiated state in a controlled environment with a regulated supply of definable nutrients and supplements (e.g., phytohormones). The tissue is axenic, and cell populations are nearly homogeneous. It is also possible, by careful manipulation of growth conditions, to bring about tissue differentiation, thus providing sterile tissue that is similar to that of the intact plant.

The interaction between a pathogen and its host plant is a multistage process (2) involving complex recognition and response mechanisms. It

From: *Methods in Molecular Biology*, vol. 6, *Plant Cell and Tissue Culture*
Edited by Jeffrey W. Pollard and John M. Walker, ©1990 by The Humana Press

has been shown that some resistance genes are able to operate in callus tissue, and this has opened up great potential for the use of tissue cultures in programs that screen for disease resistance. Dual cultures are also useful for the investigation of biochemical and physiological aspects of host–pathogen interaction, e.g., phytoalexin biosynthesis (3), and the operation of resistance mechanisms at the cellular level (4).

The protocol described here was developed in connection with a research program establishing the nature of the interaction, morphologically and biochemically, between the fungal pathogen *Septoria apiicola,* which causes late blight disease, and the crop species celery *(Apium graveolens).* There is very little resistance within celery varieties, but there is evidence that resistance is correlated with flavor compound levels in host tissues (5). The closely related and morphologically similar wild species *Levisticum officinale* (lovage) does exhibit resistance. In intact host plants, the fungus proliferates between cells of the leaf tissue, eventually forming pycnidia, which occur as characteristic brown-black lesions on the leaf surface. The basic dual culture technique, summarized in Fig. 1, could be easily manipulated to suit other source material using the same approach.

2. Materials

1. Callus induction medium—celery (CIM): MS salts (6, and *see* Appendix) supplemented with 3% sucrose; 0.5 mg/L 2,4-dichlorophenoxyacetic acid (2,4-D); 0.6 mg/L kinetin and 1% agar.
2. Callus induction medium-lovage (CIL): MS salts supplemented with 3% sucrose; 2.0 mg/L 2,4-D; 2.4 mg/L kinetin and 1% agar.
3. Callus differentiation medium (CDM): MS salts supplemented with 3% sucrose and 1% agar.
4. Growth and sporulation medium for *Septoria apiicola*: Prune-lactose-yeast medium (PLY), made up as follows: concentrated prune extract 100 mL; lactose 5 g; yeast extract (Difco) 1 g; agar 30 g; water to 1 L.

   For concentrated prune extract simmer 50 g of chopped dried prunes with 1 L water until soft. Strain the mixture through muslin, filter the solution, and make up to 1 L with water. This extract may be sterilized in convenient quantities in plugged flasks and stored at 4°C until required.
5. Lactophenol-cotton blue for microscopic examination of fungal hyphae (10 g phenol, 0.02 g aniline blue, 10 mL glycerine, 10 mL lactic acid, 10 mL distilled water).

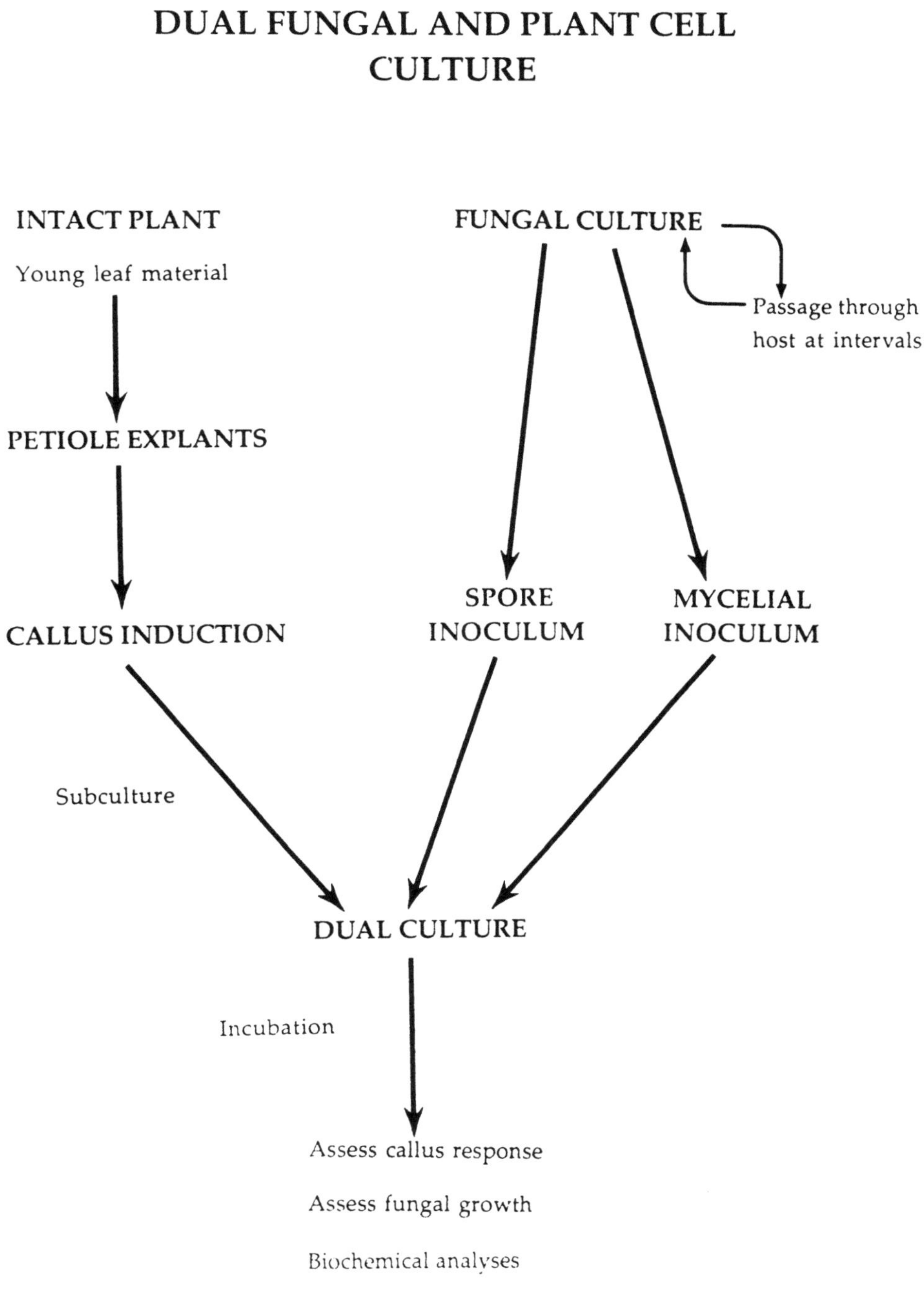

Fig. 1. Summary protocol for dual fungal and plant cell cultures.

3. Methods

3.1. Preparation of Fungal Inoculum

1. Grow *Septoria apiicola* on prune-lactose-yeast (PLY) medium in Petri dishes by placing inoculum centrally. Incubate at 20°C for 4 wk.
2. Harvest spores from 4-wk-old cultures by washing with sterile distilled water (2 x 10 mL). Determine spore concentration by counting in a hemocytometer. Place prepared suspension at 4°C for 1.5–2 h to encourage germination.
3. Alternatively, prepare mycelial inoculum by cutting disks of culture and agar (0.5 cm diameter) from the growing regions, immediately behind proliferating hyphal tips, of 4-wk-old *S. apiicola* colonies (*see* Note 1).

3.2. Inoculation of Celery with Septoria

1. At intervals, a cultured fungal pathogen should be passed through host tissue in order to maintain virulence in the culture. Make a spore suspension as above, but adjust to give a concentration of 10^6 spores/mL.
2. Inoculate young celery plant by spraying immature, unexpanded leaves with spore suspension in an atomizer or hand sprayer. Maintain plants at 15°C.
3. Isolate fungus back on to PLY medium after 2–3 wk, using single-spore isolation. Place a sample of spores into sterile distilled water and pregerminate as above. Make a dilution series and transfer samples, as appropriate, to give less than 10 spores/Petri dish. Incubate and then subculture colonies arising from single spore isolates, and maintain cultures as above.

3.3. Establishment of Callus Cultures

1. Surface sterilize young leaves and petioles of plant material. Place 1-cm petiole explants onto appropriate callus induction medium to produce undifferentiated callus material.
   a. Celery (*Apium graveolens*) var. Celebrity—CMC medium.
   b. Lovage (*Levisticum officinale*)—CML medium.
2. Differentiate callus, if required, by subculturing onto callus differentiation medium (CDM).

3. Grow in small jam jars sealed with aluminum foil at 20°C under 12 h light/dark cycle. Subculture at 4-wk intervals (*see* Note 2).

3.4. Establishment of Fungal–Plant Cultures

1. Inoculate appropriate callus (undifferentiated or differentiated; 7 d after subculture) with *S. apiicola* using either inoculum source:
   a. Spore suspension. Place 0.5 mL of suspension containing 5×10^5 spores/mL onto callus surface.
   b. Agar plug inoculum. Place plug centrally on top of callus.
2. Incubate at 20°C in 12 h light/dark cycle for 28 d. Examine frequently.

3.5. Assessment of Dual Cultures

1. Carefully slice through callus, and mount some tissue on glass slides in lactophenol, or lactophenol-cotton blue to enhance contrast, if necessary. Examine using light microscopy and assess fungal colonization of tissue subjectively (e.g., amount of surface cover; amount of aerial mycelium). Determine the extent of hyphal development within callus tissue.
2. Assess response of callus tissue subjectively, using a rating system to score the degree of browning.
3. Mount, and observe colonization and degree of hyphal penetration of both callus tissue and individual plant cells, using a scanning electron microscope. Assess the physical characteristics of the tissue and the extent of any structural changes occurring after inoculation and during incubation.

3.6. Preparation of Dual Cultures for Scanning Electron Microscopy

1. Dehydrate portions of callus inoculated with fungus by passing through an alcohol series. Place tissue into increasing concentrations of alcohol (20, 40, 60, 80, 90, 100%), allowing at least 30 min at each concentration and two changes of absolute ethanol.
2. Critical point dry material using CO_2. (A Polaron E300 critical point dryer was used here.)
3. Mount on stub for viewing, and coat with 60% gold/paladium. View in a scanning electron microscope. Figures (Fig. 2a–f) presented here were taken using a Phillips 501 B with variable accelerating voltages (7.2 or 15 kV) (*see* Notes 3–9 for analysis).

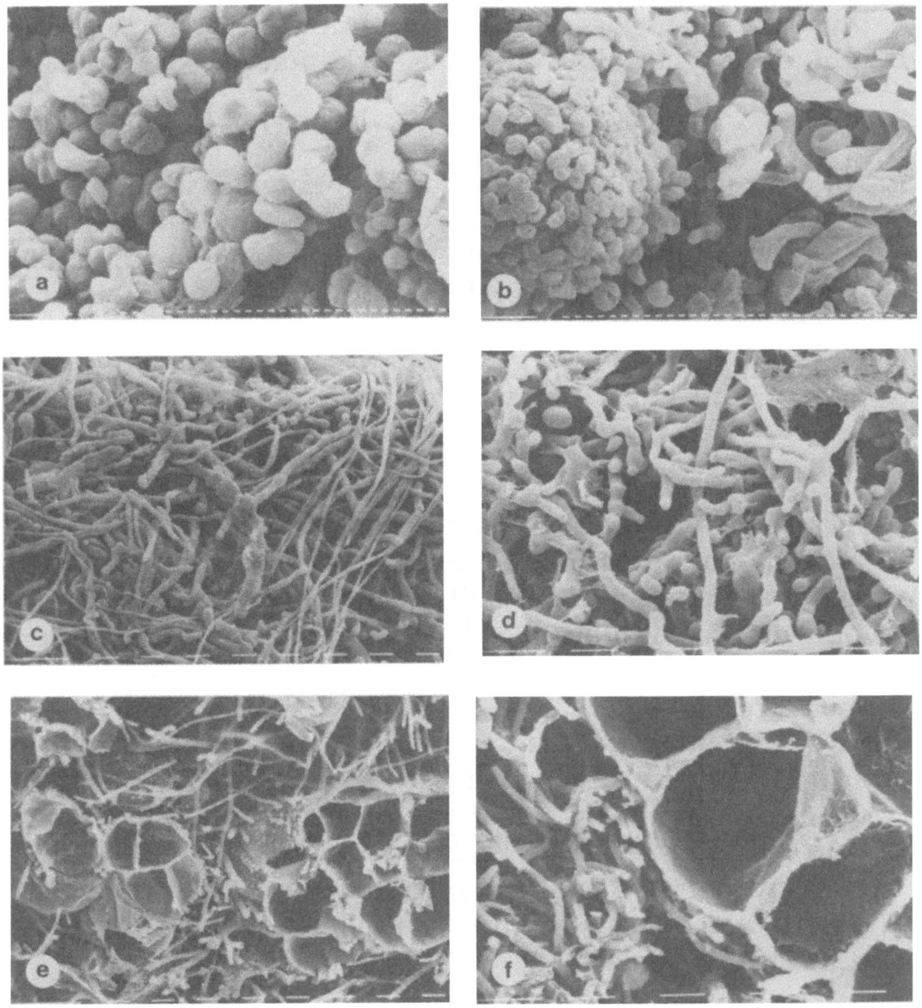

Fig. 2. Scanning electronmicrographs (SEM) of dual fungal and plant cell cultures: Uninoculated celery (a) and lovage (b) callus tissue; *Septoria* mycelium growing over celery (c) and lovage (d) callus tissue; sections through celery (e) and lovage (f) callus to show intercellular hyphal growth. Bar markers represent 10 μm (acknowledgments to K. Veltkamp for photography).

4. Notes

1. For the testing of pathogen virulence or host resistance, it is important to infect and reisolate the pathogen from initially healthy host plants, in order to maintain the aggressiveness of the isolates. Several passages through host tissue may be required if the fungus has been maintained in laboratory culture for a prolonged period.

2. In subcultured callus, chromosomal instability occurs as a consequence of prolonged culture, and structural changes to chromosomes have been reported (*7*). It is important to include sufficient replicate treatments to minimize the influence of any such aberrations.
3. Uninfected tissue cultures of celery were composed of uniform cell types of regular shape and size (Fig. 2a). Uninfected lovage callus was also composed of uniform cells, but these appeared long and thin in comparison (Fig. 2b). Inoculation of both celery (Fig. 2c) and lovage (Fig. 2d) callus with either fungal spores or mycelium resulted in the growth of fungal mycelium over the surface of callus pieces. Extensive growth into callus tissue, between the cells, was seen (Fig. 2e,f). All hyphae were intercellular, no cell penetration was observed in any preparations. No pycnidial formation occurred during the 28-d incubation period.
4. When differentiated callus was inoculated with fungus, extensive growth of mycelium occurred on the tissue surface, but no penetration between the callus cells was observed. This may be a response of the fungus to secondary compounds present in the differentiated tissues.
5. No callus cells exhibited any morphological changes as a consequence of inoculation with fungal spores or mycelium, or as a result of hyphal growth.
6. Callus cultures are useful for the study of biochemical events (particularly those that occur early after pathogen arrival, because long-term balanced cultures are not required). Extraction processes can be carried out relatively easily with such material, which is primarily composed of one major cell type. Additionally, tissues lack chlorophyll and polyphenols that can often interfere with biochemical procedures.
7. Undifferentiated callus cultures do retain the capacity for phytoalexin production. In some cases, however, the presence of plant growth substances in the callus medium may be inhibitory to the production of phytoalexins (*3*).
8. For celery, the resistance to *Septoria* has been linked to flavor compound levels (*5*). In preliminary experiments in our laboratory, healthy and infected callus tissues were shown by GLC analysis to contain similar concentrations of essential oil. However, the concentrations of the individual components differed. β-eudesmol levels were more than 20-fold higher in infected callus than in unin-

oculated material. Additionally, other compounds, normally detected at trace levels, were present in enhanced concentrations (Donovan, unpublished).

9. Mechanisms of resistance expression may not be the same as those in the intact plant, so information obtained should always be related to intact material, as far as possible.

References

1. Ingram, D. S. (1980) Tissue culture methods in plant pathology, in *Tissue Culture Methods for Plant Pathologists* (Ingram, D. S. and Helgeson, J. P., eds.), Blackwell Scientific Publications, Oxford, UK, pp. 3–9.
2. Callow, J. A. (1987) Models for host pathogen interaction, in *Genetics and Plant Pathogenesis* (Day, P. R. and Jellis, G. J., eds.), Blackwell Scientific Publications, Oxford, UK, pp. 283– 295.
3. Dixon, R. A. (1980) Plant tissue culture methods in the study of phytoalexin induction, in *Tissue Culture Methods for Plant Pathologists* (Ingram, D. S. and Helgeson, J. P., eds.), Blackwell Scientific Publications, Oxford, UK, pp. 185–196.
4. Johnson, R. (1987) The challenge of disease resistance, in *Genetics and Plant Pathogenesis* (Day, P. R. and Jellis, G. J., eds.), Blackwell Scientific Publications, Oxford, UK, pp. 311–325.
5. Donovan, A., Collin, H. A., and Isaac, S. (1986) Selection for resistance to late blight, in *Secondary Metabolism in Plant Cell Cultures* (Morris, P., Scragg, A., Stafford, A., and Fowler, M., eds.), Cambridge University Press, Cambridge, UK, pp. 244–249.
6. Murashige, T. and Skoog, F. (1962) A revised medium for rapid growth and bioassays with tobacco tissue culture. *Physiologia Plantarum* **15,** 473–487.
7. Murarata, M. and Orton, T. J. (1983) Chromosome structural changes in cultured celery cells. *In Vitro* **19,** 83–89.

Chapter 38

Mutagenesis Techniques in Plant Tissue Cultures

Joan M. Nelshoppen and Jack M. Widholm

1. Introduction

Mutant plant cell cultures can be useful in the study of the physiology and genetics of plants, as well for the improvement of crops. For recent reviews of mutant cell selection from tissue cultures, the reader is referred to Duncan and Widholm (*1*), Bright et al. (*2*), Flick (*3*), and Chapters 39–42, in this vol. Often these mutants have been obtained through selection procedures exploiting spontaneous mutations. However, the use of mutagens in culture systems has been shown to result in higher frequencies of mutant cells. Although mutagens can increase the frequency of the desired mutation in a population of cells, the frequency of undesirable mutations would also be increased. These undesirable mutations can make the produced material unusable because of properties such as slow growth, lack of plant regeneration, and sterility. Thus, one has to weigh the benefits of mutagenesis against the detrimental effects.

From: *Methods in Molecular Biology*, vol. 6, *Plant Cell and Tissue Culture*
Edited by Jeffrey W. Pollard and John M. Walker, ©1990 by The Humana Press

Plant cell cultures have been selected for resistance to amino acids, amino acid analogs, antibiotics, and herbicides at increased frequencies following chemical and physical mutagen treatments (*4,5,6,7*). It is also possible to regenerate plants from resistant calluses that carry the same resistance. In the case of Bourgin (*7*), valine-resistant tobacco plants were recovered from resistant protoplasts previously treated with ultraviolet radiation. The trait is transmissible through sexual crosses and, thus, is controlled by a stable mutation.

Cell cultures resistant to aminopterin and azaguanine have been isolated following treatment with the alkylating agents ethyl methanesulfonate (EMS) and *N*-methyl-*N*'-*N*-nitrosoguanidine (MNNG) (*8,9*). Colijn et al. (*10*) found a significant increase in the number of petunia calluses resistant to mercuric chloride and DL-6-fluorotryptophan after treatment with MNNG. The frequency of *Nicotiana sylvestris* suspension cultures resistant to the antibotics oligomycin and chloramphenicol increased 5–50x upon treatment with EMS (*11*). *Nicotiana tabacum* cell lines and plants resistant to the herbicide paraquat have been obtained from cells treated with X-rays (*12*).

Often the mutagens used in these experiments are alkylating agents, such as MNNG, *N*-ethyl-*N*-nitrosourea (ENU), and methyl methanesulfonate (MMS). A common chemical used with seeds is the promutagen sodium azide (NaN_3), which must be metabolized by plant cells to the mutagenic agent, presumably azidoalanine (*13*), to be mutagenic. There is some evidence that not all species in tissue culture (e.g., maize) are able to metabolize sodium azide to the mutagenic agent (*14,15*). Physical mutagens commonly used with plant tissue cultures are X-rays, γ rays, and ultraviolet (UV) rays. X-rays and γ rays can break the strands of DNA, whereas UV increases the occurrence of pyrimidine dimers.

The ploidy number of the tissue to be treated is also of concern in mutagen experiments. If one wishes to produce recessive mutations, haploid cultures must be used. For example, auxotrophs are normally recessive mutants. Auxotrophs are unable to synthesize necessary compounds, usually owing to the lack of an enzyme, and must be supplemented with the missing nutrient. Gebhardt et al. (*16*) found ten auxotrophic strains of *Hyoscyamus muticus* requiring histidine, leucine, tryptophan, glutamine, and asparagine, or the vitamin nicotinamide using MNNG-treated haploid protoplasts. However, other mutants can be selected using diploid or higher ploidy strains if the mutation being selected for is a dominant trait. For example, resistance to the tryptophan analog 5-methyltryptophan is normally a dominant trait and can be selected from diploid cells of several species including *Datura innoxia* (*6*).

This paper details methods for using two typical mutagens, a chemical agent (ENU) and a physical agent (γ radiation from a ^{137}C source), in the treatment of soybean embryos to initiate regenerable cultures in this lab. Safety is of great concern when working with mutagens, and some general safety guidelines for working with chemical agents and γ irradiation are given in sections 2.1. and 3.1. respectively. Preparation of a chemical mutagen stock, mutagenic treatment of tissues using proper safety procedures, and removal of the chemical mutagen are outlined in section 2. Procedures for using a γ radiation source are outlined in section 3. In section 4, the notes section, evaluation of treated tissues and the proper disposal of chemical mutagen waste are discussed. These procedures, with minor modifications, are applicable to most tissue culture systems.

1.1. Chemical Mutagenesis

An experiment using the chemical mutagen ENU in the treatment of soybean embryos is outlined below. ENU is an effective alkylating agent in plant tissue cultures (*17,18,19*) that has the added advantage of being rapidly degraded under alkaline conditions, thus leaving no toxic residue.

Though ENU is used as the chemical mutagenic agent in the experiment below, the same procedures can be followed using different chemical agents with selection of an appropriate quenching buffer. Suspension systems would obviously need more modifications than some tissue culture systems, and these adjustments are discussed.

In this experiment, immature embryos are treated with ENU in a time vs concentration experiment. A range-finding experiment, such as this, is normally used as a preliminary test to explore the effects of the mutagen and to determine a level of mutagen to be used in future experiments. A wide range of concentrations should be tested in the preliminary experiment, since a narrow range may include only concentrations much too high or low to be effective. The range can then be narrowed as one gathers more information. For example, in a prior experiment we used levels of 0.1, 10, and 100 m*M* ENU with the embryos. From the results of that experiment, it was determined that the desired level would be below 10 m*M* ENU. This experiment, which investigates the effects of 0–10 m*M* ENU on soybean embryos, was then initiated to further discern the concentration of ENU to be used in future experiments.

1.2. Safety with Chemical Mutagens

When working with mutagens, it is very important to take strict safety precautions to avoid unnecessary exposure. Unlike those for radiation, there are usually no quick tests to tell if a chemical mutagen has accidentally been left on a surface or piece of equipment. The best solution to this problem is to be very careful in the handling of the mutagen, the treated tissue, and everything that came or could have come into contact with the mutagen. Some general safety guidelines are as follows:

1. Always wear a respirator, lab coat, and disposable gloves when conducting an experiment with a mutagen. A face mask is not a respirator and does not provide adequate protection.
2. Two pairs of gloves can be worn. Besides providing the obvious increased protection, the outer pair of gloves can be removed when one wants to handle something (such as a pen) and be confident that the item has not been contaminated. Change gloves when they are contaminated, and be very aware of what items are handled and how they are handled. Gloves should never be disposed of in a normal trash can, but should be put with solid chemical waste in the hood.
3. All work should be done in a fume hood or in a closed glove box with external exhaust. This is especially important when the mutagen is volatile, as with EMS.
4. The surface area of the hood should be covered with plastic-lined absorbent paper that protects against spills and can be disposed of following the experiment.
5. Use as many disposable items as possible to avoid contaminating labware. Disposable Pasteur pipets, Eppendorf pipet tips, gloves, plastic test tubes, and so forth, should be used. A paper clip can replace a magnetic stirbar. Glassware may be used, provided it is soaked in an acid bath before washing (*see* step 9). Metal items should not be used in a mutagen experiment unless one is willing to dispose of them (e.g., a paper clip) or soak them in an acid bath. The soaking of glassware in an acid bath allows one to have some control over toxic residues.
6. It is important to have all necessary items in the fume hood or a nearby area before initiating an experiment. (If space is limited, a cart works well.) Aside from the convenience, this prevents the contamination of other lab items and of the lab in general. The proper neutralizing agent should be kept in the fume hood whenever one works with a mutagen.

7. If a spill occurs, wipe it promptly using the proper neutralizing agent. Many mutagens, such as ENU, MNNG, EMS, and MMS, degrade under alkaline conditions and can be neutralized by wiping with a paper tissue to which 1*N* KOH has been applied. (For more information and *alternatives, see* Note 6.) When cleaning a spill, keep the size of the contaminated area to a minimum. A spill on the disposable surface does not need to be wiped up, since the plastic lining will keep it contained. However, be careful not to set anything on the spot where the mutagen has been spilled.
8. Chemical mutagen waste should always be separated from normal waste. Dispose of liquid waste in a waste jug kept in the fume hood. Solid waste such as disposable Pasteur pipets, kimwipes, gloves, and so forth, should be placed in a plastic bag located in the fume hood. (An empty Petri dish sleeve propped up in an agar can work well.) Both solid and liquid waste should be disposed of by the proper authorities. Alternatively, by following the procedures in Note 6, the scientist may treat the waste and dispose of it.
9. All used glassware should be soaked in an acid bath for a minimum of 24 h before washing. A sulfuric acid bath, rather than a chromic acid bath, is recommended since chromate is toxic and mutagenic to cells. The soaking of glassware in an acid bath lowers the amount of mutagenic residue on the glassware.

2. Materials

2.1. Mutagenesis with ENU

1. *N*-ethyl-*N*-nitrosourea (ENU) stock (dissolved in DMSO) can be stored at 0°C in the dark indefinitely. ENU is light-sensitive, so it should be covered. A stock of 1*M* ENU (or 0.1*M* if very low concentrations are to be used) is sufficient for most experiments. ENU stock should be prepared beforehand according to the procedure outlined in the next section.
2. As with all tissue culture systems, sterility must be maintained. Autoclaved glass vials or sterile disposable plastic test tubes should be used to hold the tissues to be treated. Disposable Pasteur pipets and Eppendorf pipet tips should be autoclaved beforehand and kept in sterile containers that can be reclosed (e.g., a beaker with an aluminum foil covering).

3. Glass vials (scintillation vials or glass bottles with screw tops and a volume of approximately 10–15 mL) to be used for incubating the plant tissues with the mutagen, should be filled with the appropriate amount (2–5 mL) of liquid medium beforehand. (Disposable sterile plastic test tubes with screwcaps may also be used.)
4. Modified liquid Murashige and Skoog medium (*20* and *see* Appendix) with no growth hormones.
5. OR or EB media of Barwale et al. (*21*) for culturing of organogenic and embryogenic calluses and later, MSR (*21*) or TH (*22*) media (*see* Appendix).

3. Methods

3.1. Mutagenesis with ENU

3.1.1. Stock Preparation

1. Weigh 1.5-mL microcentrifuge tubes individually on a balance. The platform and surface of the balance should be covered with disposable paper or foil, to be removed when finished.
2. While working in the fume hood and wearing protective clothing and respirator, carefully add ENU to the microcentrifuge tubes, taking pains not to spill any. Add approximately 0.1 g to one tube and tightly close it.
3. Weigh the first tube on the balance, and calculate the amount of ENU contained in it. Add or remove ENU (in the hood) if necessary. Add 0.1 g to the rest of the microcentrifuge tubes and weigh them.
4. Calculate the amount of DMSO to be added to each tube to make the stock solution. The mol wt of ENU is 117.1 g, so for a 1*M* ENU stock:

$$\text{mL DMSO} = \frac{\text{g ENU in tube}}{0.1171 \text{ g}} \tag{1}$$

   Add the appropriate amount of DMSO to each tube while in the fume hood.
5. Tightly close the tubes and mix with a vortex mixer.
6. Seal each with parafilm. Keep the tubes in a small container, such as a beaker. Seal this with parafilm and cover with aluminum foil as ENU is light sensitive. Store at 0°C until ready to use.

3.2. Preparation of Embryos

Tissues to be treated should be prepared according to standard procedures. The protocol followed for soybean embryos is that of Barwale et al. (*21*).

1. Soybean pods are harvested, surface sterilized in a sodium hypochlorite solution prepared by diluting commercial bleach, and washed twice with sterile water.
2. Embryos (4–6 mm) are excised and the seed coats removed using sterile procedures in a laminar flow hood.
3. Embryos are cut into cotyledon pieces and axis pieces, which also have small amounts of cotyledonary tissue present.
4. Ten axis pieces and 15 cotyledon pieces are placed into scintillation vials containing 2 mL of liquid MS media immediately prior to the experiment (*see* Note 1).

3.3. Treatment with ENU

1. All calculations of ENU amounts to be added to vials should be done prior to the start of the experiment. In this experiment, final concentrations of 0, 0.1, 0.3, 1.0, 3.0, and 10.0 m*M* ENU are desired. For final concentrations of 0.1, 0.3, and 1.0 m*M* ENU, 2, 6, and 20 µL of 0.1*M* ENU stock will be added to the respective vials for a total vol of 2 mL. For 3.0 and 10.0 m*M* concentrations, 6 and 20 µL of 1.0*M* ENU stock will be added, respectively. (For easy reference during the experiment, make a list of the molarity and volume of ENU stock to be added to each vial, and tape this list to the outside of the fume hood.)
2. Prepare the hood completely with a disposable surface (plastic-backed absorbent paper) and a plastic bag for solid waste. (An empty Petri dish sleeve propped up in an agar can work well.) The proper neutralizing agent should also be in the hood. For ENU, which is degraded under alkaline conditions, this is 1*N* KOH. Place the frozen ENU stock in the hood to thaw approximately 20 min prior to use. The fume hood should be on.
3. Make sure that one is wearing a respirator and protective clothing before opening the ENU microcentrifuge tube. After removing the parafilm, open the tube using a paper tissue to contain droplets of ENU stock to keep from contaminating the gloves (thereby contaminating other utensils).

4. Using an Eppendorf pipet and sterile pipette tips, add the proper amount of ENU to each tissue-containing vial (*see* 1 above). Dispose of tips in the plastic solid chemical waste bag between concentrations. Inoculate all vials that will contain the same concentration (e.g., 10 m*M*) at one time, even if exposure time is to be different. Loosen the caps of all the vials before beginning pipeting of the ENU. Using an Eppendorf pipet rather than regular pipets provides for more accuracy, efficiency, and safety. If conventional pipets are used, it is very important that there be no pipeting of mutagenic liquids by mouth.
5. Sterility can be maintained in the fume hood if one uses proper techniques. Do not leave sample vials and containers for sterile tips, Pasteur pipets, and so on, open any longer than needed. Work quickly but not so quickly as to be sloppy.
6. Cap the vials containing the mutagen and plant tissues tightly and shake (100 rpm) for the required time. In many cases, it will not be possible to set up a shaker in the fume hood. The vials can be removed from the fume hood, provided one has been very careful when adding the ENU and does not suspect contamination on the outside of the vials. However, if contamination is suspected, the outside of the vial(s) should be wiped with 1*M* KOH before removal from the hood and placement on the shaker.
7. After adding ENU to the third set of samples, wrap the microcentrifuge tubes in parafilm. Cover the beaker with parafilm and foil. Store in a freezer at 0°C.

3.4. Removal of ENU and Washing of Tissue

1. After the 1-h incubation period, remove the liquid from the appropriate vials using Pasteur pipets and dispose of the liquid in a waste jug. Hold pipets firmly so as not to spill drops of ENU-laden liquid. Use a separate Pasteur pipet for each vial. The liquid is removed by suction, rather than pouring, to avoid spillage.
2. To rinse the tissue, add 2 mL MS medium to each vial (or the original treatment volume). Use a 5-mL Eppendorf pipet rather than regular pipets. It is not necessary to change tips between vials if one works quickly enough. Swirl occasionally for 10 min.
3. After 10 min, remove the first wash in the same manner as above. Repeat the washing procedure a second time.

4. After the second wash, the concentration of ENU in the vials should be down to levels safe enough to be handled outside of the fume hood. However, proper care should still be taken. Vials may now be moved to a laminar flow hood for plating on induction medium.
5. Repeat these procedures for the 2- and 3-h treatments.

3.5. Plating on Induction Medium

1. After the second wash, plate the cotyledon pieces on EB medium and the axis pieces on OR for the induction of embryoids and shoots. Prolonged exposure in liquid medium can have detrimental effects on plant tissues, so it is best to do the plating as soon as possible.
2. Use forceps to remove pieces from vials. Cap the vials tightly and place in the disposable waste bag.
3. Wrap the plates and store for 3 wk in the dark at 27–28°C, or normal culture conditions (*see* Notes 2–5).

3.6. Clean-Up

1. Do not put any equipment back into lab circulation unless it is certain that it is not contaminated. If it is possible that the equipment was contaminated, wipe with 1*M* KOH to neutralize the ENU and wash with a strong detergent.
2. Soak nondisposable glassware in a sulfuric acid bath for at least 24 h before washing.
3. Dispose of the absorbent surface in the disposable solid waste bag. Wipe the hood surface with a strong detergent. Dispose of the gloves in the solid waste bag and turn off the fume hood. Contact the proper authorities for removal of both the chemical and solid ENU waste or, if one wishes to dispose of the waste oneself, refer to Note 6 for details on the treatment and disposal of chemical mutagen waste.

3.7. Modifications for Treating Suspension Cultures

The above procedures, after minor adjustments, are suitable for conducting chemical mutagenesis experiments with most tissue culture systems. Suspension cultures are commonly treated with mutagens, and a few modifications of the above procedures are necessary.

1. Standardize the cells to be treated by always using cells in the same phase of growth. It is recommended that the investigator treat cells

that are rapidly growing and dividing, such as those in mid-log phase. This is especially important for mutagens, such as 5-bromodeoxyuridine, which must be incorporated into the DNA during DNA replication.

2. When treating cultured suspension cells, place the cells in sterile polypropylene conical test tubes with caps, which can be centrifuged.
3. It is necessary to titer the suspension, so that the same number of cells are treated in each test tube. This may be done by fresh weight, packed cell volume, cell counts, or turbidity, depending upon the methods used in the laboratory. The density of cells in a flask is determined, and medium or buffer is added until a standard number of plant cells/mL is obtained.
4. With suspensions, it is often possible to stop the action of the mutagen by the addition of ice-cold quenching buffer to the test tubes containing the plant cells and the mutagen. The quencher must, of course, be nontoxic to the plant cells. Sodium thiosulfate (0.1% in 0.1*M* sodium citrate buffer, pH 7.2), is an effective quenching buffer for an alkylating agent, such as ENU, MNNG, EMS, and MMS. Even if a quencher is used, care should still be taken when removing liquids from the test tubes.
5. To remove liquid from treated suspension cultures, it is first necessary to pellet the cells by gentle centrifugation (3 min at 140 x *g*). Liquids can then be removed from the cells, following the above procedures. When washing the cells, resuspend them in fresh liquid medium, and place the tubes on a shaker before another centrifugation and rinse step.
6. With suspension cultures, it is especially important that the cells be rotating on a shaker in such a manner so as to suspend the cells in the liquid during the 1–3-h incubation period and the washes. Not only does this provide for a more even exposure to the mutagen and washes, but the cells are aerated. If a suspension culture is thick and the cells are not aerated, cell death may occur.

3.8. Physical Mutagenesis

3.8.1. Safety with Gamma Radiation

It is beyond the scope of this chapter to list all the important safety information with which one should be familiar when working with radiation. Though some general safety recommendations will be discussed, the reader is strongly encouraged to seek further information on the type of radiation being worked with and complete safety procedures. *Radiation Protection* by Shapiro (23) is an excellent resource, providing practical information for working with many radiation sources.

The experiment outlined below uses a point source of γ radiation (^{137}Cs). γ rays are indirectly ionizing particles that lose energy through chance encounters with atoms, resulting in the ejection of electrons from these atoms. They will travel through a medium until they undergo such chance encounters and their energy is exhausted. How far a ray will travel depends on many factors, including the energy of the γ ray, the composition of the medium, and the density of the medium.

Since it is important to avoid exposing oneself to γ rays, make sure that all safety mechanisms of the source are functional before using it. γ sources have a chamber in which the material to be treated is placed. The γ source is surrounded by a thick shielding of lead. It should not be possible to bring the source out of the lead shielding up to the chamber unless the chamber door is closed. It is important that the shielding be of the appropriate thickness. Since γ rays are indirectly ionizing, a fraction of the rays penetrate the shield. As the thickness of the shield increases, the fraction becomes less. A shield appropriate for one intensity of γ rays may not be appropriate for other intensities.

Monitoring of the area in which the γ source is used should be done with a Geiger counter, which detects the emission of ionizing particles. Personal monitoring badges should also be worn when working with a γ source.

3.8.2. Planning the Experiment

It is necessary to plan the experiment according to the confines of the source one is working with. For instance, the γ chamber in the source used by this lab has an elevated, rotating platform requiring vials and test tubes to be less than 10 cm tall in order to fit into the chamber. For this reason, sterile 17 x 100 mm polystyrene test tubes (16-mL capacity) with caps are used. A special Plexiglass™ holder has been fashioned to support a test tube while allowing it to rotate on a platform in the chamber. The rotating platform is useful in providing an even exposure to all tissues.

The times of exposure should be calculated prior to the experiment. A decay table accounting for radioactive decay of the source should be consulted. Absorbed doses are expressed in units of rads or grays. The rad is the traditional unit of absorbed dose and is equal to 62.4×10^6 MeV/g tissue. The gray (gy), the unit under the International System (SI) of units, is equal to 1 joule/kilogram (1 J/kg). One gray equals 100 rads. The intensity of γ rays from a point source decreases proportionally to the square of the distance traveled. If there is more than one position in the chamber, calculations should be made for the position that will be used. A sample calculation is as follows: If the desired dose is 1 krad (1000 rad) and, according to the decay table, the γ source emits 5 krad/min at position II, the time of exposure should be 0.2 min or 12 s for tissue at this position (1/5 = 0.2).

3.9. Treatment of Tissues with Gamma Radiation

1. Prepare the tissues following standard procedures. Work should be done in a sterile laminar flow hood. Place tissues in sterile polystyrene tubes (17 x 100 mm) with caps, or small autoclaved glass vials (approximately 15 mL). Embryos are sensitive to desiccation, and so a few drops of liquid medium are added to each vial to moisten the embryos.
2. After turning on the machine, check that the safety mechanisms are functional.
3. Set the timer for the first desired dose. Dosages applied to plant tissue or cell cultures are often up to 10 krad. As with chemical mutagens, a range of different dosages should be explored in a preliminary experiment.
4. Position the rotating platform in the chamber and place the test tube onto the platform. Switch the turntable on. Close the chamber door and securely latch.
5. The γ source in use by this lab requires some hand operation of the source of radiation, the cesium rod. The cesium rod is brought up from the lead shielding to the level of the chamber using a lever. The chamber and anything in the chamber are then exposed to the cesium rod. When the time set on the timer has elapsed, the cesium rod automatically drops down into the lead shielding. When bringing the rod up to the chamber, it is best to work quickly and smoothly.
6. Repeat these procedures for all test tubes, and plate embryos on induction media in a laminar flow hood.

These procedures can be applied to any tissue culture system that can be placed in the chamber (*see* Note 7).

4. Notes

1. When modifying these procedures for use with other tissue culture systems, it is not necessary to use 25 pieces in 2 mL. Adjust the number

of tissue pieces and amount of medium so that all pieces of tissue are submerged in the medium. For the initial experiment, it may be wise to keep the number of pieces low until one is more experienced with the procedures. It is important to keep the tissues to be treated as uniform as possible. In this experiment, this is accomplished by choosing embryos of approximately the same size (4–6 mm in length). This accomplished two things: the size of the pieces to be treated is standardized, and the pieces (embryos) are of approximately the same age (immature soybean embryos of the same size are approximately the same age postanthesis). Standardization of the number, size, and total weight of the pieces to be treated will allow the same amount of mutagen to be available to each piece, and make the experiment reproducible at a later time.

For other cultures, such as organogenic shoots, weights may be used for standardization. The pieces should be approximately the same size, and the total weight/vial of the explants should be standardized. A few explants should be weighed individually to familiarize the investigator with the approximate weight of the explants. It is also important that callus pieces to be treated are uniform in terms of length of time in culture, length of time since the last transfer, and media last transferred from.

2. Treated tissues are initially evaluated at 3 wk for toxicity from the mutagen. Factors that can be scored include percentage of browning, decreased growth (measured as fresh weight of the control), percentage death, and the ability of tissues to regenerate into plants. The criteria that the investigator uses to determine the dosage for future experiments depends on the aims of the experiment. For example, if the researcher wishes to examine regenerated plants, it is important to choose a level of mutagen that does not significantly inhibit regeneration. The plant regeneration ability of maize callus (*24*) is more severely affected by γ radiation than is the growth of the same callus. If the researcher is screening for mutants resistant to inhibitory substances, the criteria for choosing the level of mutagen may be that a sufficient number of cells are viable at that level. Customarily, a level of mutagen that produces about 50% lethality is chosen, though in some cases mutants are also produced at mutagen levels that are not toxic to the tissues (*10*). If the mutagen treatment does not affect the tissue growth or viability, it is questionable if the mutagen is penetrating the tissue, and higher doses should be tested, at least to determine at what point toxicity occurs.

3. Plants regenerated from the treated tissue can be evaluated for phenotypic abnormalities. If the tissue is diploid, one may observe dominant mutations in the regenerated plants, but recessive mutations will not become apparent until progeny are grown, following self-pollination. With haploid tissue, both types of mutations may be expressed in the regenerated plants. Many tissue-culture-induced mutations expressed in regenerated plants have been found to be recessive events (25).
4. Some abnormalities in the regenerated plants and progeny may be related to epigenetic or physiological effects. Therefore, progeny from suspected mutants (termed variants until the origins of the differences are known) should be grown and examined for stable inheritance of the trait in question.
5. An alternative method for evaluating treated tissue is to plate tissues on selective medium. Mutants can be selected that are resistant to compounds normally found to be inhibitory. Before plating on screening medium, the treated cultures are grown on a normal growth medium to allow recovery and fixation of any mutations. In this type of selection scheme, there is the possibility of "escapes," calluses that appear to be resistant, but are really not and that will die upon subsequent retransfer to the selection medium. Variant cell colonies should be replated on the inhibitor medium to determine if the resistance is true.
6. Disposal of chemical mutagen waste: Both liquid and solid chemical mutagen waste should be separated from normal waste. They can be disposed of by authorities or, if the scientist prefers, treated and disposed of in the laboratory. Information on the disposal of ENU, MNNG, EMS, and MMS is presented below. For these mutagens and others, information on the hazards, reactivity, storage, and disposal of the mutagen can be found on material safety data sheets (MSDS) normally available from the chemical company from which the mutagen was ordered. Armour et al.'s *Potentially Carcinogenic Chemicals: Information and Disposal Guide* (26) is also a fine resource on the treatment and disposal of many chemical mutagens. The procedures for ENU, MNNG, EMS, and MMS waste disposal given below are taken from this reference (26) with slight modifications.
   a. Disposal of ENU and MNNG: Work should be done in the fume hood while wearing a lab coat, a respirator, goggles, and gloves. The ENU or MNNG waste (liquid, solid, or both) to be

treated should be in a beaker in the fume hood. A fresh solution of 0.3*M* potassium permanganate in 3*M* sulfuric acid should be prepared as follows: For 100 mL of this solution, slowly add 17 mL of concentrated sulfuric acid to 83 mL of cold water. Then add 4.7 g of potassium permanganate. Add a sufficient amount of this solution to fill the beaker and produce a solution of not more than 5% ENU or MNNG. Allow this to remain at room temperature for 12 h. Dilute with water, and neutralize the solution by cautiously adding 10% aqueous sodium hydroxide solution. Decant or filter the liquid into the drain with a large volume of water. Discard the solid as normal waste (*25*).

If an ENU or MNNG spill occurs on a nondisposable surface, the spill can be treated by absorbing the liquid in paper towels. Place the towels in a beaker and treat as solid waste as described above. Immediately after removing the paper towels, pour enough of a 0.3*M* potassium permanganate in 3*M* sulfuric acid solution to completely wet the contaminated area, and let stand overnight. The following day, add a 5% solution of ascorbic acid to the spill area and allow to react 15 min. Add solid soda ash to the decontaminated surface, and remove the ash using paper towels (*25*). ENU and MNNG can also be degraded using strong bases as mentioned in the text. If a spill consists of just a few drops, wipe the area with a tissue to which 1*M* KOH has been applied.

b. Disposal of EMS and MMS: Work should be done in a fume hood while wearing protective clothing, a respirator, goggles, and gloves. For disposal of liquid EMS or MMS waste, place a 100-mL, 3-necked, round-bottom flask equipped with a stirrer, condenser, and dropping funnel in the fume hood. For each 0.1 mol of EMS or MMS (12.4 g and 11.0 g respectively) to be treated, place 7.9 g (0.12 mol) of 85% potassium hydroxide pellets in the round-bottom flask, and add 32 mL of 95% ethanol rapidly. Heat the solution to gentle reflux, and add the EMS or MMS dropwise at such a rate as to maintain gentle reflux. Heat under reflux and stir for 2 h. Cool and dilute the reaction mixture with water. Wash into the drain with a large volume of water (*25*).

Solid EMS or MMS waste should be disposed of by burning. Mix the material into a combustible solvent, and burn in a furnace equipped with an afterburner and scrubber. This may have to be done by authorities, since most labs are not equipped for this (25).

Liquid or solid EMS or MMS that has spilled can be treated by covering the spill with a mixture of soda ash, clay cat litter (bentonite), and sand (1:1:1 by weight). The mixture can be disposed of by burning in a furnace with an afterburner and scrubber, or it can be treated in the fume hood. To detoxify the mixture, make a solution of 7.9 g of 85% potassium hydroxide pellets and 32 mL of 95% ethanol for each 0.1 mol of EMS or MMS (12.4 g and 11.0 g) to be destroyed. Add this solution to the ash, bentonite, sand, and EMS or MMS mixture, and heat under reflux for 8 h. Decant the cooled solution into the drain with a large volume of water and treat the solid as normal waste (25).

7. Treatment with X-rays can be done using procedures very similar to those used for γ rays. However, ultraviolet (UV) light does not penetrate tissue as deeply as γ or X-rays, and thus, only thin layers of suspension cultures or protoplasts should be treated with UV. Since UV light does not penetrate glass or plastic, the treatment would be conducted in a laminar-flow hood in open Petri dishes, using a portable UV source.

Acknowledgments

The authors gratefully acknowledge the helpful advice of Dr. U. B. Barwale and Dr. M. J. Plewa. This research was supported with funds from the American Soybean Association and the Illinois Agricultural Experiment Station.

References

1. Duncan, D. R. and Widholm, J. M. (1986) Cell selection for crop improvement, in *Plant Breeding Reviews*, vol 4 (Janick, J. ed.), AVI Publishing, Westport, Connecticut, pp. 153–173.
2. Bright, S. W. J., Ooms, G., Fouler, D., Karp, A., and Evans, N. (1986) Mutation and tissue culture, in *Plant Tissue Culture and Its Agricultural Applications* (Anderson, P. G., ed.), Butterworths, Stoneham, Massachusetts, pp. 431–449.
3. Flick, C. E. (1983) Isolation of mutants from cell culture, in *Handbook of Plant Cell Culture* (Evans, D. A., Sharp, W. R., Ammirato, P. V., and Yamada, Y., eds.), Macmillan, New York, pp. 393–441.

4. Ahmed, R., Gupta, S. D., and Ghosh, P. D. (1986) Isolation of L-ethionine resistant cell lines, *Vigna sinensis L.* after mutagen treatment. *Plant Cell, Tissue, and Organ Culture* **7,** 135–144.
5. Widholm, J. M. (1984) Induction, selection, and characterization of mutants in carrot cell cultures, in *Cell Culture and Somatic Cell Genetics of Plants,* vol 1, (Vasil, I. K., ed.), Academic, New York, pp. 563–571.
6. Ranch, J. P., Rick, S., Brotherton, J. E., and Widholm, J. M. (1983) Expression of 5-methyltryptophan resistance in plants regenerated from resistant cell lines of *Datura innoxia. Plant Physiol.* **71,** 136–140.
7. Bourgin, J. P. (1978) Valine resistant plants from in vitro selected tobacco cells. *Molec. Gen. Genet.* **161,** 225–230.
8. Shimamoto, K. and Nelson, O. E. (1981) Isolation and characterization of aminopterin-resistant cell lines in maize. *Planta* **153,** 436–442.
9. Bright, S. W. J. and Northcote, D. H. (1975) A deficiency of hypoxanthine phosphoribosyltransferase in a sycamore callus resistant to azaguanine. *Planta* **123,** 79–89.
10. Colijn, C. M., Kool, A. J., and Nijkamp, H. J. J. (1979) An effective chemical mutagenesis procedure for *petunia hybrida* cell suspension cultures. *Theor. Appl. Genet.* **55,** 101–106.
11. Durand, J. (1987) Isolation of antibiotic resistant variants in a higher plant, *Nicotiana sylvestris. Plant Sci.* **51,** 113–118.
12. Miller, O. K. and Hughes, K. W. (1980) Selection of paraquat-resistant variants of tobacco from cell cultures. *In Vitro* **16,** 1085–1091.
13. Owais, W. M., Rosichan, J. L., Ronald, R. C., Kleinhofs, A., and Nilan, R. A. (1983) A mutagenic metabolite synthesized in the presence of azide is azidoalanine. *Mut. Res.* **118,** 229–239.
14. Dotson, S. B. and Somers, D. A. (1987) Sodium azide as a tissue culture mutagen. *Maize Genet. Newsl.* **61,** 87,88.
15. Wang, A. S., Hollingworth, M. D., and Milcic, J. B. (1987) Mutagenesis of tissue cultures. *Maize Genet. Newsl.* **61,** 81–83.
16. Gebhardt, C., Schnebli, V., and King, P. J. (1981) Isolation of biochemical mutants using haploid mesophyll protoplasts of *Hyoscyamus muticus. Planta* **153,** 81–89.
17. Maliga, P. (1984) Cell culture procedures for mutant selection and characterization in *Nicotiana plumbaginifolia,* in *Cell Culture and Somatic Cell Genetics of Plants,* vol 1, (Vasil, I. K., ed.), Academic, New York, pp. 552–562.
18. Márton, L., Dung, T. M., Mendel, R. R., and Maliga, P. (1982) Nitrate reductase deficient cell lines from haploid protoplast cultures of *Nicotiana plumbaginifolia. Mol. Gen. Genet.* **182,** 301–304.
19. Cseplo, A. and Maliga, P. (1982) Lincomycin resistance, a new type of maternally inherited mutation in *Nicotiana plumbaginifolia. Curr. Gen.* **6,** 105–110.
20. Murashige, T. and Skoog, F. (1962) A revised medium for rapid growth and bioassay with tobacco tissue cultures. *Physiol. Plant.* **15,** 473–497.
21. Barwale, U. B., Kerns, H. R., and Widholm, J. M. (1986) Plant regeneration from callus cultures of several soybean genotypes via embryogenesis and organogenesis. *Planta* **167,** 473–481.
22. Barwale, U. B. and Widholm, J. M. (1988) Soybean: In vitro plant regeneration and somaclonal variation, in *Biotechnology in Agriculture and Forestry* (Bajaj, Y. P. S., ed.), Springer-Verlag, New York.

23. Shapiro, J. (1981) *Radiation Protection, Second Edition* (Harvard University Press, Cambridge, Massachusetts, and London, England), pp. 1–480.
24. Moustafa, R. A. K., Duncan, D. R., and Widholm, J. M. (1989) The effect of gamma irradiation and *N*-ethyl-*N*-nitrosourea on cultured maize callus growth and plant regeneration. *Plant Cell Tissue Organ Culture* **17,** 121–132.
25. Orton, T. J. (1984) Somaclonal variation: Theoretical and practical considerations, in *Gene Manipulation in Plant Improvement* (Gustafson, J. P., ed.), Plenum, New York, pp. 427–468.
26. Armour, M. A., Browne, L. M., McKenzie, P. A., Renecker, D. M., and Bacovsky, R. A. (1986) *Potentially Carcinogenic Chemicals: Information and Disposal Guide* (University of Alberta, Edmonton, Alberta, Canada).

Chapter 39

Mutagenesis

EMS Treatment of Cell Suspensions of *Nicotiana Sylvestris*

Jacques L. Durand

1. Introduction

Mutants are a valuable tool for solving many problems in physiology, genetics, and molecular biology. Soon after cell suspension and protoplast culture emerged as techniques in plant biology, they were applied to the isolation of selectable markers that were unavailable through classical methods using whole plant systems. This field has been pioneered by Widholm in 1972 (*1,2,* and *see* Chapter 38 and 40 this vol.) and reviewed several times (*3–8*). As far as selectable markers are concerned, mostly nuclear-coded mutations have been characterized, with the exception of chloroplast-encoded streptomycin resistance (*9,10*) and lincomycin resistance (*11*). No selectable mutations are yet known for the mitochondrial genome in higher plants.

Spontaneous mutations to give resistance to defined inhibitors are rare, even in tissue culture: their frequency is approximately 10^{-7} (*12–14*). Hence, mutagenic treatments are used to increase the frequency of desired mutations. Among chemical mutagens, EMS (Ethyl Methane Sulfonate, a common name of methanesulfonic acid ethyl ester) is widely

From: *Methods in Molecular Biology*, vol. 6, *Plant Cell and Tissue Culture*
Edited by Jeffrey W. Pollard and John M. Walker, ©1990 by The Humana Press

used with plant material. Its high efficiency has been recognized for a long time on whole plants of various species (*15–20*). It is also an efficient mutagen of mitochondrial genes in yeast (*21*), and it has been shown to induce mutations of mitochondrial genes governing cms in maize (*22*). The lethal effect of EMS on cultured plant cells has been studied in several species (*23–28*).

The primary lesion resulting from EMS in DNA is the addition of an ethyl radical (CH_3CH_2–) to either component of the macromolecule. Mutations result from erroneous restoration of DNA by repair mechanisms, as shown on bacteria and mammalian cells (*29,30*). These mechanisms also exist in plants (*31*). The probability of wrong repair is highest when the cells are engaged in DNA replication. Hence, a mutagenic treatment is most effective when applied to actively dividing cells (*32*).

The system described here was devised to meet several requirements:

1. For mutagenic treatment efficiency, the cell suspension must be finely dispersed to allow homogeneous access of the mutagen to every cell, and cells must be in the S phase of the cell cycle.
2. For efficiency of mutant cell selection, mutations must be fixed in the genome, the mutant phenotype must be expressed, wild-type cells must be eliminated by the selection, and the mutant cells recovered.
3. Control of cell density at plating is essential for a reproducible inhibition of wild-type cells, since inhibition of plating efficiency decreases when initial cell density increases (*33*).

The characteristics of plant cells make it difficult to meet these requirements: the cell wall keeps daughter cells together, leading to cell clumps of different sizes; the heterogeneity of the cell suspension is increased by differences in cell size, in cell cycle position, and even in chromosome number.

This heterogeneity can be minimized by using the properties of the cycle of cell suspension in batch culture (*34,35*). Figure 1 shows the phases of this "suspension cycle":

0: Onset of culture; a minimal cell density is necessary for growth, because of the "conditioning" effect (*35,36*); 1: Lag phase; absorption and storage of nutrients; duration depends on the inoculum, 24–48 hr in this case; 2: Beginning of cell division; smaller cell size and formation of clumps; 3: Cell enlargement and dissociation of clumps; 4: Plateau: beginning of cell decay; cell division ability decreases. These characteristics of batch culture have been used to devise a system that meets the different requirements mentioned above:

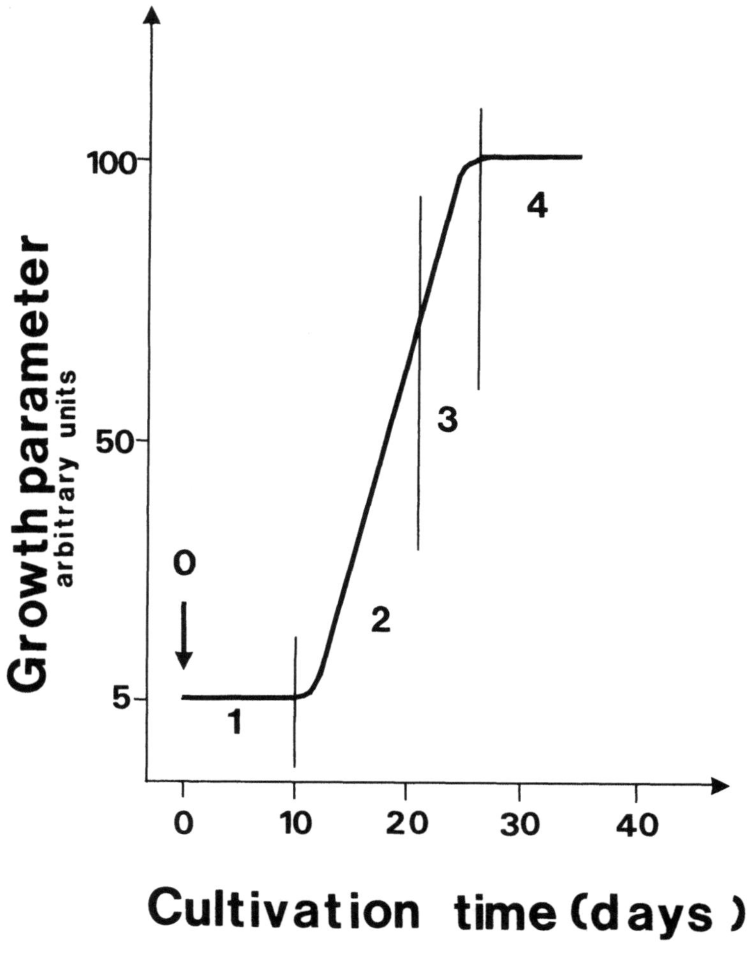

Fig. 1. The batch culture cycle. Cell growth can be monitored by different parameters: fresh weight, dry weight, or packed cell volume (*see* Chapter 2 this vol.), which give similar growth curves. After inoculation (0), growth curves show 4 different phases: (1) lag period, beginning of divisions, no increase in growth parameters; (2) exponential growth, actively dividing cells, dense cell clumps; (3) rate of division decreases, cell clumps dissociate, growth parameters still increase because of cell enlargement; (4) growth stops, beginning of cell decay.

1. Filtration in phase 3 provides a dispersed cell suspension able to divide rapidly.
2. Mutagenic treatment in phase 2: as the cells are actively growing and a large proportion of the cells are in the S phase of the cell cycle.
3. Plating the cells in phase 3 again: After the mutagenic treatment, cells are allowed to recover for a period before screening. After replication and the first cell divisions, a mixed clump of mutated and wild-type cells appears because, in plant cell cultures, daughter cells stick together. When cytoplasmic mutations are being sought, a longer period of somatic segregation should be expected because of the large number of chloroplastic and mitochondrial DNA molecules. A complete culture cycle of 3 wk was thus chosen to allow dissociation of initial cell clumps, formation of mutant clones, and expression of the mutant phenotype; the dispersion and sieving of cells favors a reproducible inhibition of wild-type cells. The importance of cell clump size for the efficiency of mutant recovery has been documented (*23,37,38*).

As a selective agent, we used oligomycin, a specific inhibitor of mitochondrial ATP-synthase. It has been used to isolate mitochondrial mutants in fungi (*39–42*) and in mammalian cells in culture (*43–45*). Mutations to oligomycin-resistance affect subunit 6 (*46*) and subunit 9 (*47*) of the mitochondrial ATP-synthase in yeast, polypeptides that are coded for by mitochondrial DNA in plants (*48,49*). Oligomycin may thus be an effective agent for the selection of mitochondrial mutants in higher plants.

2. Materials

1. Cell line: This line originates from a protoplast culture of a *Nicotiana sylvestris* doubled haploid. It is maintained by monthly subculture on gelatinized culture medium (*see below*).
2. Usual glass or disposable plastic ware, with the exception of the filter holder shown by Fig. 2, which has been specially made by us from commercial glass ground joints.
3. Empty glassware is autoclaved 20 min at 115°C, and then dried 15 min under vacuum or dry-sterilized 90 min at 150°C. Culture media are autoclaved for 20 min at 115°C.
4. Cell suspensions are incubated in 500-mL bottles containing 70 mL of culture medium on a table rotary shaker: magnetic drive; circular motion; throw diameter, 25 mm; shaking speed, 100 rpm (Model TR

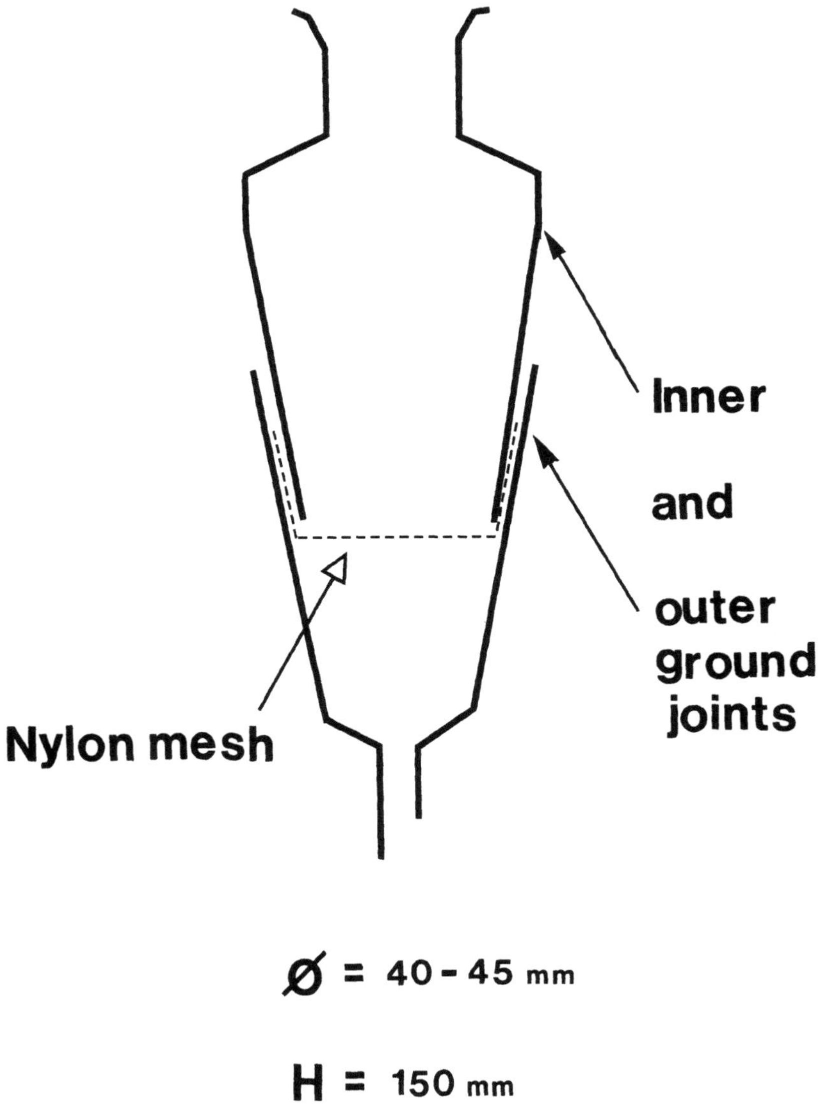

Fig. 2. Glass filter holder. The filter holder is made from commercial ground joints. Nylon mesh of various apertures is set inside, and the whole is autoclaved. It is used under aseptic conditions to separate isolated cells from larger clumps.

125 of Infors AG, Basel, Switzerland). Culture conditions: 25°C under dim light (photoperiod: 16 h).

5. Culture medium for cell suspension: Minerals as in MS medium (*50* and *see* Appendix), vitamins the same as those in ref. *51*, with the exception of thiamine which is brought to 10 mg/L; 2,4-D to 0.3 mg/L; Benzyladenine to 0.01 mg/L; sucrose to 30 g/L; pH to 5.8 before autoclaving.
6. Culture medium for callus culture: The same as above, with 7 g/L of Bacto Difco Agar added before autoclaving. When cells are embedded in agar medium, final agar concentration is 6 g/L.
7. Antibiotic stock solution: The antibiotic oligomycin is stored at –20°C as a powder or as a sterile stock solution in DMSO (1 mg/mL). This solution is prepared by dissolving a few milligrams of oligomycin in the proper volume of pure DMSO and sterile filtering with "Swinnex-13" (Millipore Corp.) equipped with Teflon™ 450 membrane (pore size 0.45 µ*M*; Gelman Sciences Inc.).
8. Selective medium: To prepare selective medium, the proper volume of stock solution is aseptically pipeted with a disposable tip and added to autoclaved medium, which is then dispensed in disposable Petri dishes.
9. EMS solution: EMS is a very active chemical, unstable in aqueous solutions and also recorded as a carcinogen (*52*, and *see* Chapter 38 this vol. for proper handling); every skin contact should be avoided. EMS is a liquid, stored under dry conditions; the flask is opened aseptically under the laminar air flow unit; a small volume is pipeted aseptically with disposable, autoclaved tips and added to autoclaved culture medium to a final concentration of 1% (v/v) = 10 µL/mL.
10. Permanganate solution: This solution is used to destroy any EMS waste after the mutagenic treatment. Introduce sequentially into a 4-L Erlenmeyer flask: distilled water—approximately 1500 mL, H_2SO_4, pure—200 mL, $KMnO_4$—20 g complete with distilled water to 2000 mL. To avoid excessive heating during the mixing of sulfuric acid and water, the flask is dipped in tap water when the acid is slowly poured into the flask.

    Immediately after use, the solution of EMS and the rinsing medium are mixed with this permanganate solution (about 1 vol to 10); every tool exposed to EMS is immersed in the permanganate solution. The experimentor must wear gloves.

3. Methods

3.1. Growth and Mutation

1. To initiate a suspension culture, a 2–3 wk-old callus is inoculated in liquid medium at 2–3 g/flask. The callus is dispersed by shaking the culture.
2. Every 3 wk, the suspension is subcultured by dilution. After a few minutes' decantation, about 10 mL of the upper cell layer (smaller aggregates) are inoculated into fresh medium. For this purpose, glass pipets with the tips cut off at the first centimeter are utilized (larger aperture). After a few subcultures, the suspensions are used to produce large amounts of cells for mutagenic treatments.
3. To prepare cells for mutagenic treatment, they are taken from 3-wk cultures. The suspension is sieved successively through nylon mesh (Fig. 2) of the following apertures: 2000 µm, 1000 µm, and 200 µm, to remove the largest clumps. The suspension is finally collected on a 60- µm aperture mesh. It consists of single cells and small clusters (mainly 2–10 cells/cluster, maximum 50 cells). The cells are counted with a Nageotte hemocytometer (50 µL), and the cell density is adjusted to 10^4 in fresh, liquid medium. The cells are cultured for a further 4 d.

   The amount of cells that can be harvested from a 70 mL flask after 3 wk of culture is, on average, two million cells (ranging from 0.5–4 million). This is point 3 of culture cycle (Fig. 1). Decanting before pouring the suspension through each mesh is important to increase the total "crop" of cells.
4. After subculture, cell division resumes, and the number of cells has multiplied by 4 when the cells are collected again on 60 µm mesh and resuspended at 5×10^5 cells/mL. EMS is then added at a final concentration of 1%. After 1/2 h of contact, the cells are rinsed and resuspended in fresh medium without EMS (treatment EMS_2). For EMS_2 treatment, the cells are not rinsed, but decanted and transferred to fresh medium. The mutagenic treatments are conducted in the laminar air flow unit under usual light conditions.
5. To determine the cell survival rate after mutagenic treatment, samples from treated and untreated suspensions are plated on solid medium (3–4 dishes/sample). Two mL of cell suspension at 10,000 cells/mL are inoculated into a disposable 5-cm Petri dish, and mixed with an equal volume of culture medium that had been autoclaved

Table 1[a]
Influence of Mutagenic Treatments
on Oligomycin-Resistant Variant Frequency

Treatment	Survival, % of control	Number of cells submitted to selection, in millions	Resistant cell lines	Observed frequency, $\times 10^{-7}$
$Control_1$	100	3	0	–
$Control_2$	100	4	1	2.5
$Control_3$	100	15	1	0.6
EMS_2	55	12	8	6
EMS_1	5	2.5	12	48

[a]After mutagenic treatment, the cells were cultured on basal medium for 3 wk, sieved, and the finest fraction (third column) was plated on solid medium that contained 1 mg/L of oligomycin. The resistant colonies were further cultured on selective medium for two passages, and then on basal medium for three passages. The lines remaining resistant were recorded after one more passage on selective medium (fourth column).

with 1.2 g/L of agar and kept at 42°C until use. The final cell density is 5000 cells/mL and agar concentration is 6 g/L. After 1 mo of culture, visible colonies are counted. The survival rate is the ratio of colony number from the mutagenized sample to colony number from the untreated sample.

3.2. Selection

The selection procedure comprises seven steps:

1. After the mutagenic treatment, cells are inoculated into fresh liquid medium at 5×10^4 cells/mL density and cultured for 3 wk until saturation is reached.
2. The suspension is then sieved, and the fraction 200 µm–6 µm is plated in solid medium as described in step 5 above, except that the cell suspension is inoculated into 8 cm dishes and 10 mL of cell suspension is mixed with an equal volume of autoclaved medium, giving a final cell density of 5000 cells/mL and an oligomycin concentration of 1 mg/L.
3. After 2 mo of culture at 25°C, visible colonies (diameter > 1 mm) are recorded, and the largest are transferred to fresh selective agar medium for 2 mo.
4. Two 2-mo subcultures on selective medium. Only lines that grow are kept for further culture.
5. Three 1-mo subcultures on basal medium.
6. The lines are inoculated on selective medium.

7. After 2 mo of culture, lines that show growth are recorded and carried on for further analysis. Table 1 shows the influence of such a mutagenic treatment on cell survival and variant line frequency.

References

1. Widholm, J. M. (1972) Cultured *Nicotiana tabacum* cells with an altered anthranilate synthetase which is less sensitive to feedback inhibition. *Biochem. Biophys. Acta.* **261,** 52–58.
2. Widholm, J. M. (1972) Anthranilate synthetase from 5-methyltryptophane-susceptible and -resistant cultured *Daucus carota* cells. *Biochem. Biophys. Acta.* **279,** 48–57.
3. Bourgin, J. P. (1978) Isolement de mutants à partir de cellules végétales en culture *in vitro. Physiol. vég.* **16,** 339–351.
4. Bourgin, J. P. (1985) Mutations et variations en culture de cellules végétales, in *Aspects industriels des cultures de cellules d'origine animale et végétale* Société, Francaise de Microbiologie. Meeting of Lyon, Mar. 7,8, 1985.
5. Maliga, P. (1976) Isolation of biochemical mutants in tissue cultures of flowering plants. *Arabidopsis Inf. Serv.* **13,** 193–199.
6. Maliga, P. (1980) Isolation, characterization and utilization of mutant cell lines in higher plants. *Internatl. Rev. Cytol.* **supp. 11 A,** 225–250.
7. Maliga, P., Menczel, L., Sidorov, V., Márton, L., Cséplö, A., Medgyesy, P., Trinh Manh, D., Lazar, G., and Nagy, F. (1982) Cell mutants and their uses, in *Plant Improvement and Somatic Cell Genetics* (Vasil, I. K., Scowcroft, W. R., and Frey, K. J., eds.), Academic, London and New York, pp. 221–237.
8. Maliga, P. (1984) Isolation and characterization of mutants in plant cell cultures. *Ann. Rev. Plant Physiol.* **35,** 519–542.
9. Maliga, P., Le Thi Xuan, Dix, P. J., and Cséplö, A. (1979) Antibiotic resistance in *Nicotiana,* in *Plant Cell Cultures: Results and Perspectives* (Sala, F., Parisi, B., Cella, R., and Ciferri, O., eds.), Elsevier/North-Holland Biomedical, pp. 161–166.
10. Yurina, N. P., Odintsova, M. S., and Maliga, P. (1978) An altered chloroplast ribosomal protein in a streptomycin resistant tobacco mutant. *Theor. Appl. Genet.* **52,** 125–128.
11. Cséplö, A. and Maliga, P. (1984) Large scale isolation of maternally inherited lincomycin resistance mutations, in diploid *Nicotiana plumbaginifolia* protoplast cultures. *Mol. Gen. Genet.* **196,** 407–412.
12. Carlson, J. E. and Widholm, J. M. (1978) Separation of two forms of anthranilate synthetase from 5-methyltryptophan-susceptible and -resistant cultured *Solanum tuberosum* cells. *Physiol. Plant.* **44,** 251–255.
13. Widholm, J. M. (1978) Selection and characterization of a *Daucus carota* L. cell line resistant to four amino acid analogues. *J. Exptl. Bot.* **29,** 1111–1116.
14. Durand, J. (1987) Isolation of antibiotic resistant variants in a higher plant, *Nicotiana sylvestris. Plant Science* **51,** 113–118.
15. Gaul, H. (1964) Mutation in plant breeding. *Radiation Bot.* **4,** 155–232.
16. Froese-Gertzen, E. E., Konzak, C. F., Nilan, R. A., and Heiner, R. E. (1964) The effect of ethyl-methane-sulfonate on the growth response, chromosome structure and mutation rate in barley. *Radiation Bot.* **4,** 61–69.
17. Jana, M. K. and Roy, K. (1975) Effectiveness and efficiency of ethyl methane sulfonate and ethylene oxide for the induction of mutations in rice. *Mutation Res.* **28,** 211–215.

18. Kaul, M. L. H. and Bhan, A. K. (1977) Mutagenic effectiveness and efficiency of EMS, DES, and gamma-rays in rice. *Theor. Appl. Genet.* **50,** 241–246.
19. Koornneef, M., Dellaert, L. W. M., and Vanderveen, J. H. (1982) EMS- and radiation- induced mutation frequencies at individual loci in *Arabidopsis thaliana* (L.) Heynh. *Mutation Res.* **93,** 109–123.
20. Gupta, P. K. and Yashvir, N. K. (1976) Induced mutations in foxtail millet (*Setaria italica* Beauv.). *Theor. Appl. Genet.* **48,** 131–136.
21. Putrament, A., Kruszewska, A., Baranowska, H., Ejchart, A., Polakowska, R., and Szczesniak, B. (1981) Mitochondrial mutagenesis in *Saccharomyces cerevisiae* V. Frequencies of different *mit*⁻ mutants and loss of their *mit*⁺ alleles in *rho*⁻clones. *Current Genet.* **3,** 57–63.
22. Cassini, R., Cornu, A., Bervillé, A., Vuillaume, E., and Panouillé, A. (1977) Hérédité et caractéristiques des sources de résistance à *Helminthosporium maydis* race T, obtenues par mutagénèse chez des Maïs à cytoplasme mâle stérile Texas. *Ann. Amélior. Plantes* **27,** 753–766.
23. Sung, Z. R. (1976) Mutagenesis of cultured plant cells. *Genetics* **81,** 51–57.
24. Horsch, R. B. and Jones, G. E. (1980) The selection of resistance mutants from cultured plant cells. *Mutation Res.* **72,** 91–100.
25. Savage, A. D., King. J., and Gamborg, O. L. (1979) Recovery of a pantothenate auxotroph from a cell suspension culture of *Datura innoxia* Mill. *Plant Sci. Let.* **16,** 367–376.
26. Horsch, R. B. and Jones, G. E. (1978) 8-azaguanine-resistant variants of cultured cells of *Haplopappus gracilis*. *Can. J. Bot.* **56,** 2660–2665.
27. King, J. and Maretzki, A. (1983) Isolation from sugarcane cultures of variants resistant to antimetabolites. *Physiol. Plant.* **58,** 457–463.
28. Caboche, M. and Muller, J. F. (1980) Use of a medium allowing low cell density growth for *in vitro* selection experiments: isolation of valine-resistant clones from nitrosoguanidine-mutagenized and gamma-irradiated tobacco plants, in *Plant Cell Cultures: Results and Perspectives* (Sala, F., Cella, R., and Ciferri, O., eds.), Elsevier/North Holland Biomedical, pp. 133–138.
29. Kimball, R. F. (1978) The relation of repair phenomena to mutation induction in bacteria. *Mutation Res.* **55,** 85–120.
30. Lehmann, A. R. and Karran, P. (1981) DNA repair. *Internatl. Rev. Cytol.* **72,** 101–146.
31. Veleminsky, J. and Gichner, T. (1978) DNA repair in mutagen-injured higher plants. *Mutation Res.* **55,** 71–84.
32. Kilbey, B. and Hunter, F. (1983) Factors affecting mutational yield from EMS exposures of yeast (*Saccharomyces cerevisiae*). *Mutation Res.* **122,** 35–38.
33. Bourgin, J. P., Hommel, M. C., and Missonier, C. (1980) Expression of resistance to valine in protoplast-derived cells of tobacco mutants in *Plant Cell Cultures: Results and Perspectives* (Sala, F., Parisi, B., Cella, R., and Ciferri, O., eds.), Elsevier/ North Holland Biomedical, pp. 167–171.
34. Stuart, R. and Street, H. E. (1969) Studies on the growth of plant cells. IV. The initiation of division in suspensions of stationary phase cells of *Acer pseudoplatanus*. *J. Exptl. Bot.* **20,** 556–571.
35. Stuart, R. and Street, H. E. (1971) Studies on the growth in culture of plant cells. X. Further studies on the conditioning of culture media by suspension of *Acer pseudoplatanus* L. cells. *J. Exptl. Bot.* **22,** 96–106.

36. Logemann, H. and Bergmann, L. (1974) Einfluss von Licht und Medium auf die "plating efficiency" isolierter Zellen aus Callus kulturen von *Nicotiana tabacum* var. "Samsum". *Planta* **121,** 283–292.
37. Murphy, T. M. (1982) Analysis of distributions of mutants in clones of plant cell aggregates. *Theor. Appl. Genet.* **61,** 367–372.
38. Grandbastien, M. A., Bourgin, J. P., and Caboche, M. (1985) Valine-resistance, a potential marker in plant cell genetics. II. Optimization of UV mutagenesis and selection of valine-resistant colonies derived from tobacco mesophyll protoplasts. *Genetics* **109,** 409–425.
39. Avner, P. R. and Griffiths, D. E. (1970) Oligomycin resistant mutants in yeast. *FEBS Lett.* **10,** 202–207.
40. Avner, P. R., Coen, D., Dujon, B., and Slonimski, P. P. (1973) Mitochondrial genetics. IV. Allelism and mapping studies of oligomycin-resistant mutants in *Saccharomyces cerevisiae. Mol. Gen. Genet.* **125,** 9–52.
41. Brunner, A. and Tuena de Cobos, A. (1980) Extrachromosomal oligomycin-resistant mutants of the petite-negative yeast *Kluyveromyces lactis*. Properties of mitochondrial ATPase and cross-resistance to inhibitors of phosphoryl transfer reactions. *Mol. Gen. Genet.* **178,** 351–355.
42. Rowlands, R. T. and Turner, G. (1973) Nuclear and extranuclear inheritance of oligomycin resistance in *Aspergillus nidulans. Mol. Gen. Genet.* **126,** 201–216.
43. Lichtor, T. and Getz, G. S. (1978) Cytoplasmic inheritance of rutamycin resistance in mouse fibroblasts. *Proc. Natl. Acad. Sci. USA* **75,** 324–328.
44. Kuhns, M. C. and Eisenstadt, J. M. (1979) Oligomycin-resistant mitochondrial ATPase from mouse fibroblasts. *Somatic Cell Genet.* **5,** 821–832.
45. Shew, J. Y. and Breen, G. A. M. (1985) Two cytoplasmically inherited oligomycin-resistant Chinese hamster cell lines exhibit an altered mitochondrial translation product. *Somatic Cell Mol. Genet.* **11,** 103–108.
46. John, U. P. and Nagley, P. (1986) Amino acid substitutions in mitochondrial ATPase subunit 6 of *Saccharomyces cerevisiae* leading to oligomycin resistance. *FEBS Lett.* **207,** 79–83.
47. Sebald, W., Wachter, E., and Tzagoloff, A. (1979) Identification of amino acid substitutions in the Dicyclohexylcarbodiimide-binding subunit of the mitochondrial ATPase complex from oligomycin-resistant mutants of *Saccharomyces cerevisiae. Eur. J. Biochem.* **100,** 599–607.
48. Dewey, R. E., Levings, C. S., and Timothy, D. H. (1985) Nucleotide sequence of ATPase subunit 6 gene of Maize mitochondria. *Plant Physiol.* **79,** 914–919.
49. Dewey, R. E., Schuster, A. M., Levings, C. S., and Timothy, D. H. (1985) Nucleotide sequence of F_o-ATPase proteolipid (subunit 9) gene of maize mitochondria. *Proc. Natl. Acad. Sci. USA* **82,** 1015–1019.
50. Murashige, T. and Skoog, F. (1962) A revised medium for rapid growth and bioassays with tobacco tissue cultures. *Physiol. Plant.* **15,** 473–497.
51. Morel, G. and Wetmore, R. H. (1951) Fern tissue culture. *Amer. J. Bot.* **38,** 141–143.
52. Nesnow, S., Argus, M., Bergman, H., Chu, K., Frith, C., Helmes, T., Mc Gaughy, R., Ray, V., Slaga, T. J., Tennant, R., and Weisburger, E. (1986) Chemical carcinogens. A review and analysis of the literature of selected chemicals and the establishment of the Gene-Tox Carcinogen Data Base. *Mutation Res.* **185,** 1–195.

Chapter 40

Techniques for Selecting Mutants from Plant Tissue Cultures

David R. Duncan and Jack M. Widholm

1. Introduction

The theory and goals of mutant selection from plant tissue cultures have been reviewed extensively over the past few years (*1–5*), and will only be dealt with in a cursory manner in this chapter. For more detail, the reader is encouraged to consult these review articles.

The ideal plant tissue culture system is useful for mutation selection because it:

1. is a defined and an easily manipulated environment of small size that facilitates the application to plant cells of selection pressures (inhibitors) that are not typically present in a field condition,
2. is a sterile environment, free of insect and pathogenic pests, which can complicate experiments and reduce the recovery of stressed mutant plant material, and
3. contains a large population (i.e., millions) of totipotent plant cells in a small space that can be screened for mutants.

From: *Methods in Molecular Biology*, vol. 6, *Plant Cell and Tissue Culture*
Edited by Jeffrey W. Pollard and John M. Walker, ©1990 by The Humana Press

To actually select for a mutant, an inhibitor is incorporated into the culture medium at a concentration that will either kill or inhibit the growth of wild-type plant cells. Resistant variant cells can be retrieved from the medium, because they will be the only cells capable of growth in the presence of the inhibitor. Ideally, these variant cells can develop into whole plants, which facilitates distinguishing between epigenetic changes in culture and true mutations since epigenetic changes, by some definitions, will not pass through meiosis. Also, the plants can be used for whole plant physiology studies, inheritance studies, or the commercial application of the selected trait.

Following this procedure, a large number of plants cells have been screened for desired traits, such as herbicide tolerance (*6–9*), salt stress tolerance (*10–15*), amino acid overproduction (*16–24*), and disease resistance (*25–27*). In some of these cases, the selected traits have been taken to the whole plant level (*6,7,9,11,13–15,17,19–22,26*). Although this procedure has produced cell lines resistant to many agents, it must be pointed out that, often for unknown reasons, not all attempts at selecting mutants have been or will be successful.

Based on reported successes (*7–12,18,20–24*) and on the ease of manipulation, the cell suspension (actually clumps of several cells) has been most commonly used for mutant selection. Consequently, the procedures described in this chapter will focus on selection in cell suspensions, with added commentary on adapting these procedures to callus cultures and plated suspensions.

The use of mutagens for increasing the probability of obtaining a mutation is covered in Chapters 38 and 39, and will not be dealt with here. Also, protoplasts can be used for selection purposes; however, they are typically grown to the stage of cloned microcalluses before they are used, and at that stage can be manipulated as either cell suspensions or callus cultures (*4*). The actual production of protoplasts is dealt with in Chapter 24, this volume.

2. Materials

2.1. Mutant Selection Using Cell Suspensions and Liquid Culture Medium

2.1.1. Determination of an Inhibitory Concentration

1. A suspension culture in mid-log phase (*see* Note 1).
2. 9.0-cm Whatman #1 filter paper.

3. A 9.0-cm (inside diameter) Buchner funnel and a 1-L Erlenmeyer suction filter flask, set up for vacuum filtration.
4. A $10^{-2}M$ (in 170 mL of medium) stock of a selection agent (inhibitor) made in culture medium (*see* Note 2).

2.1.2. Mutant Selection

Same as for determining the inhibitory concentrations, but the selection medium will contain levels of inhibitor determined by the above experiment and appropriate for the type of selection procedure used.

2.2. Mutant Selection Using Cell Suspensions and Agar-Solidified Medium

2.2.1. Determination of Inhibitory Concentration by Spreading Cell Suspensions on Top of Agar-Solidified Medium

1. Petri dishes containing 25 mL of culture medium.
2. An inhibitor stock of $2.6 \times 10^{-1}M$ in 10 mL of medium (*see* Note 2).

2.2.2. Determination of Inhibitory Concentration by Incorporating Cells in the Medium

1. Fifty mL of culture medium containing 2x the normal quantity of agar and 0.0, 2×10^{-2}, 2×10^{-3}, 2×10^{-4}, 2×10^{-5}, or $2 \times 10^{-6}M$ of inhibitor concentrations in 250-mL, aluminum-foil-covered Erlenmeyer flasks.

3. Methods

3.1. Mutant Selection Using Cell Suspensions and Liquid Culture Medium

3.1.1. Determination of an Inhibitory Concentration

1. Inhibitor concentration gradients of 0.0, 10^{-6}, 10^{-5}, 10^{-4}, and 10^{-3} are made by 10-fold serial dilutions from the $10^{-2}M$ stock (17-mL stock plus 153 mL fresh culture medium = $10^{-3}M$ inhibitor, 17-mL of the $10^{-3}M$ solution plus 153 mL fresh culture medium = $10^{-4}M$ inhibitor, and so on. These volumes will yield three 50-mL replicates with an approximate 2% margin of error. The remainder of the stock solution is used for the $10^{-2}M$ treatment).
2. Adjust the pH of each treatment to 5.8, distribute each treatment

equally among three 125-mL Erlenmeyer flasks, and sterilize by autoclaving at 121°C for 20 min (*see* Note 2).

3. Remove a 10-mL aliquot from a thoroughly suspended cell culture, and collect the cells on a wet Whatman #1 filter paper by vacuum in the Buchner funnel (*see* Note 3).
4. Scrape the cells from the filter paper and weigh them.
5. Calculate the volume of suspension required to yield 0.25 g fresh weigh (gfw) of cells, and transfer that volume from a well-suspended culture into each treatment flask (*see* Note 1).
6. Incubate the inoculated flasks for 14 d under normal growth conditions on a gyratory shaker at 130 rpm.
7. After the incubation period, harvest each flask on Whatman #1 filter paper by vacuum filtration as previously done and weigh the samples (*see* Note 4).
8. Determine the growth response of the cells to the inhibitor by calculating the growth as a percent of the growth of the control (no inhibitor added), and plot the data on semi-log graph paper (4):

$$\frac{\bar{X}_n - X_o}{\bar{X}_c - X_o} \times 100 = \% \text{ of control} \qquad (1)$$

$\bar{X}_n$ = mean weight of cells grown in inhibitor of concentration *n*; $\bar{X}_c$ = mean weight of control cells (0.0*M* of inhibitor); X_o = initial inoculum.

9. From the graph, an I_{50} (inhibitor concentration resulting in 50% inhibition of growth) or an MIC (minimum inhibitor concentration resulting in 100% inhibiton of growth) can be determined (4). These values will be used in the selection phase of this procedure (*see* Note 5).

3.1.2. Mutant Selection

Selection with cell suspensions can be done using either liquid culture medium or agar-solidified culture medium. The level of inhibitor incorporated into the medium may be either 2–3x greater than the MIC (a single-step or direct selection procedure) or a slightly inhibitory concentration that is gradually increased with subsequent subcultures (multi-step or gradual selection procedure).

The single-step method is the faster of the two procedures and is particularly useful when there is a chance that the plant regeneration capacity of a culture will be lost with prolonged culturing. Some cultures, however, do not seem to respond to single-step selection, possibly because these cultures cannot grow as single cells, or perhaps dead neighboring cells release toxic materials that kill the variant cells (*4*). Under these circumstances, the multi-step selection method may be the best procedure to use. This sometimes is the case with selections using callus cultures (*19*). Without some background experience with mutant selection using a desired plant species or with the culture habits of that species, there appears to be no *a priori* reasoning for when to use one procedure or the other. The most successful course of action would be to try both methods simultaneously.

3.1.3. Single-Step Selection in Liquid Medium (9,11,18,20–24)

1. A liter of medium containing 2–3x the MIC of inhibitor is distributed among 20 125-mL Erlenmeyer flasks in 50-mL aliquots and sterilized by autoclaving (*see* Note 1).
2. The volume of well-suspended cells containing 0.25 gfw is determined as described above, and that volume is added to each treatment flask (*see* Notes 2 and 3).
3. The inoculated flasks are incubated as described above for 8–12 wk under normal growth conditions, a time sufficient for cell growth to become evident, even if there is only one cell originally present that can divide at a moderate rate.
4. At biweekly intervals, the flasks should be examined for growth, and any growing cells should be transferred to fresh culture medium containing the same level of inhibitor.

3.1.4. Multi-Step Selection in Liquid Medium (8,10,12)

1. This procedure is similar to the single-step procedure, but the level of inhibitor used is less than the MIC, though often higher than the I_{50}.
2. Monitor the growth in inoculated flasks. When the cells have grown to approximately 1/3 the stationary phase cell mass of untreated control cultures, subculture the cells to fresh medium containing a slightly higher level of inhibitor. For instance, if the initial inhibitor level was 0.5 m*M* and the MIC was 1.0 m*M*, then with subsequent subcultures the inhibitor concentrations may be 0.6 m*M*, 0.7 m*M*, 0.8 m*M*, and so forth. If growth suddenly ceases with a single increase in the inhibitor concentration, then the increase was probably too great, and smaller concentration increases should be tried. Loss of the selected cells can be prevented by subculturing only a portion of the cells to the higher inhibitor concentration and maintaining the

remainder of the culture on fresh medium containing the inhibitor concentration on which the cells are presently growing.

3. Maintain any viable cultures on medium containing a level of the inhibitor lethal to the wild-type cells for at least two or three subcultures to ensure that only resistant cells are present.

3.2. Mutant Selections Using Cell Suspensions and Agar-Solidified Medium

Since Petri dishes can be stacked in boxes and only a minimum of shaker space is required to maintain stock suspensions, mutant selection using agar-solidified medium can be a convenient means for reducing the amount of space required and reducing the cost of the selection. Also, if a single-step selection procedure is used, there is the possibility that cell clumps growing on the inhibitor-containing medium will be of a single-cell origin, i.e., are clones.

In terms of methodology, the use of liquid medium and agar-solidified medium is the same, except for the manner in which the cells are brought into contact with the inhibitor, and possibly the I_{50} and MIC (which tend to be higher when agar-solidified medium is used).

3.2.1. Determination of Inhibitory Concentration by Spreading Cell Suspensions on Top of Agar-Solidified Medium

1. Make a serial dilution of the $2.6 \times 10^{-1}M$ stock down to $2.6 \times 10^{-5}M$ (1 mL of stock added to 9 mL of fresh medium = $2.6 \times 10^{-2}M$, 1 mL of the $2.6 \times 10^{-2}M$ solution added to 9 mL of fresh medium = $2.6 \times 10^{-3}M$, and so forth) (*see* Note 1).
2. Filter sterilize the inhibitor solutions.
3. Add 1 mL of inhibitor to each of three of the plates containing 25 mL of agar-solidified culture medium (1 mL of the $2.6 \times 10^{-1}M$ solution added to a plate of medium will make a final concentration of $10^{-2}M$; 1 mL of the $2.6 \times 10^{-2}M$ solution added to a plate of medium makes a final concentration of 10^{-3} M, and so on).
4. Let the plates stand overnight to absorb and evenly distribute the inhibitor throughout the medium.
5. Determine the volume of 0.25 gfw of suspension cells (as described above) and add uniformly to the surface of the medium in each plate. (Also add cells to three plates lacking inhibitor for control plates.) A cooled glass rod, sterilized by dipping it in 70% ethanol and re-

moving the excess ethanol by igniting it with a Bunsen burner, can be used to spread the cells over the surface of the medium.
6. Let plates stand 30 min, and then remove (with a sterile Pasteur pipet or aspirator) any excess liquid.
7. Incubate plates for 3 wk under normal growth conditions, and then examine the plates for growth.
8. Visual observation of growth is typically used to determine the inhibitory concentration, but quantitative data can be collected if the cells are carefully scraped off the agar and weighed.

3.2.2. Determination of Inhibitory Concentration by Incorporating Cells in the Medium

1. Autoclave the flasks containing the 2x level of agar and inhibitor (*see* materials 2.2.2.).
2. After autoclaving, cool flasks to 45°C and maintain in a temperature-controlled water bath. The agar will not solidify at 45°C.
3. Determine the volume of a suspension required for 1.0 gfw of cells (as described above), and place that volume in an empty, sterile flask (*see* Note 3).
4. Adjust the volume in the flask of cells to 50 mL with sterile, fresh medium.
5. Mix thoroughly a flask of cells with a flask containing 2x agar and inhibitor, and pour into four sterile Petri plates. (Mixing medium containing agar at 45°C with an equal volume of medium containing cells at ambient temperature usually is not lethal to the cells.)
6. Repeat procedure for all inhibitor concentrations and for the control containing no inhibitor.
7. After the plates have solidified, incubate them and observe growth as described for cells placed on the medium surface.

3.3. Mutant Selection by Spreading Cells on Top of Medium

1. Make 1 L of culture medium containing an inhibitor concentration 2–3x the MIC, as determined above (*see* Note 6).
2. Autoclave and pour medium into 40 sterile Petri plates.
3. Inoculate plates as described above, and incubate them for 8–12 wk under normal growth conditions.
4. Transfer any growing colonies to fresh medium containing the same level of inhibitor.

3.4. Mutant Selection by Incorporating Cells into the Medium

1. Make 500 mL of culture medium containing 2x the normal quantity of agar and an inhibitor concentration 4–6x the MIC, as determined above (*see* Note 6).
2. Divide the medium equally between ten 250-mL Erlenmeyer flasks.
3. Cover flasks, autoclave, and then cool them to 45°C.
4. Inoculate with cells as described above, pour the medium into 40 Petri plates, and then incubate them for 8–12 wk under normal growth conditions.
5. Transfer any growing colonies to fresh medium containing the same level of inhibitor.

3.5. Mutant Selection Using Callus Cultures

Mutants can be selected from callus cultures using the same techniques as described for mutant selection from cell suspensions on top of agar-solidified medium (use about 25 pieces of callus/plate, approximately 10 mgfw/callus piece). Both single-step (*6,7,17,21,25,27*) and multi-step (*7,13,16,19*) selection procedures are applicable to callus cultures.

3.6. Manipulating the Selected Variant

Once a cell line is selected that will grow at inhibitor concentrations equal to or greater than the MIC, then it must be demonstrated that the growth of the variant cells results from a mutation, and not from an adaptation of the cells to the medium. Mutations should be stable under nonselective conditions, but ultimately, whole plants must be regenerated from the variant culture and the resistance passed on to subsequent generations to prove that it is caused by a mutation.

4. Notes

1. The culture parameters mentioned in this chapter (growth phase, inoculum size, and incubation times) are based on carrot suspension cultures. These parameters may vary with the species being studied and must be empirically determined. One of these parameters, the inoculum size, can be determined by counting the cells in a culture sample using a hemocytometer. Since callus and even "cell" suspensions are actually aggregates of many cells, it is often necessary to

macerate the tissue before the individual cells can be counted. Such tissue maceration can be done by exposing a weighed quantity of tissue to a 5% chromic acid solution for 12 h, and then shaking the solution vigorously or passing it through a small syringe needle under pressure. The number of cells in the tissue can then be determined using a hemocytometer (*28*).

Inoculum size can be determined by counting the number of cells in a weighed sample of tissue because most selected resistances have occurred at a frequency near 1 in 10^{-7} cells, so to successfully select a mutant, one would expect that at least 10^8 cells should be used in the experiment. Thus, if one knows how many cells are being inoculated into a flask or Petri plate, then one can calculate how many flasks or plates need to be inoculated to obtain the 10^8 cells needed for the selection. Also, knowing the number of cells inoculated in a selection experiment will allow one to estimate the mutation (resistance) frequency of the culture.

2. Because many inhibitors are not heat or light stable or they are not soluble at high concentrations, the aseptic addition of an inhibitor to a medium can be a difficult part of selecting inhibitor-resistant mutants from cultured tissues. To avoid costly waste of medium and inhibitor, the literature or the manufacturer of an inhibitor should be consulted as to the solubility and chemical stability of the compound. For instance, heat-sensitive compounds must be filter sterilized. Also, water-insoluble compounds may need to be dissolved in organic solvents, such as ethanol or dimethylsulfoxide, prior to being added to medium. Since these solvents can be toxic, low concentrations should be used, and control treatments using only the solvent must be incorporated into the experiment in order to determine if the solvent concentrations used are toxic.
3. Some suspension cultures are composed of cell clumps of varying size. Under such circumstances, the suspension can be filtered through an autoclaved funnel lined with a layer of cheesecloth or some other sterile sieve to remove the large clumps (*4*). The smaller clumps passing through the funnel are then used in the above procedure.
4. It is useful to sample tissues exposed to the MIC (*see* step 9. of Methods) of an inhibitor and test for cell viability (as described in Chapter 3, of this vol.). How the selected variants are further cultured can depend on whether the inhibitor kills or just stops the growth of wild-type cells. For instance, wild-type cells may be present in a flask containing a selected variant. If the inhibitor just stops cell

growth of those wild-type cells, then replacing the medium in the flask with fresh medium lacking inhibitor may permit both the wild-type and the variant cells to grow. The end result would be a mixed culture that often is eventually overgrown by the more vigorous wild-type cells. In this example, the variant line should be maintained in medium containing the inhibitor for, at least, several subcultures.

5. Once an approximate inhibitory level is determined, it may be necessary to test more concentrations of the inhibitor near the previously determined inhibitory level. This is particularly useful if the transition from little to complete growth inhibition occurs over a narrow concentration range.
6. A multi-step selection can also be used with an agar-solidified medium, and the concentrations used would be similar to those used in the liquid medium procedure. Such a selection entails more effort, because many colonies will survive that must be transferred to the next higher inhibitor concentration. However, this may be the only means available for selecting mutants from some cultures.

References

1. Maliga, P. (1984) Isolation and characterization of mutants in plant cell culture. *Ann. Rev. Plant Physiol.* **35,** 519–542.
2. Widholm, J. M. (1984) Induction, selection, and characterization of mutants in carrot cell cultures, in *Cell Culture and Somatic Cell Genetics of Plants,* vol 1 (Vasil, I. K., ed.) Academic Press, NY, pp. 563–570.
3. Duncan, D. R. and Widholm, J. M. (1985) Cell selection for crop improvement. *Plant Breeding Reviews* **4,** 153–173.
4. Gonzales, R. A. and Widholm, J. M. (1985) Selection of plant cells for desirable characteristics: Inhibitor resistance, in *Plant Tissue Culture: A Practical Approach, Information Retrieval Limited,* (Dixon, R. A., ed.), New York, pp. 67–78.
5. Widholm, J. M. (1988) *In vitro* selection with plant cell and tissue cultures: An overview. *Iowa State J. Research* **62,** 587–597.
6. Chaleff, R. S. and Parsons, M. F. (1978) Direct selection *in vitro* for herbicide-resistant mutants of *Nicotiana tabacum. Proc. Natl. Acad. Sci. USA* **75,** 5104–5107.
7. Miller, O. K. and Hughes, K. W. (1980) Selection of paraquat-resistant variants of tobacco from cell culture. *In Vitro* **16,** 1085–1091.
8. Nafziger, E. D., Widholm, J. M., Steinrucken, H. C., and Killmer, J. L. (1984) Selection and characterization of a carrot cell line tolerant to glyphosate. *Plant Physiol.* **76,** 571–574.
9. Thomas, B. R. and Pratt, D. (1982) Isolation of paraquat-tolerant mutants from tomato cell cultures. *Theor. Appl. Genet.* **63,** 169–176.
10. Bennetzen, J. L. and Adams, T. L. (1984) Selection and characterization of cadmium-resistant suspension cultures of wild tomato *Lycopersicon peruvianum. Plant Cell Reports* **3,** 258–261.

11. Conner, A. H. and Meredith, C. P. (1985) Large scale selection of aluminum-resistant mutants from plant cell culture: Expression and Inheritance in Seedlings. *Theor. Appl. Genet.* **71,** 159–165.
12. Huang, B., Hatch, E., and Goldsbrough, P. B. (1987) Selection and characterization of cadmium tolerant cells in tomato. *Plant Sci.* **52,** 211–221.
13. McCoy, T. J. (1987) Characterization of alfalfa (*Medicago sativa* L.) plants regenerated from selected NaCl tolerant cell lines. *Plant Cell Reports* **6,** 417–422.
14. Nabors, M. W., Gibbs, S. E., Berstein, C. S., and Meis, M. E. (1980) NaCl-tolerant tobacco plants from cultured cells. *Z. Pflanzenphysiol.* **97,** 13–17.
15. Reddy, P. J. and Vaidyanath, K. (1986) *In Vitro* characterization of salt stress effects and the selection of salt tolerant plants in rice (*Oryza sativa* L.). *Theor. Appl. Genet.* **71,** 757–760.
16. Hibberd, K. A., Walter, T., Green, C. E., and Gengenbach, B. G. (1980) Selection and characterization of a feedback-insensitive tissue culture in maize. *Planta* **148,** 183–187.
17. Hibberd, K. A., and Green, C. E. (1982) Inheritance and expression of lysine plus threonine resistance selected in maize tissue culture. *Proc. Natl. Acad. Sci. USA* **79,** 559–563.
18. Jacobsen, E., Visser, R. G. F., and Wijbrandi, J. (1985) Phenylalanine and tyrosine accumulating cell lines of a dihaploid potato selected by resistance to 5-methyltryptophan. *Plant Cell Reports* **4,** 151–154.
19. Miao, S., Duncan, D. R., and Widholm, J. M. (1988) Selection of regenerable maize callus cultures resistant to 5-methyl-DL-tryptophan, S-2-aminoethyl-L-cysteine and high levels of L-lysine plus L-threonine. *Plant Cell Tissue and Organ Culture,* **14,** 3–14.
20. Ranch, J. P., Rick, S., Brotherton, J. E., and Widholm, J. M. (1983) Expression of 5-methyltryptophan resistance in plants regenerated from resistant cell lines of *Dature innoxia. Plant Physiol.* **71,** 136–140.
21. Wakasa, K. and Widholm, J. M. (1987) A 5-methyltryptophan resistant mutant, MTR1, selected in tissue culture. *Theor. Appl. Genet.* **74,** 49–54.
22. Widholm, J. M. (1974) Cultured carrot cell mutants: 5-methyltryptophan resistance trait carried from cell to plant and back. *Plant Sci. Letts.* **3,** 323–330.
23. Widholm, J. M. (1976) Selection and characterization of cultured carrot and tobacco cells resistant to lysine, methionine, and proline analogs. *Can. J. Bot.* **54,** 1523–1529.
24. Widholm, J. M. (1978) Selection and characterization of a *Daucus carota* L. cell line resistant to four amino acid analogues. *J. Exptl. Bot.* **29,** 1111–1116.
25. Gengenbach, B. G. and Green, C. E. (1975) Selection of T-cytoplasm maize callus cultures resitant to *Helminthosporium maydis* race T pathotoxin. *Crop Sci.* **15,** 545–649.
26. Gengenbach, B. G., Green, C. E., and Donovan, C. M. (1977) Inheritance of selected pathotoxin resistance in maize plants regenerated from cell culture. *Proc. Natl. Acad. Sci. USA* **74,** 5113–5117.
27. Rines, H. W. and Luke, H. H. (1985) Selection and characterization of toxin-insensitive plants from tissue cultures of oats (*Avena sativa*) susceptible to *Helminthosporium victoriae. Theor. Appl. Gent.* **71,** 16–21.
28. Brown, R. and Rickless, P. (1949) A new method for the study of cell division and cell extension with some preliminary observations on the effect of temperature and of nutrients. *Proc. Roy. Soc. B.* **136,** 110–125.

Chapter 41

Selection of Antimetabolite Resistant Mutants

Philip J. Dix

1. Introduction

Mutants resistant to chemicals that inhibit growth (antimetabolites) are the most readily selected in plant cell cultures. A number of such mutants have been isolated, with resistance to amino acids and their analogs, base analogs, toxins from pathogenic microorganisms, herbicides, salts, and antibiotics. The mutants can be used for fundamental investigations into cellular metabolism, as markers for plant genetic manipulation, and in efforts to improve crop tolerance to diseases and agrochemicals. Progress in mutant selection has been reviewed extensively in recent years (*1–4*). Detailed considerations of the technical aspects of in vitro selection are available (*1*), and some protocols have been published (*4–7*).

At the outset of any program of mutant selection in vitro, several basic decisions have to be made, including the type of culture system to be used, whether a mutagenic treatment is necessary or desirable, and, if so, what kind and at what stage it should be applied. A detailed consideration of these questions is beyond the scope of this chapter, but useful guidelines can be found elsewhere (*1*). Generally, the decision regarding starting

From: *Methods in Molecular Biology*, vol. 6, *Plant Cell and Tissue Culture*
Edited by Jeffrey W. Pollard and John M. Walker, ©1990 by The Humana Press

material depends on the range of culture procedures that are routine for the plant species under investigation, and the ultimate use to which the mutants will be put, i.e., whether or not it is important that fertile plants can be regenerated.

The choice of plant species and culture systems for the protocol described here reflects some of the considerations outlined above. *Nicotiana plumbaginifolia* is one of a number of species in the family *Solanaceae* for which a fairly full range of tissue and protoplast culture techniques is available (*7*). Furthermore, it is widely viewed as a model species for investigations into genetic manipulation of plants by somatic hybridization and transformation.

This protocol incorporates the use of two culture systems: established cell suspension cultures, which have a relatively poor regeneration potential, and mesophyll protoplasts, from which plants can be readily regenerated. It relates specifically to *N. plumbaginifolia*, but should also work for *N. tabacum*, tobacco. For other species for which suitable tissue and protoplast cultures have been reported, the protocol can be appropriately modified.

The choice of hydroxyproline as the selective agent is based upon a particular case study, aimed at producing variants for use in investigating the cellular physiology of salt stress (*8*). Any other chemical known to inhibit the growth of cells can be substituted, although the chances of success depend upon the potential of the cells for the required genetic or epigenetic change. The procedure should work particularly well when selecting for resistance to amino acid analogs.

2. Materials

1. MS plant salt mix (*see* Appendix).
2. The following vitamins and plant growth hormones (*see* Appendix), concentrations of aqueous stock solutions are given:
   a. Thiamine HCl. 1 mg/mL.
   b. Meso-inositol. No stock, used as powder.
   c. 2,4-dichlorophenoxyacetic acid. (2,4-D) 0.1 mg/mL.
   d. 6-furfurylaminopurine (kinetin). 0.05 mg/mL.
   e. Benzylaminopurine (BA). 0.5 mg/mL.
   f. Naphthalene acetic acid (NAA). 1 mg/mL.
   g. Gibberellic acid (GA_3). 0.5 mg/mL.

   All stock solutions are stored in the refrigerator.

3. All components of K_3 medium and W5 salts (*see* Appendix).
4. 1*M* cis-4-hydroxy-L-proline stock solution.
5. 10 m*M* *N*-nitroso-*N*-ethylurea (NEU; 1.17 mg/mL) or 10 m*M* *N*-nitroso-*N*-methylurea (NMU; 1.03 mg/mL) freshly prepared in RMP medium and filter sterilized.
6. 20% (v/v) Domestos or other proprietary bleach.
7. RM medium: 4.6 g MS plant salt mix, 20 g sucrose, and 6.5 g Difco Bactoagar/L.
8. RMP medium: RM medium, but with 30 g sucrose, 100 mg mesoinositol, 1 mg thiamine-HCl, 0.1 mg 2,4-D, and 0.1 mg kinetin/L.
9. RMOP medium: RM medium, but with 100 mg mesoinositol, 1 mg thiamine-HCl, 1 mg BA, and 0.1 mg NAA/L.
10. *Nicotiana plumbaginifolia* seeds.
11. Nylon bolting cloth and sewing thread.
12. Culture room facilities. Preferred conditions: 25°C, 1500 1x, 16-h day.

3. Methods

The individual procedures that together make up the protocol are dealt with under appropriate subheadings below. They include the sterilization and germination of seeds and the initiation of shoot, callus, and cell suspension cultures; the determination of the appropriate selective level of the antimetabolite; and mutagenesis and the selection of resistant colonies, and the recovery of plants from them.

3.1. Initiation of Shoot, Callus, and Cell Suspension Cultures

1. Cut circles of nylon bolting cloth 3–4 cm in diameter and make into small parcels containing 100–200 seeds, tied with sewing thread.
2. Surface sterilize the seeds by placing in 70% alcohol (5 s), followed by 20% Domestos (10 min) and two washes in sterile distilled water (*see* Note 1).
3. Cut open parcels and transfer seeds to the surface of 20 mL RM medium in 9-cm plastic Petri dishes. Seal with parafilm and incubate in the culture room until the seeds germinate.
4. Thin out seedlings at 0.5–1 cm in height, transferring to RM medium in Plantcon (Flow Laboratories) containers. At a height of 4–5 cm, transfer to individual containers. When doing so, remove the roots and the basal 2 mm of the stem with a single clean cut, and insert the

cut surface into the fresh medium. This encourages vigorous growth of adventitious roots.

5. When the seedling approaches the top of the container, establish shoot cultures by dissecting out nodes (with approx. 1 cm internode remaining above and below) and transferring individually to RM medium in fresh containers, with the lower, cut surface embedded in the agar. Axillary buds will grow, and the process can be repeated with the resulting shoots at 4–8 wk intervals.
6. To initiate callus cultures, remove leaves from shoot cultures and cut into strips (3–4 mm wide) with sterile scalpel and forceps. Transfer to the surface of 20 mL RMP medium in 9-cm Petri dishes, 5 explants/dish, lower epidermis downwards. Seal with parafilm, and culture in culture room or in an incubator at 25°C (illumination not necessary).
7. When sufficient callus has developed from the cut surface of the explants (4–6 wk), excise small pieces (0.3–0.5 cm in diameter), transfer to fresh RMP medium, and culture under the same conditions as for callus initiation. Callus stocks so obtained can be maintained indefinitely by subculturing at intervals of approximately 1 mo.
8. To initiate cell suspension cultures, transfer about 1 g callus to 50 mL RMP medium (without the agar) in a 250-mL Erlenmeyer flask. Development of finely dispersed cell suspensions may be accelerated by cutting the callus into pieces first. Incubate the cultures at 25°C on a rotary shaker at 100 rpm.
9. Subculture the newly formed suspensions after 3 wk by swirling each flask and pouring, at a 3–4 x dilution, into flasks of fresh RMP medium.
10. Subculture the cell suspension cultures in the same way, but at about 5 x dilution, at 10–14-d intervals. After several such culture passages, the suspensions should become more finely dispersed and rapidly dividing. At this stage, the culture passage can be shortened to 7 d, to eliminate lag and stationary phases of the growth curve. (For further comments on the suspension culture procedure, *see* Note 2.)

3.2. Determination of Selective Level of the Antimetabolite (See also *Note 3*)

1. Prepare RMP medium containing the following concentrations of hydroxyproline (mol wt = 131): 0, 0.1, 0.3, 1, 3, 10, 30, and 100 m*M* (*see* Note 4). The media are supplemented with 0.65% Difco Bactoagar and dispensed, 20 mL/dish, into 9 cm sterile Petri dishes. Hydroxyproline, from freshly prepared filter sterilized stock solutions, is

added to flasks of autoclaved medium held at 45°C in a water bath. As addition of stock will in some cases result in up to 33% dilution of the medium, appropriate adjustments must be made to the volumes of medium in some of the flasks prior to autoclaving. Two stock solutions, which should be made up in liquid RMP medium and subsequently pH adjusted to 5.6, are recommended: 300 m*M* (39.3 mg/mL) for 100 and 30 m*M* final concentrations, and 100 mM (13.1 mg/mL) for the remaining concentrations. The volumes of media and stock solution required to give final volumes of 250 mL (sufficient for 12 dishes) are summarized in Table 1.
2. Place small (20–30 mg) pieces of healthy callus, taken from cultures about 3 wk after subculture, on the surface of the media described above. Five inocula should be evenly spaced on each dish. Use 10 dishes/hydroxyproline concentration, to give a total of 50 calluses/treatment. Seal dishes with parafilm and incubate at 25°C.
3. After 4–6 wk, weigh the calluses, determine the mean and standard deviation for each treatment, and, after subtracting the initial inoculum fresh weight, calculate the percentage of the control (no hydroxyproline) fresh weight increase obtained at each hydroxyproline concentration. Plot these values against hydroxyproline concentration (log scale).
4. Select the lowest concentration at which total inhibition is found as the selective concentration for mutant isolation. (This is likely to be 10 or 30 m*M*.)

3.3. Mutagenesis and Selection of Resistant Cell Lines

The following protocol commences with cell suspension cultures. If protoplasts are preferred, *see* Note 5.

1. To ten flasks of freshly subcultured cell suspension cultures, on a 7-d passage cycle, add filter sterilized NEU or NMU to give a final concentration of 0.1 m*M* (*important: see* Note 6).
2. Culture the treated cell suspensions and nonmutagenized controls in the usual way, for 7 d.
3. For each suspension culture, prepare RMP medium; levels of all components and agar for 250 mL, and the selective level of hydroxyproline, stock concentrations, and volumes for nine of the ten flasks are again deduced from Table 1. In this case, however, make each

Table 1
Preparation of Media
Containing a Range of Hydroxyproline Concentrations

Final concentration, mM	Stock solution, mM	Volume of autoclaved medium*, mL	Volume of stock solution added, mL
100	300	167	83
30	300	225	25
10	100	225	25
3	100	242.5	7.5
1	100	250	2.5
0.3	100	250	0.75
0.1	100	250	0.25
0	–	250	0

*Each flask contains agar for 0.65% concentration in a final volume of 250 mL (i.e., 1.63 g).

flask up to a final volume of 200 mL, not 250 mL (omit 50 mL distilled water when preparing the medium). Place these flasks in a water bath at 45°C, to keep the agar molten, for 30–60 min prior to plating the cells.

4. Filter the contents of each flask of cell suspension through two layers of sterile muslin to remove the large aggregates, and pour the fine suspension so obtained (approximately 50 mL) into one of the flasks of molten medium. Mix by swirling, and pour into 12 sterile plastic 9 cm Petri dishes. Over 100 dishes containing the selective level of hydroxyproline (and 12 control dishes without hydroxyproline) should be obtained.

   Cell aggregates arising from mesophyll protoplasts (Note 5 and Fig. 1) are plated in the same way, except that filtration through muslin is not necessary, and the medium should include 0.2*M* glucose or sucrose. In this case, pipet (using a sterile Pasteur pipet) the contents of five Petri dishes directly into each flask of molten medium with or without hydroxyproline, swirl, and pour as above. The number of dishes of protoplast-derived cell aggregates, and hence, the number of flasks of medium required for plating, will depend on the initial yield of protoplasts, which is difficult to standardize.
5. Allow the agar to set, wrap each dish with a strip of parafilm, and incubate at 25°C in the dark.

Freshly isolated protoplasts cultured at 10^5 ml^{-1} in 5 cm petri dishes, 5 ml per dish in K_3 with 0.4 M glucose and 0.1 mM NEU.

↓ 7-10 days

Contents of each dish transferred to a 9 cm dish and diluted with K_3 with 0.4 M glucose to a final volume of 10 ml.

↓ 7-10 days

Dilution with an equal volume of K_3 with 0.3 M glucose. Transferred to fresh 9 cm dishes at 10 ml per dish.

↓ 7-10 days

Dilution with an equal volume of K_3 with 0.2 M glucose. Transferred to fresh 9 cm dishes at 10 ml per dish.

↓ 7-10 days

Contents of 5 dishes pipetted into RMP with 0.2 M glucose, the selective level of hydroxyproline, and agar to give a total volume of 250 ml. Poured into 12 x 9 cm dishes.

↓ 5-8 weeks

Remove resistant colonies.

Fig. 1. A flow diagram for isolation of hydroxyproline resistant colonies from mesophyll protoplasts of *N. plumbaginifolia*. For experimental details *see* protocols in this Chapter and Chapter 42.

6. Examine the dishes at weekly intervals. After 5–8 wk, remove any colonies from selective plates and transfer individually to dishes of fresh RMP medium supplemented with the selective level of hydroxyproline.
7. After 4–7 wk, transfer small pieces (20–30 mg) of callus, from lines continuing to grow on medium containing hydroxyproline, to RMOP medium in Petri dishes. Seal with parafilm and incubate in the culture room at moderate (1000–2000 lx) illumination.
8. After 5–8 wk, excise any adventitious shoots with normal leaves, and transfer to RM medium in Plantcon containers. Incubate for a further 4–6 wk, until a vigorous root system is established. Transfer calluses that did not produce adventitious shoots on RMOP medium to fresh RMB medium. Shoots may be regenerated on RMB, or after several alternating culture passages on RMB and RMOP (*see also* Note 7).
9. Shoot cultures can either be maintained in vitro by transferring nodal cuttings to fresh medium as already described, or transferred to the greenhouse for growth to maturity and genetic analysis. In the latter case, gently remove agar from the roots, and plant individually in 12-cm pots 2/3 full of wet potting compost. To avoid desiccation, cover tightly with cling film for 1–2 wk, until the plantlets are established. When growing vigorously, transfer them to larger pots, where they should flower within 2–3 mo.
10. Check resistance of selected plants by initiating callus from shoot cultures and testing for growth on medium containing selective levels of hydroxyproline as already described (*see also* Note 8).

4. Notes

1. It is important that the seeds should all be properly exposed to the sterilizing solution. A convenient way of ensuring this is to perform the sterilization steps in open, sterile Petri dishes placed close to the sterile air flow. The bags can then periodically be stirred and gently squeezed with sterile forceps, to expel air pockets. In the event of poor germination, a gibberellic acid treatment can be included. Sterilized seeds are placed in an aqueous solution of GA_3 (0.5 mg/mL) for 1 h before transferring to Petri dishes.
2. The object of this protocol is to provide a straightforward set of instructions that will give a good chance of obtaining the desired mutants. For this reason, detailed monitoring of the growth of the

suspension cultures has been excluded, in favor of a simple empirical approach that has proven effective in our hands. If quantitative data on mutation frequency, or direct comparisons between a number of experiments are required, it may be desirable to monitor the growth of the suspensions to ensure that comparable cultures are always used, and to perform cell counts prior to plating to ensure that a uniform cell density (about 10^4 cells/mL) is used in all experiments. Such measurements can pose their own problems in highly aggregated suspension cultures, but some useful guidelines and procedures can be found in references *9* and *10* and Chapter 3, this vol.

3. Callus is the most convenient material for producing a dosage–response curve, as described in the protocol. In the case of hydroxyproline, there is a good correlation between the sensitivity of callus and that of plated cells or protoplasts. For some chemicals, this correlation can be less precise, and it may be wise, if sufficient material is available, to use, as additional selective levels, concentrations above and below that determined by the callus test. A more rigorous alternative is to measure dose response by plating cells or protoplasts at the complete range of antimetabolite concentrations, using exactly the same procedure described in section 3.3. In this case, instead of fresh weight, colony number after a suitable period (4–6 wk for plated cell suspensions, 6–8 wk for plated protoplasts) should be plotted against antimetabolite concentration.
4. Storage times of culture medium: All media described here, including those containing hydroxyproline, can be stored for 3–4 wk, preferably in the cold room, without serious deterioration. Medium in Petri dishes, however, is subject to concentration by evaporation, and should be used within 2 wk of preparation. Even for this period of storage, it is important that they are sealed in bags (e.g., those in which the sterile dishes are obtained from the suppliers) and kept in the cold room. NMU and NEU should be added to the medium only immediately before use.
5. If mesophyll protoplasts are the desired starting material, the protocol for isolation, mutagenesis, and culture of the protoplasts is exactly as described for isolation of chloroplast mutants in Chapter 42, this vol. When the colonies are at the stage where they are ready to be plated in solid medium, they can be handled in the same way as mutagen-treated cell suspension cultures in the current protocol, except that 0.2*M* glucose or sucrose should be included in the RMP medium

used for plating. For clarification, the steps in the isolation of hydroxyproline-resistant mutants from mesophyll protoplasts are outlined in the flow diagram, Fig. 1.

6. Both NMU and NEU, which are closely related, are dangerous carcinogens, and must be handled with extreme care. For details of the precautions required, see Chapters 38 and 42, this vol. Both mutagens are also unstable in aqueous solution, with the consequent advantage that the low levels used here do not have to be washed from the cells. This does mean, however, that mutagen solutions have to be prepared immediately before use. The rate of breakdown is extremely pH-dependent, and it is important that stock solutions are prepared using culture medium at its usual pH (5.6). Further information about these mutagens can be found in ref. *11*. It is also possible to obtain spontaneous mutants, with the mutagenesis treatment omitted, but a larger number of cultures may be required.
7. Shoot regeneration from the resistant cell lines is a critical phase of the procedure. The efficiency with which morphogenesis can be induced by manipulating the hormonal composition of the medium (i.e., using RMOP or RMB) is variable, and in the case of suspension cultures, may be quite low. If difficulties are encountered, regeneration can often be stimulated by the inclusion in the medium of Ag^+ ions in the form of silver nitrate, at 10 or 50 mg/L (*12*).
8. Resistance to the antimetabolite can also be checked in seedling progeny and callus derived from them.

References

1. Dix, P. J. (1986) Cell line selection, in *Plant Cell Culture Technology* (Yeoman, M. M., ed.), Blackwell Scientific, Oxford, pp. 143–201.
2. Bright, S. W. J., Ooms, G., Foulger, D., Karp, A., and Evans, N. (1986) Mutation and tissue culture, in *Plant Tissue Culture and Its Agricultural Applications* (Withers, L. A. and Alderson, P. G., eds.), Butterworths, London, Wellington, pp. 431–450.
3. Maliga, P. (1984) Isolation and characterization of mutants in plant cell culture. *Ann. Rev. Plant Physiol.* **35,** 519–542.
4. Flick, C. E. (1983) Isolation of mutants from cell culture, in *Handbook of Plant Cell Culture,* vol. 1, *Techniques for Propagation and Breeding* (Evans, D. A., Sharp, W. R., Ammirato, P. V., and Yamada, Y., eds.), Macmillan, New York, pp. 393–441.
5. Gonzales, R. A. and Widholm, J. M. (1985) Selection of plant cells for desirable characteristics: Inhibitor resistance, in *Plant Cell Culture, a Practical Approach* (Dixon, R. A., ed.), IRL Press, Oxford, pp. 67–78.
6. Widholm, J. M. (1984) Induction, selection and characterization of mutants in carrot cell cultures, in *Cell Culture and Somatic Cell Genetics of Plants,* vol 1, *Laboratory Pro-*

cedures and Their Applications (Vasil, I. K., ed.), Academic Press, New York, pp. 563–570.

7. Maliga, P. (1984) Cell culture procedures for mutant selection and characterization in *Nicotiana plumbaginifolia*, in *Cell Culture and Somatic Cell Genetics of Plants,* vol 1, *Laboratory Procedures and Their Applications* (Vasil, I. K., ed.), Academic Press, New York, pp. 552–562.
8. Dix, P. J., McLysaght, U. A., and Plunkett, A. (1986) Salt stress: Resistance mechanisms and *in vitro* selection procedures, in *Plant Tissue Culture and Its Agricultural Applications* (Withers, L. A. and Alderson, P. G., eds.), Butterworths, London, Wellington, pp. 469–478.
9. King, P. J. (1984) Induction and maintenance of cell suspension cultures, in *Cell Culture and Somatic Cell Genetics of Plants,* vol. 1, *Laboratory Procedures and Their Applications* (Vasil, I. K., ed.), Academic Press, New York, pp. 130–138.
10. Gilissen, L. W. J., Hänisch ten Cate, C. H., and Keen, B. (1983) A rapid method of determining growth characteristics of plant cell populations in batch suspension culture. *Plant Cell Rep.* **2,** 232–235.
11. Hageman, R. (1982) Induction of plastome mutations by nitrosourea compounds, in *Methods in Chloroplast Molecular Biology* (Edelman, M., Hallick, R. B., and Chua, N. H., eds.), Elsevier/North-Holland, Biomedical Press, Amsterdam, p. 119.
12. Purnhauser, L., Medgyesy, P., Czakó, M., Dix, P. J., and Márton, L. (1987) Stimulation of shoot regeneration in *Triticum aestivum* and *Nicotiana plumbaginifolia* viv. tissue cultures using the ethylene inhibitor $AgNO_3$. *Plant Cell Rep.* **6,** 1–4.

Chapter 42

Selection of Chloroplast Mutants

Paul F. McCabe, Agnes Cséplö, Aileen M. Timmons, and Philip J. Dix

1. Introduction

The chloroplast genome encodes a number of proteins, including thylakoid proteins and the large subunit of ribulose biphosphate carboxylase, associated with the structure and function of the chloroplast (*1,2*). In addition, many components of the chloroplast translational machinery, such as all of the RNAs and some of the ribosomal proteins, are coded by the chloroplast DNA. Although there have been numerous investigations into the genetics of algal chloroplasts, similar studies with higher plants have been hampered by the uniparental (maternal) pattern of transmission of chloroplasts observed in most species, and the shortage of suitable genetic markers (*3,4*).

Protoplast fusion has provided a means whereby cells with mixed cytoplasm (cybrids) can be obtained (*5,6*), and the recombination of chloroplast DNA (*7*) and transfer of the organelles (*5,6*) investigated. Tissue culture procedures have also been applied to selecting an increasing number of genetic markers, including chlorophyll-deficient (*8*) and antibiotic-resistant (*4,9*) and herbicide-resistant (*10*) mutants.

From: *Methods in Molecular Biology*, vol. 6, *Plant Cell and Tissue Culture*
Edited by Jeffrey W. Pollard and John M. Walker, ©1990 by The Humana Press

Quantitiative mutagenesis of chloroplasts is complicated by the large number of copies of ptDNA in the cell, but, in the absence of a plastome mutator gene, the rate of mutagenesis has been estimated to be as low as 10^{-9} (*11*). Workers in this field have, therefore, turned to the use of efficient plastome-targeted mutagens, such as nitroso-methylurea (*12*). This has facilitated selection of chloroplast mutants in *Nicotiana* (*9,11*) and *Lycopersicon* (*13*).

Here we describe detailed protocols for the isolation of such mutants, starting from protoplasts or leaf strips. The protoplast procedure is effective for *Nicotiana plumbaginifolia* and, with modifications, can be used to obtain herbicide-resistant as well as antibiotic-resistant (streptomycin, lincomycin) mutants. The leaf strip procedure has now been used successfully for several species (*Nicotiana tabacum, N. plumbaginifolia, N. sylvestris, Solanum nigrum,* and *Lycopersicon peruvianum*), but so far, only to obtain streptomycin- and spectinomycin-resistant and chlorophyll-deficient mutants.

2. Materials

1. Shoot cultures of the species to be used (*see above*). The procedures for obtaining and maintaining these cultures, starting from seed, are as described in Chapters 11 and 41, this vol.
2. Protoplast enzyme solution: Modified K3 medium (*see* Appendix) containing 0.4*M* sucrose and 0.5% Driselase (Kyowa Hakko Kogyo Co., Tokyo, Japan (*see* Note 1). K3 can be filter sterilized, or autoclaved in a pressure cooker, and stored for up to 4 wk, preferably in the cold room. Driselase must be added immediately before use, and the resulting enzyme solution filter sterilized.
3. Protoplast wash solution, W5 (*see* Appendix; 14): 150 m*M* NaCl, 125 m*M* $CaCl_2$, 5 m*M* KCl, 5 m*M* glucose, pH 5.6. W5 may be autoclaved and stored in the cold room for up to 2 mo.
4. Protoplast culture medium: Modified K3 supplemented with 0.4*M* glucose, 0.1 mg/L 2,4-dichlorophenoxyacetic acid (2,4-D), 0.2 mg/L 6-benzylaminopurine (BA), and 1 mg/L naphthaleneacetic acid (NAA). Also, the same medium with glucose levels reduced to 0.3 and 0.2*M*.
5. Regeneration medium (RMOP). Contains (per liter): 4.6 g MS salts (*see* Appendix), 20 g sucrose, 100 mg meso-inositol, 1 mg thiamine-HCl, 1 mg BA, 0.1 mg NAA, pH 5.6, solidified with 0.65% Difco Bactoagar.

6. Leaf strip medium for *L. peruvianum*: Contains (per liter) 4.6 g MS salts, 20 g sucrose, 100 mg meso-inositol, 1 mg thiamine-HCl, 0.5 mg nicotinic acid, 0.5 mg pyridoxine-HCl, 0.2 mg indoleacetic acid (IAA), and 2 mg zeatin (Sigma), pH 5.6, solidified with 0.65% Difco Bactoagar.
7. Leaf strip medium for *S. nigrum:* As for *L. peruvianum*, except zeatin is replaced by 1 mg/L BA.
8. RM medium: Contains (per/liter) 4.6 g MS salts and 20 g sucrose, pH 5.6, solidified with 0.65% Difco Bactoagar.
9. RM solution: as for RM medium, but without the agar.
10. 60 µm mesh nylon bolting cloth.
11. Mutagens: *N*-nitroso-*N*-methylurea (NMU); *N*-nitroso-*N*-ethylurea (NEU) (*see* Note 2).
12. Protective clothing: plastic or rubber apron and disposable gloves, and industrial organic vapor cartridge respirator with cartridges, and prefilters (obtainable from "Sa-fir," East Hoathly, East Sussex, England).
13. Large sheets of absorbent paper backed with foil.
14. 5*M* NaOH.
15. 5% (w/v) streptomycin sulfate in H_2O, filter sterilized.
16. 5% (w/v) lincomycin hydrochloride in H_2O, filter sterilized.
17. Terbutryn (Chem. Service, West Chester, PA, USA).
18. Culture room facilities. Preferred conditions: 25°C, 1500 1x, 16-h day.
19. 0.4*M* glucose.

3. Methods

Mutant isolation is described below, using two different starting materials, leaf strips and mesophyll protoplasts. The use of leaf strips is technically more straightforward and is described first. Healthy leaves from shoot cultures of any of the species mentioned in the introduction can be used in exactly the same way, the only differences being in the culture medium required. In our experience, *S. nigrum* gives substantially higher yields of streptomycin-resistant mutants than the other species. Albino mutants can readily be obtained in the same experiments, as described in Note 3.

The rest of this methods section deals with the isolation of streptomycin- and lincomycin-resistant mutants of *N. plumbaginifolia* from cultures initiated from mesophyll protoplasts. Modifications of this procedure, necessary to obtain mutants resistant to terbutryn and other photosynthetic herbicides, are described in Note 4.

3.1. Isolation of Streptomycin-Resistant Mutants from Leaf Strips

1. Prepare mutagen solutions after carefully reading Note 2. Prepare a stock solution containing 80 mg NMU in 20 mL of RM solution, and filter sterilize. For a 1 m*M* solution, add 2.6 mL of stock to 97.4 mL of autoclaved RM solution in a wide-necked 250 mL Erlenmeyer flask. For a 5 m*M* solution, add 12.9 mL to 87.1 mL of RM solution. It is important that the RM solutions used have pH values of 5.5–6.0 to enhance the stability of the mutagen.
2. Remove leaves from shoot cultures and cut into small strips (2–3 mm wide by 5–15 mm long, depending on leaf size). Add 250 strips to 100 mL of each (1 and 5 m*M*) NMU solution and to 100 mL of RM solution (nonmutagenized control).
3. Incubate the flasks on a rotary shaker (about 50 rpm) at 25°C for 90 min.
4. Decant the mutagen solution, and wash the leaf strips 4–5x in a large excess of RM solution of sterile distilled water (pH adjusted to 5.5.–6.0).
5. For each treatment, place 200 of the strips on the surface of selective medium in 40 plastic Petri dishes (5 strips/dish). The selective media are RMOP for *Nicotiana* species, and the media described in Materials for *S. nigrum* and *L. peruvianum*, in each case supplemented with 500 mg/L streptomycin sulfate (1000 mg/L can also be used for *Nicotiana* spp. and *S. nigrum*). Streptomycin sulfate is highly soluble, and the medium is prepared by addition of a small aliquot from the concentrated, filter-sterilized stock solution to the autoclaved culture medium prior to pouring into dishes. Place the remaining 50 strips on the same medium without streptomycin. Seal all dishes with parafilm and incubate in the culture room.
6. After about 40 d, the leaf strips on control dishes should show prolific shoot regeneration, whereas those on medium containing streptomycin should be bleached and show little morphogenesis. Green nodules will appear at the edges of some of the bleached leaf strips (from one or both mutagen treatments), and most of these will differentiate into shoots. When shoots have at least two leaves that are normal, not "vitreous" (glassy and translucent), in appearance, remove cleanly with a scalpel, and transfer, with cut stem embedded in the agar, to RM medium for rooting. Continue to incubate in the culture room.

7. After 4–8 wk, plantlets with vigorous roots are obtained. Remove a single leaf, cut into strips, and test for streptomycin resistance on the same medium as that originally used for mutant selection. Typically, resistance is expressed in the mutants by the retention of chlorophyll and the appearance of numerous green adventitious shoots, although growth is usually slower than in cultures grown in the absence of streptomycin.
8. Rooted plantlets can be either propagated in vitro by nodal cuttings, or transferred to soil for growth to maturity and genetic analysis. Both these procedures are described in Chapter 41, this vol. Freshly potted plantlets must be covered with cling film or an inverted glass or plastic container for about 10 d, to prevent desiccation.

3.2. Isolation of Streptomycin or Lincomycin-Resistant Mutants from Protoplast Cultures of N. plumbaginafolia

1. Remove healthy, fully expanded leaves from shoot cultures, slice finely using a sterile razor blade (in a holder) or scalpel and forceps, and transfer to sterile protoplast enzyme solution in 9 cm Petri dishes. Typically, 10 mL of solution in a dish should be sufficient for 3–4 leaves, and three such dishes should provide enough protoplasts for an experiment. Protoplast yields can be variable, however, so it may be necessary to start with more material.
2. Incubate the dishes overnight (12–16 h) at 25°C in the dark.
3. Swirl the dishes several times to liberate protoplasts from leaf debris, remove the solution with a Pasteur pipet, and filter through 60 µm nylon mesh into a 100 mL Erlenmeyer flask.
4. Transfer the protoplast preparation to 10 mL capped glass centrifuge tubes (sterile) and spin at about 300 x *g* for 3 min.
5. Intact protoplasts float and form a tight green band at the surface of the medium. Remove this carefully with a Pasteur pipet and transfer to a clean tube (not more than 1 mL/tube). Fill the tubes with W5 solution, cap, invert to ensure thorough mixing, and spin at about 50 x *g* for 2 min.
6. Pipet off the supernatant, and resuspend the protoplast pellet in a small volume of protoplast culture medium (K3) supplemented with 0.4*M* glucose. Mix the contents of the tubes, count the intact (spherical, with an uninterrupted plasma membrane) protoplasts, and dilute to 10^5m/L with the culture medium.

7. Transfer to 5 cm Petri dishes, 5 mL/dish, using either previously calibrated Pasteur pipets, or graduated pipets with wide-bore tips.
8. To individual dishes, add (to a final concentration of 0.1 or 0.3m*M*) NEU from a concentrated stock prepared in the culture medium and filter sterilized. (*Important:* read Note 2 carefully before using the mutagen.) Wrap all dishes with parafilm and incubate in the culture room at low light intensity (ca. 100 lx).
9. After 7–10 d, providing divisions can be observed in the protoplasts, dilute the protoplasts 2x with fresh K3 medium with 0.4*M* glucose. To do this, Pasteur pipet the contents of each dish into a 9 cm dish and add 5 mL of medium. Seal and incubate as before. There is no need to wash out the mutagen, since it is unstable and breaks down within 48 h at the pH used.
10. Monitor the development of the protoplast-derived cell aggregates and make dilutions at suitable intervals (*see* Note 5). In a good preparation, these intervals should be of 7–10 d. Each dilution should be 2x, and lead to a doubling of the number of dishes containing 10 mL culture. For the first dilution, use K3 medium with the glucose reduced to 0.3*M*, and for the second, glucose reduced to 0.2*M*. Within 10–14 d of the latter dilution, numerous microcolonies (about 1 mm in diameter) should be visible, and the cultures are then ready for plating into solid medium.
11. For every four (more if there is a low colony density) dishes, prepare 500 mL of RMOP medium with 0.2*M* glucose (instead of 2% sucrose), 0.65% agar, and 1000 mg/L streptomycin sulfate or lincomycin hydrochloride. The antibiotics are added from the concentrated filter-sterilized stock solution to autoclaved medium, held at 45°C in a water bath.
12. Using a fine-tipped Pasteur pipet, remove the excess K3 medium from the cultures to be plated. This is best done by tilting the dishes slightly and allowing the microcolonies to settle, so that the medium can be removed from above them. Then, using a broad-tipped pipet, wash the contents of four dishes into the 500 mL molten medium by repeatedly transferring small amounts of the medium to the dishes and sucking up again, together with the colonies.
13. Mix well, and pour into 9 cm Petri dishes (about 20 mL/dish). Allow the agar to set, wrap the dishes with parafilm, and incubate in the culture room at 1000–1500 lx illumination.
14. Numerous white colonies should appear after 1–2 mo. Streptomycin-

or lincomycin-resistant colonies are green and are easily selected visually. Pick them off when large enough and transfer to the same medium, but with sucrose reduced to 0.1*M*.

15. Resistant colonies will continue to grow and remain green. Subculture them onto RMOP medium, without the antibiotic, for shoot regeneration. Regenerated shoots are handled as described for mutants isolated from leaf strips.

4. Notes

1. Driselase powder is a crude preparation containing much insoluble material that can quickly block the millipore filter. This can be prevented by either spinning for a few minutes in a bench centrifuge, or filtering, to obtain a clean solution prior to filter sterilization.
2. Both mutagens, which are closely related, are dangerous carcinogens and must be used with great care. We recommend the use of a respirator, and protective gloves and apron, during all manipulations involving the mutagens. It is important to avoid skin contact. All working surfaces, balances, and the like, where spillage might have occurred, should be washed down immediately after use. Placing large sheets of absorbent paper backed with foil on the laminar-flow work surface helps to contain any spillage. It is a good idea to exclude other workers from the work area while manipulations with NEU or NMU are in progress. In the event of skin contact, and routinely after use, wash hands in soap and water gently, avoiding excessive rubbing of the skin. Do not use the mutagens in alkaline solutions because they are very unstable. After filter sterilizing mutagen solutions, do not remove the syringe from the millipore unit immediately, since the pressure that has built up in the syringe can result in the release of an aerosol of the mutagen.

   Mutagen solutions should be inactivated before disposal. Add an excess of 5% NaOH in the fume hood, and leave open overnight, before pouring down the sink and chasing with a large volume of tap water. Contaminated apparatus should also be treated with 5% NaOH, but in this case, after an overnight soaking, a second wash (greater than 1 h) is recommended, followed by a thorough rinse in running tap water. Additional advice on the use of these mutagens is given in ref. *12*.

3. For all the species mentioned, albino mutants have also been obtained from leaf strips. These arise in response to the mutagenesis treatment, on the dishes from which the selective agent (antibiotic or herbicide) has been excluded. Among the normal green shoots differentiating from the leaf strips, some albino or variegated shoots are frequently obtained. Albino shoots can be rooted and maintained on RM medium, in the same way as other shoot cultures. Albino shoots can be obtained from variegated ones by dissecting out white sectors and culturing on the appropriate regeneration medium.
4. The procedure for isolating antibiotic-resistant mutants from protoplast cultures of *N. plumbaginifolia* can be applied to the selection of mutants resistant to herbicides that inhibit photosynthesis, providing a selective medium permitting photomixotrophic growth is used. This is achieved by lowering to 0.3% the glucose level in the RMOP medium in which the microcolonies are plated. In order to reduce the osmotic stress resulting from plating in such a low-sugar medium, an additional dilution step (with K3 medium plus 0.1*M* glucose) is introduced into the protoplast (e.g., atrazine, simazine) a suitable selective level is $10^{-4}M$, and selection is based on the greening of colonies, exactly as in the case of antibiotic-resistant mutants. After retesting for resistance on selective medium, shoots are regenerated by transfer of small callus pieces to RMOP medium without terbutryn. Mutants resistant to metobromuron and bromoxynil have also been obtained in this way.
5. The instructions for the gradual dilution of protoplast cultures with fresh medium of decreasing glucose concentration are given as accurately as possible. The 7–10 d interval should work, but careful monitoring of the cultures is desirable. If the growth rate of the colonies seems to be slow, a longer interval must be used. On the other hand, rapid growth rates, especially if accompanied by browning or the appearance of dead cells, indicates a requirement for more rapid dilution.

References

1. Dyer, T. A. (1985) The chloroplast genome and its products, in *Oxford Surveys of Plant Molecular and Cell Biology*, vol 2 (Miflin, B. J., ed.), Oxford University Press, New York, pp. 147– 177.
2. Borner, T. and Sears, B. B. (1986) Plastome Mutants. *Plant Molecular Biology Reporter* 4, 69–92.

3. Maliga, P. (1986) Cell fusion to introduce genetic information coded by chloroplasts and mitochondria in flowering plants, in *Molecular Development Biology* Liss, New York, pp. 45–53.
4. Maliga, P. Breznovits, A., and Márton, L. (1973) Streptomycin resistant plants from haploid callus culture of tobacco. *Nature* (*New Biol.*) **244,** 28–30.
5. Galun, E. and Aviv, D. (1983) Cytoplasmic hybridization: Genetic and breeding applications, in *Handbook of Plant Cell Culture*, vol 1 (Evans, D. A., Sharp, W. R., Ammirato, P. V., and Yamada, Y., eds.), Macmillan, New York, pp. 358–392.
6. Pelletier, G. (1986) Plant organelle genetics through somatic hybridization in *Oxford Surveys of Plant Molecular and Cell Biology*, vol 3 (Miflin, B. J., ed.), Oxford University Press, pp. 96–121.
7. Medgyesy, P., Fejes, E., and Maliga, P. (1985) Interspecific chloroplast recombination in a *Nicotiana* somatic hybrid. *Proc. Natl. Acad. Sci. USA* **82,** 6960–6964.
8. Svab, Z. and Maliga, P. (1986) *Nicotiana tabacum* mutants with chloroplast encoded streptomycin resistance and pigment deficiency. *Theor. Appl. Gent.* **72,** 637–643.
9. Cséplö, A. and Maliga, P. (1984) Large scale isolation of maternally inherited lincomycin resistance mutations in diploid *Nicotiana plumbaginifolia* protoplast cultures. *Mol. Gen. Genet.* **196,** 407–412.
10. Cséplö, A., Medgyesy, P., Hideg, E., Demeter, S., Márton, L., and Maliga, P. (1985) Triazine resistant *Nicotiana* mutants from photomixotrophic cell cultures. *Mol. Gen. Genet.* **200,** 508–510.
11. Fluhr, R., Aviv, D., Galun, E., and Edelman, M. (1985) Efficient induction and selection of chloroplast-encoded anitbiotic resitant mutants in *Nicotiana. Proc. Natl. Acad. Sci. USA* **82,** 1485–1489.
12. Hagemann, R. (1982) Induction of plastome mutations by nitroso-urea-compounds, in *Methods in Chloroplast Molecular Biology* (Edelman, M., Hallick, R. B., and Chua, N. H., eds.), Elsevier Biomedical Press, Amsterdam, pp. 119–127.
13. Hosticka, L. P. and Hanson, M. R. (1984) Induction of plastid mutations in tomatoes by nitrosomethylurea. *J. Hered.* **75,** 242–246.
14. Maliga, P. (1984) Cell culture procedures for mutant selection and characterization in *Nicotiana plumbaginifolia*, in *Cell Culture and Somatic Cell Genetics of Plants,* vol 1 (Vasil, I. K., ed.), Academic, New York, pp. 552–562.

Chapter 43

Large-Scale Culture of Plant Cells

Alan H. Scragg and Michael W. Fowler

1. Introduction

The large-scale or mass cultivation of plant cells is the growth of plant cell suspensions at volumes above those normally produced in shake flasks, that is, above 1 L. Attempts to grow plant cells in fermenters or bioreactors started in the early 1960s with converted carboys. The area has developed steadily, such that today bioreactors in excess of 5000 L have been used successfully for large-scale plant cell culture (*1,2*). Much of the early work was carried out using bioreactors designed for microbial culture. It was soon found, however, that although plant cell suspensions appear to be similar in many ways to microbial cultures, there are, in fact, key differences that can have a significant influence on large-scale cultivation. Plant cells are large, 20–40 μM in diameter, and up to 100 μM in length; further, they rarely occur as single cells, but as aggregates of up to 2 mm in diameter (Fig. 1). The individual plant cell soon after division is typically rounded, containing considerable amounts of cytoplasm; however, as it ages, the cell expands and becomes dominated by a large vacuole. In consequence, the overall metabolic activity is low compared with microbial cells, which in turn gives a very slow growth rate (measured in days,

From: *Methods in Molecular Biology*, vol. 6, *Plant Cell and Tissue Culture*
Edited by Jeffrey W. Pollard and John M. Walker, ©1990 by The Humana Press

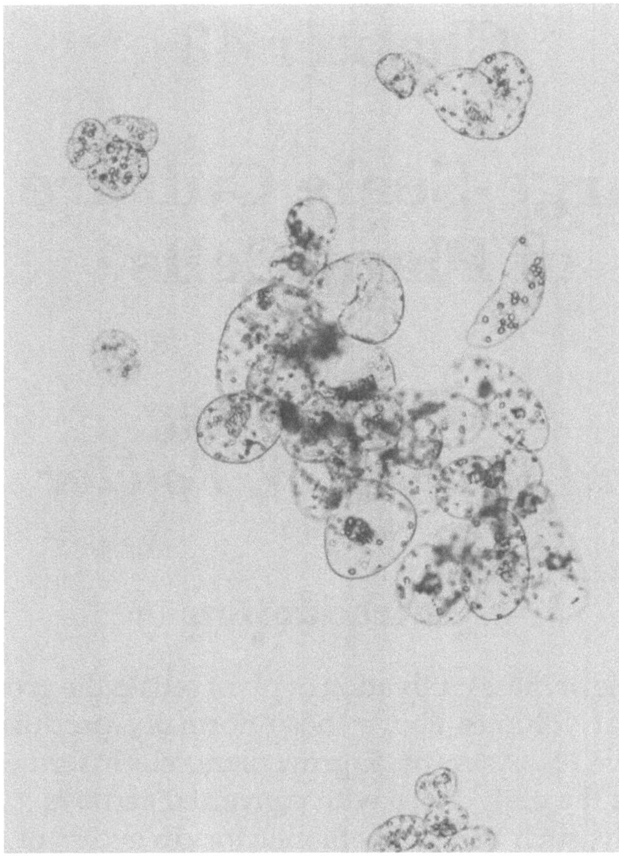

Fig. 1. Cell aggregates in a culture of *Catharanthus roseus.*

typically 2–3 d doubling time) and a low oxygen demand. The large size, rigid cellulose cell wall, and extensive vacuole have also been used to explain the apparent sensitivity to shear off plant cell suspensions. Quite obviously, these distinctive properties influence approaches to the cultivation of plant cell suspensions in bioreactors. The consequences of these properties on cultivation are as follows: mixing is as important as aeration, mixing has to be achieved under low shear conditions, and the bioreactor runs will be long (10–30 d). The long bioreactor runs mean that a strict observance of sterility is required and that water loss can be a problem. Because of the long run times, the number of experiments that can be performed with one bioreactor are also limited. The cell aggregates can cause an additional problem with sampling, blocking the ports, or forming a nonhomogeneous sample as a result of rapid settling.

In the face of these problems, various alternative bioreactor designs have been considered. Many of the original studies were carried out in stirred-tank bioreactors using slow agitation speeds, but the development of the airlift bioreactor design for single-cell protein in the 1970s, with its low shear characteristics, offered an important alternative for plant cell suspensions (*3*). The design has subsequently proved successful in the culture of a number of plant cell lines, and has also been used for animal cell cultures. Recent research has shown, however, that plant cell suspensions are not as sensitive to shear as was first thought, and that the stirred-tank bioreactor cannot be ignored, in particular when considering scale-up.

There are perhaps two reasons why plant cells should be grown in a bioreactor. First, a bioreactor allows close control and monitoring of culture conditions, particularly the gassing regime, which are not possible in shake flasks and it also allows large and numerous samples to be taken for enzyme or product analysis. Secondly, if a large-scale plant cell culture process is to be developed, the response of the culture to scale-up must be determined, since this cannot be predicted from shake-flask experiments. The types of processes for which plant cell culture might be considered as a process system include the production of biomass (e. g., ginseng), the production of certain secondary products (*see* Chapter 48, this vol.) for the food and pharmaceutical industry (e. g., quinine), and the production of enzymes. In this chapter, we will consider the growth and alkaloid formation (ajmalicine and serpentine), by a suspension culture, of *Catharanthus roseus* in a 30-L (working volume) airlift bioreactor.

2. Materials

1. There are no general rules for the formulation of a medium that will support both growth and product formation in plant cell cultures, other than that secondary product formation is, in general, encouraged by removal of the auxin 2,4-dichlorophenoxyacetic acid (2,4-D). However, a medium (M3) has been developed (*4*) that supports both growth and production formation in *C. roseus*. The medium is based on the medium of Murashige and Skoog (*5; see* Appendix), but contains 20 g-L sucrose, 0.1mg-L kinetin, and 1.0 mg-L Naphthaleneucetic acid (NAA), and lacks KH_2PO_4. The other most popular medium, Gamborg's B5 (*6, see* Appendix), is supplemented with 0.1 mg-L kinetin and 2 mg-L 2,4-D.
2. YEPG Plates: These consist of 1% yeast extract, 1% peptone, 1% glucose, and 0.8% agar.

3. Methods

3.1. Inoculum

The *C. roseus* cultures are subcultured every 2 wk as is typical for most plant cell suspensions. Twenty milliliters ofculture are added to 100 mL of fresh medium, M3 in this case, in a 250-mL shake flask. The flask is closed with a double layer of aluminum foil, incubation is at 25°C, and the culture is shaken at 150 rpm in a gyrorotatory shaker. Normally the 1:5 ratio of inoculant to new medium is reduced for the bioreactors to 1:10, which means that 3 L are required for the 30-L bioreactor. The 3-L inoculum is grown in 2-L shake flasks containing 1 L of medium and inoculated with 100 mL of cells (1:10 ratio). Therefore, as shown in Fig. 2, preparation of the inoculum requires starting some 2 wk before seeding the vessel. Despite using laminar flow cabinets and strict sterile techniques, cultures can become contaminated or grow poorly. If this happens, one can be set back 2–4 wk, depending on the stage of growth of the stock lines when problems occur. Thus, it is prudent to include two extra 2-L flasks; someone will always find a use for them after 2 wk!

3.2. Preparation of Bioreactors

One week before the 3-L inoculum is available (Fig. 2), the bioreactor should be cleaned and assembled. The bioreactor used in this study is an L. H. Engineering 30-L airlift bioreactor as shown in Fig. 3. A schematic diagram of its construction is shown in Fig. 4. The bioreactor is constructed from two Q. V. F. borosilicate glass sections, 1500 cm and 30 cm in length, joined by a stainless-steel (316) neck flange. The neck flange is arranged with 5 ports (5 x 40 mm) approximately 170 mm apart, designed to take pH, temperature, and other probes. The upper glass section (30 cm) forms an expansion area. The top and base are provided with stainless-steel plates, into which are placed another series of ports. The base plate has a single port, taking the sample valve. The sample valve is screwed up into the vessel for sampling, and has an independent supply of steam, so that it can be sterilized before and after sampling. The headplate has eight ports fitted with blue rubber septa, through which needles can be inserted for inoculation or pH control. The needles as supplied had diameters of 3.5 mm, but these have been enlarged to 6.5 mm, in order to accommodate aggregated plant cell suspensions. When not in use, the septa are sealed with stainless-

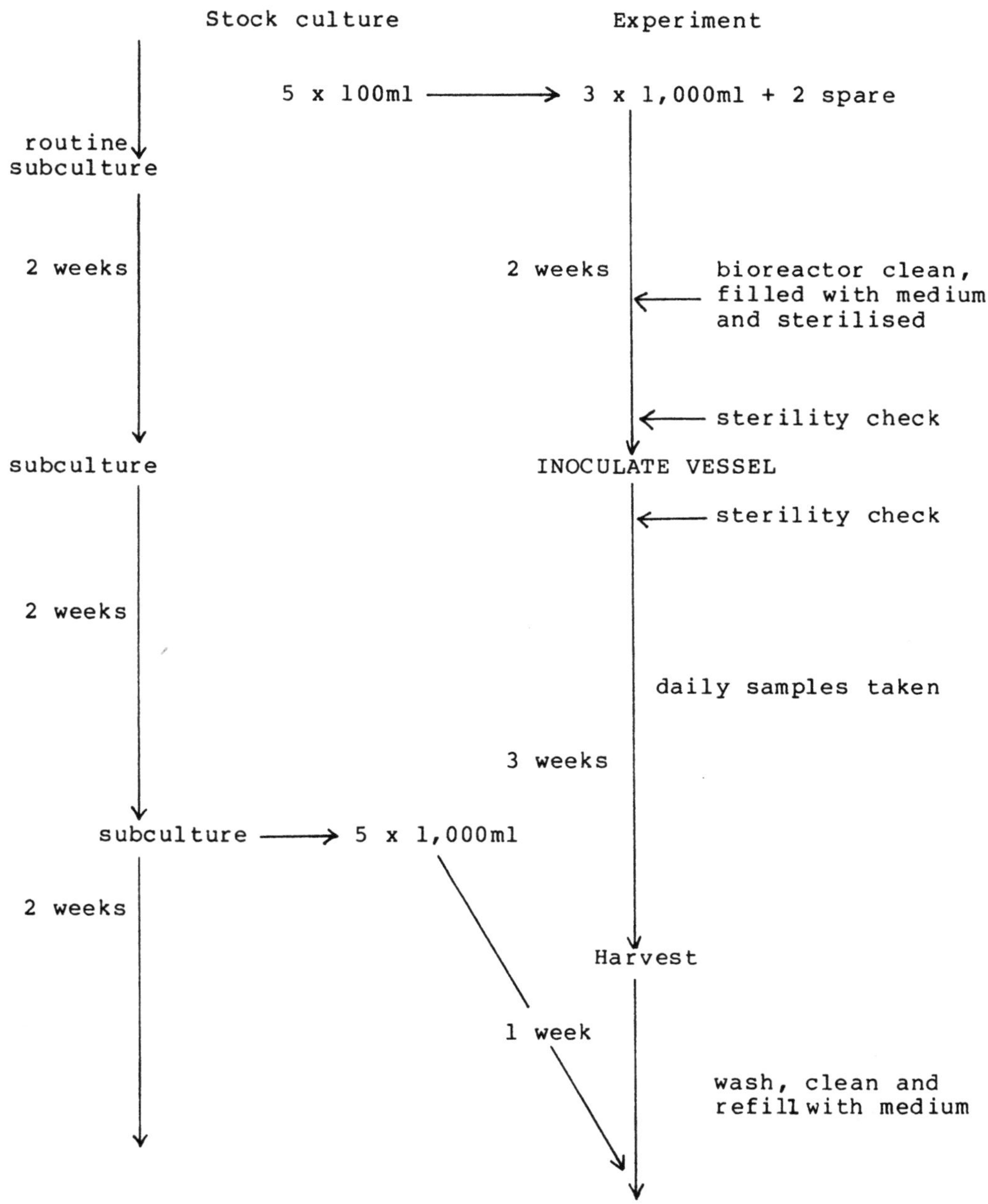

Fig. 2. Sequence of events during the preparation and running of a 30-L airlift bioreactor.

Fig. 3. The glass vessel on the far right is a 30-mL L. H. Engineering airlift bioreactor with associated controls.

steel end caps. In addition, the headplate has a port for an antifoam probe (not used), pressure gage, and pressure release valve. The draft tube is constructed of stainless-steel (double skinned), which acts as a heat exchanger and air sparger. The sparger is connected to the draft tube, through the head plate, to a ring sparger with 20 holes, 3/4 mm in size, angled to the center. Three further connections are made to the draft tube through the head plate, two for water inlet and outlet, and the third as a stabilizer. The exit gas leaves the bioreactor through the top plate via a

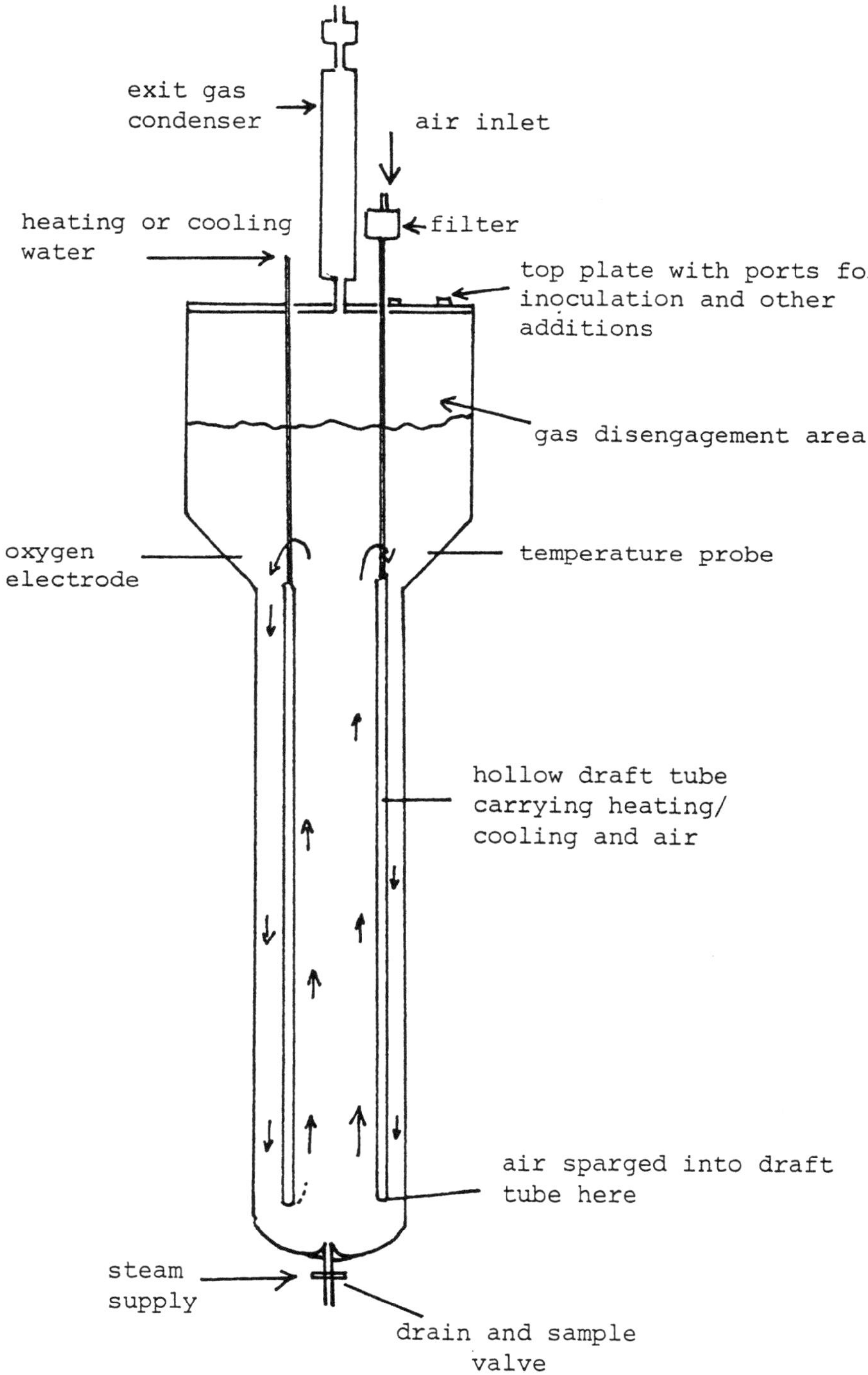

Fig. 4. A schematic diagram of the L. H. Engineering 30-L airlift bioreactor.

multitube condenser and PAL filtration unit (to stop grow-back). The inlet air is passed through a PAL cartridge filter connected by silicon tubing to a quick connector on the top plate.

As the vessel is cleaned and assembled, the various septa and "O"-rings are inspected for wear, and replaced if worn. Before each run, the vessel is pressure tested and, if pressure is maintained, filled with medium. Before the medium is added, a temperature probe and a dissolved oxygen probe (Ingold) are inserted into the neck flange. Other probes are not normally included, since antifoam addition is not carried out, and the culture pH is allowed to drift.

3.3. Sterilization

The bioreactor is sterilized by passing superheated water from a thermocirculator (130°C) through the draft tube to raise the temperature of the bioreactor to 121°C. The water supply to the condenser is turned off and the sparger turned to bleed into the headspace. The inlet PAL filter and needle are sterilized separately (121°C for 40 min). The temperature of the bioreactor is maintained at 121°C for 1 h and then turned down to 25°C. When the bioreactor has cooled to 90°C, the sterile PAL filter is connected and, as the temperature drops further, air is introduced into the vessel, normally at a rate of 6 L/min (0.2 vvm).

Once the bioreactor has reached 25°C, the dissolved oxygen probe (dO_2) is connected to the controller and the value set at 100%. A portion of the exit gas (200 mL/min) is diverted to determine exit gas carbon dioxide and oxygen; the same is done for the inlet gas. The bioreactor is then incubated for at least 3 d, in order to detect any contamination. The medium after sterilization should be very pale yellow and clear; any microbial contamination will show as cloudiness and an accompanying drop in dissolved oxygen, as well as increase in exit carbon dioxide. If the M3 medium is left too long (1–2 wk), a precipitate can develop that can be mistaken for contamination.

3.4. Inoculation

When the inoculating cells are ready (day 14), 3 L are transferred to a large 5-L inoculating flask (Fig. 5) fitted with a side arm, connected by silicon tubing to an inoculating needle and stainless-steel cover. The flask and needle have previously been sterilized (121°C for 40 min), and the flask is fitted with a muslin-covered cotton-wool bung. The 3-L inoculation,

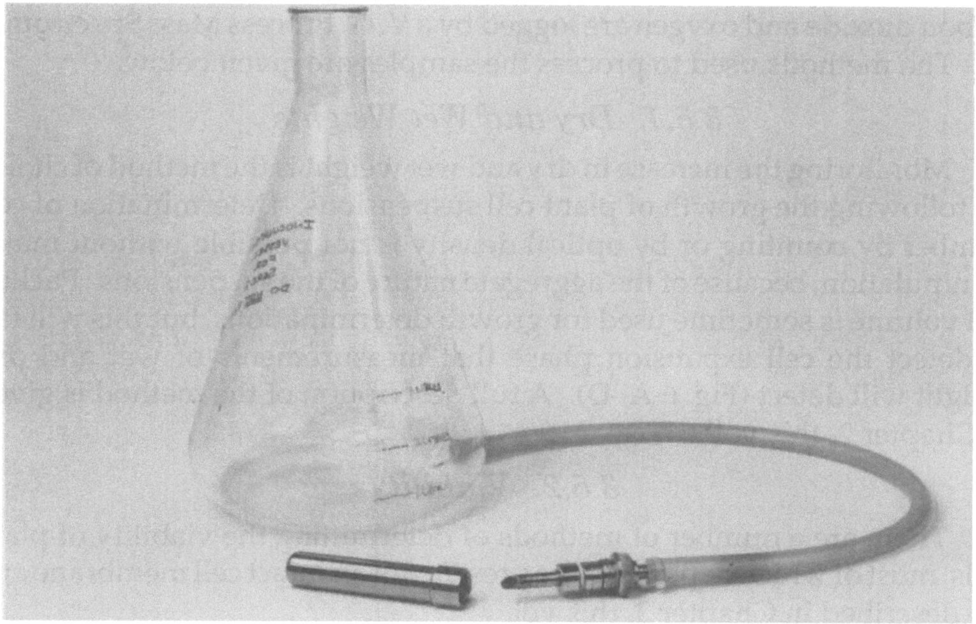

Fig. 5. An inoculation flask with needle and cover.

once mixed, is sampled for what is known as the preinoculation sample. The needle is inserted through a blue sepum on the top plate, and the cells run into the vessel. The sparging air is turned off to reduce back-pressure, and the flask is swirled to stop settling out, which can block the silicon tubing. Once the cells have been transferred, the needle is removed and replaced rapidly by a presterilized end cap. The air is turned on again and the vessel sampled for the various parameters, and for contamination, by plating out some of the culture on YEPG plates (*see* 2, under Materials).

3.5. Sampling

The advantage of a bioreactor of 30-L working volume is that reasonably sized samples (up to 250 mL) can be taken, and therefore, a wide range of parameters can be followed. Traditionally, because of the slow growth of plant cells, samples were taken every 3 d for up to 30 d, but with the growth rates achieved in bioreactors, samples are taken each day. The samples are normally used to determine dry weight, wet weight, viability, total carbohydrate, product (serpentine and ajmalicine), and pH. The dissolved oxygen is noted at the time of sampling, and the inlet and outlet carbon dioxide and oxygen are logged by a V. G. Process Mass Spectrometer. The methods used to process the samples are given below.

3.5.1. Dry and Wet Weights

Monitoring the increase in dry and wet weight is the method of choice for following the growth of plant cell suspensions. Determination of cell number by counting or by optical density is not possible without major manipulation, because of the aggregate nature of the suspensions. Packed cell volume is sometime used for growth determinations, but this will fail to detect the cell expansion phase that measurements of wet and dry weight will detect (Fig. 6 A–D). A full description of the method is given in Chapter 2, this vol.

3.5.2. Viability

There are a number of methods of determining the viability of plant cells, most of which depend on the presence of an intact cell membrane and are described in Chapter 3, this vol.

3.5.3. Total Carbohydrate

Since sucrose is used as the carbon source for *C. roseus*, the commercial kits used for glucose determination cannot be used. Total carbohydrates are determined using the anthrone reagent, by a procedure described in Chapter 2, this vol.

This method is being replaced by the determination of sucrose, glucose, and fructose by HPLC analysis, a more accurate method with less interference from other media components. HPLC analysis also does not use concentrated H_2SO_4. In addition, it has been observed that the early media samples, when assayed by anthrone, provide a lower reading than anticipated. However, the values rise after 2–3 d, and it is suspected that this may be related to the presence of interfering substances as mentioned above. This is not observed in HPLC analysis (Fig. 7).

3.5.4. Alkaloid Analysis

Some 50–100 mg of dried cells are required for alkaloid analysis; therefore, quite large samples will be needed in the early stages of culture growth (200 mL).

1. The suspension cells are harvested by filtration in a Hartley funnel using Miracloth. For small samples (> 20 g wet wt), the cells are placed in small plastic bags and frozen rapidly in liquid nitrogen. Larger

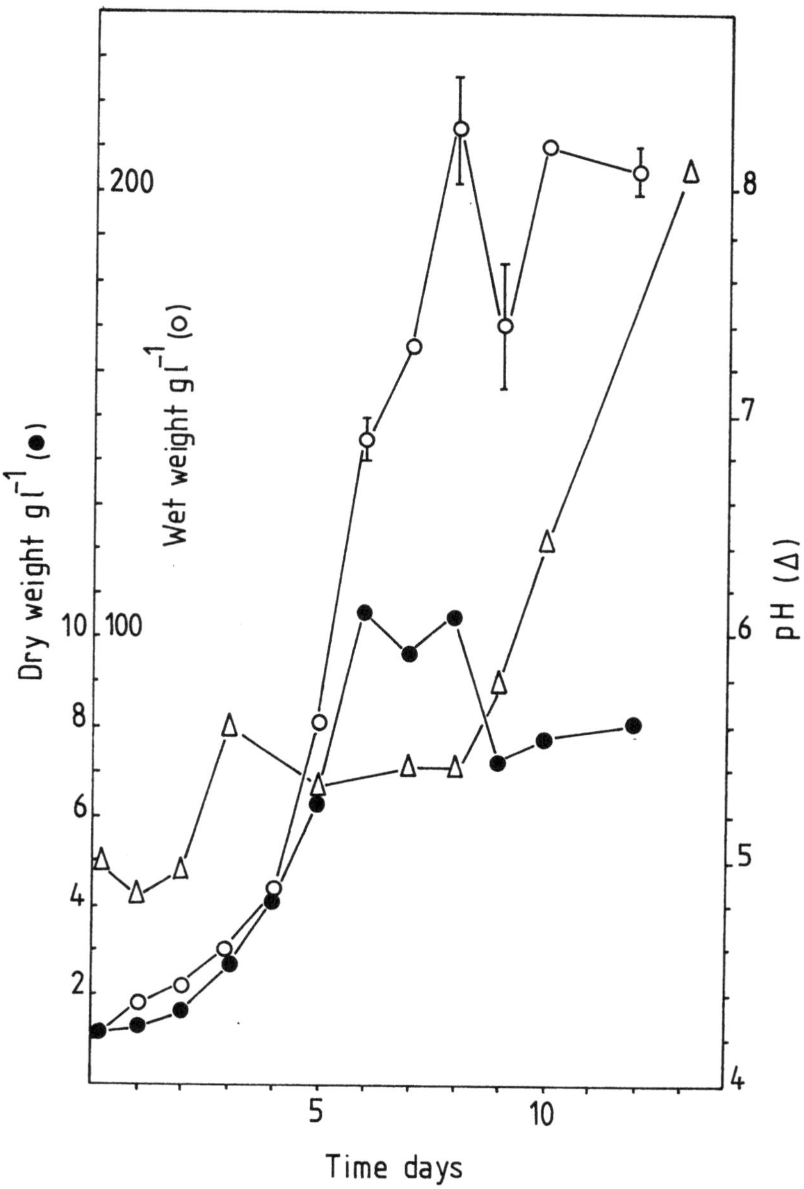

Fig. 6A. The growth of *Catharanthus roseus* ID1 in a 30-L L. H. Engineering bioreactor. Dry weight g/L, wet weight g/L, and pH.

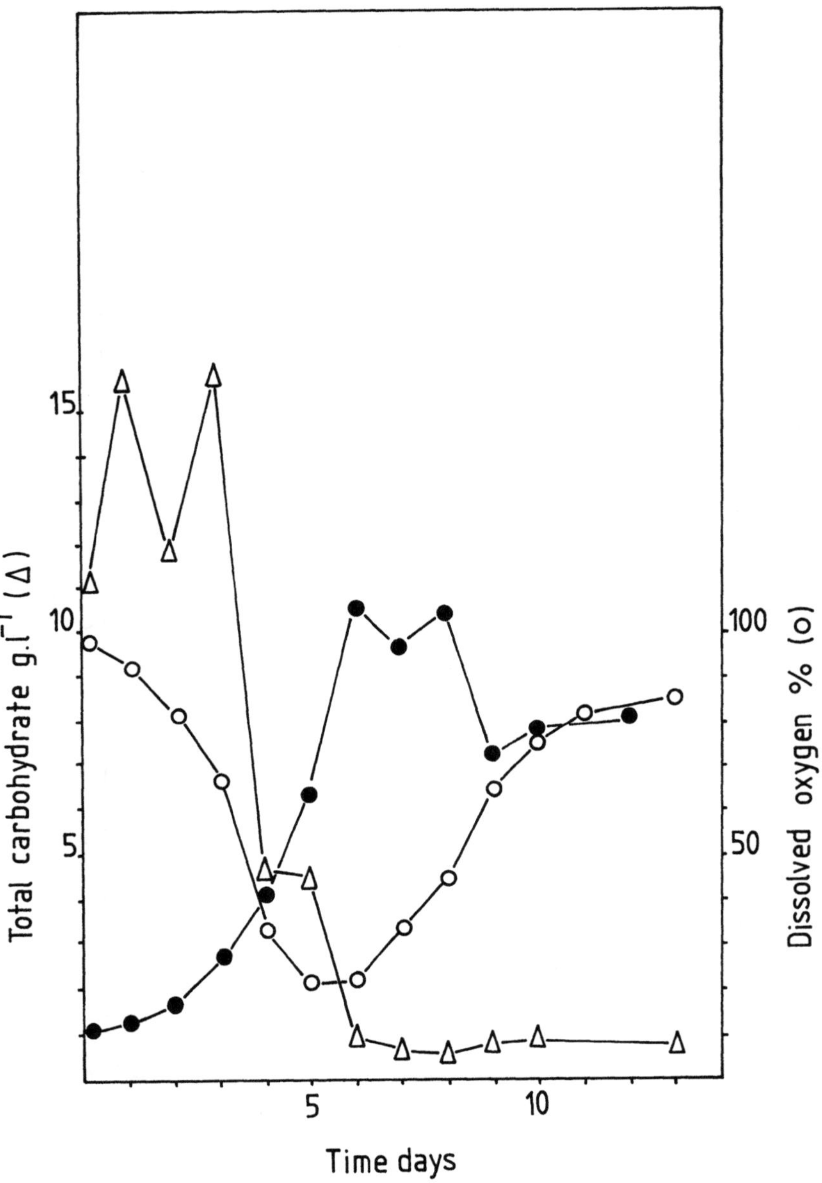

Fig. 6B. Dry weight g/L, dissolved oxygen % and total carbohydrate g/L.

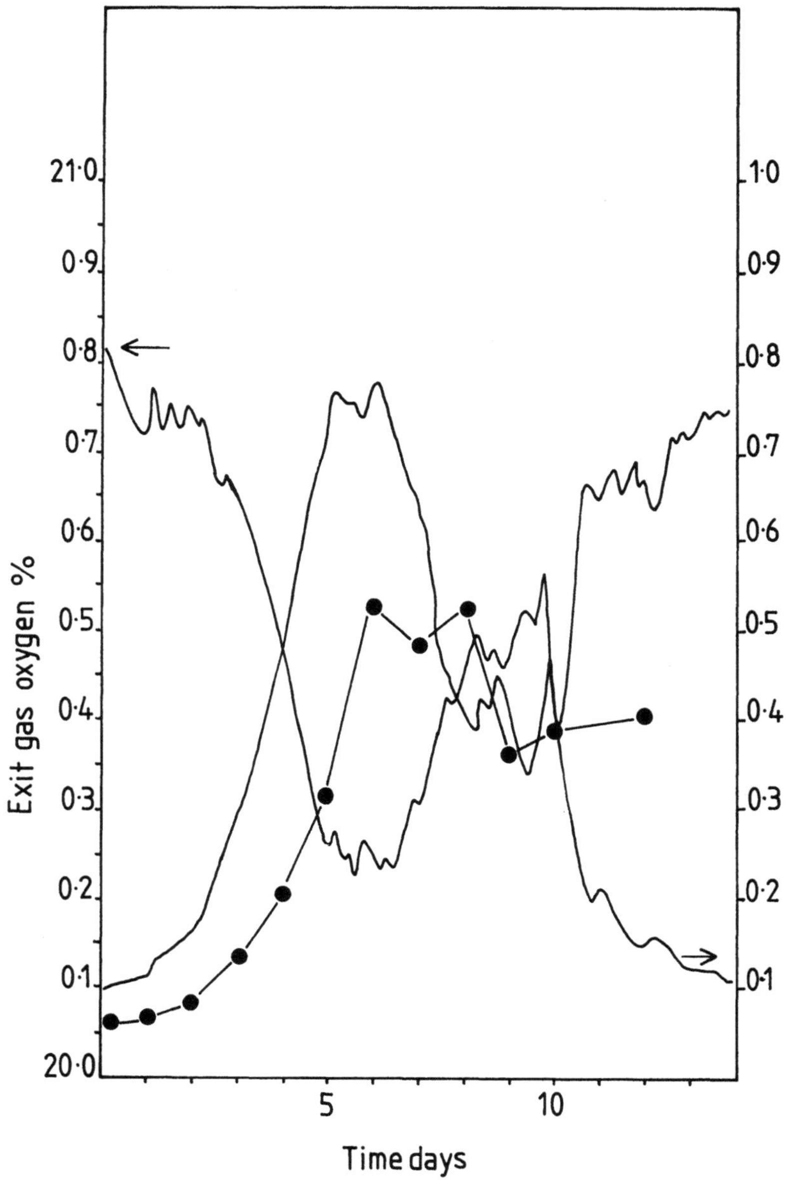

Fig. 6C. Dry weight g/L, exit gas oxygen %, and exit gas carbon dioxide %.

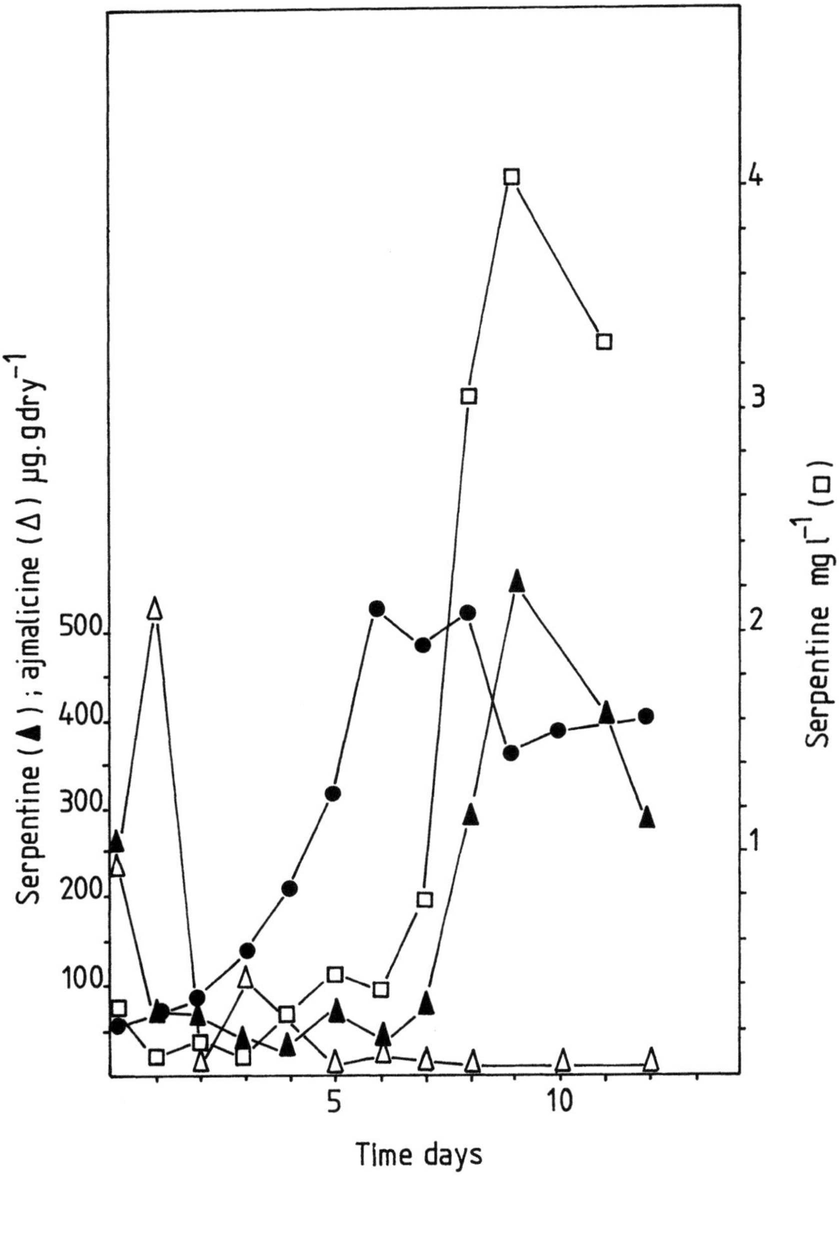

Fig. 6D. Dry weight g/L, ajmalicine μg/g dry wt, serpentine μg/g dry wt, and serpentine mg/L.

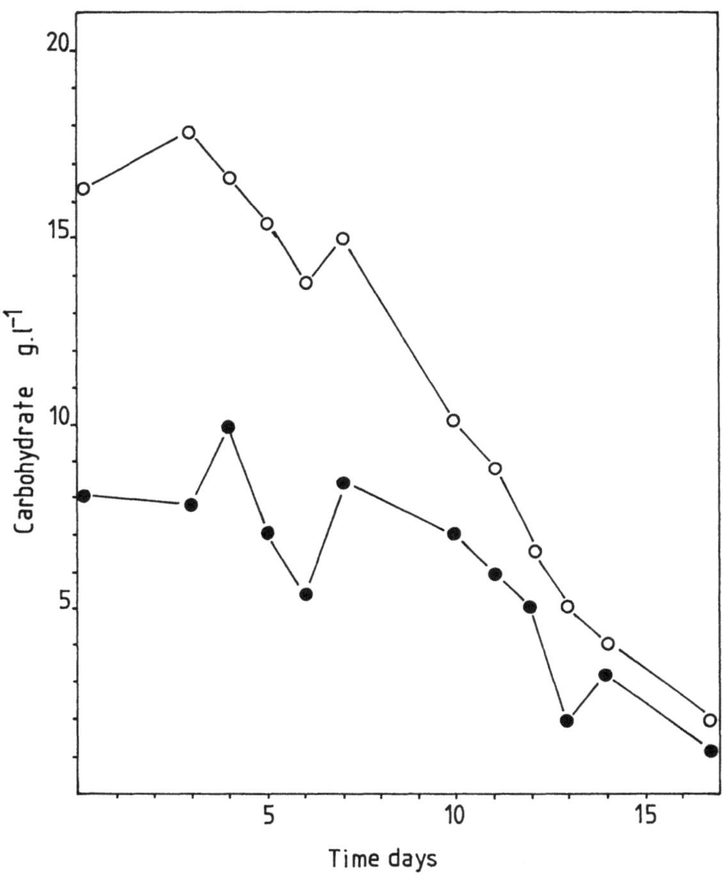

Fig. 7. The measurement of total carbohydrate from a *Catharanthus roseus* culture with time using; anthrone method (●), HPLC method (○).

samples are placed in round-bottomed flasks, and then frozen using liquid nitrogen.

2. Once frozen, the flasks or bags are freeze-dried on a New Brunswick freeze-drier.
3. The freeze-dried cells are weighed out into Soxhlet thimbles (10–50 mm) and extracted using 50 mL of methanol for 2 h (temperature

60°C). Any remaining dried cells can be stored in plastic boxes at 4°C for a number of months.

4. The methanolic extract is reduced to dryness, using a rotary evaporator.
5. The residue is taken up in 1 mL of methanol and analyzed by HPLC.

3.5.5. Other Samples

The pH is noted for each sample taken. Usually, pH is not controlled with plant cell suspensions, but rather allowed to drift (Fig. 6). Other parameters that can be determined are the concentrations of phosphate, nitrate, and ammonia in the medium, cell number, and mitotic index.

3.6. Growth

The *C. roseus* cultures generally grow rapidly, reaching a maximum biomass of 10.5 g/L dry weight after 6 d. This represents a growth rate of 0.46 d^{-1} and a doubling time of 1.54 d. Growth rates are determined over the period of increasing dry weight, using linear regression analysis of a log linear plot. Doubling time is calculated using the equation $td = 0.693/\mu m$.

The culture should start out being pale yellow in color, and thicken appreciably as growth continues. The onset of growth is accompanied by a drop in dissolved oxygen and a peak in carbon dioxide evolution (Fig. 6C). Any rapid browning in the culture is an indication of cell death or contamination. Another problem that may occur is the formation of a crust or meringue (Fig. 8). This appears to start as a stable foam in which cells are trapped, and then increases with time, until it spreads across the top of the bioreactor. Under normal circumstances, this causes no real problems other than an underestimate of the final biomass, but if this is allowed to build up to fill the head space, a number of problems ensue. First, the level of the medium may be lowered sufficiently to stop the flow in the bioreactor. Secondly, the meringue can block the condenser and exit filters, causing back-pressure. The addition of antifoam will reduce foaming and, hence, meringue formation.

3.7. Harvesting

The aggregate nature of plant cell suspensions causes them to sediment rapidly, so that only low-speed centrifugation is required for harvesting. An alternative is filtering through Miracloth, but this is suitable only for volumes up to 5 L. Thirty liters of *C. roseus* cells represent some 300 g

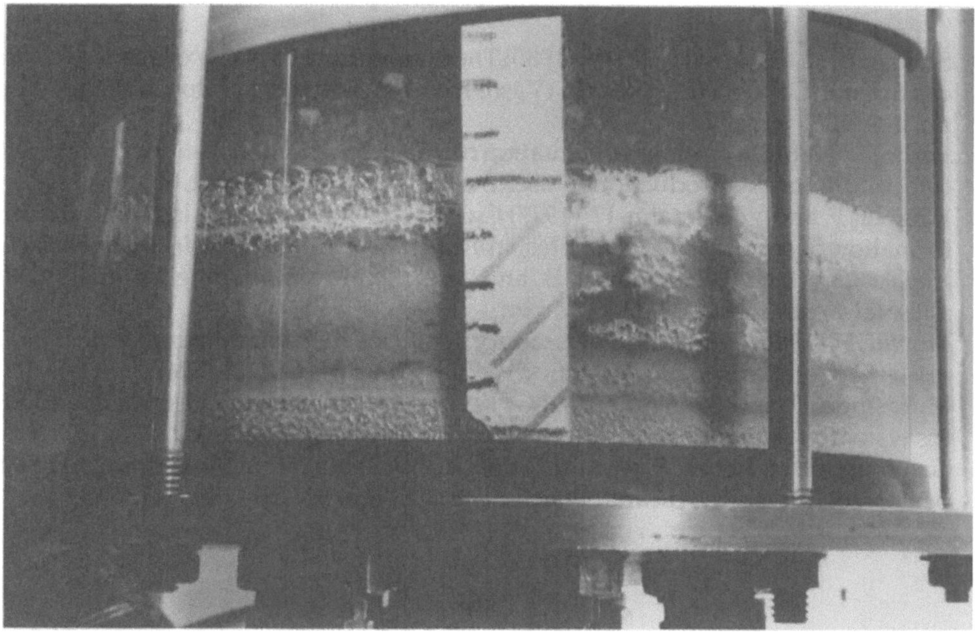

Fig. 8. Meringue forming just above the bubbles on the surface of a *Catharanthus roseus* culture growing in a 7-L airlift bioreactor.

dry cells and between 2–3 kg wet cells. We have found that a nylon bag in an ordinary domestic spin dryer is sufficient to harvest 30 L of cells.

4. Notes

HPLC analysis of carbohydrates: The carbohydrates sucrose, glucose, and fructose can be analyzed in culture medium using a LiChrosphor 100 NH_2, 5-μm column (MERCK) run with 75% acetonitrite. The detector used is a Waters differential refractometer 410.

Acknowledgments

The authors would like to thank P. Bond and S. Ashton for supplying some of the data, and M. Tune for the photographs. We would also like to thank the SERC Biotechnology Directorate and Plant Science Ltd. for their financial support.

References

1. Scragg, A. H. and Fowler, M. W. (1985) The mass culture of plant cells, in *Cell Culture and Somatic Cell Genetics of Plants* (Vasil, I. ed.), Academic, London and New York, pp. 103–128.
2. Berlin, J. (1986) Large scale fermentation of transformed and non-transformed plant cell culture for the production of useful compounds. *Symbiosis* **2,** 55–65.
3. Wagner, F. and Vogelmann, H. (1977) Cultivation of plant tissue in bioreactors and formation of secondary metabolites, in *Plant Tissue Culture and Its Biotechnological Applications* (Barz, W., Reinhard, E., and Zenk, M. H., eds.), Springer-Verlag, Berlin and New York, pp. 245–252.
4. Morris, P. (1986) Regulation of product synthesis in cell cultures of *Catharanthus roseus.* II. Comparison of production media. *Planta Medica* 121–126.
5. Murashige, T. and Skoog, F. (1962) A revised medium for rapid growth and bioassays with tobacco tissue cultures. *Physiol. Plantarum* **15,** 473–497.
6. Gamborg, O. L., Miller, R. A., and Ojima, K. (1968) Nutritional requirement of suspension cultures of soybean root cells. *Exp. Cell Res.* **50,** 151–158.

Chapter 44

Batch Suspension Culture

Pamela A. Bond

1. Introduction

Batch culture involves the inoculation of a known volume or mass of cells into a volume of (normally) defined medium. Growth is allowed to proceed, and the resulting biomass is harvested at some stage during the growth cycle. Throughout the growth period, no additions are made to the media, with the exception of acid and alkali, if needed, for pH control. The scale of operation depends on the requirements.

The growth profile of plant cells in suspension culture has been described in Chapter 2, this vol. There is an initial lag phase, during which the cells are adjusting to the new culture conditions, followed by an exponential growth period, when the cells are rapidly dividing. As the culture ages, the rate of cell division declines, until a stationary phase is reached (rate of production is equal to rate of cell death), and finally there is a senescent phase when the cells are dying.

The growth rate of plant cell cultures can be assumed to follow Monod kinetics. This allows an estimation of growth rates and doubling times. The growth rate of a suspension culture of plant cells can be calculated over the exponential growth period. If the natural log (ln) of the dry weight is plotted against time, the gradient of the straight line produced on the graph is a measure of the growth rate (Fig. 1). The nonlinear parts of the graph show the lag phase and the stationary phase, respectively. The doubling time represents the time taken for one cell to divide into two cells, or two cells into four, and so on. This is obtained by dividing ln 2 by the growth rate. For a discussion on the applications of Monod kinetics, *see* Bailey and Ollis, 1986 (*1*).

From: *Methods in Molecular Biology*, vol. 6, *Plant Cell and Tissue Culture*
Edited by Jeffrey W. Pollard and John M. Walker, ©1990 by The Humana Press

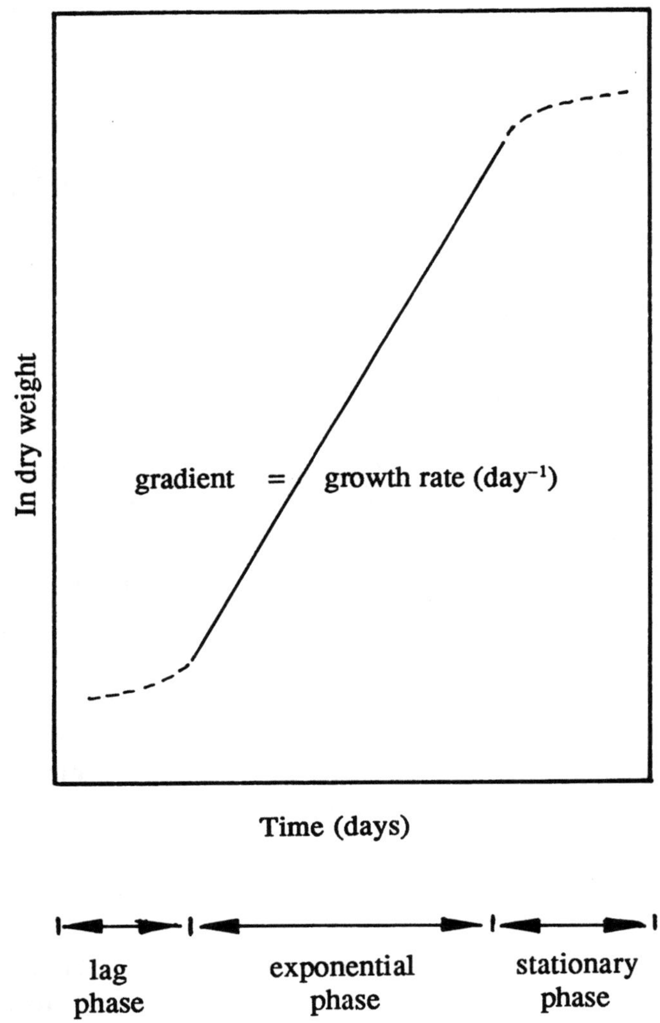

Fig. 1. Calculation of the growth rate using Monod kinetics.

Typically, the doubling times of microbes ranges from 20 min (*E. coli*) to several hours, and animal cells are able to double in number every 12–24 h. Plant cells, however, have doubling times measurable in days (48–96 h), highlighting the slow growth of plant cells in culture compared to other suspension systems. Despite this, cell culture maintains a distinct advantage over growth of the whole plant in the field. A batch culture may take from 10 d to 1 mo to complete. In comparison, growth of, for example, the Cinchona tree, from which quinine is extracted for use both as an antimalarial drug, and as a bittering agent in soft drinks (tonic water), takes several years.

This chapter investigates links between the growth cycle and such features as fresh and dry weights of cultures, cell viability, carbohydrate use, and metabolic activity (oxygen uptake and carbon dioxide evolution, respiratory quotients and so forth).

2. Materials

2.1. Batch Culture in an Airlift Bioreactor

1. Airlift bioreactor.
2. Inoculation culture of *Catharanthus roseus* (M3 medium; ref. 2 and *see* Appendix).
3. Dissolved oxygen probe: polarographic or galvanic.
4. G MM80 Gas Analyzer (or similar) to monitor metabolic activity.
5. Equipment for culture analysis (*see* Chapter 2, this vol.).

3. Methods

1. An airlift bioreactor is set up as described in Chapters 43 and 45, this vol.
2. The bioreactor is autoclaved together with a suitable volume of M3 medium (2% sucrose) for at least 30 min at 121°C.
3. The bioreactor is inoculated with a culture of *C. roseus*. The following parameters are monitored daily: fresh weight, dry weight, viability, pH, total carbohydrate, dissolved oxygen levels.
4. During the experiment, the inlet and outlet concentrations of oxygen, carbon dioxide, nitrogen, and argon in the air streams are monitored using the Gas Analyzer. This allows the oxygen uptake rate (OUR), carbon dioxide evolution rate (CER), and respiratory quotient (RQ) to be calculated.

4. Notes

1. Good aseptic technique must be observed when taking samples, to reduce the risk of contamination of the culture with microorganisms.
2. The lag phase, logarithmic phase, and stationary phase can be seen in the fresh and dry weight curves obtained (Fig. 2). As in this case, the lag phase may be short, or it may not be apparent.
3. The initial viability of the cells is high; viability begins to fall only as the cells senesce (Fig. 3).
4. As the cells lose viability, the pH increases (Fig. 3). This is a result of the release of cellular compounds into the medium.

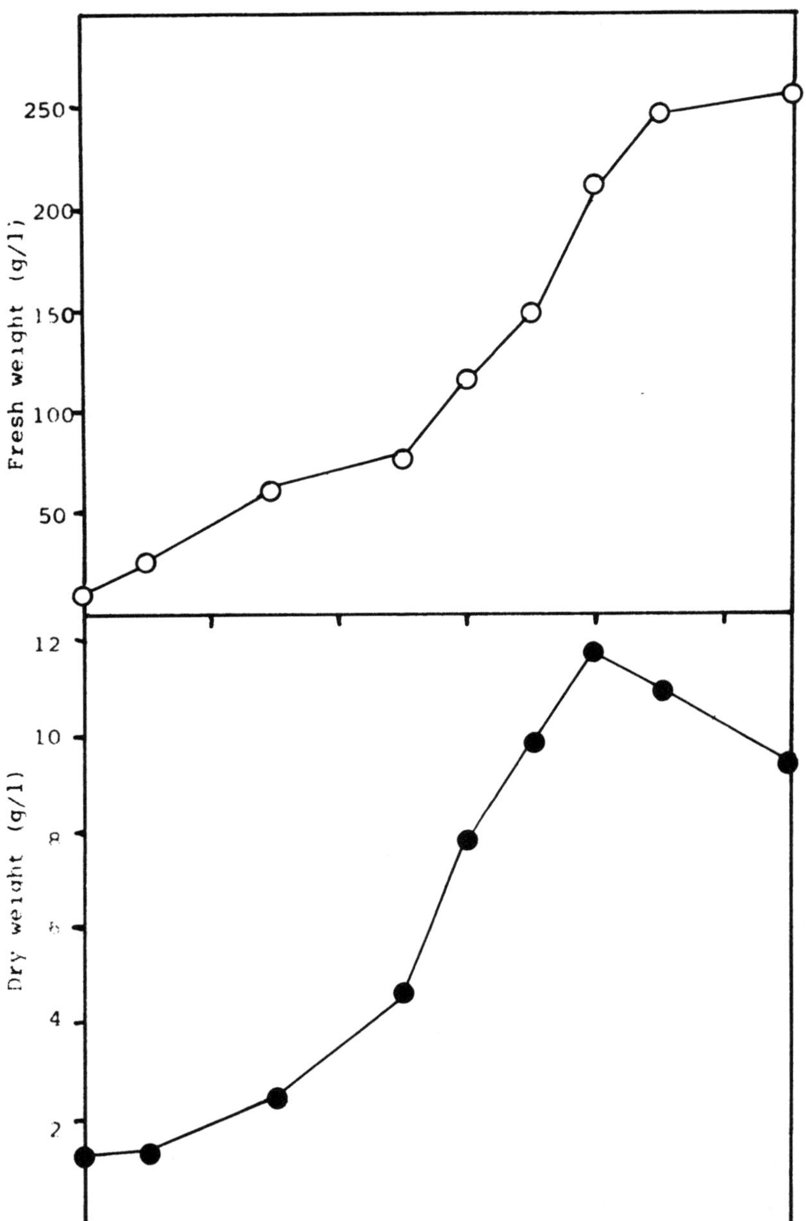

Fig. 2. Graph to show fresh and dry weight levels during a batch culture.

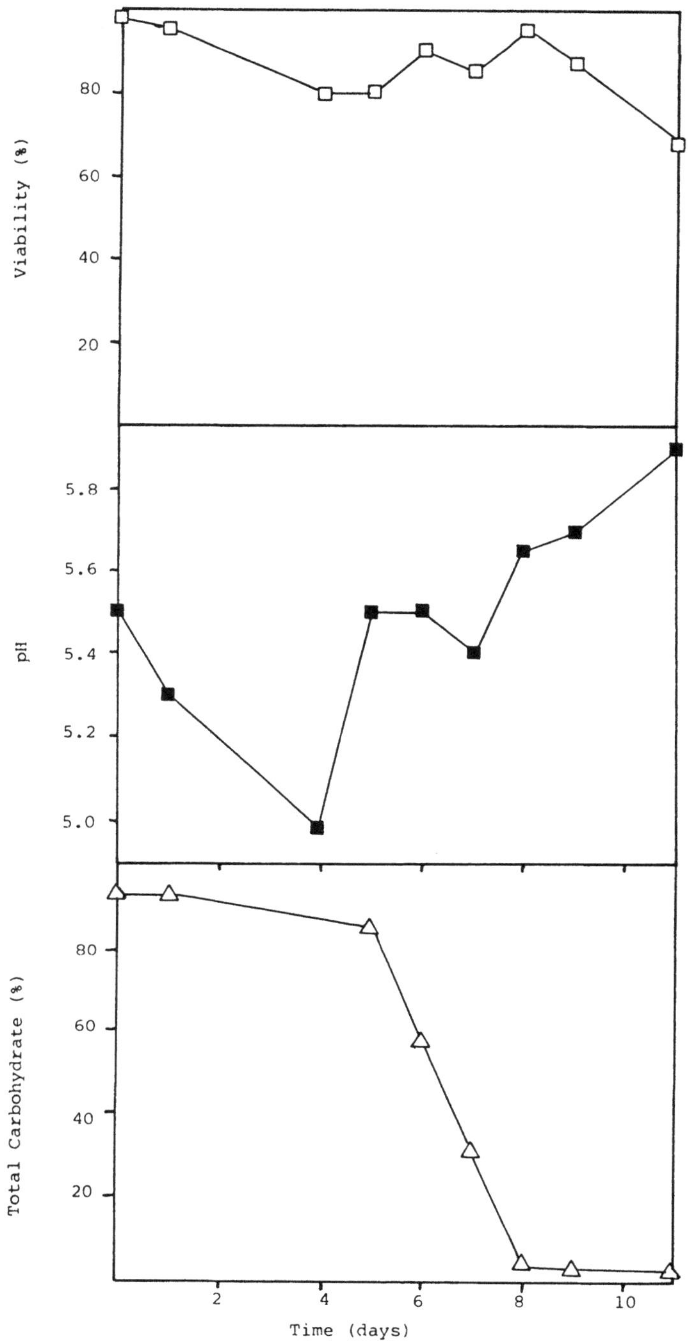

Fig. 3. Changes in cell viability, media pH, and carbohydrate levels during batch culture.

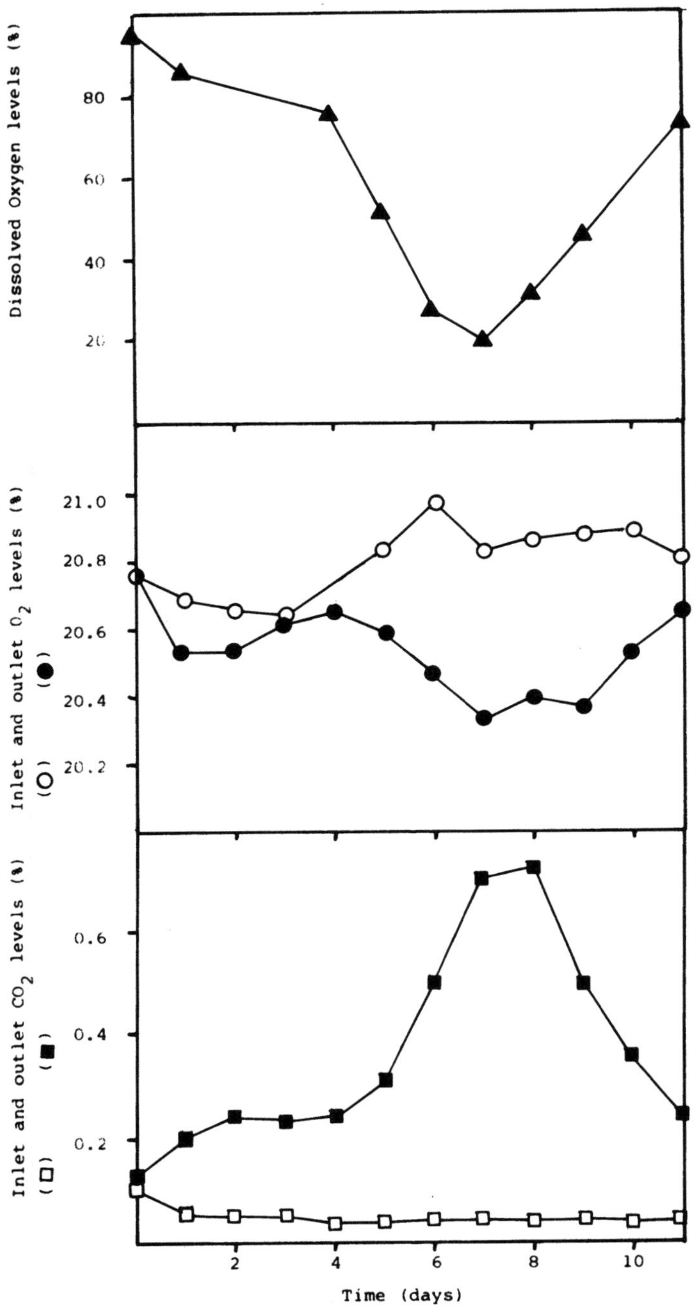

Fig. 4. Gaseous O_2 and CO_2 levels compared to dissolved oxygen levels, during batch culture.

Table 1
Oxygen Use and Carbon Dioxide Evolution by a Batch Culture of *C. roseus*

	Oxygen utilization rate		Carbon dioxide evolution rate	
Day no.	g/g dry wt/h	g/L/h	g/g dry wt/h	g/L/h
0	–	–	4.60×10^{-3}	1.16×10^{-3}
1	1.60×10^{-3}	1.16×10^{-3}	2.90×10^{-3}	2.10×10^{-3}
5	4.00×10^{-2}	8.66×10^{-3}	5.94×10^{-2}	1.28×10^{-2}
6	8.00×10^{-2}	1.02×10^{-2}	1.00×10^{-1}	1.28×10^{-2}
7	8.00×10^{-2}	8.06×10^{-3}	1.46×10^{-2}	1.48×10^{-2}
8	7.52×10^{-2}	6.41×10^{-3}	1.51×10^{-1}	1.29×10^{-2}
9	7.52×10^{-2}	6.41×10^{-3}	1.51×10^{-1}	1.29×10^{-2}
10	8.16×10^{-2}	7.42×10^{-3}	1.01×10^{-1}	9.14×10^{-3}
11	5.76×10^{-2}	–	6.91×10^{-2}	–

5. As the cells grow and divide, uptake of carbohydrate is rapid. The media carbohydrate is depleted before growth has ceased, indicating that the cells are able to store sugar for future use.
6. A good correlation exists between oxygen use and carbon dioxide evolution (Fig. 4). Maximum oxygen use also corresponds to the decrease in media-dissolved oxygen levels.
7. The RQ calculated for the *C. roseus* cells in this experiment is 1.2 (RQ = carbon dioxide produced/oxygen used).
8. Maximum metabolic activity/cell dry weight occurs at the midlogarithmic or midexponential growth phase. This is a period of rapid cell division and has a high energy requirement (Table 1).
9. Maximum activity/liter of culture does not necessarily coincide with this midexponential phase. A large mass of cells of low metabolic activity may show a higher oxygen use/unit volume than a culture of very high metabolic activity, but a low dry weight.

References

1. Bailey, J. E. and Ollis, D. F. (1986) *Biochemical Engineering Fundamentals,* 2nd Ed. (McGraw-Hill, Singapore.)
2. Morris, P. (1986) Regulation of product synthesis in cell cultures of *Catharanthus roseus* II comparison of production media. *Planta Medica* **2,** 121–126.

Chapter 45

Production Systems

Pamela A. Bond

1. Introduction

This chapter ties together ideas introduced in previous articles of this volume in the search for ways to use plant cell cultures as industrial production systems. The economics of a process must always be carefully considered when examining the suitability of plant cell culture for the manufacture of target compounds. The object is to produce the maximum amount of required product in as short a time as possible, using the most cost-effective method. New processes can then be compared with existing systems, and their feasibility assessed. This also requires a knowledge of the market value of the product under consideration, the projected stability (or growth) of the market, and the potential market share that your product could hope to attain.

When examining the actual cost of production, the list of considerations is extensive:

1. Initial cost of equipment, e.g., bioreactors, and their depreciation in value with age and use.
2. Cost of materials and backup for each production run, e.g., media, electricity, heating, lighting.
3. Inoculum production costs and maintenance of stock cultures.
4. Cost of harvesting the biomass and extraction or purification of the product.
5. Downtime of the bioreactors.

From: *Methods in Molecular Biology*, vol. 6, *Plant Cell and Tissue Culture*
Edited by Jeffrey W. Pollard and John M. Walker, ©1990 by The Humana Press

The "downtime" of the bioreactors is the time spent cleaning, making new media, and sterilizing the vessel. This exerts a very real cost. Time is equivalent to money, and if the vessel is inactive for long periods of time, it is not generating revenue. The aim, therefore, is to maximize production and to minimize costs.

There are several methods that can be considered when trying to locate the most cost-effective production system; these systems are available for the production of plant cell, animal cell, and microbial cultures, but their suitability for each varies.

1. Batch culture: Cells are inoculated into the media, growth takes place, and the cells are harvested when product levels (or biomass levels) are at a maximum (*see* Chapter 44, this vol.).
2. Fed batch: Initially, the growth phase is identical to that in batch culture. However, growth-limiting factors (normally the carbohydrate supply) are added to the bioreactor contents at various stages during growth. The medium volume is therefore increased during the culture period. The name for this method of culture, "fed batch," was first introduced by Yoshida et al. in 1973 (*1*).
3. Two-stage batch: The cells are grown to a high biomass level in a batch mode on a medium that has been optimized for growth. The medium is then removed, and the cells are transferred to a production medium for incubation before the final harvest. The advantage of this system is that both stages can be run at their maximum capacity; optimum conditions for growth are not necessarily the same as optimum conditions for product formation, but, under this system, the two can be separated and controlled individually.
4. Draw–fill: This technique is an extension of the batch cultivation mode. The cells are allowed to divide; in the middle of the exponential growth phase, all but 10–20% of the cells are harvested. The vessel is then refilled with medium, and the remaining cells act as the inoculum for the next batch growth stage. Problems encountered using this method include a high risk of contamination (minimized by taking great care over correct aseptic techniques) and the fact that immature cells are harvested—if the compound required is a secondary metabolite, production will not be initiated. This can be overcome by use of a second "production" stage (cf two-stage batch systems).
5. Continuous cultivation: An extension of the fed-batch system, continuous culture is used widely in the cultivation of microorganisms.

The cells are first grown in a batch mode until they reach the mid-log phase of growth, and then, based on a knowledge of the growth rate, spent medium containing cells is continuously removed from the culture and replaced with fresh medium, allowing an equilibrium to be reached. This method is difficult to employ with plant cell cultures, because of their slow growth rates and doubling times. The amount of culture that must be removed is small, and it is difficult to obtain pumps that will cope with the situation. Obviously, as larger vessels are used, the problem is eased, but most laboratories do not have the facility for this scale of cultivation. Another solution is to use a semicontinuous culture system in which the medium is added in small, discrete doses and an equivalent amount of spent medium is removed at fixed time intervals.

6. Immobilization: This technique is most useful when the product is a compound that is excreted from the cell into the medium. Cells are entrapped either in a mesh or in a gel substance; medium is provided (continuous or batch), and the product is removed with the medium. The use of immobilized plant cells is discussed in Chapter 47, this vol.

In this chapter, two production systems will be considered. A fed-batch system conducted in 250-mL shake flasks will be compared to the use of a high initial sucrose concentration, and a draw–fill experiment using a 5-L airlift bioreactor is described.

2. Materials

2.1. High Sucrose Concentration/Fed-Batch Experiment

1. 250-mL shake flasks.
2. Orbital shaker at 150 rpm maintained at 25°C.
3. Material for growth analysis (*see* Chapter 2, this vol.).
4. Stock culture of *Catharanthus roseus*.
5. M3 media (2 and *see* Appendix).

2.2. Draw–Fill Comparison Experiment

1. 5-L bioreactor.
2. Material for growth analysis.
3. Inoculation cultures of *C. roseus* prepared from stock culture.
4. M3 media.

3. Methods

3.1. *Fed-Batch Method* (See *Notes 1–4*)

1. Prepare at least three 250-mL shake flasks containing 250 mL M3 medium (2% sucrose).
2. Inoculate the flasks with 10–20 mL of stock culture of *C. roseus*. Incubate at 25°C on an orbital shaker in continuous, diffuse light.
3. Monitor the flasks daily for fresh and dry weight, viability, pH, medium carbohydrate levels, and alkaloid production (*see* Chapters 2 and 48, this vol.).
4. At intervals (when the culture reaches mid-log phase) add sterile, concentrated solutions of sucrose (40% w/v) using a Gilson pipet with a sterile tip, so the carbohydrate level is increased from 2–4%, then 4–6%, and so forth.
5. A parallel experiment is conducted, using an initial sucrose concentration of 6%, as a comparison.

3.2. *Draw–Fill Method* (See *Notes 5–8*)

1. The 5-L (or similar) airlift vessels (Figs. 1 and 2), containing 5 L of M3 medium, are sterilized in an autoclave at 121°C, 15 psig for 35–40 min.
2. The vessels are each inoculated via a sterile port (Fig. 3) with 750 mL to 1 L of *C. roseus* culture (prepared from the stock line 14 d earlier by subculturing 100 mL of cells into a 2-L shake flask containing 1 L of medium). Ports can be kept sterile after sterilization by the use of cotton wool and tin foil prior to autoclaving (Fig. 3). Figure 2 shows the vessel in place in the laboratory after inoculation with *C. roseus*.
3. The culture is monitored daily for fresh and dry weight, viability, pH, medium carbohydrate levels, and alkaloid production (*see* Chapters 2 and 3, this vol.).
4. All but 1 L of the medium is withdrawn from the vessel, at the mid-log phase of the growth cycle. The cells that are removed can then be harvested directly or placed in a second "holding vessel" with medium, to allow alkaloid production.
5. The vessel is refilled with fresh, sterile medium, and the culture parameters are monitored.
6. Inoculation with fresh medium as described above can be repeated over several "subcultures."

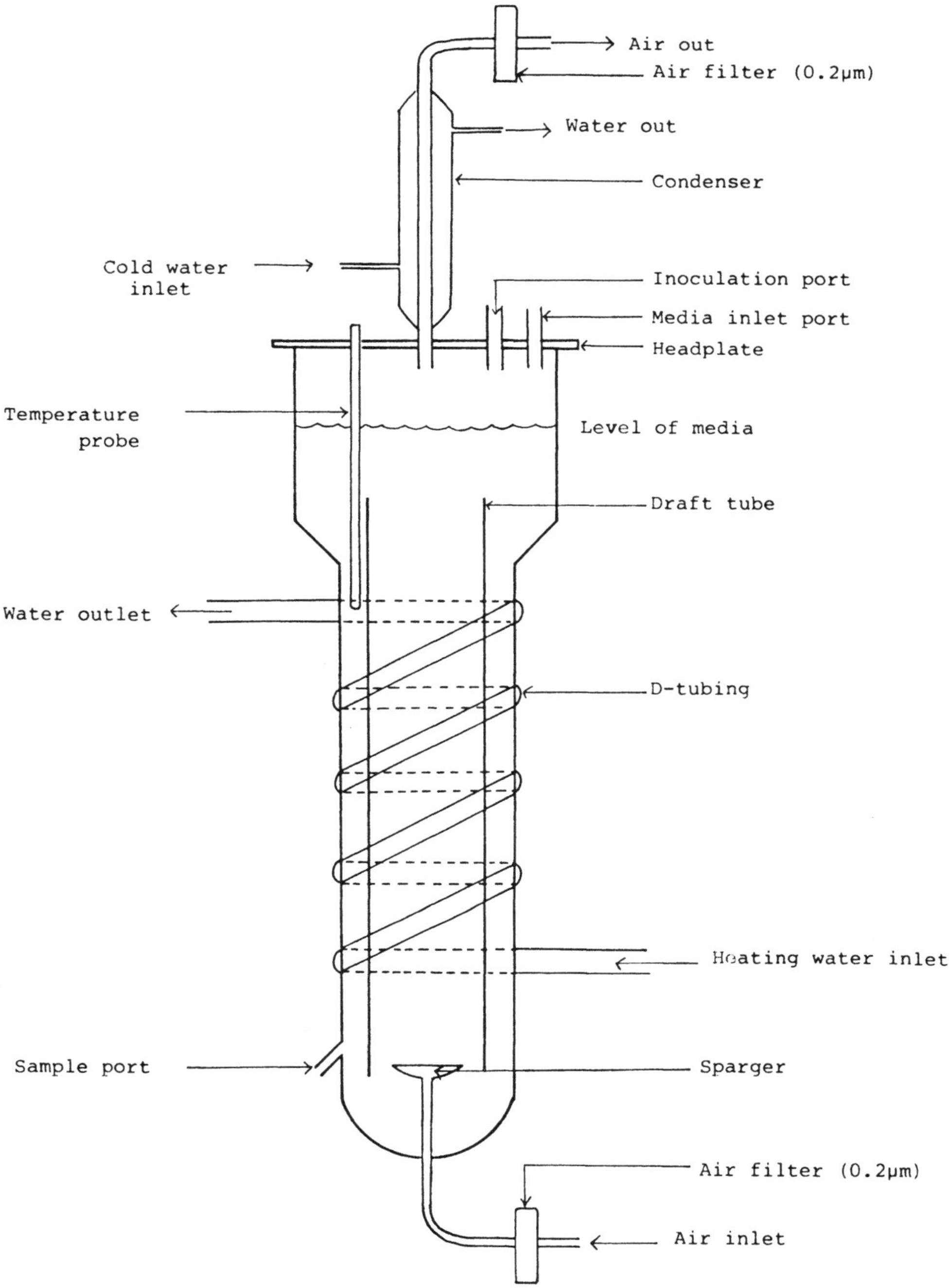

Fig. 1. The 5-L airlift bioreactor used in the draw–fill experiment. Temperature control is provided by the use of D-tubing and a water bath at 25°C.

Fig. 2. The 5-L airlift bioreactor in use in the laboratory.

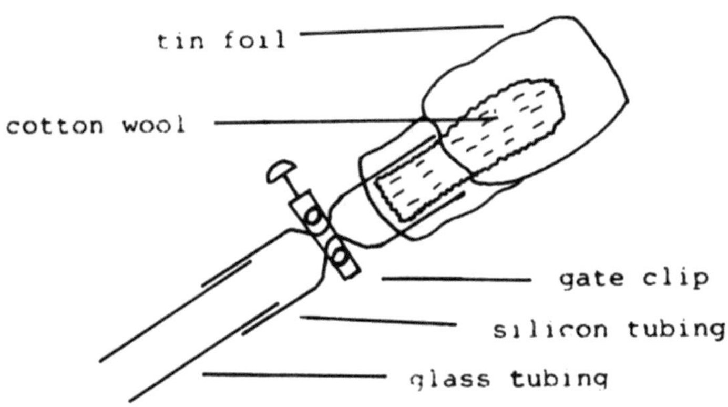

Fig. 3. The use of tin foil and cotton wool to maintain sterility—the cotton wool acts as a filter and the tin foil as a barrier to prevent entry of contaminants.

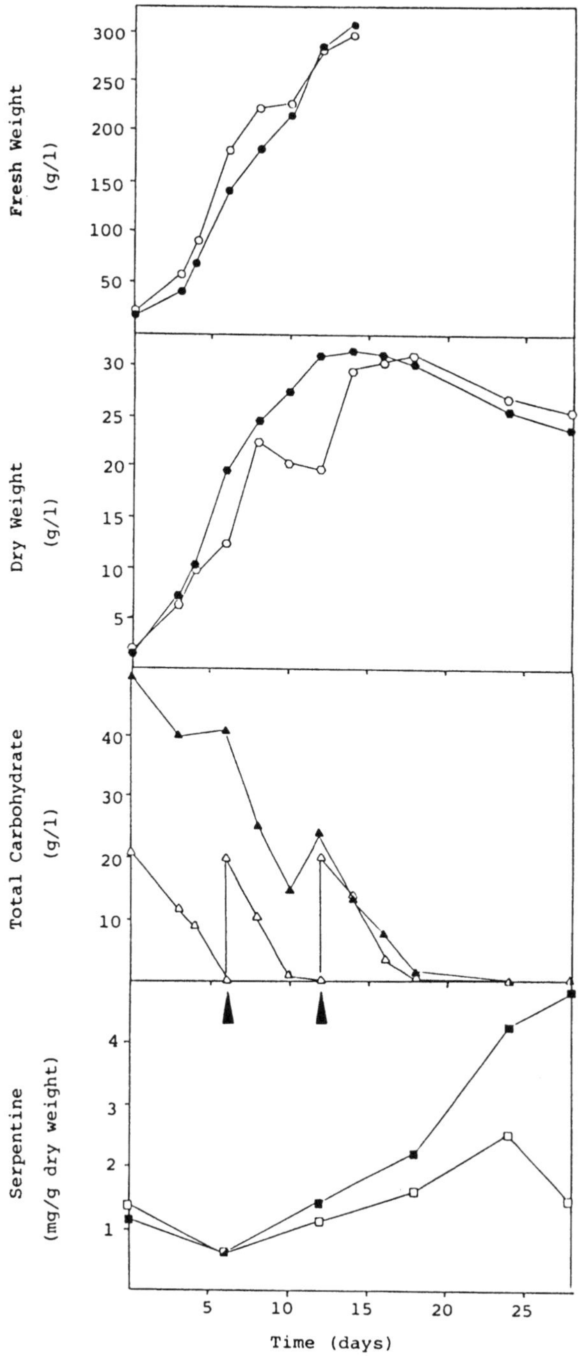

Fig. 4. Growth profiles of cells grown in 250-mL shake flasks. Open symbols represent cells fed with 2% sucrose at days 6 and 12. Closed symbols represent cells grown in 6% sucrose.

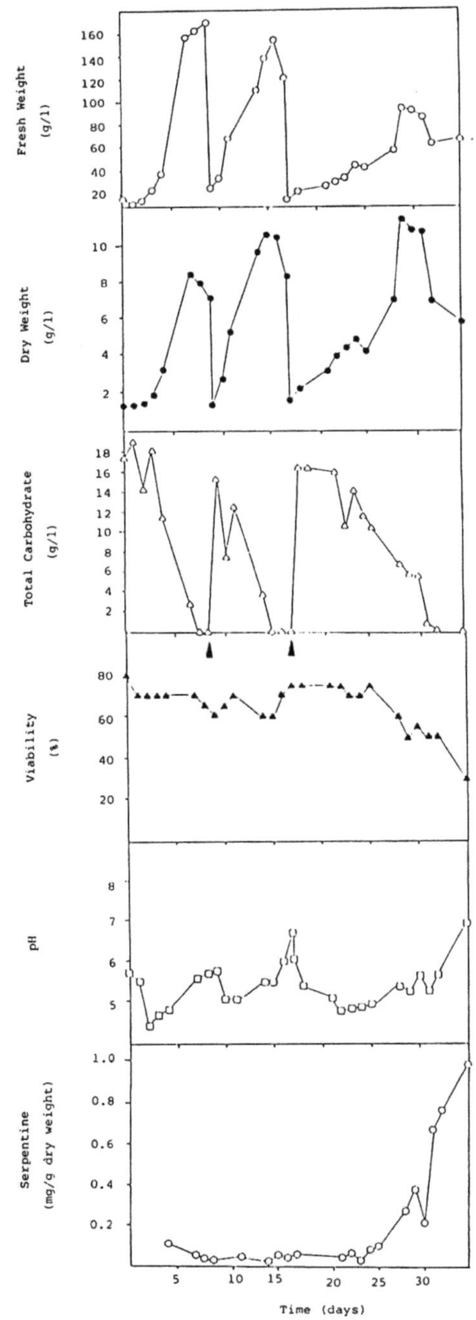

Fig. 5. Growth of *C. roseus* in a 5-L airlift bioreactor. Medium was withdrawn (draw–fill) at days 9 and 17.

4. Notes

1. In the fed-batch method (Fig. 4), fresh weight increases more rapidly when the initial sucrose concentration is low (2%). If the initial sucrose concentration is high (e.g., 6%, as shown in Fig. 4), there is an osmotic effect, and water cannot be accumulated by the cells.
2. At high fresh weights, accurate measurement becomes difficult, and the standard deviation is above 10%. These results are, therefore, not recorded on Fig. 4.
3. The dry weight accumulates rapidly in a high initial sucrose concentration. This is because of the conversion of excess carbohydrate to starch.
4. The production of alkaloids is enhanced by high sucrose concentrations. This has also been noticed in other *C. roseus* cell lines (*3*).
5. The draw–fill method (Fig. 5) allows for a programmed supply of cells from the vessel.
6. Aseptic technique must be adhered to continuously to reduce the risk of contamination.
7. Figure 5 does not show an "ideal" draw–fill experiment. In this case, one of the common problems is illustrated. The culture has been allowed to reach stationary phase before the harvesting/reinoculation stage. After 3–4 harvests, the growth of the culture is reduced, and the cells lose viability. Eventually, the culture is lost.
8. Under ideal conditions, the culture can be continued indefinitely. If large amounts of biomass are needed, a fed-batch system is ideal. However, if the product required is a secondary metabolite, a second stage is essential to allow formation of the compounds required. Figure 5 shows that serpentine levels remain low (below 200 µg/g dry weight), until the cells are allowed to mature after the final draw–fill stage.

References

1. Yoshida, F., Yamane, T., and Nakamoyo, K. (1973) Fed-batch hydrocarbon fermentations with colloidial emulsion feed. *Biotech. Bioeng.* **15,** 257–270.
2. Morris, P. (1986) Regulation of product synthesis in cell cultures of *Catharanthus roseus* II comparison of production media. *Planta Medica* **2,** 121–126.
3. Bailey, C., Nicholson, H., Morris, P., and Smart, N. J. (1985) A simple model of growth and product formation in cell suspensions of *Catharanthus roseus* G Don. *Appl. Biochem. Biotechnol.* **11,** 207–219.

Chapter 46

Immobilization of Cells by Spontaneous Adhesion

Peter J. Facchini, Frank DiCosmo, Laszlo G. Radvanyi, and A. Wilhelm Neumann

1. Introduction

Various immobilization strategies have been developed to optimize the biosynthetic potential of cultured plant cells. Immobilization involves the retaining of suspension-cultured plant cells on, or within, a physical barrier that promotes cell aggregation and separates the cells from the surrounding media. The advantages of immobilization include the formation of diffusionary gradients around and between the cells that increase intercellular biochemical communication conducive to the coordinated expression of secondary metabolism. In effect, immobilization simulates the physiological conditions within large aggregates of cells in a manner amenable to manipulation for fermentation purposes.

One of the more promising immobilization processes is the entrapment of suspension-cultured cells within inert, reusable, and relatively inexpensive reticulate polymer matrices. Reticulated matrices of polyure-

From: *Methods in Molecular Biology*, vol. 6, *Plant Cell and Tissue Culture*
Edited by Jeffrey W. Pollard and John M. Walker, ©1990 by The Humana Press

thane and polyester have been used for this purpose, resulting in an increased biosynthetic potential of the immobilized cells compared to freely suspended cells (*1*). The process involves the invasion by the cells into the reticulated network, their adhesion to the polymer fibers, and the subsequent growth of the plant cells into the surrounding space within the matrix. In order to optimize this method of immobilization, the ideal polymer matrix material must be determined.

The initial interaction of the plant cells with various substrate surfaces can be studied and manipulated according to well-defined surface thermodynamic principles. Various thermodynamic properties can be defined in terms of Legendre transforms of the surface energy and can be used as an index of the hydrophobicity of the surface. A thermodynamic model of particle adhesion has been used to predict the extent of adhesion of suspension-cultured plant cells to various substrates (2). The extent of adhesion depends on the relative values of the substrate (γ_{sv}), liquid (γ_{lv}), and cellular (γ_{cv}) surface tensions. Briefly, when:

$$\gamma_{lv} > \gamma_{cv} \tag{1}$$

cell adhesion increases with decreasing surface tension γ_{sv} of the substrate. Alternatively, when:

$$\gamma_{lv} < \gamma_{cv} \tag{2}$$

the opposite pattern occurs. For the case of equality:

$$\gamma_{lv} = \gamma_{cv} \tag{3}$$

adhesion is negligible for conditions described herein and is independent of the value of γ_{sv}. The model provides a method of determining the cellular surface tension (γ_{cv}) for suspension-cultured plant cells, in order to select the optimum liquid medium (γ_{lv}) and substrate (γ_{sv}) surface tensions for maximizing the spontaneous adhesion of the cells to the substrate. Immobilization systems that involve the attachment of cultured cells to substrate matrices can be improved by increasing the initial level of cell adhesion, and hence, biomass loading as well as strengthening the adhesive bond.

The thermodynamic approach to the adhesion phenomenon assumes electrical charge interactions to be constant and considers the process to be governed by an interfacial free energy balance involving van der Waals forces. The changes in the free energy of adhesion (ΔF^{adh}) have been found to be proportional to variations in the extent of adhesion as predicted by the thermodynamic model:

$$\Delta F^{adh} = \gamma_{sc} - \gamma_{cl} - \gamma_{sl} < 0 \tag{4}$$

where γ_{sc} is the solid-cellular interfacial tension, γ_{cl} is the cellular–liquid interfacial tension, and γ_{sl} is the solid–liquid interfacial tension. The distinguishing cases implied by this model in terms of the relative values of the cellular (γ_{cv}), liquid (γ_{lv}), and subtrate (γ_{sv}) surface tensions are represented in Fig. 1.

The methods described herein allow for the determination of the cellular surface tension (γ_{cv}), the most difficult of the three surface tension measurements. A discussion follows for the control and modification of adhesion using the thermodynamic model. A summary of some substrate materials and their respective surface tensions (γ_{sv}) is also included.

2. Materials

1. A series of nylon mesh filters in the range of 100–500 µm pore size. These should be mounted onto some sort of vacuum filtration apparatus.
2. Smooth polymer films of a wide range of surface tensions (γ_{sv}). Examples of some polymers and their associated γ_{sv} values are given in Table 1.
3. l-propanol; *n*-hexane; 95% ethanol.
4. Teflon™ blocks (approximate dimensions: 9.5 x 3.1 x 1.3 cm) with three 1-cm diameter holes bored through (Fig. 2).
5. Silastic™ gaskets with an inner diameter of 1 cm to fit between the Teflon™ block and the polymer surface to prevent leakage.
6. Standard glass microscope slides.
7. A short-focus telescope equipped with a goniometer eyepiece to measure contact angle (θ) of sessile water drops on surfaces.
8. Standard compound microscope or image analysis system.

3. Methods

3.1. Preparation of Polymer Surfaces

1. Polymers that are received as smooth films should be cleaned by sonicating in 95% ethanol for 15 min or, in the case of sulfonated polystyrene (SPS), dipping for 10 s in *n*-hexane.
2. Allow substrates to air dry, and mount on ethanol-cleaned glass microscope slides, using small pieces of autoclave tape at the ends.

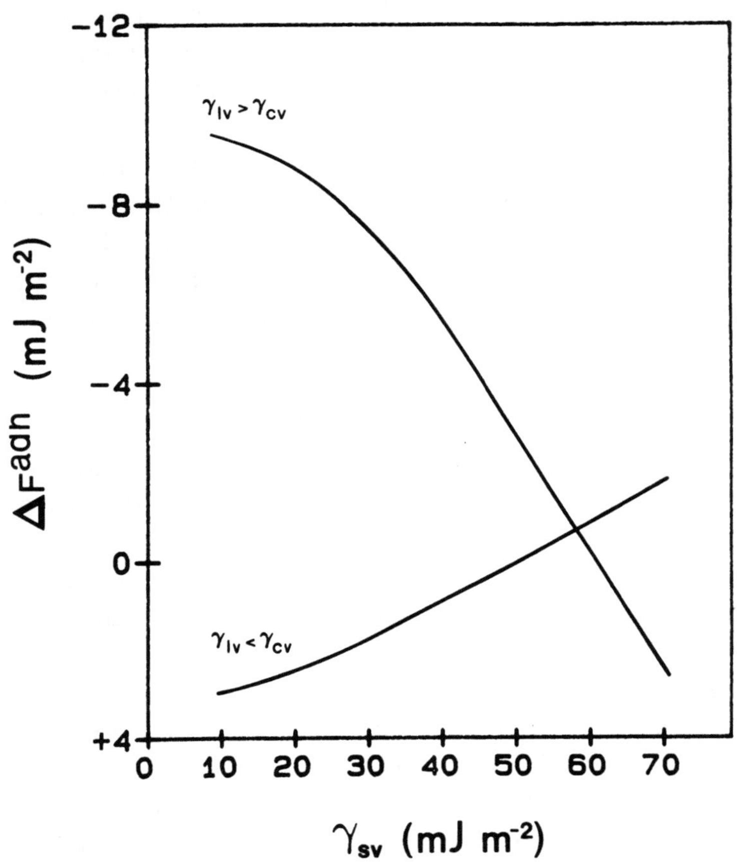

Fig. 1. Dependence of the change in the free energy of adhesion (ΔF^{adh}) of particles as a function of substrate surface tension γ_{sv}. The plant cells are assumed to have a surface tension of γ_{cv} = 55 mJ m^{-2}. For $\gamma_{lv} > \gamma_{cv}$, γ_{lv} = 72.5 mJ m^{-2} was chosen, whereas for γ_{lv} , γ_{cv}, γ_{lv} = 44.5 mJ m^{-2} was used.

The polymer sample should be cut slightly smaller than the dimensions of the glass slide.

3. Polymers, such as fluorinated ethylene-propylene (FEP), low-density polyethylene (LDPE), and acetal resin (Ac), must be heated to their plastic transition temperature and pressed to produce smooth surfaces. This can be done by pressing ethanol-cleaned pieces of the polymer between two ethanol-cleaned glass microscope slides in a heated

Table 1
Polymer Substrates Used in Adhesion Experiments

Polymer	Method of preparation	Source	Contact angle θH_2O degrees	Surface tension mJ m^{-2}
Sulfonated polystyrene (SPS)	Used as received	Dow Chemical Co.	24 ± 3	66.6
Polyethylene terephthalate (PET)	Used as received	Celanese, Ltd Toronto	60 ± 3	47.0
Acetal resin	Heat press	Commercial Plastics, Toronto	64 ± 1	44.6
Low density polyethylene (LDPE)	Heat press	Commercial Plastics, Toronto	84 ± 4	32.5
Polystyrene (PS)	Used as received	Dow Chemical Co.	95 ± 2	25.6
Fluorinated ethylene-propylene (FEP)	Heat press	Commercial Plastics, Toronto	110 ± 3	16.4

vise (*see* Note 1). Clean the pressed polymers by sonication in ethanol, and mount on glass slides as described above.

4. Determine that the polymer surfaces have the desired surface tension (γ_{sv}) by observing the advancing contact angle of a drop of double-distilled water on several representative areas of each surface, using the goniometer telescope (Fig. 3). An advancing contact angle is produced by increasing slightly the vol of the sessile water drop just prior to making a measurement of the contact angle (3). An observed contact angle that falls within the error limits reported in Table 1 (*see* Note 2) indicates an acceptable surface of desired surface tension (γ_{sv}).

3.2. Preparation of Suspending Liquid Medium

1. The suspending liquid used is distilled water, with various concentrations of 1-propanol added to produce a wide range of liquid surface

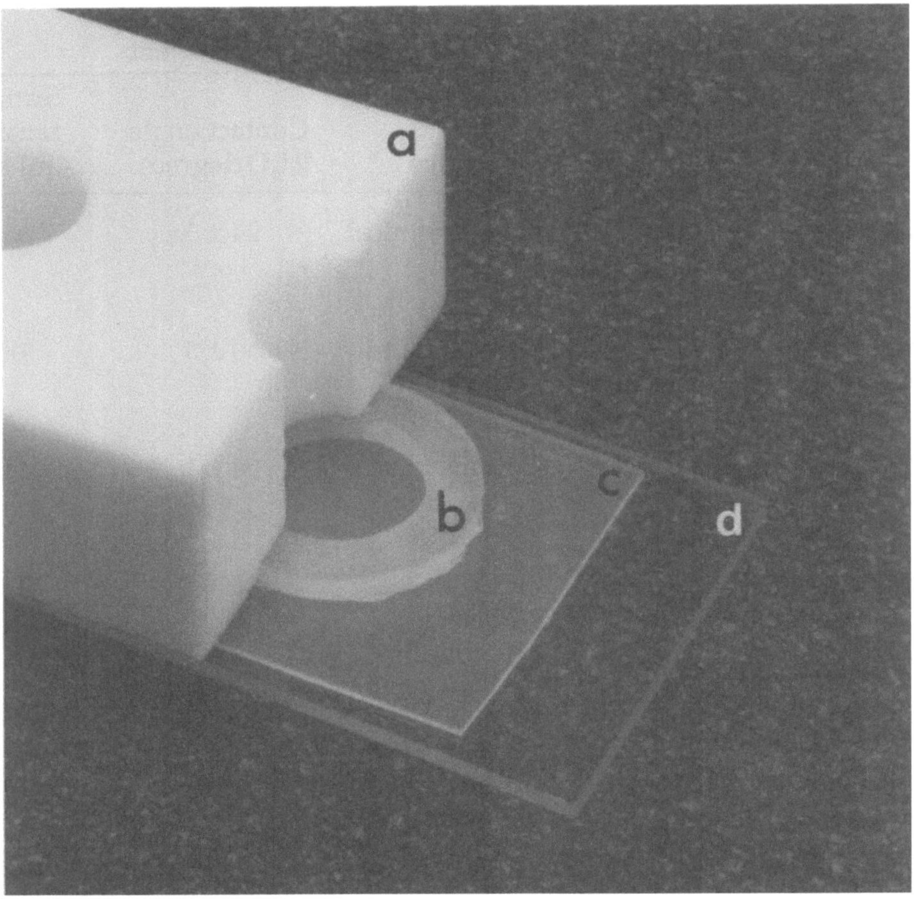

Fig. 2. Cross-section of the Teflon™ well system used in the adhesion experiments. The Teflon™ block (a), with three 1-cm diameter holes bored through, is secured to the polymer surface to be tested (c), which is retained on a standard glass microscope slide (d). Silastic™ gaskets (b) are used between the Teflon™ block and the polymer surface to prevent leakage. Elastic bands can be fastened between wells to secure the entire system.

tensions (γ_{lv}) (*see* Note 3). The exposure times of the cells to the 1-propanol are short, and low concentrations of l-propanol are used, thus assuring viability. The effect of l-propanol in modifying the surface tension (γ_{lv}) of water is shown in Fig. 4.

2. A series of l-propanol–water mixtures should be prepared; for example, liquid mixtures with surface tensions that decrease from 72.5 mJ m^{-2} (the surface tension of water) in 5 mJ m^{-2} increments are recommended.

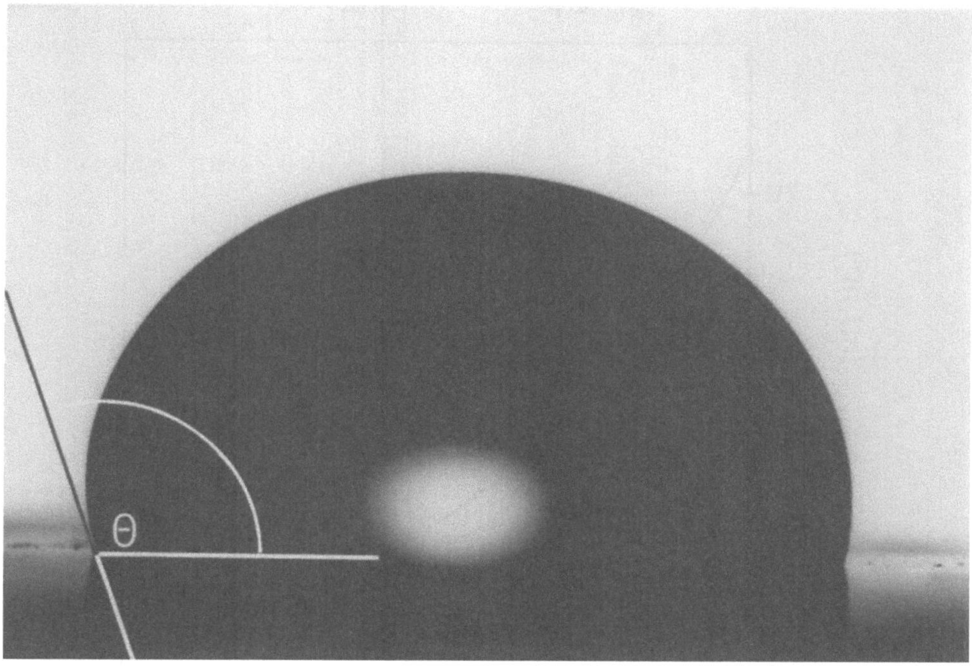

Fig. 3. Sessile drop of double-distilled water on fluorinated ethylene propylene (FEP) demonstrating an advancing contact angle (θ). The angle is measured through the drop from a tangent to the drop at the three-phase point. The angle represented here for FEP is θ = 110 degrees.

3.3. Cultured Plant Cell Preparation

1. Cell suspensions are cultivated according to any of several standard methods. Dilute approximately 100 mL of cell suspension culture with 6 vol of distilled water, and filter under gentle vacuum through a filter series (e.g., 500, 350, 210, and 105 µm) to obtain a fine cell suspension. The final cell suspension should consist of unicells or small aggregates of 2–5 cells. Repeat filtration as required.
2. Wash the filtered suspension by centrifugation at 600 x *g* for 3 min. Repeat 3x, resuspending the pellet in fresh distilled water each time.
3. This procedure produces cell suspensions (e.g., *Catharanthus roseus*) consisting of greater than 97% small aggregates (2–5 cells) with low levels of extracellular polysaccharides and proteins in the suspending liquid.
4. Adjust the suspension to a final 1% packed cell vol (PCV) concentration in each of the chosen l-propanol–water mixtures.

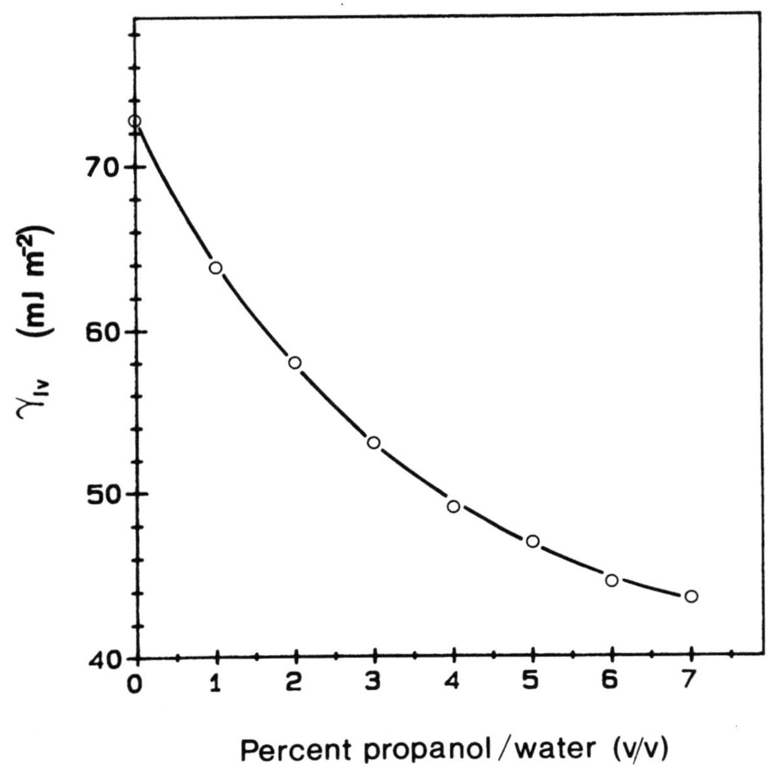

Fig. 4. Reduction in the surface tension γ_{lv} of distilled water by the addition of l-propanol.

3.4. Adhesion Protocol

1. Pipet 1 mL of the 1% PCV plant cell suspension, in the appropriate test liquid of known γ_{lv}, into the 1-mL capacity Teflon™ wells secured by elastic bands to the polymer surface, which is retained on a glass microscope slide. Silastic™ gaskets must be used between the polymer surface and the Teflon™ block to prevent leakage. Test a number of replicate wells/polymer surface and test liquid mixture (*see* Note 4).
2. Incubate the cells for 20 min at 25°C. Then submerge the blocks in a distilled water bath at 25°C, and subsequently invert them for 10 min to remove nonadherent cells.
3. Return the submerged blocks to the upright position before removing them from the bath, and allow to stand at room temperature for 20 min before carefully separating the Teflon™ blocks from the polymer

films. Drain the wells without greatly disturbing the adherent layer of cells. Allow the polymers to air dry with the Silastic™ gaskets still attached.

4. The extent of cultured plant cell adhesion can be assessed in a number of ways. The simplest method is to count the number of adherent cells adhered/unit surface area, using a standard compound microscope. Alternatively, if an image analysis system is available, the percentage of the surface area covered by cells can be used as the means of quantifying the relative extent of adhesion on each surface.

3.5. Determination of Cellular Surface Tension

1. After determining the relative extent of adhesion of the cultured plant cells to various polymer substrates (e.g., 3–5 polymers of a wide range of γ_{sv} values) in various 1-propanol–water suspending liquid mixtures (e.g., 5–10 liquid surface tension values covering a wide range), the data should be plotted as the extent of adhesion (e.g., cell number; percentage of surface area covered by cells) vs the substrate surface tension (Fig. 5).
2. The slopes of each of the lines should be plotted vs liquid surface tensions (γ_{lv}). Since adhesion becomes independent of γ_{sv} when $\gamma_{lv} = \gamma_{cv}$, the intercept with the γ_{lv} axis, according to the thermodynamic model, is equal to the surface tension of the cultured plant cells (Fig. 6).

4. Notes

1. After heat pressing, some of the polymers have a tendency to stick to the glass slides; do not force these apart. Immersing these in distilled water for 24 h will facilitate separation.
2. Polymers that produce reproducible water contact angles that are different from those listed in Table 1 might have a different surface tension (γ_{sv}); tables are available to convert contact angle values to solid surface tensions in Neumann et al. (*4*).
3. The surface tension of the various 1-propanol–water mixtures can be checked if a Whilhelmy balance or any other form of tensiometer is available. This is unnecessary if the solutions are freshly prepared prior to each experiment. A tensiometer is useful, however, to measure the surface tension of various unknown liquids of interest such as growth media.

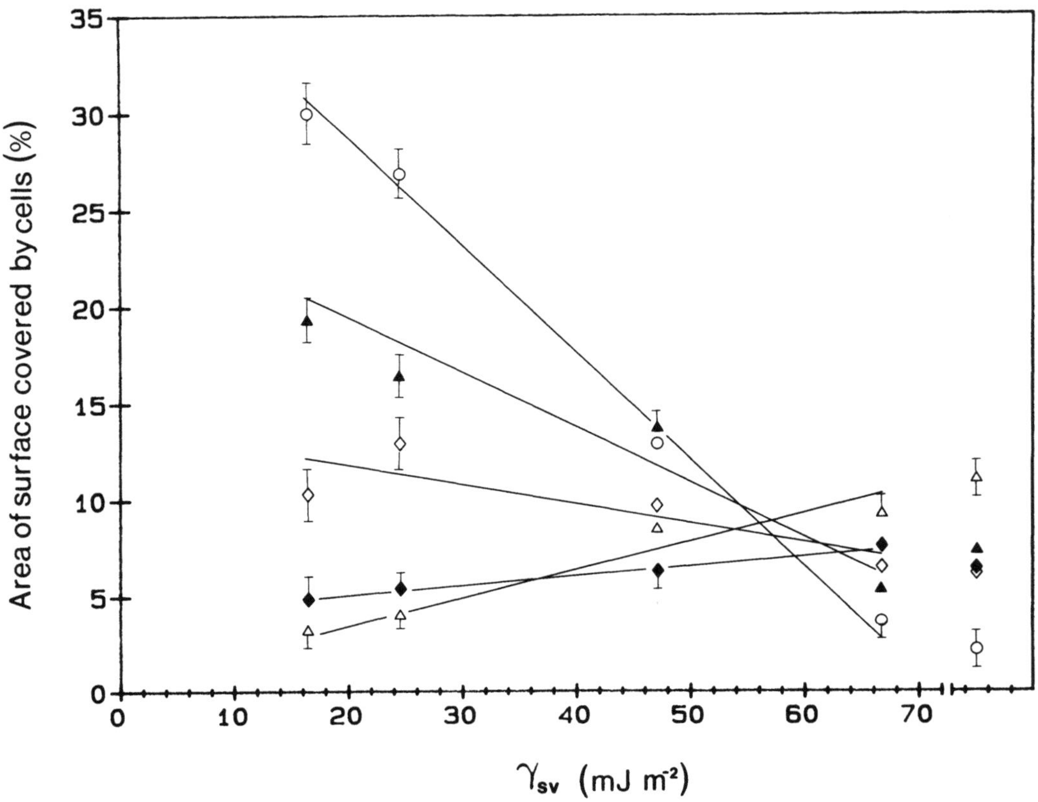

Fig. 5. Adhesion of suspension-cultured *Catharanthus roseus* cells as a function of surface tension γ_{sv} for the various l-propanol concentrations. Lines were plotted using a least squares estimate. The points for glass (γ_{sv} > 72.5 mJ m^{-2}) were not used to fit lines, because its precise surface tension was unknown. The liquid surface tensions used were: (0) 72.5 mJ m^{-2}; (▲) 63.0 mJ m^{-2}; (◇) 58.0 mJ m^{-2}; (◆) 53.0 mJ m^{-2}; (Δ) 44.5 mJ m^{-2}.

4. The choice of immobilization substrate material when spontaneous adhesion is involved will depend on the relative values of the suspending liquid and cellular surface tensions as discussed in the Introduction. The γ_{sv} value for Murashige and Skoog (MS) medium with 3% sucrose, 2.4 μm α-naphthalenacetic acid, and 1 μm kinetin is 47.2 mJ m^{-2}. Therefore, when this growth medium is used as the suspending liquid, a γ_{cv} value higher than 47.2 mJ m^{-2} will indicate that a hydrophilic substrate material (i.e., with a high surface tension) should be used to maximize adhesion. A hydrophobic substrate (i.e., one with a low γ_{sv} value) should be chosen under the opposite circumstances. The surface tension of the liquid medium used for the immobilization procedure can be lowered by increasing the concentration of organic compounds in the solution. Nonorganics, such as mineral salts, have

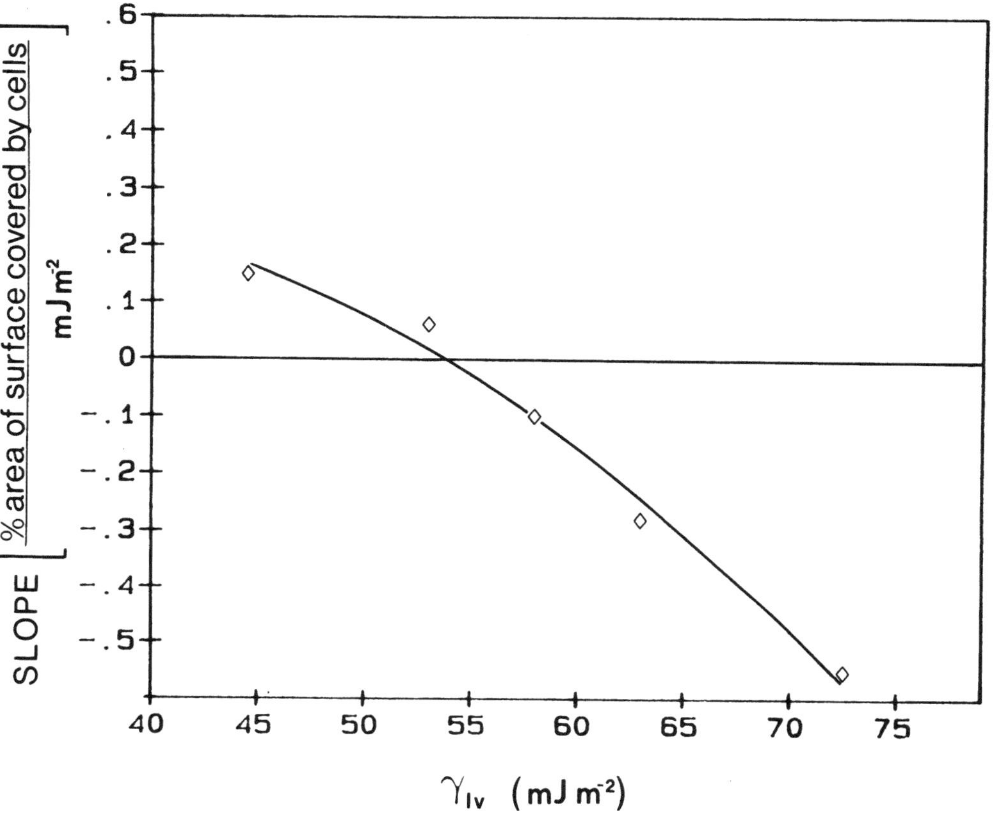

Fig. 6. Slopes of the straight lines in Fig. 5 vs γ_{lv}. The slope here is equal to 0 for $\gamma_{lv} = \gamma_{cv} = 54$ mJ m^{-2}. The line was plotted by means of a second order computerized curve fit.

little effect on liquid surface tension. Again, the γ_{lv} value of the suspending medium used should be checked with a tensiometer. The greater the difference between γ_{lv} and γ_{cv}, the more effective an extremum in γ_{sv} will be in increasing adhesion.

References

1. Lindsey, K., Yeoman, M. M., Black, G. M., and Mavituna, F. (1983) A novel method for the immobilization and culture of plant cells. *FEBS Letters* **155,** 143–149.
2. Facchini, P. J., Neumann, A. W., and DiCosmo, F. (1988) Thermodynamic aspects of cultured plant cell adhesion to polymer surfaces. *Applied Microbiology and Biotechnology* **29,** 346–355.
3. Neumann, A. W. (1974) Contact angles and their temperature dependence: thermodynamic status, measurement interpretation and application. *Advances in Colloid and Interface Science* **4,** 105–191.
4. Neumann, A. W., Absolom, D. R., Francis, D. W., and van Oss, C. J. (1980) Conversion tables of contact angles to surface tension. *Separation and Purification Methods* **9,** 62–163.

Chapter 47

Plant Cell Immobilization in Alginate and Polyurethane Foam

Mohamed T. Ziyad-Mohamed and Alan H. Scragg

1. Introduction

Plant cell culture has great potential as an alternative system for the production of phytochemicals normally extracted from whole plants, many of which have commercial value. Suspension cultures of plant cells have been shown to produce compounds characteristic of the original plant, and in some cases the yield of these phytochemicals has been higher than that from the plant.

Immobilized plant cells offer an alternative to batch culture for the production of secondary metabolites. Immobilization of plant cells confers a number of advantages; it encourages product release, it protects the cells from shear stress, and it gives good cell-to-cell contact, known to affect secondary metabolite production.

From: *Methods in Molecular Biology*, vol. 6, *Plant Cell and Tissue Culture*
Edited by Jeffrey W. Pollard and John M. Walker, ©1990 by The Humana Press

Two types of matrices have been found to be successful for the immobilization of plant cells. These are (a) entrapment in polysaccharide polymers, such as alginate (*1*), and (b) entrapment in porous matrices, such as polyurethane foam (*2,3*). Although alginate has been widely used for plant cells, it has its drawbacks, for example, rupture of beads with cell growth.

In this chapter, we describe the immobilization of *Catharanthus roseus* (Madagascan Periwinkle) cells, both in polyurethane foam and calcium alginate, and the production of the alkaloids ajmalicine and serpentine. Since we are looking at secondary metabolites, a growth-limiting medium is used in the method with polyurethane foam. With alginate, by controlling the bead load, cell load, and the media composition, rupture of beads is avoided.

2. Materials

1. Polyurethane foam, 60 pores/inch (Declon Ltd., Northants UK): Rhodes et al. (*4*) reported the phytotoxicity of the foam obtained from various manufacturers, but in our study with *C. roseus, Dacaus carota, Theobroma,* and *Cinchona ledgeriana,* no toxic effect was noticed with the foam supplied by Declon Ltd., UK.
2. Sodium alginate is obtained from BDH Chemicals UK (*see* Note 1).
3. P5 medium: MS medium (*see* Appendix) containing 8.5 mg/L KH_2PO_4, 0.2 mg/L naphthaleneacetic acid (NAA), 0.1 mg/L kinetin, and 2% w/v sucrose, pH 5.8.
4. Stock cultures of *C. roseus* cell line ID 1 are grown on M3 medium (*see* Appendix) at 25°C under diffuse light (5 W/m^2) in 250-mL conical flasks containing 100 mL medium. Cells are subcultured every 14 d at a dilution ratio of 1:5.
5. M3 medium: MS medium (*see* Appendix) containing 1 mg/L NAA, 0.1 mg/L kinetin, and 2% w/v sucrose, pH 5.8.

3. Methods

3.1. Immobilization in Polyurethane Foam

1. Cut the polyurethane foam sheet into 1-cm cubes, weigh it, wash it in distilled water, and autoclave it in distilled water for 30 min (*see* Note 2).
2. Inoculate flasks containing 100 mL P5 medium and 5 preweighed foam cubes with 20 mL of 14-d-old culture of *C. roseus* cells.

3. Agitate the cultures on a rotary shaker at 120 rev/min[1], at 25°C. At intervals, flasks can be harvested and determinations made of fresh and dry weight of suspended cells, carbohydrate, and dry weight of immobilized cells (*see* Chapter 2, this vol.).
4. To estimate the cells in the foam cubes, the excess water is removed from the foam by suction on a Buchner funnel, and the cubes are frozen in liquid nitrogen and freeze-dried. The foam cubes are weighed after drying to determine the weight of cells present.
5. The alkaloids, from the freely suspended cells and from the medium, can be extracted as described in Chapter 48, this vol. Alkaloids are extracted from foam-immobilized cells by placing the dry cubes in a Soxhlet thimble and extracting with methanol in the normal way (*see* Chapter 48, this vol.).
6. Figure 1 shows *C. roseus* cells immobilized in polyurethane foam cubes. Growth kinetics of *C. roseus* cell line ID 1 are given in Fig. 2. It can be seen from Fig. 2 that each foam cube entraps about 47.5 mg dry weight of cells, and once the foam cubes are full of cells (by day 7), there is a rapid increase in the dry weight of cells in suspension, reaching a maximum biomass of about 6.2 mg/mL.
7. Alkaloid analysis (Fig. 3) shows that the entrapped cells produce about 4.2 mg/g dry weight ajmalicine and 2.3 mg/g dry weight serpentine, compared to 0.20 mg/g ajmalicine and 0.80 mg/g serpentine produced by cells in suspension. The entrapped cells also release 8.5 µg/mL ajmalicine and 3.0 µg/mL serpentine to the medium, which is much higher than the maximum of about 1.0 µg/mL ajmalicine and 0.30 µg/mL serpentine normally released by suspension cells.

3.2. Immobilization in Alginate Beads

1. *C. roseus* cells are collected from a 7-d old, late logarithmic phase culture on a 50 µm pore nylon net and washed with M3 medium.
2. The washed cells (20 g fresh weight) are resuspended in sterile 4% (w/v) sodium alginate solution in MS medium, in order to reduce osmotic shock to the cells.
3. The mixture is added dropwise (Fig. 4) with the aid of a sterile syringe and large bore needle (diameter 1 mm) to 200 mL of stirred culture medium containing 0.2*M* $CaCl_2$. The $CaCl_2$ solution is stirred using a magnetic stirrer, and the beads that form (Fig. 5) are stirred for a further 2 h, after which the beads are removed and washed twice with culture medium (*see* Note 3).
4. Beads (15 g wet weight) containing immobilized cells are finally sus-

Fig. 1. *C. roseus* cell line ID 1 immobilized in polyurethane foam.

pended in 100 mL of M3 medium in 250 mL shake flasks and shaken at 120 rev/min at 25°C under diffuse light.

5. The medium can be separated from the beads by filtration and analyzed for alkaloid and carbohydrate content (Chapters 2 and 48, this vol., and *see* Note 4).
6. For the analysis of the fresh weight, dry weight, and the alkaloid accumulated by the immobilized cells, the beads are dissolved in 100 mL of 0.1*M* trisodium citrate, 0.1*M* EDTA, or 0.5*M* KH_2PO_4.
7. Growth kinetics of *C. roseus* cell line ID 1 immobilized in calcium alginate are shown in Fig. 6. After a longer initial lag phase than that in polyurethane foam (5 d), the cells in the beads grow well. The longer lag phase may be the result of diffusional limitation. Once the beads are almost full (day 10), fine cells are released into the medium as observed by Morris and Fowler (*5*). The maximum suspension cell concentration of about 2.9 mg/mL is found by day 20, whereas the weight of the immobilized cells has reached its maximum of about 420 mg (in 15 g beads) by day 15. The beads begin to disintegrate after day 15, and the total weight of cells/flask is about 675 mg.

The immobilized cells produce about 2.02 mg serpentine/g dry wt of cells by day 15 (Fig. 7), whereas suspension cells produce about 1.86

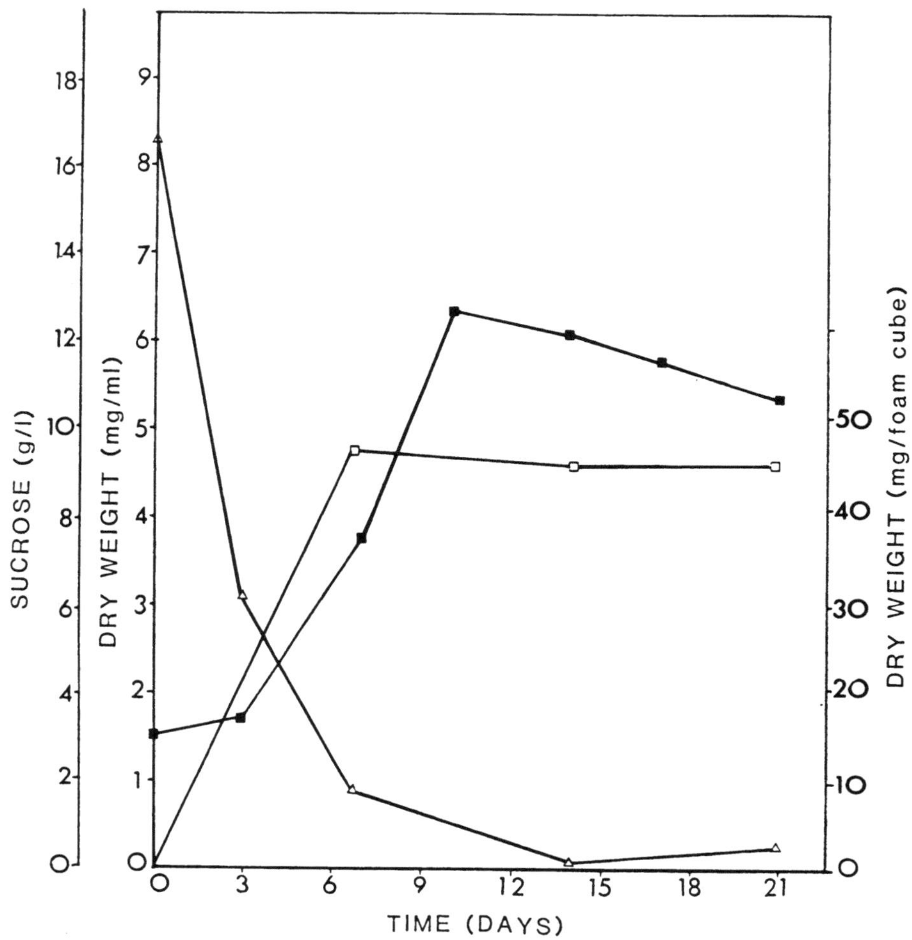

Fig. 2. Growth kinetics of *C. roseus* cell line ID 1 immobilized in five polyurethane foam cubes using P5 medium in 250-mL conical flasks: (□) dry wt of cells in foam; (■) dry wt of cells in suspension; (△) sucrose concentration.

mg/g dry wt serpentine. Whereas the polyurethane foam method produced high ajmalicine levels, quite low levels were produced both in suspension and immobilized cells (about 0.08 and 0.26 mg/g dry wt) respectively. In addition, the immobilized cells released 0.9 μg/mL serpentine to the medium and, as with the cells, the ajmalicine level is very low (*see* Notes 5–7).

4. Notes

1. Alginates are heteropolymer carboxylic acids coupled by 1,4 glycosidic linkages of β-D-mannuronic acid (M) and α-L-guluronic acid

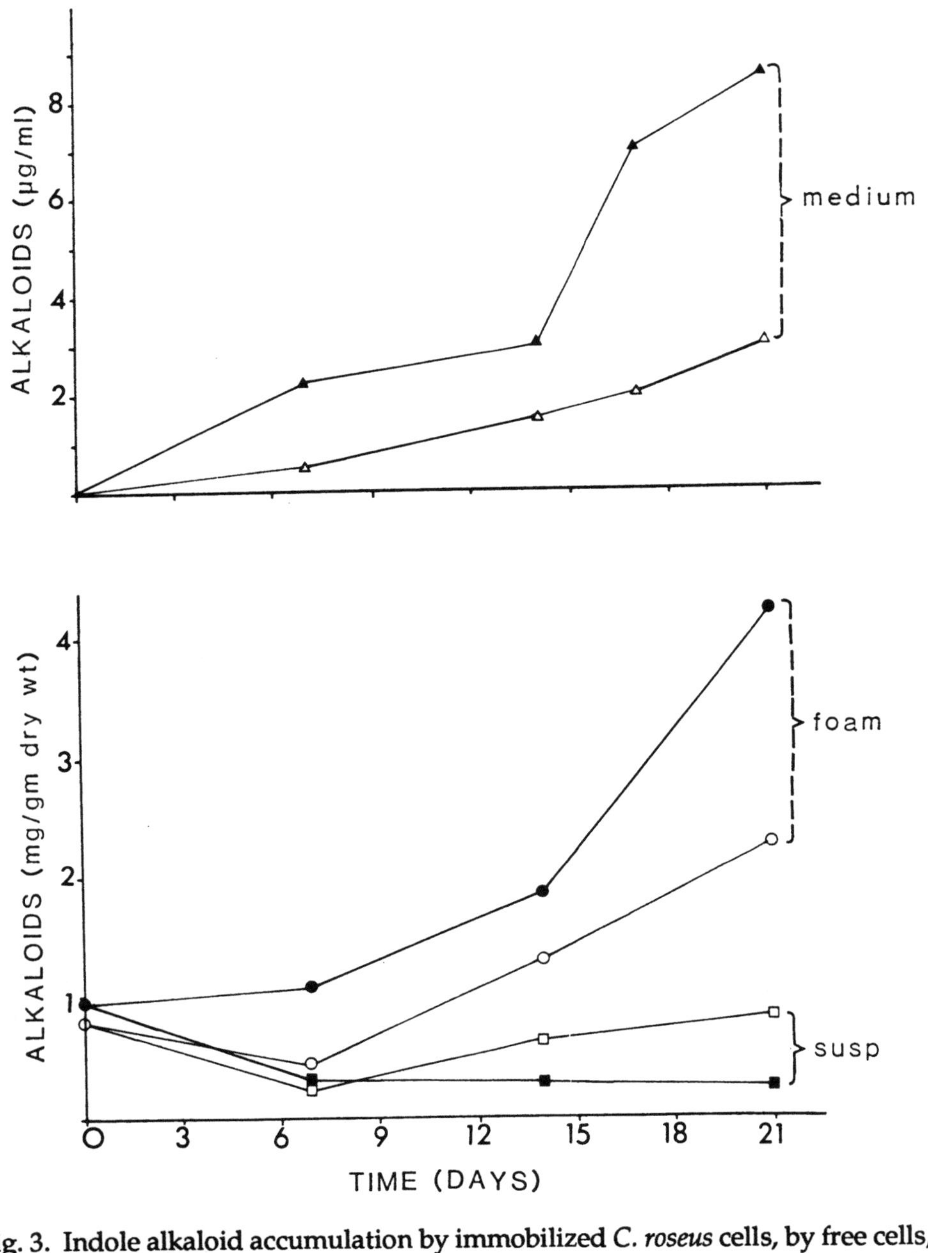

Fig. 3. Indole alkaloid accumulation by immobilized *C. roseus* cells, by free cells, and in the medium: (●) (■) (▲) ajmalicine accumulation; (○) (□) (Δ) serpentine accumulation.

(G) units. This polymer contains blocks of M_n and G_n, together with blocks of alternating sequence (MG_n). Alkali and magnesia alginates are soluble in water, whereas alginic acids and the salts of polyvalent metal cations are insoluble. Dropping a sodium alginate solution into a solution containing polyvalent cations (except Mg^{2+})

Fig. 4. Immobilization of *C. roseus* cells on calcium alginate beads.

forms gels. The alginate we use is extracted from the algae *Laminaria hyperborea*. The viscosity of a 4% (w/v) solution of alginate can vary from 25–1400 mPa, depending on the species from which it is extracted. The viscosity, in turn, can affect the strength and diffusional properties of the beads formed. The strength and texture of the alginate gels are mainly dominated by the guluronic acid (G_n) sequence.

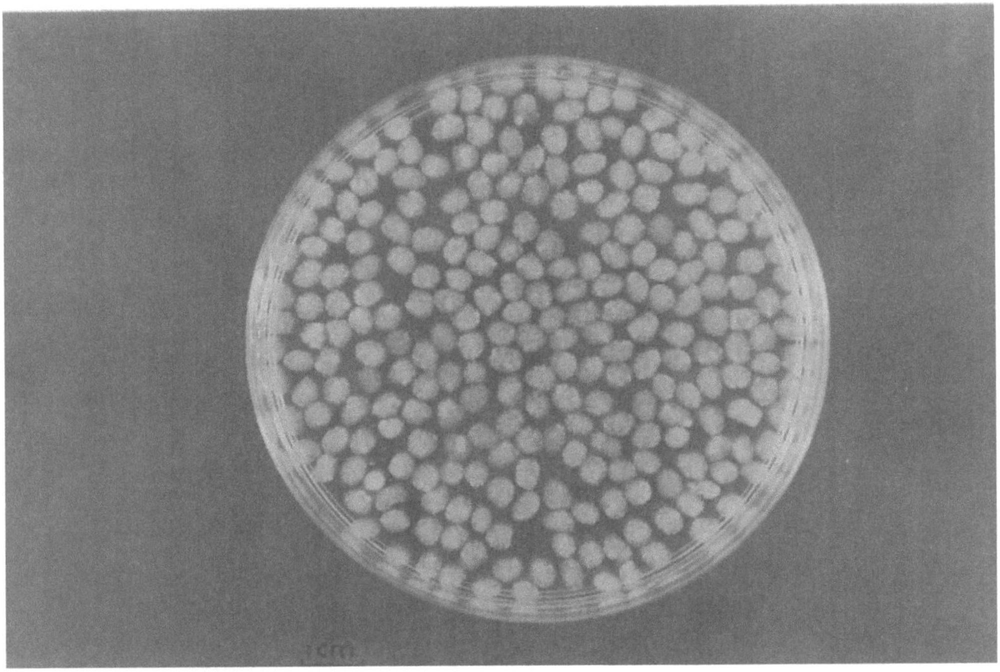

Fig. 5. Calcium alginate beads containing immobilized *C. roseus* cells.

2. This helps to reduce the hydrophobicity of the foam. Plant cells in suspension form aggregates, and therefore, the foam porosity should be selected to match the average aggregate size. Since immobilization in polyurethane foam is a passive adsorption process, the most effective immobilization is achieved when the foam particles are stationary. Immobilization can be improved by increasing the number of foam cubes or by threading them onto a stiff stainless-steel wire loop (*3*). The foam is about 95% void, and when full of cells, can present diffusion problems that may require the reduction of the size of the foam cubes.
3. In our alginate experiment, we have used the simple method of needle and syringe for the formation of beads, since we were working with small volumes (up to 120 mL). For larger volumes, Morris and Fowler (*5*) employed a perisaltic pump. An alternative method of bead formation, based on vibration technique, has been used by Hulst et al. (*6*) to give more spherical beads. Nilsson et al. (*7*) employed a different technique, using a hydrophobic phase, such as soy oil, to produce uniform beads.
4. Brodelius et al. (*8*) observed that product removal from the medium

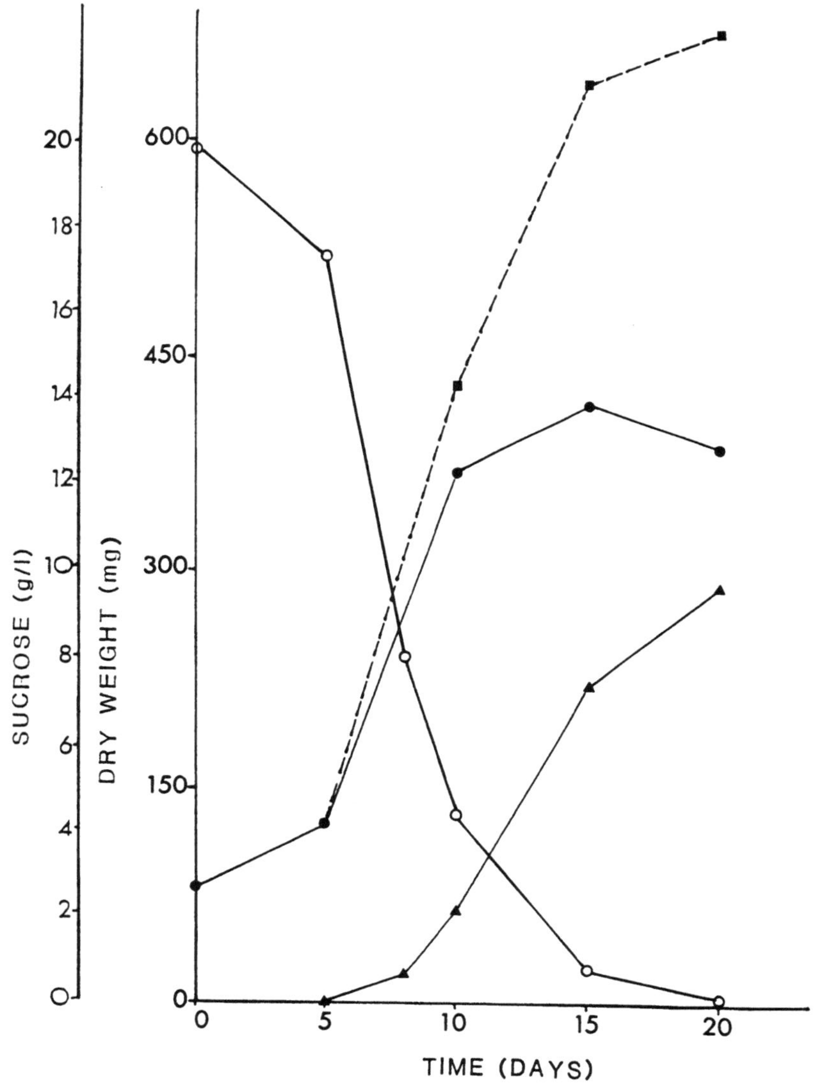

Fig. 6. Growth kinetics of *C. roseus* cells immobilized in calcium alginate beads: (●) dry wt of cells in beads; (▲) dry wt of cells in suspension; (■) total dry wt of cells per flask; (○) sucrose concentration.

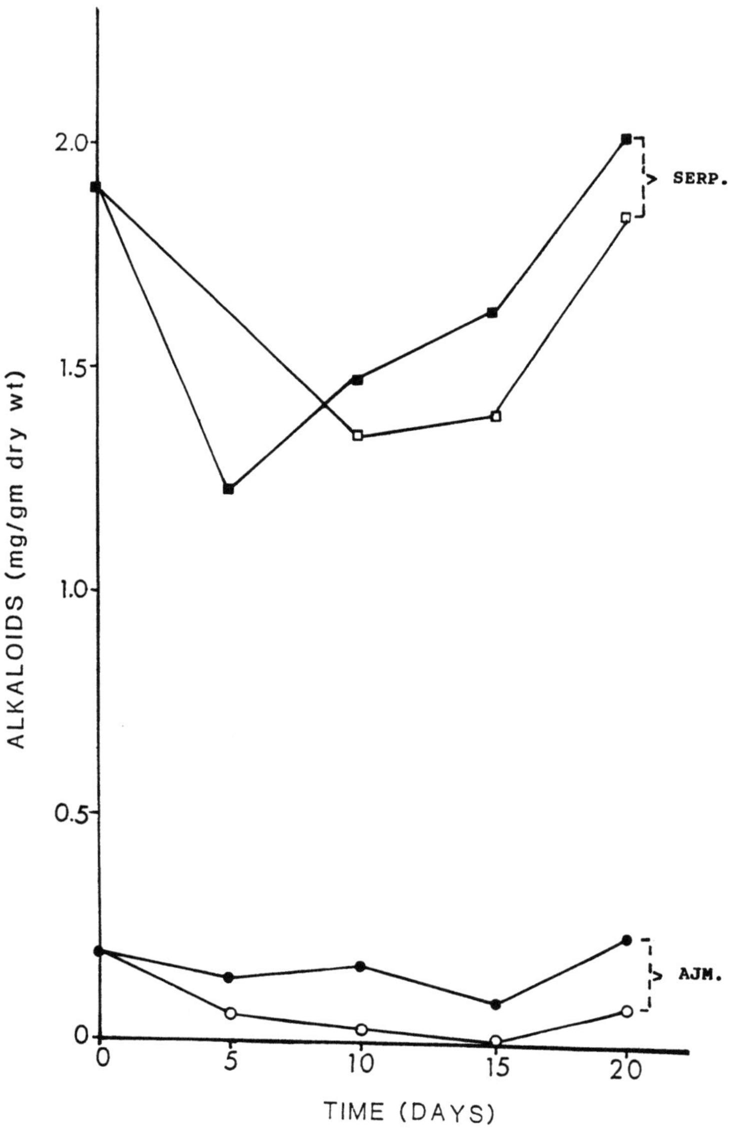

Fig. 7. Indole alkaloid accumulation by calcium alginate-immobilized cells and free cells: (■) (●) by immobilized cells; (□) (○) by suspension cells.

could enhance the formation of secondary products. For a commercial process, it would be advantageous if the product could be isolated from the medium, rather than extracted from the biomass. The additional advantages would be

a. The biomass need not be destroyed,
b. The productive capacity of the system would not depend on the cellular accumulative capacity, but on the biosynthetic rate and rate of release to the medium,
c. Product could be removed continuously or semicontinuously from the medium.

Brodelius and Nilsson (*9*) have claimed improved product release by permeabilization of cells with dimethyl sulfoxide (DMSO); however, other workers (*10*) have observed that chemical or electrical permeabilization leads to cell lysis.

5. Solid-phase product recovery (e.g., using XAD-7 resin) would be preferred to liquid-phase (using chloroform) recovery, since the former gives a higher concentration and easier mode of operation. Knorr et al. (*11*) were successful in using chitosan (a polyelectrolyte) as an immobilizing and permeabilizing agent for plant cells.
6. The major objective of immobilization is enhancement of product formation. In suspension culture, the product formation starts when the growth ceases (i.e., in the late stationary phase). Generally in plant cell cultures, growth limitation induces secondary product formation. In our experiment with polyurethane foam, reduced phosphate was used to limit growth.

Secondary product formation can be induced by stress, such as nutritional limitation, which has been used to explain product accumulation in aggregates (*12*). The type of gel, gel concentration, and cell loading can affect the diffusion rate of substrate in or products out, and therefore may affect secondary product formation.

References

1. Brodelius, P. (1983) Immobilised plant cells, in *Immobilised Cells and Organelles* (Mattiason, B., ed.), CRC Press, Boca Raton, pp. 27–55.
2. Lindsey, K., Yeoman, M. M., Black, G. M., and Mavituna, F. (1983) A novel method for the immobilisation and culture of plant cells. *FEBS Lett.* **155,** 143–149.
3. Mavituna, F. and Park, J. M. (1985) Growth of immobilised plant cells in reticulate polyurethane foam matrices. *Biotech. Letters* **7,** 637–640.
4. Rhodes, M. J. C., Smith, J. I., and Robins, R. J. (1987) Factors affecting the immobilisation of plant cells on reticulated polyurethane foam particles. *Appl. Microb. and Biotech.* **26,** 28–35.

5. Morris, P. and Fowler, M. W. (1981) A new method for the production of fine plant cell suspension cultures. *Plant Cell Tissue Organ Culture* **1,** 15–24.
6. Hulst, A. C., Tramper, J., Van't Riet, K., and Westerberk, J. M. M. (1985) A new technique for the production of immobilised biocatalyst in large quantities. *Biotech. Bioeng.* **27,** 870–876.
7. Nilsson, K., Birnbaum, S., Flygare, S., Linse, L., Schroder, U., Jeppson, U., Larsson, P., Mosbach, K., and Brodelius, P. (1983) A general method for the immobilisation of cells with preserved viability. *Eur. J. Appl. Microbiol. Biotech.* **17,** 319–326.
8. Brodelius, P., Deus, B., Mosbach, K., and Zenk, M. H. (1979) Immobilised plant cells for the production and transformation of natural products. *FEBS Lett.* **103,** 93–97.
9. Brodelius, P. and Nilsson, K. (1983) Permeabilization of immobilised plant cells, resulting in release of intracellularly stored products with preserved cell viability. *Eur. J. Appl. Microb. Biotech.* **17,** 275–280.
10. Parr, A. J., Robins, R. J., and Rhodes, M. J. C. (1984) Permeabilisation of *Cinchona ledgeriana* cells by dimethylsulfoxide. Effects on alkaloid release and long term membrane integrity. *Plant cell report* **3,** 262–265.
11. Knorr, D., Miazga, S. M., and Teutonico, R. A. (1985) Immobilisation and permeabilisation of cultured plant cells. *Food Technology* **October,** 135–142.
12. Shuler, M. L. and Hallsby, G. A. (1985) Bioreactor considerations for chemical production from plant cell cultures, in *Biotechnology in Plant Science* (Zaitlin, M., Day, P., and Hollaender, A., eds.), Academic Press, London, pp. 191–205.

Chapter 48

Alkaloid Secondary Products from *Catharanthus roseus* Cell Suspension

Alan H. Scragg

1. Introduction

Plant cell culture, the growth of plant cells on solid medium or in liquid, was originally used to study the physiology and biochemistry of plants without the complication of the whole plant. However, it was soon found that plant cell cultures were often capable of producing compounds characteristic of the original plant. Of particular interest were those compounds known as secondary products. Secondary products are poorly defined, but in general are products that are not essential for growth, but may confer some advantage. The plant kingdom produces a vast range of secondary products, many of which, such as morphine and diosgenin, are of commercial interest. Since plant cell cultures are capable of producing such compounds (*1–3*), they have been of interest for the development of a biotechnological process to produce these compounds. Initially, the accumulation of secondary products was fortuitous, but with selection and medium changes, high levels of the products have been achieved that in some cases exceed those found in the whole plant (*2–4*).

From: *Methods in Molecular Biology*, vol. 6, *Plant Cell and Tissue Culture*
Edited by Jeffrey W. Pollard and John M. Walker, ©1990 by The Humana Press

The methods used to induce secondary product formation are empirical, since often the pathways and controls involved are unknown. However, a number of general rules on secondary product accumulation appear to exist. A change in the cytokinin and auxin (for example, the removal of 2,4-dichlorophenoxyacetic acid (2,4-D)) tends to stimulate product formation (*5*), as does a change in the cultural condition that stresses the cells, for example, the removal of all culture components other than the carbon source (*6*). Small changes in medium and cultural conditions can also affect the levels of product formation (*7*), but the effects are limited. The idea of using a growth medium followed by a change to a production medium has prompted the development of a two-stage process (*5,6,8*), although the alternative, a single growth and production medium, has also been developed (*9*). In this chapter, the formation, extraction, separation, and identification of the alkaloids ajmalicine and serpentine from cultures of *Catharanthus roseus* will be described.

2. Materials

1. Suspension cultures of *C. roseus* are developed, and maintained in B5 medium (*10* see Appendix), as described in Chapter 43. The induction medium can be B5 medium in which 2,4-D is replaced by indole acetic acid (IAA), or the culture can be developed in a growth and production medium, such as M3, as described previously in Chapter 43, in which 2,4-D was replaced by napthaleneacetic acid (NAA).
2. TLC plates: 20 x 20 cm, 0.25 mm silica 60_{F254} fluorescence activation (E. Merck, Damstradt); activate by heating to 100°C for 1 h.
3. Iodoplatinate stain: 5% w/v platinum chloride (10 mL), conc. HCl (5 mL), 2% w/v KI (240 mL).
4. Ceric ammonium sulfate stain: ceric ammonium sulfate (1 g) in 100 mL of 85% *o*-phosphoric acid.
5. 0.8 x 10 cm μm Bondpak C18 HPLC column.

3. Methods

3.1. Growth

An example of the growth cycle and alkaloid accumulation by a culture of *C. roseus* ID 1 grown in M3 medium is shown in Fig. 1 (*see* Note 1). The culture has been grown in 250-mL shake flasks and samples taken

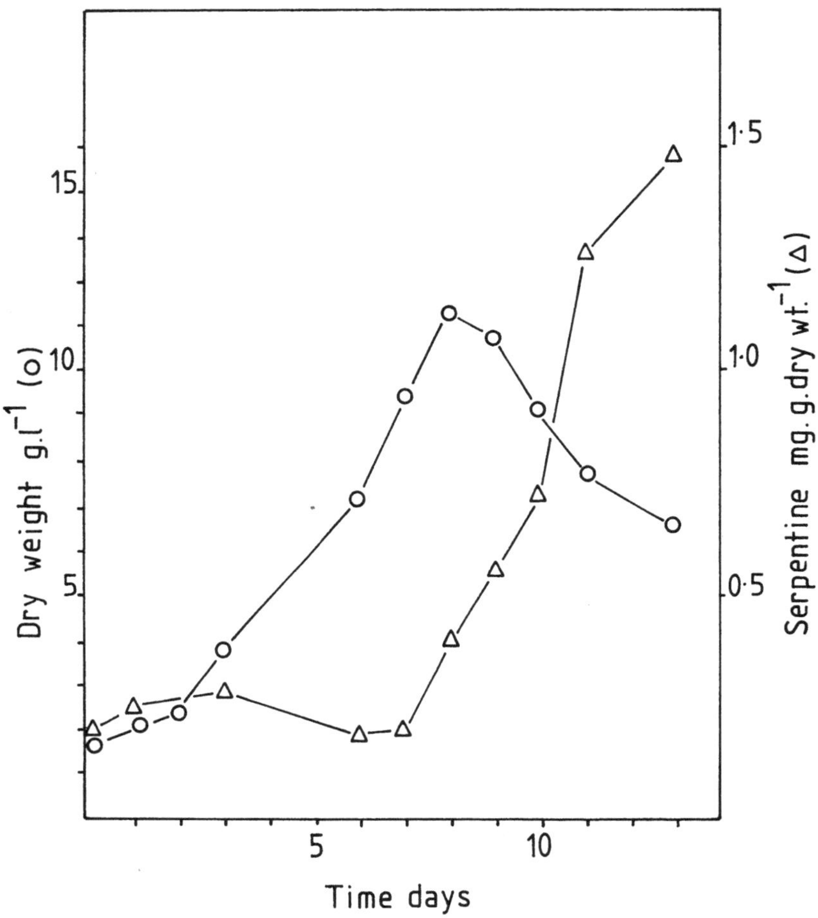

Fig. 1. The growth of and product formation by *Catharanthus roseus* ID 1 grown on M3 medium.

for dry and wet weight determination at intervals. The cultural conditions and sampling methods have been described in Chapter 43.

3.2. Alkaloid Extractions and Analysis

1. Samples of between 50–100 mg dry weight are required for alkaloid extraction, so the samples will be in the region of 0.5–1 g wet wt. At the end of growth, when the culture has reached maximum biomass, this will normally represent a volume of 5–20 mL of culture (wet weights are between 100–200 g/L). Therefore, 1–3 samples can be removed from a single 120-mL culture in a 250-mL shake flask before the reduction in volume becomes critical. However, if alka-

loid accumulation is to be followed early in the culture (at low cell densities) and duplicate samples taken, considerably larger volumes (50–100 mL) are required. Under these circumstances, a large number of flasks are set up from a single inoculum and whole flasks used in the early stages for product analysis. The number of flasks used will depend upon the frequency of sampling and rate of growth of the cultures.

2. The samples, once taken, are filtered to remove medium using a 9-cm Hartley Funnel fitted with a Miracloth Filter (Calbiochem-Behring, La Jolla, USA). The cells are washed with 100 mL of distilled water, placed in a small plastic bag, and rapidly frozen using liquid nitrogen (*see* Note 2). The frozen cells can be stored at –20°C or, more usually, freeze-dried.
3. The method used for the extraction of alkaloids will depend on whether one is extracting from cells or the growth medium. For cells, between 0.05–0.1 g of freeze-dried *C. roseus* cells are broken up and placed in a small thimble (10 x 50 mm, Whatman cellulose) that is then plugged with nonabsorbant cotton wool. The actual amount of dried cells in the thimble is estimated by weighing before and after addition of cells.
4. The thimble is placed in a small Soxhlet apparatus (Fig. 2), and refluxed with 100 mL of methanol for 2 h at 85°C. The extract (in round–bottomed flask) is reduced to dryness by rotary evaporation (Rotavapor R110, BuchiLabs) at 60°C under vacuum.
5. The residue is resuspended in 1 mL of methanol using a 1-min exposure in a Decon Frequency Sweep sonicator bath (Decon Ultrasonic Ltd., Hove, UK) to help resuspension. Two further 1-mL methanol washes are used, and these are combined and stored in sealed bottles at 4°C (*see* Note 3).
6. Alkaloid accumulation in the medium can be estimated by direct extraction of the alkaloids by addition of chloroform (50 mL of $CHCl_3$ to 100 mL of medium).
7. However, a more convenient method involves the use of SEP-PAK C18 cartridges (Waters Associates). The cartridges are preactivated with 4 mL of acetonitrile, followed by 4 mL of 95% methanol.
8. Identification can be carried out either by HPLC or TLC.
   a. HPLC: In general, the identification of plant products is achieved using high-performance liquid chromatography (HPLC). In this system, the peaks that were eluted from the column were identified on the basis of their elution time, and sample amounts were calculated using a response factor de-

Fig. 2. The Soxhlet extraction of alkaloids from *Catharanthus roseus*.

rived from the peak areas of known quantity of alkaloid standards.

For the identification of the *C. roseus* alkaloids ajmalicine and serpentine, the following system was used. The column was a 0.8 x 10 cm µm Bondpak C18 reverse phase, run isocratically at 55:45:5 methanol:water:*n*-heptane sulfonate for 14 min and then with a linear gradient to 70:25:5 (methanol:water:*n*HS). The sample volume was 10–30 µL, with a flow rate of 1.5–2 mL/min, and the wavelengths of 254 nm and 280 nm were used for detection. An example of such a separation and some of the standards run is shown in Fig. 3. The standards are run after every mix run, and the retention times are updated.

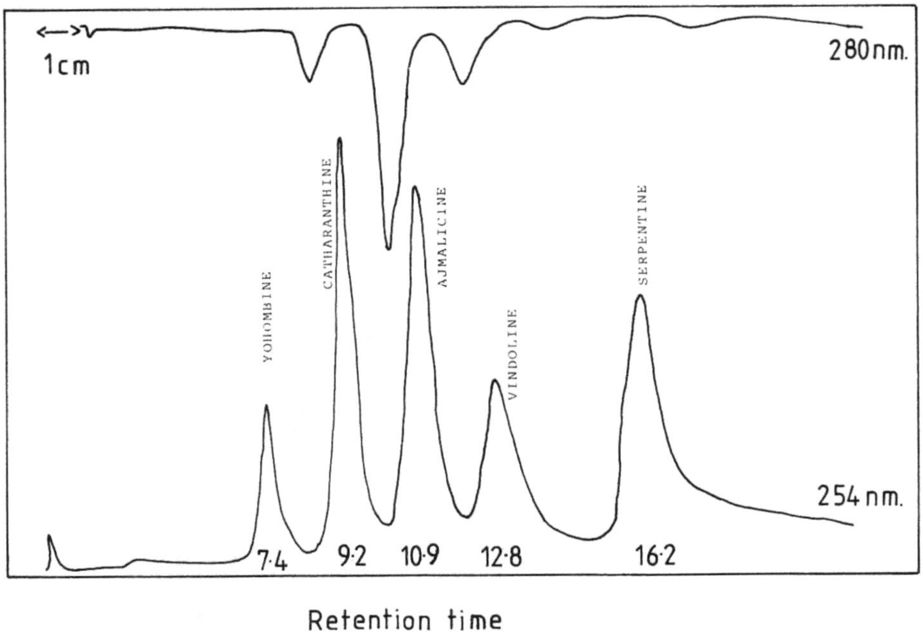

Fig. 3 A typical elution pattern for *Catharanthus roseus* alkaloid from an HPLC system.

b. TLC: Although the peaks eluted from the HPLC system are monitored at two wavelengths, and thus both retention time and wavelength ratio can be used to identify the peak, a second method is really required for confirmation. Perhaps the best method is to use a preparative HPLC analysis, in which the particular peaks can be collected and analyzed using mass spectrometry. This method requires considerable effort and time, however. Thus for routine analysis, thin-layer chromatography (TLC) is used.

There are a number of methods for the separation of alkaloids by TLC and sprays that can detect them once run. The following systems have been found suitable for serpentine and ajmalicine.

1. Chloroform, acetone, diethylanine 5:4:1.
2. Ethylacetate, absolute ethanol, NH_4OH 3:1:1.

The plates are run in these solvents in normal TLC tanks for 45–50 mins and dried at 60°C.

Depending upon the solvent system used, a number of detection methods can be used.

Table 1
TLC of *Catharanthus roseus* Alkaloids

	Rf values			
Alkaloids	Chloroform: acetone: diethylamine	Ethylacetate: ethanol	UV, 254mm	CAS
Ajmalacine	68	80	–	Gray
Catharanthine	70	70	–	Yellow-pink
Serpentine	31	2–5	Intense blue	–
Vindoline	62	72	–	Pink
Yohombine	59	61	Intense light blue	–

(1) The dried plates can be examined under UV light at 254 nm, where the alkaloids ajmalicine and serpentine fluoresce (Table 1), and their Rf values can be compared.
(2) At a UV wavelength of 366 nm, the alkaloids will quench, and therefore form dark spots.
(3) When sprayed with iodoplatinate, serpentine will give a purple spot; with all others red-purple or pink.
(4) The color obtained from spraying with ceric ammonium sulfate (CAS) and using the solvent in point 2 is described in Table 1.

Serpentine and ajmalacine can be detected by their intense fluorescence and subsequent response to the CAS spray. The same extract can be used for TLC and HPLC analysis.

4. Notes

1. In the cell line *C. roseus* ID 1, the alkaloids are produced after growth has ceased, as one would find in a two-stage process. This contrasts with the results found with a related *C. roseus* line C87, in which alkaloid production was growth-related.
2. The use of small plastic bags allows a large number of samples to be freeze-dried at one time using a large metal container. Also, it reduces the loss of limited samples that can occur when glass flasks are used and is cheaper than using glass flasks.
3. If the extracts contain high levels of nonalkaloidal material, such as phenolics, the SEP-PAK cartridges can be used to reduce these. The residue after rotary evaporation is resuspended in methanol and di-

luted to a concentration of 20% methanol. This solution should be run though a preactivated cartridge, the phenolics eluted with 4 mL of 25% methanol, and the alkaloids by 4 mL of 95% methanol.

References

1. Berlin, J. (1986) Secondary products from plant cell cultures, in *Biotechnology* vol 4 (Pape, H. and Rehm, H.-J., eds.) V. C. H., Weinheim, pp. 629–658.
2. Dougall, D. K. (1984) Factors affecting the yields of secondary products in plant tissue culture, in *Plant Cell and Tissue Cultures. Principles and Application* (Sharp, W. E., ed.), Ohio State University Press, Columbia, pp. 727–743.
3. Barz, W. and Ellis, B. E. (1981) Potential of plant cell culture for pharmaceutical production. *Ber. Deutsch. Bot. Gas.* **94,** 1–26.
4. Fowler, M. W. (1983) Commercial applications and economic aspects of plant cell culture, in *Plant Biotechnology* (Mantell, S. H. and Smith, H., eds.) Cambridge University Press, London, pp. 3–37.
5. Zenk, M. H., El-Shagi, H., Arens, H., Stockigt, J., Weiler, E. W., and Deus, B. (1977) Formation of the indole alkaloids serpentine and ajmalicine in cell suspension cultures of *Catharanthus roseus,* in *Plant Tissues Culture and Cyts Biotechnological Application* (Barz, W., Reinhard, E., and Zenk, M. H., ed.) Springer-Verlag, Berlin, pp. 27–43.
6. Ulbrich, B., Wilsner, W., and Arens, H. (1985) Large-scale production of resmarinie acid from plant cell cultures of *Coleus blumei* Benth, in *Primary and Secondary Metabolism of Plant Cell Cultures* (Deus-Neumann, B., Barz, W., and Reinhard, E., eds.) Springer-Verlag, Berlin, pp. 293–303.
7. Mantell, S. H. and Smith, N. (1983) Cultural factors that influence secondary metabolite accumulations in plant cell and tissue cultures, in *Plant Biotechnology* (Mantell, S. H. and Smith, H. eds.) Cambridge University Press, London, pp. 75–94.
8. Curtin, M. E. (1983) Harvesting profitable products from plant tissue culture. *Biotechnology* **1,** 649–657.
9. Morris, P. (1986) Regulation of product synthesis in cell cultures of *Catharanthus roseus. Plants Medica,* 121–126.
10. Gamborg, O. L., Miller, R. A., and Ojuma, K. (1986) Nutritional requirements of suspension cultures of soybean root cells. *Exp. Cell. Res.* **50,** 151–158.

Chapter 49

Betanin Production and Release In Vitro from Suspension Cultures of *Beta vulgaris*

Christopher S. Hunter and Nigel J. Kilby

1. Introduction

The diverse group of compounds known as plant secondary products includes many compounds with pharmaceutical activity (e.g., morphine, vincristine), fragrances, pigments, latex, enzymes, and carbohydrates (*1*). The commercial production of such plant secondary metabolites has, until recently, always been via the harvesting of field-grown plants followed by the extraction and/or direct utilization of the active principle(s). The proportion of the plant that is the active ingredient is invariably only a few percent of the total plant dry weight, (for example, the quinine content of *Cinchona* bark averages 6–14%) (*2*). Although annual crops are common for the production of such compounds, it is also often the case that the time required for field growth of a plant may be several years, 7–16 yr in the case of *Cinchona*, 5–7 yr in the case of the shikonin-yielding root of *Lithospermum erythrorhizon* (*3*). Field-grown

From: *Methods in Molecular Biology*, vol. 6, *Plant Cell and Tissue Culture*
Edited by Jeffrey W. Pollard and John M. Walker, ©1990 by The Humana Press

crops are also subject to the vagaries of the weather, pests, diseases, and nutrient availability. These, with other biotic and abiotic factors (*4*), have combined to encourage the evaluation of in vitro systems for the production of plant secondary products as an economically sound alternative to conventional growth and processing systems.

The in vitro systems investigated include the growth of free cells and aggregates suspended in stirred-tank chemostats and turbidostats (*5*), in air-lift fermenters (*6*), in various immobilized systems (*7*), and as calluses, both in flatbed bioreactors (*8*) and on agar-based media in flasks (*9*). However, the only in vitro scheme to succeed to date in a commercial sense is that of the Mitsui Petroleum Company in Japan for the production of the antibacterial, anti-inflammatory, napthoquinone compound shikonin (*3*). The Mitsui process uses two media sequentially in stirred-tank reactors: the first is designed for the rapid growth of the cell population from the initial inoculum during a 7 d period, and the second is designed for the production of shikonin by those cells during a further 14 d incubation period. The cells produced in this way are harvested and extracted destructively for the red product, which has a published retail value of approximately $4500/kg (*10*).

With a cell doubling time ranging between 16 h and several days, the production of cells and secondary products are expensive processes (*10*), and it has been variously estimated (*10,11*) that the final product must have an end value above $5,000/kg to enable the process to be economically viable. Attempts have been made to harvest secondary products by non-destructive techniques, thereby allowing the cells to produce second and subsequent "crops" of secondary product(s), and thus to capitalize on the initial investment in cell growth. Because few cell lines release the product directly into their growth medium, permeabilization to allow the controlled release of the product by various applied techniques has been investigated. The essential feature of such permeabilization is that it should be reversible. The use of dimethylsulfoxide (*12*), glycerol (*13*), media of low pH value (*13*), and other treatments (*12*) have been only of limited value, principally because there has been either limited release of the metabolite, or cell death/lysis. Two new experimental techniques for the non-destructive release of secondary products are being evaluated on the release of the pigment betanin from suspension cultures of *Beta vulgaris* (*14,15*). One technique uses an ultrasonic treatment for up to 60 s, followed by an incubation period. There is a steady release of a proportion of the vacuole-located betanin into the surrounding medium during the initial 30 min of this post sonication incubation.

The other technique uses a DC pulse of up to 750V/cm to electropermeabilize cells. Following electropermeabilization, betanin is released into the surrounding medium within 1–2 min. With both systems, either the individual treated cells or the treated population can regrow.

Two other secondary metabolite production systems should also be mentioned as having interesting potential for continuous production. One is the production and use of genetically transformed roots (*see* Hunter and Neill, Chapter 27, this vol.) in bioreactors, and the other is the submerged growth of shoot cultures (*16*). One alleged advantage of both these systems is associated with increased genetic stability of differentiated/organized cultures in comparison to those based on callus or cell suspensions and the similarity of the profile of secondary product(s) produced to that in the parent plant; the latter is particularly important for the production of flavor and fragrance compounds, which are invariably complex mixtures whose identity and relative proportions are of paramount importance for their commercial value. A further advantage with transformed roots is their rapid rate of growth.

1.1. Red Beetroot: A Model System

Although of low financial value, the red pigment from *B. vulgaris* (red beetroot)—a commerically important food colorant (E162) used in such products as salami and soups—has been used as a "model" system for in vitro studies on the production of secondary products. Its particular merit lies in its obvious color and the consequent relative simplicity by which it can be quantitatively assayed colorimetrically, in comparison with colorless compounds, which usually have to be assayed by gas chromatography or HPLC (*see* Chapter 48, this vol.). The principal pigments in beetroot are two red betacyanins (betanin and prebetanin) and two yellow betaxanthins (vulgaxanthin I and II). In most of the recent studies on pigment prodution by *B. vulgaris* in vitro, attention has focused on cultures selected for their ability to produce the dominant red pigment betanin (*13*). The beetroot system has been used in hairy root (*17*), immobilization (*18*), and permeabilization (*19*) investigations.

The following protocol enables:

1. Axenic beetroot cultures to be initiated;
2. Callus to be grown on agar-based media;
3. Suspension cultures to be initiated and subcultured;
4. Pigment to be extracted using both destructive (whole-cell disruption) and nondestructive (electropermeabilization) methods.

2. Materials

1. *Beta vulgaris* cv Boltardy. Seed is widely available from seeds merchants or direct from Suttons Seeds PLC, Ltd., Torquay, Devon, UK.
2. Tissue culture media: all media are based on that of Gamborg's B5 (*20*, and *see* appendix). The basic medium without plant growth regulators, sucrose or agar can be purchased.
3. Suspension culture (S1) medium: Gamborg's B5 medium Flow Labs 3.87 g, sucrose 20 g, kinetin (6-furfurylaminopurine) 0.1 mg, 2,4-D (2,4-dichlorophenoxyacetic acid) 0.02 mg. Medium is made up to 1 L with single distilled water and then the pH adjusted to 5.5 prior to autoclaving at 106 kPa at 121°C for 15 min.
4. Agar-based medium (A2) for callus growth: as for S1, except that 6 g/L agar is added prior to pH adjustment.
5. Callus induction medium (A1): as for A2 except that the 2,4-D concentration is 1 mg/L.
6. Seed germination medium (A3): as for A2 except that it contains neither kinetin nor 2,4-D.
7. Erlenmeyer flasks (250 mL) with the necks closed by a double layer of aluminum foil and oven sterilized (dry) at 160°C for 90 min are used for seed germination, callus initiation, growth, and suspension cultures. Disposable 9-cm diameter styrene Petri dishes are used for feeder cultures.
8. Culture incubation: all cultures are grown at 25° C in white fluorescent lighted conditions (the intensity and spectral conditions are not critically important). Suspension cultures are incubated on an orbital shaker at constant speed (100–125 rpm).
9. Electropermeabilization equipment: this is based on the electrofusion apparatus of Watts and King (*21*), except that no signal generator is used. The electropermeabilization chamber is constructed from a 75 x 24 mm glass microscope slide to which are epoxy-bonded two parallel platinum electrodes (0.2 mm diameter) 5 mm apart (Fig. 1).

3. Methods

3.1. Callus Initiation

3.1.1. From Seed

1. Soak the seeds in a sterile flask containing an aqueous solution of sodium hypochlorite (2% available chlorine) plus 1–2 drops of Tween 20 for 20 min (*see* Note 1).

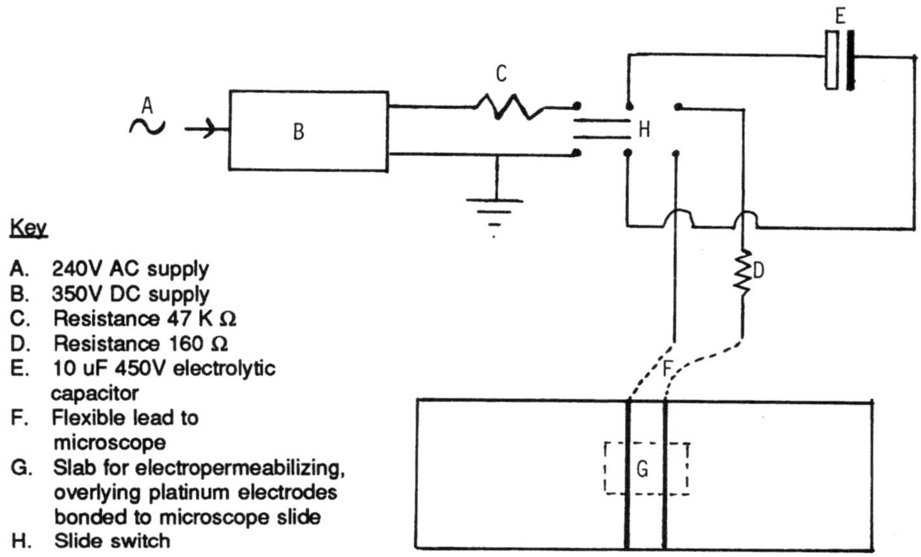

Fig. 1. Electropermeabilization equipment: circuit diagram.

2. Wash the seeds (4x) in sterile distilled water (SDW) and aseptically transfer them to the surface of A3 media: use 5 seeds/flask (*see* Note 2).
3. When the seeds have germinated and seedlings grown to produce the second true leaf, aseptically remove individual seedlings, dissect petioles and hypocotyls into sections approximately 15 mm long, and implant onto the surface of A1 medium.
4. Within 21 d, callus will have initiated and should have grown into structures several millimeters in diameter. Excise these calluses and transfer to A2 medium.

3.1.2. From the Flowering Stems of Mature Plants

1. Remove the flowering stem, trim off any leaves and flowers, and then surface sterilize them in a sodium hypochlorite solution (1% available chlorine + 1 drop Tween 20) for 20 min.
2. Wash the stems (4x) in SDW, and then section them aseptically into 20-mm lengths.
3. Bisect each 20-mm length longitudinally and place, cut sides down, onto A1 medium.
4. Incubate the flasks: within 21 d, callus will be initiated, and should be treated as described above for the seed-derived callus and grown on into A2 medium.

3.2. Callus Culture

1. When the calluses have grown to an approximate diameter of 20 mm, they should be subcultured by transfer with a sterile loop to fresh A2 medium. The quantity transferred at subculture is not critical, but a lump 5–10 mm diameter is recommended. A fast-growing callus will require subculture at 21– 28 d intervals.
2. Transfer approximately 5 g of callus to 100 mL of S1 medium and incubate on the orbital shaker.
3. Subculture at 7–10 d intervals to fresh S1 medium by transfering 50% of the total culture into 50 mL of fresh medium.

3.3. Pigment Production, Extraction, and Analysis

Most calluses and suspension cultures of beetroot contain a mixture of cells that can visually be classified as "colorless, pink, or red." For the production of betanin it is clear that, at least in the short term, one must select for "red" cells.

1. To select for red cells, prepare several flasks of A2 medium, and onto the surface of each pour approximately 2 mL of suspension culture, swirl the suspension over the A2 surface to distribute the cells (single cells and aggregates), and incubate the flasks.
2. From each flask, many calluses will develop that can be assessed visually for their color and subcultured to fresh A2 and later, after sufficient growth, to S1 medium. It is our experience that selected cell lines will require reselection after many subcultures. Also, we have found that it is prudent to have a stock of "backup" callus cultures in reserve to overcome the seemingly inevitable problems of a shaker power-supply failure or other such disaster!
3. The medium, cells, and aggregates from a suspension culture flask are homogenized in a Waring blender to produce a slurry.
4. Centrifuge the homogenate at 600 x *g* for 5 min at 15–21°C, remove, and centrifuge the supernatant at 600 x *g* for 5 min.
5. Rehomogenize the cell debris in 50 mL of S1 medium, and centrifuge twice, as before. Discard the cell debris and combine all the supernatant extracts: this should then be filtered through glass microfiber filter paper (Whatman, GF/C) to yield the pigment extract for assay.
6. Although the extract contains a mixture of pigments, their separation is beyond the scope of this chapter; however, if intensely red cells have been selected as the cell line for these experiments, the major red pigment will be betanin. The λmax of betanin varies with

pH of the solution (22). To minimize both the pH effect and the spectral overlap of the other betalain pigments, the pH of the pigment extract should be adjusted to a standard value of pH 5.5. Spectrophotometric assay should be at 537 nm: there is a straight-line relationship between betanin concentration and absorbance at this wavelength in S1 medium (Fig. 2). For quantification of the pigment, *see* (22,23).

3.4. Electropermeabilization

1. Prepare A2 medium, and cool it to 40°C.
2. Aseptically sieve a suspension culture through a 0.6 mm mesh to collect small aggregates and single cells, and remove 0.25 mL (settled cell volume) using a graduated glass pipet that has had its distal end sawn off to allow free ingress and egress of cells.
3. Add these cells to 25 mL of the warm media and, after rapid mixing by swirling, dispense the suspension onto the bases of several Petri dishes to form layers about 1 mm thick. After the agar has set, remove slabs 10 x 10 mm when required for permeabilization.
4. Feeder cultures are to be prepared containing suspended cells in A2. A feeder culture must be used to support the long-term growth of the cells in the slab after permeabilization, because they are at a concentration below the minimum effective cell density for independent survival. In the same manner as above, obtain 4 mL (SCV) of beetroot cells and suspend them in 25 mL of A2 medium. Dispense the suspension into Petri dishes to form layers approximately 4–5 mm deep; allow the agar to set.
5. With great care (*see* Note 3), connect up the electropermeabilization apparatus and position the chamber firmly onto a microscope stage, place the cell slab to bridge the two electrodes, and select an area of the slab, viewing with the lowest magnification possible (10x objective), to locate discrete cells or small aggregates of intensely red cells situ-ated between the electrodes.
6. Adjust the equipment to hold a charge in its capacitor at 350 V and, when ready, switch the equipment to allow the capacitor to discharge across the electrodes.
7. Watch the cells selected, and observe the dissipation of the red pigment into the agar during the next 2 min. If there is no such dissipation, either give a further capacitor discharge or select another area on the slab for treatment. This is a straightforward technique, but does require skill and practice, so it is advisable to prepare

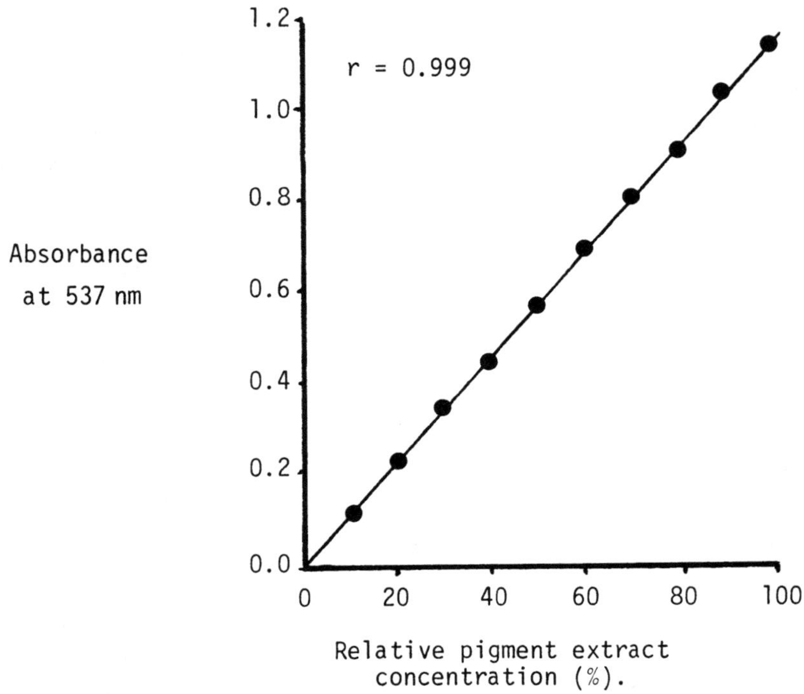

Fig. 2. Absorbance of betanin at 537 nm when dissolved in S1 medium.

several slabs and feeder layers. If possible, photograph the cells both before and after treatment for an objective record.

8. Immediately after treatment, transfer the slabs to the feeder layers. Subsequent weekly microscopic observation of the treated cells will show the recovery and regrowth of pigment-depleted cells; here also, photographic records are most helpful.

4. Notes

1. Surface sterilization of seeds can sometimes result in explant death or insufficient killing of contaminating organisms. If either of these occurs, adjust the time of soaking or the concentration of sodium hypochlorite until satisfactory surface sterilization is achieved.
2. Normal aseptic techniques must be adopted for work with the cultures.
3. Care should be taken in the connection and operation of the electropermeabilization apparatus because of the high voltages involved.

If the circuit is identical to ours, then at no time is the power supply connected across the electropermeabilization chamber. If in doubt, take qualified electrical safety advice.

References

1. Vickery, M. L. and Vickery, B. (1981) Secondary plant metabolism (Macmillan Press, London).
2. McHale, D. (1986) The cinchona tree. *Biologist* **33, (1)**, 45–53.
3. Curtain, M. E. (1983) Harvesting profitable products from plant tissue cultures. *Biotechnology* **1**, 649–657.
4. Mantell, S. H., Matthews, J. A., and McKnee, R. A. (1985) Principles of plant biotechnology (Blackwell Scientific Publications, Oxford).
5. Hashimoto, T., Azechi, S., Sugita, S., and Suzuki, K. (1982) Large scale production of tobacco cells by continuous cultivation, in *Plant Tissue Culture 1982* (Fujiwara, A., ed.) Maruzen, Tokyo, pp. 403, 404.
6. Smart, N. J., Morris, P., and Fowler, M. W. (1982) Alkaloid production by cells of *Catharanthus roseus* grown in airlift fermenter systems, in *Plant Tissue Culture* (Fujiwara, A., ed.), Maruzen, Tokyo, pp. 397, 398.
7. Lindsey, K. (1986) The production of secondary metabolites by immobilized plant cells, in *Secondary Metabolism in Plant Cell Cultures,* (Morris, P., Scragg, A. H., Stafford, A., and Fowler, M. W., eds.), Cambridge University Press, Cambridge, UK, pp. 143–155.
8. Lindsey, K. and Yeoman, M. M. (1983) Novel experimental systems for studying the production of secondary metabolites by plant tissue cultures, in *Plant Biotechnology,* (Mantell, S. H. and Smith, H., eds.), Cambridge University Press, Cambridge, pp. 39–66.
9. Ikuta, A. and Itokawa, H. (1982) Studies on the alkaloids from tissue culture of *Nandina domestica* in *Plant Tissue Culture,* (Fujiwara, A., ed.), Maruzan, Tokyo, pp. 315, 316.
10. Scragg, A. H. (1986) The economics of mass cell culture, in *Secondary Metabolism in Plant Cell Cultures,* (Morris, P., Scragg, A. H., Stafford, A., and Fowler, M. W., eds.), Cambridge University Press, Cambridge, UK, pp. 202–207.
11. Sahai, O. and Knuth, M. (1985) Commercializing plant tissue culture processes: economics, problems, and prospects. *Biotechnology Progress* **1**, 1–9.
12. Parr, A. J., Robins, R. J., and Rhodes, M. J. C. (1986) Product release from plant cells grown in culture, in *Secondary Metabolism in Plant Cell Cultures* (Morris, P., Scragg, A. H., Stafford, A., and Fowler, M. W., eds.), Cambridge University Press, Cambridge, UK, pp. 173–177.
13. Kilby, N. J. (1987) An investigation of metabolite release from plant cells *in vitro* to their surrounding medium. Ph. D. thesis, Bristol Polytechnic, UK.
14. Kilby, N. J. and Hunter, C. S. (1986) Ultrasonic stimulation of betanin release from *Beta vulgaris* cells *in vitro*: a non-thermal, cavitation-mediated effect, in *Abstracts of VI International Congress of Plant Cell and Tissue Culture* (Somers, D. A., Gengenbach, B. G., Biesboer, D. D., Hackett, W. P., and Green, C. E., eds.), University of Minnesota, Minneapolis, USA, p. 352.
15. Hunter, C. S. and Kilby, N. J. (1988) Electropermeabilization and ultrasonic techniques for harvesting secondary metabolites from plant cells *in vitro*, in *Manipu-*

lating Secondary Metabolism in Culture (Robins, R. J. and Rhodes, M. J. C., eds.), Cambridge University Press, Cambridge, UK.

16. Charlwood, B. V., Brown, J. T., Moustou, C., and Charlwood, K. A. (1988) Pelargoniums: flavors, fragrances and the new technology. *Plants Today* **1 (2),** 42–46.
17. Hamill, J. D., Parr, A. J., Robins, R. J., and Rhodes, M. J. C. (1986) Secondary product formation by cultures of *Beta vulgaris* and *Nicotiana rustica* transformed with *Agrobacterium rhizogenes*. *Plant Cell Reports* **5,** 111–114.
18. Rhodes, M. J. C., Smith, J. I., and Robins, R. J. (1987) Factors affecting the immobilization of plant cells on reticulated polyurethane foam particles. *Appl. Microbiol. Bio-technol.* **26,** 26–28.
19. Kilby, N. J. and Hunter, C. S. (Submitted) Repeated harvest of vacuolar-located secondary product from *in vitro* grown plant cells using 1.02MHz ultrasound.
20. Gamborg, O. (1970) The effects of amino acids and ammonium on the growth of plant cells in suspension culture. *Plant Physiology* **45,** 372–375.
21. Watts, J. W. and King. J. M. (1984) A simple method for large-scale electrofusion and culture of plant protoplasts. *Bioscience Reports* **4,** 335–342.
22. Nilsson, T. (1970) Studies into the pigments in beetroot. *Lantbrukshogskolans annaler* **36,** 179–219.
23. Saguy, I., Kopelman, I. J., and Mizrahi, S. (1978) Computer-aided determination of beet pigments. *J. Food Science* **43,** 124–127.

Chapter 50

Enzyme Extraction and Assessment of Enzyme Production

Debbie Grey

1. Introduction

There have been numerous reports of enzyme synthesis in cultured plant cells and the presence of certain enzymes in the culture medium. Most of these studies have been performed in the context of the characterization of these enzymes from cell cultures. However, cell cultures—in particular those secreting enzymes into the medium—can be used as a process system for enzyme production. Surprisingly, there is very little information regarding the potential of plant cell cultures as a tool for enzyme productivity.

At present, the industrial production of enzymes is based to a large extent on the use of microorganisms. The obvious advantages of these systems in mass culture (fast growth rates, high productivity, rapid turn around, and so forth) leave little room for competition by plant cell cultures. However, plants contain a vast repertoire of enzymes that catalyze specifically plant reactions. In addition, there is a growing requirement in the food industry, prompted by possible legislative guide-

From: *Methods in Molecular Biology*, vol. 6, *Plant Cell and Tissue Culture*
Edited by Jeffrey W. Pollard and John M. Walker, ©1990 by The Humana Press

lines, to use plant-derived aids in the processing of food products. In this context, mass production of plant enzymes would be very advantageous. In particular, if the target enzyme is secreted into the nutrient medium, the downstream processing and recovery of the enzyme protein will exclude the laborious and costly process of cell tissue breakage.

Although the method described in this chapter relates to small-scale extractions, the same rationale can be applied to a vessel of any size in terms of assessing productivity. In this chapter, productivity is described in terms of the contribution made by both intra- and extracellular enzyme activities.

2. Materials

2.1. For Culture Fractionation

1. Filter unit: stainless-steel filter bed (*see* Note 1) or small Hartley funnel and Whatman's No. 1 filters to fit unit.

2.2. For Intracellular Enzyme Extraction

1. Prechilled acid-washed sand.
2. Ice-cold extraction buffer (*see* Note 2).
3. Centrifuge and prechilled tubes (*see* Note 3).

2.3. For Protein Determination

1. Coomassie Brilliant Blue G-250.
2. 95% (v/v) ethanol.
3. 85% (w/v) phosphoric acid.
4. Standard protein solution (100 mg bovine serum albumin/100 mL extraction buffer). This solution should be made fresh each time (*see* Note 4).

3. Methods

3.1. Culture Fractionation

1. Position a Whatman's No. 1 filter paper disk in the filter unit, and place this on a Buchner flask fitted with a rubber stopper under vacuum.
2. Filter a known volume of culture (e.g., 10 mL). To ensure that a representative sample is taken, the culture should be mixed well before sampling. Maintain the vacuum until the cells appear dry.

3. Note the weight of the filtered cells and the volume of the medium (filtrate) using the graduated test tube (Fig. 1, columns B and C).
4. Place the test tube containing the medium in ice, and use the contents for enzyme assays.

3.2. Intracellular Enzyme Extraction

1. Position a prechilled mortar and pestle firmly in the ice.
2. Place the weighed cells in the prechilled mortar along with approximately 10% (w/w) prechilled acid-washed sand.
3. Add ice-cold extraction buffer to the cells in the approximate ratio of 0.5:1 with the weight of the cells.
4. Disrupt the cells for approximately 3 min (*see* Note 5).
5. Remove the slurry to prechilled centrifuge tubes, using a Pasteur pipet.
6. Wash the mortar and pestle with a minimal volume of the prechilled extraction buffer and combine with the slurry.
7. If using an Eppendorf system, spin the slurry for 2 min to remove the sand and cell debris. Transfer the resulting supernatant to clean tubes and spin for a further 5 min.

   If an Eppendorf system is not available, spin down the sand and cell debris with a short spin (3 min) in a bench centrifuge before transferring the supernatant to a high-speed centrifuge. Centrifuge at 44,000 x *g* and 4°C for 10 min.
8. Carefully remove the supernatant from the centrifuge tubes, and measure the volume using the graduated test tube. This becomes the extraction volume.
9. Place the graduated test tube in ice, and use the contents for enzyme assays.

3.3. Protein Determination

3.3.1. Preparation of Bradford Reagent (1 *and* see *Vol. 3*)

1. Add 100 mg of Coomassie Brilliant Blue G-250 to 50 mL of 95% ethanol. Leave to stir on a magnetic stirrer for 15 min.
2. To this solution, add 10 mL of 85% phosphoric acid. Leave to stir on a magnetic stirrer for at least 15 min.
3. Make the volume of the resulting solution up to 1 L and store in a dark bottle. This reagent is stable for 1 mo at room temperature.
4. Before using this reagent, it is necessary to ensure that it is well mixed.

Date____ Enzyme____	I Extraction				II Assay													
	A	B	C	D	E	F	G	H	I	J	K	L	M	N	O	P	Q	R
Source____	Cult. vol. (ml)	Tiss. wt. (g)	Med. vol. (ml)	Extr. vol. (ml)	C/M	Aliq. vol. (ml)	Absorb. (/min)	Volume act. (U/ml)	Mean volume act. (U/ml)	Total C or M act. (U)	Total cult. act. (U/ml)	Mean cult. act. (U/ml)	Weight act. (U/g)	Protein (mg/ml)	Spec. act. (U/mg)	Mean Spec. act. (U/mg)	Protein in cells (mg/g)	
										IxD/C	J_C+J_M:A		J:B		H:N		ND:B	

Fig. 1. Suggested chart layout for recording enzyme productivity data from plant cell cultures. The abbreviations used are explained in the text. Column R is left blank for the calculation and recording of an additional parameter.

3.3.2. Calibration Curve

It is advisable to construct the calibration curve prior to any sample analysis (*see* Note 6). Under normal conditions, it is sufficient to construct one standard curve/batch of reagent.

1. Pipet 20, 40, 60, 80, and 100 µL of standard protein solution into duplicate sets of test tubes. Make up all volumes to 100 µL with extraction buffer, and mix well. The blank will contain 100 µL of extraction buffer.
2. At timed intervals, add 5 mL of Bradford reagent and mix well. Incubate at room temperature for exactly 10 min.
3. After exactly 10 min, zero the spectrophotometer at 595 nm against the blank assay, and at timed intervals thereafter, read calibration tubes against this setting (*see* Note 7). The absorbance at 595 nm of the tube containing 100 µL of standard protein solution should be between 0.6 and 0.7, approximately.
4. Construct a calibration curve of protein concentrations (mg/mL) against absorbance readings (*see* Note 8).

3.3.3. Analysis of Protein Samples

1. Pipet into duplicate sets of test tubes 100 µL of each sample to give absorbance readings that lie within the range of the standard curve. Dilution of cell extracts with extraction buffer may be necessary. The blank will contain 100 µL of extraction buffer.
2. Repeat steps 2 and 3 of section 3.3.2. (calibration curve).
3. To calculate protein concentration in samples, use the standard curve constructed above (*see* Note 8).

3.4. Calculation of Enzyme Activity

1. Figure 1 represents a sample of a chart used to record essential information and calculations of enzyme productivity. The chart is divided into two main sections. The first (I) deals with the culture fractionation and extraction details, and the second (II) is concerned with calculations of enzyme activity.
2. Enter the details from culture fractionation and enzyme extraction into section I (columns A-D).
3. Columns E and F specify the provenance (cell extract [C] or culture medium [M]) and volume of the sample aliquot used in the assay.

4. The change in absorbance as a result of enzyme activity/unit time (A/min) is recorded in column F and calculated/min and entered in column G.
5. For most enzymes, the volume activity (U/mL) is calculated according to equation 1 (*see* Note 9):

$$\frac{\text{Change in absorbance (A/min)} \times \text{Total assay vol (mL)}}{\text{mol extinction coefficient (cm}^2\text{/mol)} \times \text{light path (cm)} \times \text{aliq. vol (mL)}} \quad (1)$$

   A unit of enzyme activity is defined as µmol of substrate utilized or product formed/min. The volume activity of each assay (U/mL) is recorded in column H, with the mean values for cell extract and medium fraction entered in column I (*see* Note 10).
6. Column J records the total number of enzyme activity units in cellular and medium fractions. This is calculated by multiplying the volume activity (column H) by the extraction volume (column D) for the cells and the volume activity (column H) by the medium volume (column C).
7. This information is now used to calculate the total number of units/mL of culture (cells + medium). It is essential to assay an equal number of replicates from each culture fraction, i.e., cells or medium. The values in column J for the first replicates, cell and medium, are added together, and the sum is then divided by the total culture volume (column A). This calculation is then repeated for each subsequent replicate. The results are entered into column K. The mean and standard errors for these values are calculated, and the results recorded in column L.
8. The intracellular activity of the enzyme/gram fresh weight (column M) is calculated by dividing the cellular activity values in column J by the tissue weight in column B.
9. The protein concentrations of the cell extract and culture medium, and calculations based on them, are recorded in columns N–Q. The protein concentrations (mg/mL), determined according to the method outlined above, are recorded in column N. The specific activity of the enzyme (column O) in each fraction is calculated by dividing the column activity of each replicate expressed in column H by the appropriate protein concentration. The mean and standard errors of these values are recorded in column P. Finally, the protein content of the cells (mg protein/g fresh weight) is calculated by multiplying the

protein concentration of the cell extract (column N) by the extract volume (column D), and dividing the result by the tissue weight (column B).

4. Notes

1. For details of the stainless-steel filter unit, refer to Fig. 3 in Chapter 2, this vol.
2. A review of the literature should indicate the ionic strength, pH, and buffering species best suited for the stability of a given enzyme extraction. For certain enzymes, detergents such as Tween 20, or salts, such as KCl, are added to the extraction buffer to increase the degree of enzyme release. Certain sulfur-containing compounds (e.g., 2-mercaptoethanol) are added to the extraction buffer to enhance enzyme stability.
3. As mentioned in the introduction, this method is intended for culture screening and, consequently, for small-scale work. Ideally, therefore, an Eppendorf centrifuge system should be used. However, if this is unavailable, it is possible to use a bench centrifuge in conjunction with a high-speed centrifuge. Drawbacks of this latter system include a requirement for a larger sample volume and a lengthy procedure.
4. To prevent frothing of the protein standard solution, it is advisable to weigh the protein directly into a 100-mL volumetric flask and slowly add distilled water up to volume before shaking the flask to dissolve the protein.
5. During the first few attempts, it is recommended that one check the degree of cell disruption under a microscope. Ideally, over 90% of cells should be disrupted.
6. Care must be taken when constructing the protein standard curve, particularly soon after the preparation of the reagent. There is a tendency for the standard curve to be unreliable, since it may not go through the origin and can be biphasic in nature. To overcome these problems, it is recommended that the standard curve be constructed at least 24 h after the reagent has been prepared.
7. Use a 3-mL glass cuvet for protein absorbance readings. The reagent dye tends to stain the glass. The stain can be removed by repeated rinsing with ethanol or methanol, and the cuvet should be stored in a container filled with either ethyl or methyl alcohol . It is advisable to monitor the build up of stain on cuvet surfaces during serial assays,

and rinse the cuvet with alcohol if necessary. Alternatively, disposable plastic cuvets may be used.

8. When a straight line standard curve which passes through the origin has been obtained, a constant can be determined that, when multiplied by the absorbance of an assay sample, will give the protein concentration of that sample. This constant is calculated by finding the reciprocal of the slope of the standard curve.
9. Equation 1 is used when the enzyme activity is monitored by measuring the initial rate of the reaction. However, in some cases enzyme activity is assessed by colorimetric methods. In these cases, it is usual to construct a standard curve from which the amount of substrate used, or product formed, is determined.
10. In order to have statistically meaningful results, it is essential to assay the enzyme in both cell extract and medium in at least five replicates. This justifies the calculation of the standard error of the mean.

Acknowledgment

The author is grateful to Dr. G. Stepan-Sarkissian for helpful suggestions and critical reading of the script.

Reference

1. Bradford, M. M. (1976) A rapid and sensitive method for the quantitation of microgram quantities of protein utilizing the principle of protein–dye binding. *Anal. Biochem.* **72,** 248–254.

Chapter 51

Micromachines to Automate Plant Tissue Culture

Brent Tisserat

1. Introduction

This chapter is concerned with the use of mechanical and electronic techniques (i. e., the Automated Plant Culture System [APCS]) to aid in the sterile cultivation of plants for the purpose of prolonging culture life and/or increasing yields over that obtained from conventional technology (*1,2*). The use of machines to facilitate the culturing of plants in vitro is common (e. g., shakers, roller bottle apparatuses, incubators, magnetic stirrers, and roller drums). However, current techniques employed in plant tissue culture (i. e., agar or liquid culture), developed about 30 yr ago, are highly labor intensive because of the requirement for numerous hand–plant interactions in the culturing of plants in a sterile environment. Culture of any plant in vitro requires periodic transferring to fresh medium. Nutrient medium is either exhausted and/or altered by plant cultures, thus requiring the constant transfers to fresh medium (usually every 4–8 wk or sooner). The system presented in this chapter is meant to reduce dependence on labor through substitution of hand–plant interactions with machine–plant interactions via automatic medium replenishment (Fig. 1). The APCS culture vessel is, therefore, large enough to accommodate the increasing plant culture size associated with long-term growing of plants in a single culture vessel.

From: *Methods in Molecular Biology*, vol. 6, *Plant Cell and Tissue Culture*
Edited by Jeffrey W. Pollard and John M. Walker, ©1990 by The Humana Press

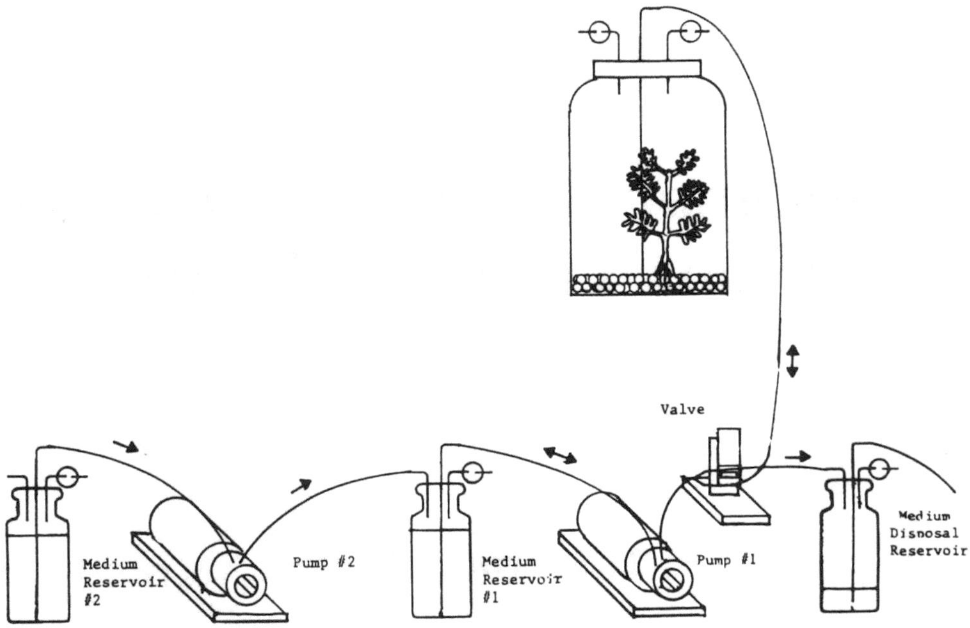

Fig. 1. Diagrammatic representation of the components employed in the autoclavable automated plant culture system.

2. Materials

1. Microcomputer. Microcomputers (e. g., Commodore Vic-20, C-64, or Atari 400) can be purchased from local retail or mail-order outlets.
2. Interface assembly. This can be purchased ready made from local or mail-order sources, but it can be easily constructed as well (Fig. 2). Interface assembly consists of the following parts:
   a. Relays, miniature 5 VDC SPDT or DPDT types (Jameco Electronics, Belmont, CA).
   b. Transformers, +5 and +12 VDC types with bridge rectifier and capacitor assemblies (Jameco Electronics, Belmont, CA).
   c. Relay board with cables, to connect the assembly to the computer, and the assembly to the pumps and valves (Jameco Electronics, Belmont, CA).
3. Sterile Culture System (SCS). This can be constructed of a variety of materials, depending on the needs of the investigator. The SCS consists of the following components:
   a. Culture chamber, 1 quart (0.946 L) or 1/2 gallon (1.89 L) glass mason jar filled to a depth of 2 cm with 5-mm diameter (dia.)

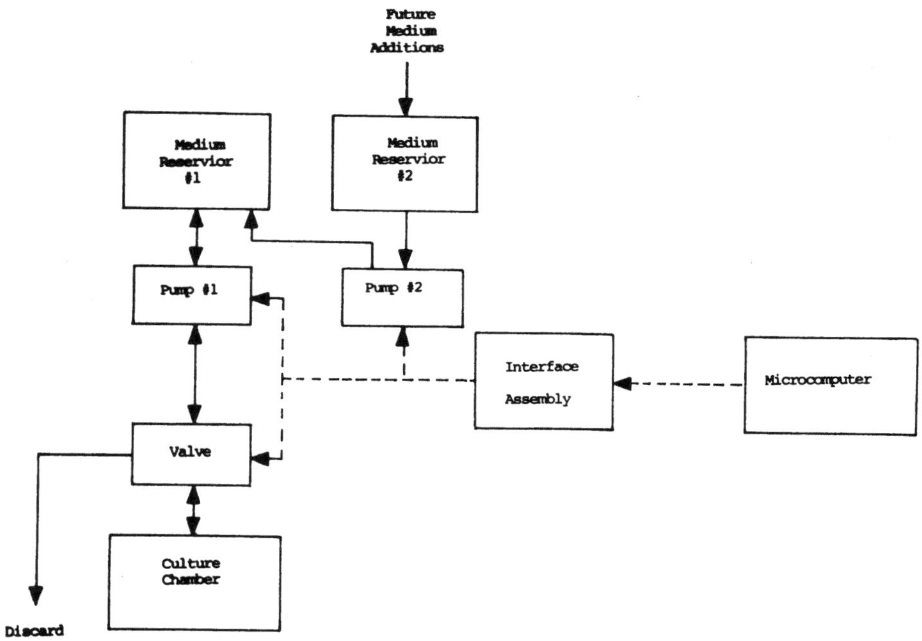

Fig. 2. Block diagram of the automated plant culture system. Electrical lines are represented by broken lines (---) and liquid lines by solid lines (—).

solid glass beads and fitted with an autoclavable No. 7 (30 mm dia.); "Twistit" rubber stopper containing three pre-drilled, countersunk holes of 8-mm dia. (American Scientific Products, McPaw Park, IL; Bellco Glass, Inc., Vineland, NJ).

b. Peristalic pumps, variable-speed type, with a 100-rpm motor capable of delivering a flow range of 5–80 mL/min with size 16 tubing, to be equipped with Lexan polycarbonate heads and cold-rolled steel rotor (Cole-Parmer Instrument Company, Chicago, IL).

c. Silicone tubing, dimensions: size 16; 3.1-mm inside dia. (i. d.); 3.2-mm hose barb size; and 7-mm outer dia. (o. d.). Employ 0.318-mm dia. polypropylene quick-disconnect Luer taper miniature plastic fittings as needed to aid in tubing connections (Cole-Parmer Instrument Company, Chicago, IL).

d. Media reservoirs, 2 L capacity Pyrex™ or borosilicate glass media storage bottles (101 mm o.d. x 225 mm height [h] x 45 mm screw cap size). One 2 L capacity Pyrex™ or borosilicate glass media-storage bottle (138 mm o.d. x 246 mm h x 45 mm screw cap size), to be fitted with linerless polypropylene screw caps (Bellco Glass, Inc., Vineland, NJ).

e. Pinch clamp valve, three-way type 12 VDC, to provide at least 20 psi pressure (Bio-chem Valve Corp., East Hanover, NJ).

f. Pyrex™ (borosilicate) standard wall glass tubing, dimensions: 6 mm o.d., 1 mm wall thickness (American Scientific Products, McPaw Park, IL).

g. Sterilizing vent filters, ACRO 37 TF 37 (Gelman 4464) equipped with hydrophobic PTFE membrane, 0.2-micron membrane, dimensions: 0.64–0.95 cm hose barb connections; 37 mm dia. Five required/system (American Scientific Products, McPaw Park, IL).

3. Methods

3.1. Construction of the Sterile Culture System

1. Culture chamber: Position the rubber stopper in the center of the mason jar lid by drilling a 2.86 cm dia. hole using a keyhole drill. Into two of the three holes in the stopper, insert two 120 mm length (l.) x 6 mm o.d. glass tubes, bent in half at a right angle. Connect a 3 cm l. x 6 mm o.d. piece of silicone tubing attached to an air filter to each of these tubes. Insert the medium inlet line tube (260 mm l. x 6 mm o.d.) into the third stopper hole, with a right angle bent 50 mm from the end that is exposed to the atmosphere. This tube must reach the bottom of the chamber. Proper positioning of this tube is important to ensure adequate medium evacuation following culture soaking. Connect a 70 cm l. x 6 mm o.d. piece of silicone tubing to this medium inlet line, and connect the other end to the leg of a 6 mm o.d. barbed polypropylene Y connector. To one of the arms of the Y connection a 50 cm l. x 6 mm o.d. silicone tubing line is connected to the medium inlet line of the media drain reservoir. To the other arm of the Y connection, connect a 50 cm l. x 6 mm o.d. silicone tubing line that proceeds to the culture medium reservoir inlet line. The arm lines bifurcating from the Y are positioned in a pinch clamp value. The line to the culture chamber is in the normally open position, whereas the line to the media drain reservoir is in the normally closed position.
2. Culture-media reservoirs: Drill a 2.86 cm dia. hole using a keyhole drill in the twist top closure to accommodate the stopper. Insert three

glass tubes as previously described, except omit the filter to one line and connect a 90 cm l. x 6 mm o.d. piece of silicone tubing. The other end of this line connects to the medium inlet line of the second reservoir.

3. Culture-media drain reservoir: Prepare as described for the culture media reservoirs, but connect the inlet line to the line in the normally closed position proceeding from the pinch clamp valve. The 260-mm glass tubing line referred to as the medium inlet line for the culture chamber and media reservoirs will be used for extraction of waste medium at later times. This will be plugged at this time.

3.2. Construction of Microcomputer and Interface System

1. A diagram for the construction of the relay board employed in the interface system is shown in Fig. 3.
2. The culture program is written in BASIC, using appropriate POKE commands to access the microcomputer's output lines. These output lines control the relays, which in turn control the operation of pumps and valves to control medium flow.

3.3. Culturing Procedures

1. Connect the SCS system, as described, to media bottles. Add 950 mL of an appropriate medium to each media reservoir bottle. Use pinch clamp closures to seal all silicone connections for medium reservoirs during autoclaving. Autoclave at 15 psi for 15 min at 120°C.
2. Adjust the input flow of medium into the chamber to allow the culture to be only half submerged in solution. Thereafter, program the computer for this time. Program entry of medium into the chamber every 2 h.
3. Program medium removal from the media reservior #1 (e. g., 15 mL) to occur every 48 h, at which time an equal amount of medium from reservoir #2 will replace the discarded media. Reservoir #2 should last 3 mo using 950 mL of medium, or 6 mos using 1900 mL of medium.

4. Notes

1. The length of time that the culture soaks in medium is critical. This time will vary for each plant employed in the system and needs to be determined empirically.

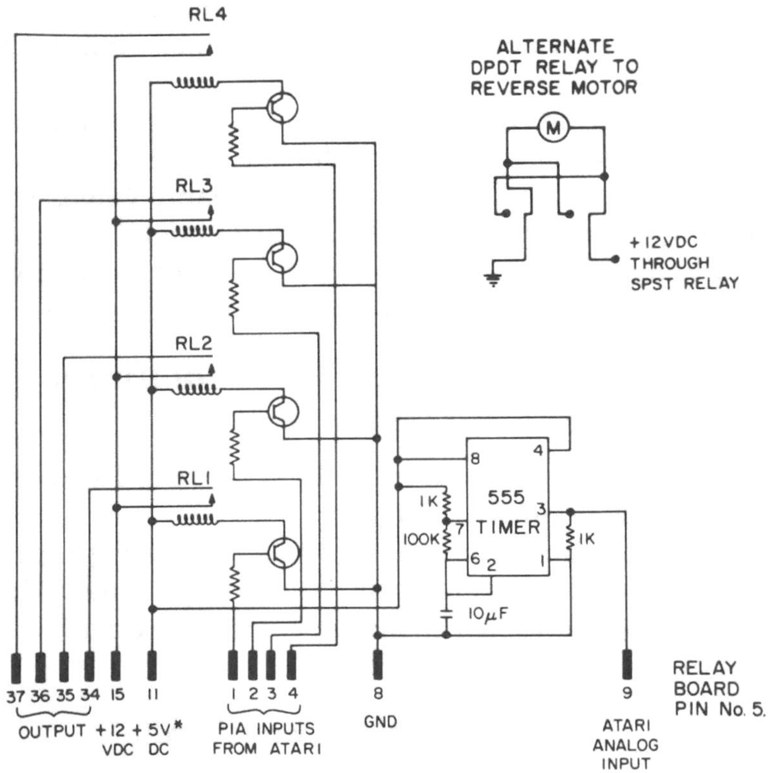

Fig. 3. Schematic representation of a relay board wiring with its connections to the Atari 400 microcomputer.

2. Because of its complexity, this system is not intended as a small-scale experimental system, but rather as an aid in the prolonged culture of well established cultures. It is recommended that only established cultures, grown on medium that has been previously determined to give optimum growth, be employed in this system. This will avoid wasting time on contamination problems and nutritional studies.
3. Larger reservoirs and culture chambers may be employed, depending on the desires of the investigator.
4. A variety of microcomputers or microprocessor-controller systems, and even some multichannel timers, can be used for this system. Microcomputers are employed because they are less expensive, easier to interface, more versatile, and more easily programmed than micro-

processors or programmable timers. Appropriate POKE commands are found in each microcomputer's programmer's reference manual.

5. The SCS described in this paper is completely autoclavable. A variety of peristalic or impeller pumps can be employed. Systems can be constructed to be either totally autoclavable or partially autoclavable (e. g., media bottles and culture chamber) and partially ethylene oxide sterilizable (e. g., impeller pumps and culture chamber).
6. Growth rates of 2–4x that obtained in agar medium are plausible. In some cases, depending on the culture species, even higher growth rates are obtainable (e. g., embryogenetic carrot callus can give 10x the growth that is obtained in agar medium).

Acknowledgment

I wish to thank C. E. Vandercook for his help in the early construction of this system.

Disclaimer

Mention of a trademark name or proprietary product does not constitute a guarantee or warranty of the product by the US Department of Agriculture and does not imply its approval to the exclusion of other products that may be suitable.

References

1. Tisserat, B. and Vandercook, C. E. (1985) Development of an automated plant culture system. *Plant Cell Tissue Organ Culture* **5,** 107–117.
2. Tisserat, B. and Vandercook, C. E. (1986) Computerized long-term culture for orchids. *Amer. Orchid Soc. Bull.* **55,** 35–42.

Chapter 52

Automation of the Surface Sterilization System Procedure

Brent Tisserat and Carl E. Vandercook

1. Introduction

Reliable methods for obtaining sterile explants (i. e., that part of the parent plant introduced to in vitro conditions) are critical in tissue culture. Normally, explants are soaked in disinfectants in order to eliminate the coating layer of microorganisms ubiquitously found on plants. This chapter is concerned with the use of mechanical and electronic techniques (i. e., an automated Surface Sterilization System [SSS]) to aid in the sterile establishment of explants. The automated SSS reduces labor input and increases the effectiveness and reproducibility of the surface sterilization treatment (*1*) (*see* Fig. 1). Numerous disinfectant types and techniques have been developed to surface sterilize plants (2). However, all procedures can be summarized as follows:

1. Explants are washed in a detergent to remove excess debris.
2. Explants are wrapped in cheesecloth and immersed in a container containing a disinfectant that is usually corrosive (i. e., containing a halogen compound).

From: *Methods in Molecular Biology*, vol. 6, *Plant Cell and Tissue Culture*
Edited by Jeffrey W. Pollard and John M. Walker, ©1990 by The Humana Press

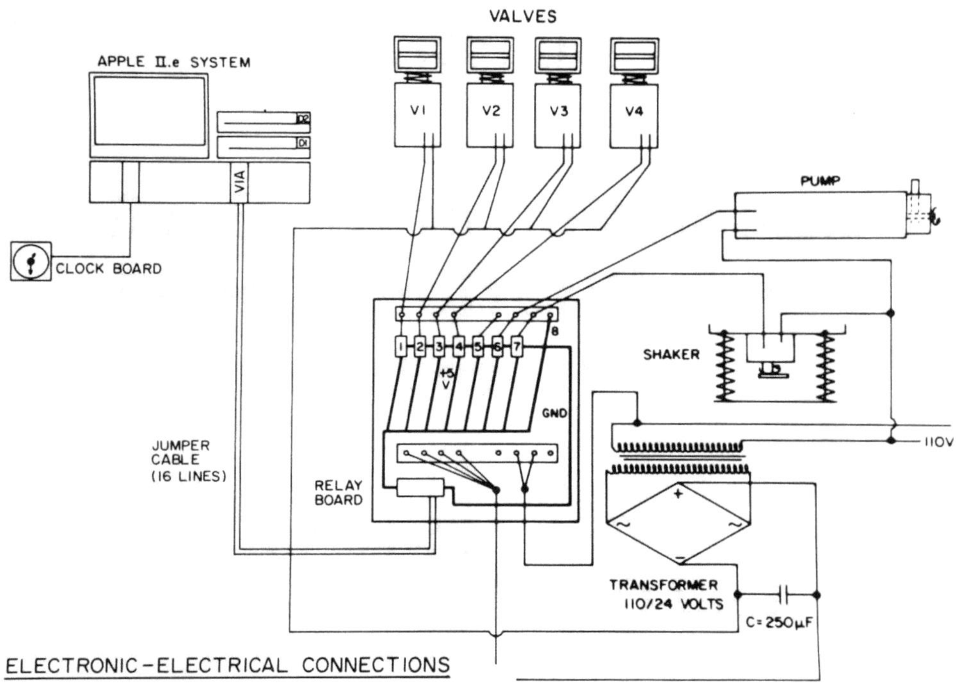

Fig. 1. Schematic representation of a relay board wiring with its connections to the Apple II+ microcomputer for the automatic surface sterilization system. V_1 to V_4 represent valves 1–4.

3. Explants are periodically agitated to disperse the air bubbles adhering to explants and facilitate an even distribution of the disinfectant.
4. Following a set soaking period (e. g., 5–30 min), explants are rinsed 3x with water, and then prepared for planting on nutrient medium.

The system presented in this chapter substitutes hand–plant interactions with machine–plant interactions via automatic solution washing and removal, timing, and agitation (Figs. 1 and 2).

2. Materials

1. Microcomputer. Microcomputers (e. g., Apple II+, Atari 400, or the Commodore Vic-20 or C-64) can be purchased from local retail or mail-order outlets.
2. Interface assembly. This can be purchased ready-made from local or mail-order sources, but it can be easily constructed as well (Fig. 3). Interface assembly consists of the following parts:

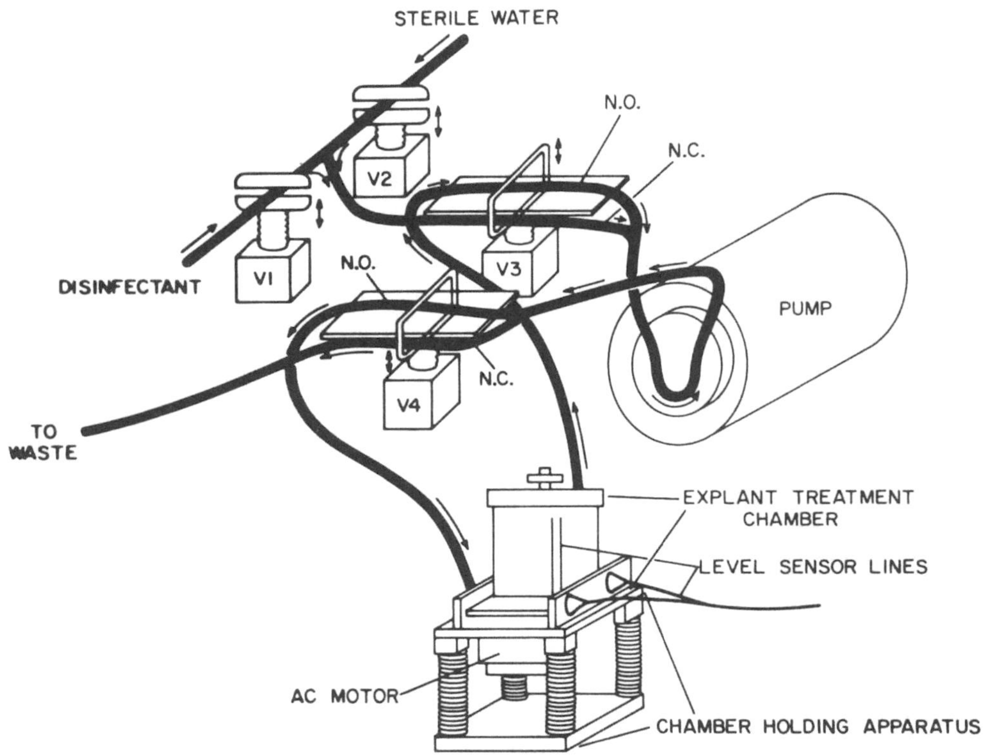

Fig. 2. Diagrammatic representation of the components employed in the autoclavable automated surface sterilization system. N.O. = normally open; N.C. = normally closed.

a. Relays, miniature 5 VDC SPDT or DPDT types (Jameco Electronics, Belmont, CA).
b. Transformers, +5 and +12 VDC types with bridge rectifier and capacitor assemblies (Jameco Electronics, Belmont, CA).
c. Relay board with cables, to connect the assembly to the computer, and the assembly to the pumps and valves (Jameco Electronics, Belmont, CA).

3. SSS. This can be constructed of a variety of materials, depending on the needs of the investigator (Fig. 2). The SSS consists of the following components:
   a. Explant treatment chamber, (0.125 L) glass jar fitted with an autoclavable No. 7 (30-mm diameter [dia.]) "Twistit" rubber stopper containing three predrilled, countersunk holes of 8 mm dia. (American Sci. Products, McPaw Park, IL).
   b. Peristalic pumps, variable-speed type, with a 100-rpm motor

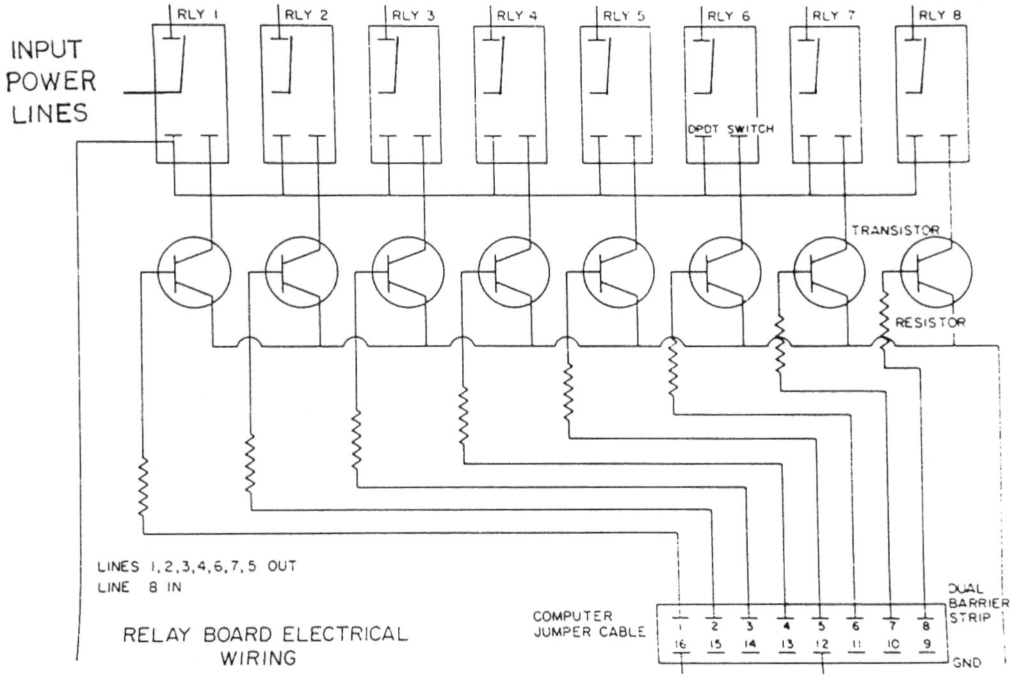

Fig. 3. Schematic representation of a relay board wiring with its connections to a microcomputer.

capable of delivering a flow range of 5–80 mL/min with size 16 tubing, to be equipped with Lexan polycarbonate heads and cold-rolled steel rotor (Cole-Parmer Instrument Company, Chicago, IL).

c. Silicone tubing, dimensions: size 16; 3.1-mm inside dia. (i. d.); 3.2-mm hose barb size; and 7-mm outer dia. (o.d.). Employ 0.318-mm dia. polypropylene quick-disconnect Luer taper miniature plastic fittings as needed to aid in tubing connections (Cole-Parmer Instrument Company, Chicago, IL).

d. Solution reservoirs, three 2-L capacity Pyrex™ or borosilicate glass media-storage bottles (138-mm o. d. x 246-mm height [h] x 45-mm screw cap size), to be fitted with linerless polypropylene screw caps (Bellco Glass, Inc., Vineland, NJ).

e. Chamber holding assembly and agitator apparatus, consists of two Plexiglas™ platforms connected by four stainless-steel springs with dimensions: 10-mm o. d. x 70-mm h. Top platform is fitted with a 120 VAC motor with an offset weight attached to its shaft (Fig. 2).

f. Optical liquid sensing system consisting of Levelite controller, quartz probe, 15 mm length x 9-mm o. d., and Levelite probe assembly (Ryan Herco Products Corporation, Burbank, CA). Alternatively, an inexpensive liquid level sensor can be constructed using stainless-steel probes and an optical isolator integrated circuit (Fig. 4).

g. Pinch clamp valve, two-way and three-way types, 12 VDC, to provide at least 20 psi pressure (Bio-chem Valve Corp., East Hanover, NJ). However, as an alternative, inexpensive spring-powered pinch clamp valves can be constructed using Plexiglas™, wood, springs, and solenoid valves (Guardian Electrical Manufacturing Co., Chicago, IL) (Fig. 5).

h. Pyrex™ (borosilicate) standard wall glass tubing, dimensions: 6-mm o. d., 1-mm wall thickness (American Scientific Products, McPaw Park, IL).

i. Sterilizing vent filters, ACRO 37 TF 37 (Gelman 4464) equipped with hydrophobic PTFE membrane, 0.2 micron membrane; dimensions: 0.64–0.95 cm hose barb connections, 37 mm dia. Five are required/system (American Scientific Products, McPaw Park, IL).

3. Methods

3.1. Construction of the Automated SSS

3.1.1. Explant Treatment Chamber

1. Position the rubber stopper in the center of the jar's lid by drilling a 2.86 cm dia. hole using a keyhole drill. Into two of the three holes in the stopper, insert two 50 mm length (l.) x 6 mm o. d. glass tubes, bent in half at a right angle. These tubes should penetrate only 1 mm into the chamber.
2. Connect a 30 mm long x 6 mm o. d. piece of silicone tubing attached to an air filter to one of these tubes. To the other tube (solution inlet line tube), connect a 400 cm long x 6 mm o. d. piece of silicone tubing. Connect the other end of this tubing to one of the arms of a 6 mm o. d. barbed polypropylene Y connector (Fig. 2).

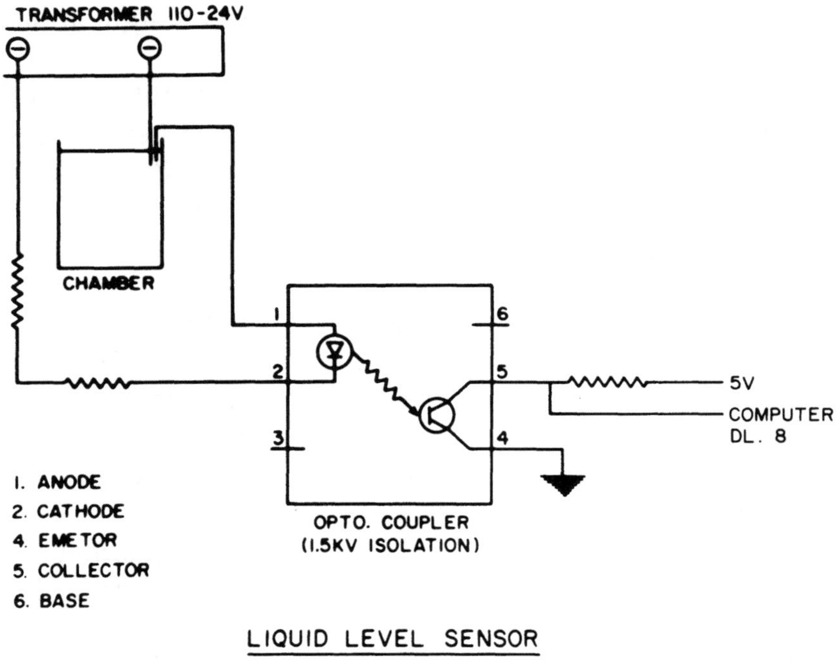

Fig. 4. Schematic representation of the liquid level sensor employed in the explant treatment chamber.

3. Insert the medium outlet line tube (120 mm long x 6 mm o. d.) into the third stopper hole, with a right angle bent 40 mm from the end that is exposed to the atmosphere. This tube must reach the bottom of the chamber. Proper positioning of this tube is important to ensure adequate solution evacuation following explant soaking.
4. Connect a 250 cm long x 6 mm o. d. piece of silicone tubing to this medium inlet line, and connect one of the arms of another Y connector.
5. To one of the arms of the Y connection attached to the solution chamber inlet line, attach a 200 mm long x 6 mm o. d. silicone tubing line and connect to a polypropylene T connector, which in turn connects tubing to the inlet lines of the solution reservoirs.
6. To the other arm of the Y connection attached to the explant treatment chamber inlet line, connect a 200 mm long x 6 mm o. d. silicone tubing line that proceeds to the Y connector attached to the outlet line of the explant chamber.
7. To the remaining arm of this Y connector, attach 400 mm long x 6 mm o. d. silicone tubing and attach to waste reservior.
8. The arm lines bifurcating from the Y are positioned in a pinch lamp valve.

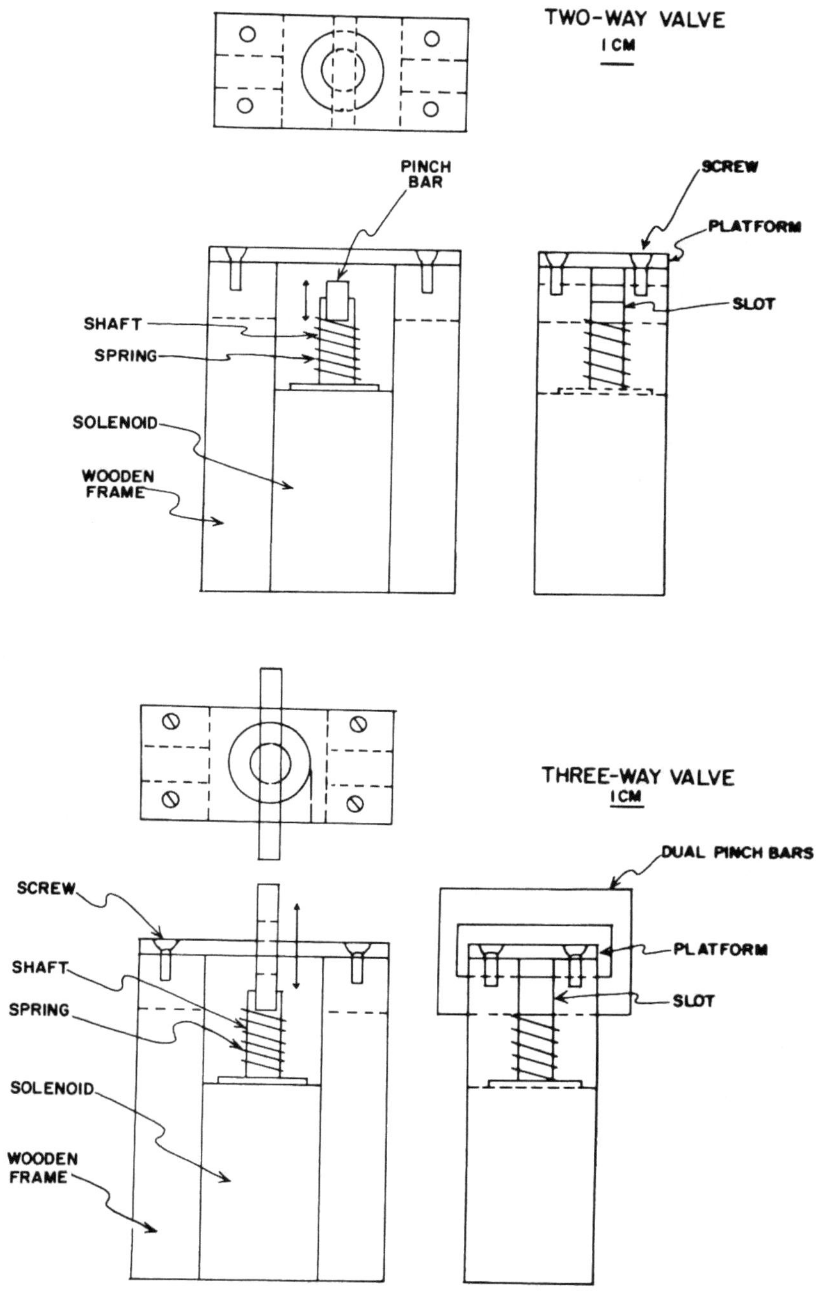

Fig. 5. Diagrammatic representation of the components employed in construction of pinch clamp valves.

The solution inlet line to the explant chamber is in the normally open position, whereas the line to the media drain reservior is in the normally closed position. The solution outlet line from the explant chamber is in the normally open position, whereas the solution line from the solution reservoirs is in the normally closed position (Fig. 2). Insert level sensor lines 2 mm into chamber (Fig. 4).

3.1.2. Solution Reservoirs

1. Drill a 2.86-cm dia. hole using a keyhole drill in the twist top closure to accommodate the stopper.
2. Insert two glass tubes of 40-mm long x 6-mm o. d. and connect air filters as described.
3. The solution outlet glass tube is 260-mm long x 6-mm o. d. This tube is positioned almost to the bottom of the reservoir. A 300-mm long x 6-mm o. d. silicone tube connects the solution outlet line to the T connection via the normally closed valve.

3.1.3. Drain Reservoir

1. Prepare as described for the culture media reservoirs, but connect the inlet line to the line in the normally closed position proceeding from the pinch clamp valve.
2. The 260-mm glass tubing line referred to as the medium inlet line for the culture chamber and media reservoirs will be used for extraction of waste solutions at later times. This will be plugged at this time.

3.2. Construction of Microcomputer and Interface System

1. A diagram for the construction of the relay board employed in the interface system is shown in Fig. 3. Connections of electrical lines are illustrated in Fig. 1.
2. The culture program is written in BASIC, using appropriate POKE commands to access the microcomputer's output lines. These output lines control the relays, which in turn control the operation of pumps and valves to control medium flow.

3.3. Sterilization Procedures

1. Connect the SSS system, as described in figures and text, to solution reservoir bottles. Add 1800 mL deionized distilled water to its reservoir bottle. Use pinch clamp closures to seal all silicone connections

for solution reservoirs during autoclaving. Autoclave at 15 psi for 15 min at 120°C. Following autoclaving, add disinfectant solution to its reservoir.

2. Flow lines for automated SSS illustrates the operation of the system (Fig. 2).

4. Notes

1. This system is not intended for the novice or for use in small-scale experimental laboratories. It should be useful as an aid in the mass establishment of cultures or when critical control of the disinfectant conditions is sought. It is recommended that disinfectant and times used in manual disinfectant techniques be initially employed in this system. Such conditions should be suitable for the automated system. The length of time that explants soak in medium is critical; this time will vary for each plant and explant type, and needs to be optimally determined empirically.
2. Larger reservoirs and sterilizing chambers may be employed, depending on the desires of the investigator. Also, additional solutions (i. e., detergent and water) can be added for additional treatments as desired. Minimize the length of tube connections to reduce the time required to fill the explant treatment chamber and unnecessary utilization of test solutions.
3. A variety of microcomputers or microprocessor-controller systems, and even some multichannel timers, can be used for this system. Microcomputers are employed because they are less expensive, easier to interface, more versatile, and more easily programmed than microprocessors or programmable timers. Appropriate POKE and PEEK commands are found in each microcomputer's programmer's reference manual.
4. It is desirable to autoclave lines prior to intensive use, to minimize introduction of contaminates. A variety of peristalic or impeller pumps can be employed. Systems can be constructed to be either totally autoclavable, or partially autoclavable (e. g., media bottles and culture chamber) and partially ethylene oxide sterilizable (e. g., impeller pumps and culture chambers).
5. Rates of surface sterilization using this system are comparable to those obtained from the manual procedure. In some cases, depending on the species and explant type, higher sterilization rates are obtained using the automated SSS (*1*).

Acknowledgment

We wish to thank V. Maurice for his help in the initial construction of this system.

Disclaimer

Mention of a trademark name or proprietary product does not constitute a guarantee or warranty of the product by the US Department of Agriculture, and does not imply its approval to the exclusion of other products that may be suitable.

References

1. Maurice, V., Vandercook, C. E., and Tisserat, B. (1985) Automated plant surface sterilization system. *Physiol. Veg.* **23,** 127–133.
2. Tisserat, B. (1985) Embryogenesis, organogenesis, and plant regeneration, in *Plant Tissue Culture: A Practical Approach.* (Dixon, R. A., ed.), Information Retrieval Limited Press, Oxford, pp. 76–106.

Appendix

Plant Cell Culture Medium

Compiled by

Jeffrey W. Pollard

Table 1
Simple Tissue Culture Salts and Media[a]

Ingredient	BE or CPW salts, mg/L	Murashige & Skoog salts,[b] mg/L	Murashige & Skoog medium,[b,c] mg/L	OR[d] medium, mg/L	Gamborg's B5 medium, mg/L	White's modified medium, mg/L	Nitsch's medium, mg/L
$CaCl_2 \cdot 2H_2O$	1480	440.0	440.0	440.0	150.0		166.0
$Ca(NO_3)_2$anh						208.5	
KCl						65.0	
KH_2PO_4	27.2	170.0	170.0	170.0			68.0
KNO_3	101.0	1900	1900	1900	3000	80.0	950.0
$MgSO_4 \cdot 7H_2O$	246.0	370.0	370.0	370.0	500.0	720.0	185.0
Na_2SO_4						200.0	
$NaH_2PO_4 \cdot 2H_2O$					150.0	18.7	
NH_4NO_3		1650	1650	1650			720.0
$(NH_4)_2SO_4$					134.0		
$CoCl_2 \cdot 6H_2O$		0.025	0.025	0.1	0.025		
$CuSO_4 \cdot 5H_2O$	0.025	0.025	0.025	0.1	0.025	0.001	0.025
$FeSO_4 \cdot 7H_2O$[e]		27.8	27.8	27.8	27.8	3.48	27.8
Na_2EDTA[e]		37.3	37.3	37.3	37.3	4.66	37.3
H_3BO_3		6.2	6.2	24.8	3.0	1.5	10.0
KI	0.16	0.83	0.83	3.32	0.75	0.75	
$MnSO_4 \cdot 4H_2O$		22.3	22.3	89.2	13.2	7.0	25.0
MoO_3						0.0001	
$Na_2MoO_4 \cdot 2H_2O$		0.25	0.25	1.0	0.25		0.25
$ZnSO_4 \cdot 7H_2O$		8.6	8.6	34.4	2.0	3.0	10.0
Sucrose			30,000	30,000	20,000		
Myo-Inositol			100.0	100.0	100.0		100.0
Folic acid							0.5

Nicotinic acid			0.5	1.0	1.0	0.5	5.0
Thiamine •HCl			0.1	13.0	10.0	0.1	0.5
Pyridoxine•HCl			0.5	1.0	1.0	0.1	0.5
Biotin							0.05
Glycine			2.0			3.0	2.0
Proline				1381			
pH	5.8	5.8	5.8	5.8	5.8	5.8	5.8
Reference	*1*	*2*	*2*	*3*	*4*	*5*	*6*

[a]Details of media preparation are given in Chapter 1, this volume.[f]

[b]Details of the manufacture and storage of Murashige and Skoog's (MS) medium is given in Chapter 6, this volume.

[c]MS medium comes with many minor variations (*see* text chapters and Linsmaier and Skoog [1965], *Physiologia Plantarum* **18,** 100–127). Some of these are commercially available (e.g., *see* the Flow Laboratories catalog). In Chapter 38, this volume, EB medium is MS medium plus 43 μ*M* NAA, 5.0 μ*M* additional thiamine•HCl, 30 μ*M* additional nicotinic acid, and 0.6% bactoagar; MSR medium is MS medium containing 1.7 μ*M* BAP, 0.025 μ*M* indolebutyric acid (IBA), and 0.6% bactoagar; R5 medium is MS basal containing 9.8 μ*M* IBA, 5.0 n*M* BAP, and 5 μ*M* Gibberellic acid.

[d]*See* Chapter 38, this volume.

[e]$FeSO_4 \cdot 7H_2O$ and Na_2EDTA are added from the following stocks: $FeSO_4 \cdot 7H_2O$ (557 mg) and Na_2EDTA (745 mg) dissolved in 100 mL by boiling, 5 mL being added to I L of medium except for White's medium, where 0.625 mL is added.

[f]The amount of individual hormones depends on the type of growth required and plant species (*see* Chapter 5). These values will be varied appropriately. Stocks of plant hormones can be made up at 0.1 or 1 mg /mL: 2, 4-D (2, 4-dichlorophenoxyacetic acid), NAA (1–Naphthalanacetic acid), Kinetin (6–furfurylaminopurine) in H_2O with 1*M* NaOH added dropwise until they dissolve, BAP (benzylaminopurine) is made up in 0.1*M* HCl. All stocks can be stored at 4°C for up to 3 mo.

Table 2
Refined Tissue Culture Medium

Ingredient	K_3 medium, mg/L	Modified K_3 medium,[a] mg/L	NT medium, mg/L	SH medium, mg/L	Kao's medium,[b] mg/L	Yehia medium,[c] mg/L
$CaCl_2 \cdot 2H_2O$	900.0	900.0	220.0	200.0	600.0	44.0
$CaHPO_4$	50.0					
KCl					300.0	
KH_2PO_4			680.0		170.0	17.0
KNO_3	2500	2400	950.0	2500	1900	190.0
$MgSO_4 \cdot 7H_2O$	250.0		1233	400.0	300.0	37.0
$NaH_2PO_4 \cdot 2H_2O$	150.0	120.0				
NH_4NO_3	250.0	240.0	825		600.0	
$NH_4H_2PO_4$				300.0		
$(NH_4)_2SO_4$	134.0	130.0				
$CoCl_2 \cdot 6H_2O$	0.025	0.025		0.1	0.025	0.0025
$CoSO_4 \cdot 7H_2O$			0.030			
$CuSO_4 \cdot 7H_2O$	0.025	0.025	0.025	0.2	0.025	0.0025
$FeSO_4 \cdot 7H_2O$	27.8	27.85	27.8	15.0		27.8
Na_2EDTA	37.3	37.3	37.3	20.0		37.3
Sequestrene® 330Fe					28.0	
H_3BO_3	3.0	3.0	6.2	5.0	3.0	0.62
KI	0.75	0.75	0.83	1.0	0.75	0.083
$MnSO_4 \cdot 4H_2O$		6.7	22.3	10.0	10.0	2.23
$MnSO_4 \cdot H_2O$	10.0					
$Na_2MoO_4 \cdot 2H_2O$	0.25	0.24	0.25	0.1	0.25	0.025
$ZnSO_4 \cdot 7H_2O$	2.0	2.3	8.6	1.0	2.0	0.86
Sucrose[d]	30,000		10,000	30,000	250.0	107,000
Glucose					68.4	

p-Aminobenzoic acid					0.02	
Ascorbic acid					2.0	
Choline chloride					1.0	
D-Ca-pantothenate					1.0	
m-Inositol	100.0	100.0	100.0	1000	100.0	5600
Folic acid					0.4	
Nicotinamide					1.0	
Nicotinic acid	1.0	1.0		5.0		1.0
Pyridoxine•HCl	1.0	1.0		0.5	1.0	1.0
Riboflavin					0.2	
Thiamine•HCl	10.0	10.0	1.0	5.0	1.0	10.0
Vitamin A					0.01	
Vitamin D_3					0.01	
Vitamin B12					0.02	
Biotin					0.01	
Zeatin					0.5	
D-maniton			1.275×10^5		250.0	5700
Sorbitol					250.0	5700
Xylitol						4700
Cellobiose					250.0	
Fructose					250.0	
Mannose					250.0	
Rhamnose					250.0	
Ribose					250.0	
Xylose	250.0	250.0			250.0	
Agar[d]				6000		
Agarose[d] (low M. P.)						10,000 or 0

(continued)

Table 2 *(continued)*

Ingredient	K_3	Modified K_3 medium,[a] mg/L	NT medium, mg/L	SH medium, mg/L	Kao's medium,[b] mg/L	Yehia medium,[c] mg/L
Sodium pyruvate				20.0		
Citric acid				40.0		
Malic acid				40.0		
Fumaric acid				40.0		
L-amino acids (all except Gln, Ala, Glu, Cys)				0.1 each		
Alanine				0.6		
Cysteine				0.2		
Glutamic acid				0.6		
Glutamine				5.6		
Adenine				0.1		
Guanine				0.03		
Thymine				0.03		
Uracil				0.03		
Hypoxanthine				0.03		
Xanthine				0.03		
Vitamin free casamino acid				250.0		
Coconut water (from mature fruits heated to 60°C for 30 min)				20 mL/L		

Casein hydrolysate						50
pH	5.6	5.6	5.8	5.9	5.6	5.6
Reference	*7*	*7*	*8*	*9*	*10*	*11*

[a]This is a slight modification of the K_3 medium used for *N. plumbaginifolia* protoplast culture as described in Chapter 41, this volume. It is convenient to make up 5 L of the medium double strength omitting only the sucrose or glucose and storing at –20°C. It is generally recommended to filter K_3 medium, but for *N. plumbaginifolia* protoplasts, good results can be obtained using autoclaved medium but the pH should be 5.7 or higher.

[b]This is medium K8p from ref. *10* for growth of a single colony of protoplasts. This reference should be consulted for other minor variants of this medium.

[c]This is the medium that allows for cell wall regeneration; *see* Chapter 20, this volume.

[d]Amounts may be variable as indicated in the text chapters.

References

1. Banks, M. S. and Evans, P. K. (1976) A comparision of the isolation and culture of mesophyll protoplasts from several Nicotiana species and their hybrids. *Plant. Sci. Lett.* **7,** 409–416.
2. Murashige, T. and Skoog, F. (1982) A revised medium for rapid growth and bioassays with tobacco tissue cultures. *Physiol. Plant.* **15,** 473–497.
3. Barwale, U. B., Kerns, H. R., and Widholm, J. M. (1986) Plant regeneration from callus cultures of several soybean genotypes via embryogenesis and organogenesis. *Planta* **167,** 473–481.
4. Gamborg, O. L., Miller, R. A., and Ojima, K. (1968) Nutrient requirements of suspension cultures of soybean root cells. *Exp. Cell. Res.* **50,** 151–158.
5. White, P. R. (1954) The cultivation of animal and plant cells (Ronald Press, New York).
6. Nitsch, J. P. and Nitsch, C. (1969) Haploid plants from pollen. *Science* **163,** 85–87.
7. Nagy, J. I. and Magliga, P. (1976) Callus induction and plant regeneration from mesophyll protoplasts of Nicotiana sylvestris. *Z. Pflanzenphysiol. Bd.* **78,** S, 446–452.
8. Nagata, T. and Takebe, I. (1971) Planting of isolated tobacco mesophyll protoplasts on agar medium. *Planta (Berl.)* **99,** 12–20.
9. Schenk, R. U. and Hildebrandt, A. C. (1972) Medium and techniques for induction and growth of monocotyledonous and dicotyledonous plant cell cultures. *Can. J. Bot.* **50,** 199–204.
10. Kao, K. N. and Michayluk, M. R. (1975) Nutritional requirements for growth of *Vicia hajastana* cells and protoplasts at a very low population density in liquid media. *Planta (Berl.)* **126,** 105–110.
11. Ochatt, S. J. and Caso, O. H. (1986) Shoot regeneration from leaf mesophyll protoplasts of wild pear (*Pyrus communis,* var. *Pyraster L.*) *J. Plant Physiol.* **122,** 243–249.

Index

MIX
Papier aus verantwortungsvollen Quellen
Paper from responsible sources
FSC® C105338

If you have any concerns about our products,
you can contact us on
ProductSafety@springernature.com

In case Publisher is established outside the EU,
the EU authorized representative is:
Springer Nature Customer Service Center GmbH
Europaplatz 3, 69115 Heidelberg, Germany

Printed by Libri Plureos GmbH
in Hamburg, Germany